The Moral Authority of Nature

The Moral Authority of Nature

Edited by Lorraine Daston and Fernando Vidal

THE UNIVERSITY OF CHICAGO PRESS
CHICAGO AND LONDON

The University of Chicago Press, Chicago 60637
The University of Chicago Press, Ltd., London
© 2004 by The University of Chicago
All rights reserved. Published 2004
Printed in the United States of America

13 12 11 10 09 08 07 06 05 04   1 2 3 4 5

ISBN: 0-226-13680-9 (cloth)
ISBN: 0-226-13681-7 (paper)

Library of Congress Cataloging-in-Publication Data

The moral authority of nature / edited by Lorraine Daston and Fernando Vidal.
    p.   cm.
Includes bibliographical references (p.   ) and index.
ISBN 0-226-13680-9 (cloth : alk. paper)—ISBN 0-226-13681-7 (paper : alk. paper)
    1. Philosophy of nature—History. 2. Nature—Moral and ethical aspects—History.
I. Daston, Lorraine, 1951– II. Vidal, Fernando.

BD581.M78 2004
113'.09—dc21

2003008351

❧ CONTENTS ❧

Doing What Comes Naturally

Lorraine Daston and Fernando Vidal

The oration of the worlds is untiring.
I can repeat all of it from the beginning
with a pen inherited from a goose and Homer
with a diminished spear
stand in front of the elements
I can repeat all of it from the beginning
the hand will lose to the mountain
the throat is weaker than the spring
I will not outshout the sand
not with saliva tie a metaphor
the eye with a star
and with the ear next to a stone
I won't bring out stillness
from the grainy silence.
      ZBIGNIEW HERBERT, "Mr. Cogito Considers the Difference between the
      Human Voice and the Voice of Nature"

This book is about how humans use nature to think about standards of the
good, the beautiful, the just, and the valuable. The authors of these essays
concentrate less on the question of whether nature is or should be used in
this fashion than on the how, why, and what of nature's authority in the
human realm. The "is" can be easily established by following stories in the
daily media on topics ranging from genetically manipulated organisms to
surrogate motherhood;[1] the "should be" is already the subject of a vast

---

1. Other examples could be listed at length. To stay with the two mentioned in the text,

I

critical literature, dating back at least to David Hume's warning against the conflation of "is" and "ought."[2] Instead, we examine how nature's authority works in different times and places, why it is a force to wield or to escape, and to which domains it does (or does not) apply. Our aim is to advance discussion about the interactions of the natural and the human beyond controversies over social constructionism versus realism, questions of transgressive boundary crossings between the categories of nature and culture (or society, nurture, civilization, artifice), and exposures of ideologies and fallacies (naturalistic, pathetic).[3] We are indebted to these earlier explorations for identifying the problem of nature's authority *as* a problem, with ample documentation and critical analysis; we hope to contribute a comparative dimension, both historical and cross-cultural, and to shift the focus of inquiry from the existence and (il)legitimacy of nature's authority to its jurisdictions and workings.

Key to this project is understanding what "nature" means and has meant in a range of contexts, and what kinds of authority these meanings exert. Consider the term "naturalization," in current English usage perhaps the most common way of describing nature's authority in human affairs. "Naturalization" in this sense means to shore up a social convention (for example, reading from left to right) or political arrangement (the disenfranchisement of slaves and women, for instance) by asserting that it is dictated by nature and is therefore either irrevocable or optimal or both. Friedrich Engels's remarks on social Darwinism offer a clear example of this strategy:

> The whole Darwinist teaching of the struggle for existence is simply a transference, from society to living nature, of Hobbes' doctrine bellum omnium contra omnes and of the bourgeois-economic doctrine of competition, together with Malthus' theory of population. When this sleight of hand has been performed . . . , the same theories are transferred back again from organic nature into history, and it is now claimed that their validity as eternal laws of human society has been proved.[4]

---

media reporting on GMOs and reproductive issues do not necessarily take sides, but often convey antagonisms between groups who claim to speak for science and groups who oppose "tampering" with nature.

2. David Hume, *Treatise on Human Nature,* ed. L. A. Selby-Bigge, 2nd ed. with text rev. and notes by P. H. Nidditch (1739–40; Oxford: Clarendon Press, 1978), III.i.1, 469–70.

3. See the bibliography for a sample of relevant studies.

4. Friedrich Engels, letter to Pjotr Lawrowitsch Lawrow, 12–17 November 1875, in Karl Marx and Friedrich Engels, *Werke* (Berlin: Dietz Verlag, 1966), 34:170.

More recent accounts of naturalization have enlarged the range of objects to which it can be applied—from homosexuality to motherhood, intelligence to madness, wilderness to gardens—but the image of the critic debunking a sleight of hand endures.

As Engels explained, the trick consists in smuggling certain items (in his example, competition) back and forth across the boundary that separates the natural and the social. Critics, like customs inspectors, return items to their rightful categories, extraditing the natural from the social, and especially, the social from the natural. Naturalization in this form assumes the existence of distinct categories of nature and society, of well-drawn boundaries between them, and of a certain asymmetric advantage in dwelling in one territory over another, nature being the land of choice for immigration. Naturalization imparts universality, firmness, even necessity—in short, authority—to the social.

These metaphors of smuggling, customs inspection, extradition, and immigration may seem fanciful, but they in fact recall the original sense of the word "naturalization," which was embedded in a very different semantic field. In both English and French the word "naturalization" (*naturalisation,* cp. Italian *naturalizzare*) entered the language in the sixteenth century, and referred to extending to an alien the rights and social position of a native-born subject or citizen, or to adopting a foreign word or custom on equal footing with what was native. By the seventeenth century, it had acquired the further meaning of becoming accustomed to a new locale, and by the early eighteenth century, it was applied metaphorically to imported plants and animals successfully cultivated in new surroundings.[5] By modern lights, this earlier usage seems oddly confused, itself an instance of the more recent definition of "naturalization" as a covert conflation of the natural and the social. Why use the word "naturalize" to refer to the obviously conventional status of subject or citizen? The application of the term to the more "natural" (in our view) plants and animals fails to illuminate modern sensibilities: didn't these strains flourish equally "naturally," or still more so, in their lands of origin? Moreover, why use the term "naturalize" in conjunction with expressly local conditions, when

5. See "Naturalization," *Oxford English Dictionary,* 2nd ed (Oxford: Clarendon Press, 1989); "Naturalisation," Emile Littré, *Dictionnaire de la langue française,* 7 vols. (Paris: Gallimard/Hachette, 1957); Alain Rey, ed., *Dictionnaire historique de la langue française,* 3 vols. (Paris: Le Robert, 1998); "Naturalizzare," Carlo Battisti and Giovanni Alessio, *Dizionario Etimologico Italiano,* 5 vols. (Florence: G. Barbera, 1954).

nature stands for what holds everywhere and always? These queries testify to the gap between early modern and modern European understandings of nature and its sources of authority. Their nature was bounded not so much by "society" as by "art" on one side and "divinity" on the other; their nature referred to local circumstances as well as to universal regularities. It is of course still possible to imagine (and find) early modern instances of "naturalization" *avant la lettre,* but the authority of nature differs crucially according to the meanings of nature and its opposites.

Hence an inquiry into how, why, and where nature's authority is called on to buttress or subvert human norms requires a careful sorting out of the meanings of both nature and authority in specific contexts. The derivatives of the Latin *natura* in modern European languages have notoriously long and rich definitions, and their common Latin root itself derives many of its connotations from the Greek *physis,* which has its own convoluted semantic history.[6] Even the most cursory survey of other cultural traditions greatly complicates an already labyrinthine situation, which stands as a warning against unthinking attempts to universalize often idiosyncratic modern Western intellectual traditions to history and the world at large. Despite their important continuities with European scientific, philosophical, and religious traditions, archaic Greek and classical Arabic sources offer no single word that corresponds to "nature" as the sum total of the entire universe, although they do possess cognates to "nature" as the essence of an individual or class.[7]

The ancient Sanskrit word *dharma* refers both to the orderliness of the world and that which dictates the appropriate behavior of every kind of thing: it is, for example, the dharma of a snake to bite and of an owl to be nocturnal.[8] Dharma implies that "should" and "is" are one—that one

6. In addition to the entries for "Nature" and "*Natura*" in the sources cited in note 5, see "Natuur," A. Kluyver, A. Lodewyck, J. Heinsius, and J. A. N. Knuttel, *Woordenboek der Nederlansche Taal,* 29 vols. ('s-Gravenhage: Martinus Nijhoff/A.W. Sijthoff, 1913); "Natur," Jacob and Wilhelm Grimm, *Deutsches Wörterbuch,* 33 vols. (Munich: Deutscher Taschenbuch Verlag, 1991); "PHYSIS," Franz Passow, *Handwörterbuch der Griechischen Sprache,* neu bearbeitet und zeitgemäss umgestaltet von V. C. R. Rost, F. Palm, O. Kreussler, K. Keil, F. Peter, und G. E. Benseler, 2 parts in 4 vols., 5th ed. (Leipzig: F. C. W. Vogel, 1857).

7. Felix Heinemann, *Nomos und Physis. Herkunft und Bedeutung einer Antithese im griechischen Denken des 5. Jahrhunderts* (1945; Darmstadt: Wissenschaftliche Buchgesellschaft, 1965); Seyyed Hossein Nasr, "The Meaning of Nature in Various Perspectives in Islam," in his *Islamic Life and Thought* (London: George Allen & Unwin, 1981), 96–101; Friedrich Dieterici, trans., *Die Naturanschauung und Naturphilosophie der Araber im zehten Jahrhundert aus den Schriften der lautern Brüder* (1869; Hildesheim: Georg Olms, 1969), 141–60.

8. Wendy Doniger O'Flaherty, *The Origins of Evil in Hindu Mythology* (Berkeley: University of California Press, 1980), 46, 81–95.

should do what one's nature inclines one to do. Confronted with the difficulties of translating Western philosophical and scientific texts peppered with references to nature, intellectuals in nineteenth- and twentieth-century China and Japan had creative recourse to a web of words: in Japanese, these included *manbutsu* ("the myriad objects") as well as *honzen no sei* (human nature) and the neologism *shizen*;[9] in Chinese, there was *tiandi* (heaven and earth) and *zaohua* (creative transformation in the universe), as well as the term *ziran,* which like the Japanese *shizen* became standard only in the late nineteenth century.[10] Nothing even approximating a full-dress account of the multiple meanings and histories of the word "nature" and its cognates (or lack thereof) in other languages can be provided here.[11] Instead, we wish to highlight how these varied and complex meanings have been enlisted to endow nature with authority in human affairs.

The three sections of the book allude to principal sources of nature's authority (and resistance thereto): as a source of values and a value in itself; as a means of thinking about necessity and freedom; and as a permeable and moving boundary that demarcates individuals and categories. Each section begins with a concise statement of the problematic that links its essays together. Here, we draw out for special consideration themes concerning the workings of nature's authority that reappear throughout the volume.

The most inexorable authority is that of necessity, against which human will strives in vain. Nature's necessity may operate at several levels: it

9. See Julia Adeney Thomas's essay in this volume; see also Tessa Morris-Suzuki, "Concepts of Nature and Technology in Pre-Industrial Japan," *East Asian History* 1 (1991): 81–97.

10. See Fa-ti Fan's essay in this volume; see also Joseph Needham, "Human Laws and the Laws of Nature," in his *The Grand Titration: Science and Society in East and West* (Toronto: University of Toronto Press, 1969), 299–331; Heiner Roetz, "Etymologische Vorbemerkung zum chinesischen Begriff der 'Natur'," in his *Mensch und Natur im alten China* (Frankfurt am Main: Peter Lang, 1984), 113–17.

11. On the concept of nature accross time in the Western intellectual tradition, see "Natur" and related articles in Joachim Ritter and Karlfried Gründer, eds. (new ed. by Rudolf Eisler), *Historisches Wörterbuch der Philosophie* (Darmstadt: Wissenschaftliche Buchgesellschaft, 1984), vol. 6; R. G. Collingwood, *The Idea of Nature* (New York: Oxford University Press, 1945); Clarence J. Glacken, *Traces on the Rhodian Shore: Nature and Culture in Western Thought from Ancient Times to the End of the Eighteenth Century* (Berkeley: University of California Press, 1967); C. S. Lewis, "Nature," in *Studies in Words* (1960; Cambridge: Cambridge University Press, 1990); Arthur O. Lovejoy, "Nature as an Aesthetic Norm," in *Essays in the History of Ideas* (Baltimore: Johns Hopkins University Press, 1948), 69–78; John Torrance, ed., *The Concept of Nature* (Oxford: Clarendon Press, 1992); Raymond Williams, "Nature," in *Keywords: A Vocabulary of Culture and Society,* rev. ed. (New York: Oxford University Press, 1983). See the bibliography for studies of particular periods.

may govern the cosmic order; it may be indelibly inscribed in the body; or it may unfold over time as a result of (and often retaliation against) human actions. Laura Slatkin (essay 1) describes how in early Greek literature the cosmic and human orders were so tightly bound to one another as to be inextricable, both orders defined by the principles of due measure and the seasonable: justice can be summoned to harry the sun for transgressing its measures as well as to punish humans for violating fair exchange and reciprocity. The cosmic order depends in turn on sorting out the inventory of the universe into "natural kinds," as Joan Cadden (essay 8) points out in the context of Chaucer's fourteenth-century poem "Parliament of Fowls": overall order is ensured by that among categories (in this case, species of birds), each of whose members runs "true to type" and which are arranged in a hierarchy of dignity and merit. The imperative to adhere to one's type is at once a duty and a necessity, again blending the realms of "is" and "ought."

Within these natural kinds—the human species, for example— anatomy and physiology outline still further categories that allegedly determine character and conduct. In medieval Galenic medicine, as Valentin Groebner (essay 14) explains, an individual's nature was thought to be fixed by a balance of bodily humors known as a "complexion," which came in endless varieties but which lay beyond the power of individual choice to change—hence its utility as a means of penetrating impostures and disguises of all kinds. Superimposed on these cyclic and classificatory orders of nature's necessity are those of cause and effect, or rather, reward and punishment. Slatkin quotes Hesiod's pronouncement: fertile lands, woolly sheep, and children who resemble their parents are the lot of the just. Conversely, nature strikes back at those who flaunt its dictates: Fernando Vidal (essay 10) recounts how Enlightenment physicians threatened masturbators with nature's "immanent justice," a cruel and early death; Matt Price (essay 7) reports the doomsday predictions of ecological economists who warn that we squander nature's finite energy sources at our peril; and Gregg Mitman (essay 17) charts the picaresque career of ragweed (and hay fever) as nature's revenge for greedy land development.

Yet these necessities rarely prove ironclad; if they did, nature's moral authority would be superfluous. No one deliberates the rights and wrongs of obeying the law of gravity; vices necessitated by nature are not blameworthy. But humans often evade, overrule, or even defy nature's purported authority, particularly in the domain of sexual desire and reproduction. Sometimes it is nature's own contradictions that undermine its authority, as when the quirks of individual nature subvert that of the species and cosmos. Cadden notes that within the medieval Christian tradition, Nature, as God's vicar, oversaw the replenishment of the world by ensuring that like reproduced like, but that she also planted unruly desires within individuals that could wreak havoc with her scheme: the worthy might mate with the less worthy; male might couple with male, female with female. Examining the use of the phrase "crimes against nature [*contra naturam*]" in late medieval and early modern sodomy trials, Helmut Puff (essay 9) remarks that within this legal framework, only individuals could act *contra naturam,* a phrase here resonant of both natural and divine order. When they did so, the peculiarities of their own individual natures could be mustered in their defense: nature rebelling against Nature.

Sometimes nature's all-too-oppressive authority is confronted by another, such as that of political freedom. Michelle Murphy (essay 13) shows how twentieth-century feminists sought control over women's reproductive biology as the *sine qua non* of their liberation. So vigorous was the language of control that nature came to be seen as more malleable than culture, sex more readily alterable than gender. Writing in the aftermath of World War II, Japanese political theorist Maruyama Masao also attacked the tyranny of nature in the name of political emancipation. In contrast to Frankfurt School intellectuals who traced German fascism to the ruthless mastery of nature, Maruyama, as Julia Adeney Thomas (essay 12) observes, blamed Japanese fascism on enslavement to nature, understood as the dead hand of tradition and the indolence of sensual pleasures.

The categories of the natural and the nonnatural must be constantly construed and filled with content: is the opposite of nature art? nurture? culture? history? the supernatural? the unnatural? The work of construal and instantiation turns out to be widely and interestingly distributed: sages, poets, physicians, jurists, philosophers, theologians, and scientists all have their say about what nature is and is not, as do political theorists and activists of the most diverse stripes. Nonetheless, there is a distinct tendency in the European tradition from the late Middle Ages to the present to create experts in the natural—at first physicians and natural philosophers in the early modern period, and later scientists. Such experts were (and are) relied on to interpret nature, especially in controversial cases involving politically charged issues such as race, gender, sexuality, colonialism, and labor. Indeed, the very existence of such controversies helped call into existence both nature as an allegedly neutral judge and experts as allegedly disinterested interpreters of nature's verdicts. Nature's neutrality and scientific disinterestedness are thus intertwined, both doctrines emerging at the turn of the nineteenth century, in stark contrast to Enlightenment assertions about the normative dictates of nature and the moral vocation of scientists. But there is another class of experts whose construal of nature is perhaps less visible but no less powerful: the historians charged with explaining how, for good or ill, human arrangements came to diverge from a state of nature. Whether their story is one of the advance of civilization from a primitive state of animality or one that recounts the loss of an Edenic golden age, historians help define the meaning and morality of nature.

Conflicts between nature's authority and its rivals create oppositions that depend on a logic of hypostasis and polarization: nature versus freedom, nature versus art, nature versus civilization, nature versus history, nature versus nurture, nature versus society. The terms in this far-from-exhaustive list overlap but do not coincide; each is the product of a distinct historical context and shifts the authority of nature in a somewhat different direction. Nature versus civilization, for example, opposes the wild to the refined, while nature versus history contrasts the eternal with the mutable. Indeed, nature's authority often depends on this multiplicity and elasticity. Vidal draws attention to the all-inclusiveness of nature for Enlightenment *philosophes,* capacious enough to embrace if not to reconcile opposites. In his account of the fortunes of pedagogical naturalism in nineteenth-century Prussia, Eckhardt Fuchs (essay 6) documents the suppleness of the word *Natur* in educational debates, sometimes opposed to religion and at other times allied with it, in competition with *Bildung* but also supporting it. Nature's authority works by paradox and obscurity as well as by the clash of clear-cut positions. Herein may lie the importance of allegories, personifications, and emblems in representing nature: figurative language and images offer a rich and plastic repertoire for depicting nature's traits, which can be juxtaposed without necessarily being made consistent with one another. Katharine Park (essay 2) follows allegorical and visual traditions of personified nature in medieval and Renaissance Europe, noting how nature can be depicted as at once regal but fragile, a goddess-like figure with a torn gown, or simultaneously bounteous but bestial, her many lactating breasts resembling bovine udders. Perhaps this is why poetry— with its imagery, displacements, and condensations that permit superimposed meanings and oblique implications—plays as prominent a role as treatises do among the sources that proclaim nature's authority.

Whether nature is personified or merely hypostasized, writ small or large, located in the world order or in the makeup of an individual, its authority in the foregoing examples has been explicit and imposed. Hence the tone of admonition and exhortation to obey (or defy) nature's dictates:

nature appears as an external authority, even if its imperatives are lodged deep in the body or psyche. Nature's authority can also be internalized, made "natural" in the sense of seeming inevitable or effortless. Danielle Allen (essay 3) analyzes the scandal provoked by Bernard Mandeville's *Fable of the Bees* among eighteenth-century readers in terms of the "cunning" that disguises the second nature of upbringing as the first nature of inborn human nature. For those who excoriated and burned his book, Mandeville's crime lay not so much in reducing virtue and good manners to self-interest, but rather in exposing allegedly natural conduct as in fact acquired.

Cultivated forms of experience—aesthetic, erotic, sentimental—can also convert nature's authority into a felt reality, in contrast to an imposed command. Lorraine Daston (essay 4) describes how regimens of attention in Enlightenment natural history saturated lowly insects with value—and brought human and natural productions within a common framework of utility. The fierce devotion to insects persisted among nineteenth- and twentieth-century myrmecologists, as Abigail J. Lustig (essay 11) shows, and grounded their recommendations for the future of human societies in observations of ants. In retelling Goethe's epiphany of the *Urpflanze,* Robert J. Richards (essay 5) argues that the poet's erotic awakening paved the way for the recognition of natural archetypes. Negative emotions can work just as powerfully as positive ones to establish nature's authority at a visceral level: both Puff and Vidal emphasize how literary styles attaching expressions of disgust and terror to sodomy and onanism created a reflex response of revulsion among early modern jurists and Enlightenment physicians, respectively. Nature's authority can thus circumvent the discourse of legitimation to penetrate directly into habits, perceptions, judgments, and feelings.

Both arguments of legitimation and self-certifying intuitions make nature their final court of appeal: cases rest when they reach nature, whether led there by evidence or self-evidence. But nature can also be deployed as a means as well as an end of thought. Here its authority derives from the way it eases thinking about otherwise intractable problems. One such problem is how to imagine the structure and functioning of society, an entity at once all-pervasive but invisible. As Price observes in his analysis of the impasse reached in current debates over the economic and ethical valuation of nature, the social is oddly absent in a world divided up into the subjective and the objective, the psychological and the physical. Allen suggests that the principal attraction of the beehive in the history of Western political thinking has been precisely its utility as a way of imagining society, otherwise so elusive: compact and surveyable, the hive serves as an exemplar for the integration of parts into a whole. Nor did Darwinian theory put an end to this long tradition of using social insects to think about human societies, as Lustig notes in her account of how ants were enlisted as models (good and bad) for European and American societies after the devastation of World War I. However distant ants and bees may be from humans by phylogenetic measures, their societies now appeared as alternative solutions to the same problems faced by human societies—all of which had been subsumed under the rubric of the "natural" by the theory of evolution.

Since the Enlightenment, nature's authority has been closely associated with universalism and uniformity, whether nature's claims are made on behalf of the Rights of Man or the metric system. Locality, however, can exert an equally magnetic pull. Fa-ti Fan (essay 16) explores the centrality of place to the efforts of early-twentieth-century Chinese intellectuals to root a new brand of nationalism in nature, which elevated local civil and natural history into the "geobody of the nation." When Maruyama criticized Japanese ultranationalism for its thralldom to nature, Thomas explains, he had in mind among other things the cult of the distinctively Japanese landscape. Mitman's story of ragweed's migrations from wilderness to urban wasteland evoke place both literally and metaphorically: ragweed, like all weeds, is a plant out of place—a weed being anything judged unfit for a garden; ragweed could also symbolize the displaced immigrant in the xenophobic American imagination. Groebner points out that medieval and early modern theories of human types derived from climatic zones, an intellectual tradition that persisted well into the eighteenth century in the form of what Londa Schiebinger (essay 15) calls "environmental racism." The peculiarities of place forged natural identities that could be transmuted into nations and races, natives and foreigners, insiders and outsiders.

Human nature is the point at which the human and the natural intersect, and it is therefore the flashpoint of endless controversies over whether, why, and when nature's authority may be hauled into human affairs. Because human nature implies a species type, essence, or program (depending on one's preference for old-fashioned or newfangled metaphors) that is at once a description and a prescription for how to act, think, and feel, discussions of human nature are inevitably efforts to draw boundaries. These boundaries are almost always simultaneously hierarchies: between ensouled humans and brutes, between hominids and Homo sapiens, between superior and inferior races. How are the boundaries marked? Groebner discerns a shift in the sixteenth century from an internal balance of bodily humors to external visible signs, from Galenic *complexio* to the modern complexion of cosmetic companies. In the context of eighteenth-century medical experimentation, Schiebinger describes how corporeal uniformity could efface social markers, but only up to a point: the bodies of royalty and criminals, slaveholders and slaves, rich and poor were deemed interchangeable for the purposes of testing a new drug or therapy—but not those of men and women. Reporting on current research on human origins, Robert Proctor (essay 18) detects the lingering influence of the 1952 UNESCO Statement on Race in the reluctance of paleoanthropologists to admit the existence of multiple coexisting species of humans. On this view, the unity of human nature, tragically shattered by a now discredited racism, must be preserved at all costs—although the definition of humanness flits from criterion to criterion: upright posture, language, tool use. Each quiddity of the human becomes a standard to uphold, not only with respect to other species, but also within our own: that which is essentially human becomes the trait to promote, be it crafty intelligence or a handy thumb.

The sheer variety and flexibility of nature's meanings and uses has long aroused the suspicions of the analytically minded. From Robert Boyle to Arthur Lovejoy,[12] there have been no end of critics who have made careful lists of the multiple senses of "nature," and some have concluded that the word should be abolished altogether. How can such a protean entity mean anything at all? Cultural critics have swelled the chorus: "nature" is not only unclear; it is dangerous. Almost every ideology seeks to sign up nature for its cause, to bolster its shaky political credentials with nature's authority. Just because that authority has been so widely commandeered, critics imagine nature as a kind of blank screen or mirror on which the most diverse human fantasies may be projected. Paradoxically, the most obdurate authority—nature's necessity—has been joined in these accounts to the most malleable of cultural constructs—nature as anything you care to make it. Following this line of thought, it is not surprising that

12. Robert Boyle, *A Free Inquiry into the Vulgarly Received Notion of Nature* [1685–86], in Thomas Birch, ed., *The Works of the Honourable Robert Boyle* [1772], 6 vols. (Hildesheim: Georg Olms, 1965–66), 5:167–71; Arthur O. Lovejoy and George Boas, *Primitivism and Related Ideas in Antiquity*, with supplementary essays by W. F. Albright and P. E. Dumont (Baltimore: Johns Hopkins University Press, 1935), 447–56. Boyle lists eight senses of the word "nature"; Lovejoy made it to sixty-six.

critical attention has shifted from nature conceived as mute and tractable to those who claim to speak for nature: doctors, scientists, jurists, theologians, politicians, activists. Although there may be excellent reasons for investigating these authoritative voices quite independently of their rhetoric of the natural, to unmask them as nature's ventriloquists does not explain why they go to the trouble. If nature is merely a dummy, whence its authority?

This book suggests that no amount of well-intentioned intellectual and political hygiene is likely to get rid of nature. At the dawn of the "biotech century," as some hail genes as the new stars for human destiny and others revile the geneticizing of human existence as intellectually simplistic and economically exploitative,[13] nature stays in place—and in familiar guise. In 1988, a $3 billion initiative was launched, aimed at mapping and sequencing the human genome. Arguably the grandest scientific enterprise of the late twentieth century, the Human Genome Project (HGP) has given rise to fierce debates that are still partially shaped by older categories. In 1874, for example, the British polymath and eugenicist Francis Galton made up "nature and nurture" as "a convenient jingle of words" to differentiate what he viewed as the two main determinants of each person's physical and psychological makeup and destiny.[14] Its echoes have reverberated throughout the twentieth century—and into the HGP—in debates over the inheritance of intelligence, the origins of sexual orientation and racial differences, the possibilities and limits of education, and the causes of crime. The controverted questions remain unresolved, even if the HGP seemed to herald victory for nature over nurture. Such a victory, however, would also signal nature's demise: nurture as control of persons and environments would become the means of creating and maintaining an order of normality whose only claim to naturalness could lie in the objects it genetically engineers. For those who fear such a prospect, the line between gene surgery and gene manipulation looks perversely thin.

13. Jeremy Rifkin, *The Biotech Century: Harnessing the Gene and Remaking the World* (New York: Jeremy P. Tarcher/Putnam, 1998). James D. Watson, codiscoverer of the structure of DNA and first director of the Human Genome Project, claimed, "We used to think our fate was in the stars. Now we know, in large measure, our fate is in our genes" (quoted by Leon Jaroff, "The Gene Hunt," *Time,* 20 March 1989, 67). Critics include, among others, Ruth Hubbard and Elijah Wald, *Exploding the Gene Myth; How Genetic Information Is Produced and Manipulated by Scientists, Physicians, Employers, Insurance Companies, Educators, and Law Enforcers* (Boston: Beacon Press, 1993); and Richard C. Lewontin, *It Ain't Necessarily So: The Dream of the Human Genome Project and Other Illusions* (New York: New York Review of Books, 2000).

14. Francis Galton, *English Men of Science: Their Nature and Nurture* (London: Macmillan, 1874), 12.

Issues about the dignity of nature and humans—about their being considered not as means, but as ends in themselves—continue to be examined in fiction, philosophy, and moral theory. In connection with the HGP, however, they are most concretely threshed out in the abstruse domain of patents and trademarks. Here questions about the value of nature in its most hardheaded economic form are front and center. Since the 1980s, courts have ruled that it is possible to patent human cell lines, arguing for the manufactured character of cell products (for example, antibodies), asserting the nonidentity and discontinuity of cell lines and the providers of the primary cells, attributing property rights to an individual, institution or company, and in effect redefining biological objects as inventions.[15]

Central to the patenting of life (from cell lines to organisms) are issues of exploitation, economic justice, and the ethics of ownership; so is the vision of nature as a self-standing source of value and authority that ought to remain absolutely distinct from human invention. Resistance to the Human Genome Diversity Project rapidly developed when in 1998 the Icelandic Parliament licensed one company to create a database connecting the population's genealogical and health records to a database of individual DNA genotypes.[16] That same year, a UNESCO declaration proposed a ban on the commercial uses of the human genome "in its natural state."[17] Indigenous peoples, already hostile to bioprospecting as an instance of bio-colonialism, concurred. Their opposition to the patenting of "all natural genetic materials" stems from their wish to ensure "the continuity of the natural order," "hold precious all life in its natural form," and refusal to "violate the principles of Creation by manipulating and changing" it.[18] For the Roman Catholic Church, intervention is permissible "not in order to modify nature but to favor its development;" and it is imperative to avoid the "demiurgic degeneration of research."[19] Patenting "raw" nature stands condemned here because it violates the boundaries that define humanity's

15. Hannah Landecker, "Between Beneficence and Chattel: The Human Biological in Law and Science," *Science in Context* 12 (1990): 203–25.

16. See Association of Icelanders for Ethics in Science and Medicine, http://www.mannvernd.is/english/home.html.

17. "Universal Declaration on the Human Genome and Human Rights" (1998), art. 4, http://www.unesco.org/ibc/en/genome/projet/.

18. "Declaration of Indigenous Peoples of the Western Hemisphere Regarding the Human Genome Diversity Project" (1995), http://www.indians.org/welker/genome.htm.

19. John Paul II, "Dangers of Genetic Manipulation" (1983), widely available on the Web; and Pontificia Academia Pro Vita, *Reflexions on Cloning* (1997), § 4, http://www.vatican.va/roman_curia/pontifical_academies/acdlife/documents/rc_pa_acdlife_doc_30091997_clon_en.html.

place in a universal order where the authority of nature is second only to God's.

Indeed, the biotech era comes close to blurring the distinction between Aristotle's two efficient causes of things, nature and art.[20] On the front cover of this book, Charles Sheeler's 1943 painting *The Artist Looks at Nature* archly conflates these two ancient categories: an artist paints from nature, but the landscape that surrounds his easel is a domesticated one of walls and other human handiwork in a manicured green setting; and the image on his canvas represents a further step away from nature toward art by depicting the interior of a house in tones of black and white. Yet if all art is really nature (the motto of the scientific revolution of the seventeenth century) or all nature really art (the promise of the genetic revolution of the twenty-first century), then the ironies of Sheeler's painting dissolve into a plain statement of fact. At once unbiquitous and elusive, nature remains as protean as ever.

Nature may be a quick-change artist, but it is a hard-working one. These essays document the diversity of nature's work, and its peculiar efficacy. In the most varied contexts, nature stands for order, sometimes to the point of tyranny: the division of labor among insects and humans, the great circle of the constellations and seasons, the fixity of types, the balanced give-and-take of exchange, the conservation of energy, the hierarchy of kinds, the permanence of place, the truth of bodies, the immanent justice of cause and effect. Within the framework of nature's moral authority, even the disorder wrought by earthquakes and floods becomes part of a scheme of vengeance for human malfeasance. Whether this order is perceived as harmonious or oppressive, ineluctable or changeable, just or unjust, it is oddly insidious. The natural is synonymous with the self-evident, melding habit with duty. The "is" and "ought" blur together, despite strenuous efforts to hold them apart. Nature's order seems to reconcile autonomy and obedience, the strait and narrow path to virtue with the lazy path of least resistance. Hence the steady tug of nature's authority, despite centuries of closely argued criticism. "Doing what comes naturally" holds out the dream of the self-enforcing rule.

Authority presumes voice, and centuries' worth of allegories, personifications, and hypostatized abstractions bear witness to the urge to give nature a voice that could exhort, reproach, and command: "The oration of

20. Efficient causes concern who or what made an object; the other three causes Aristotle distinguishes in *Physics* II.3 are the formal (what is it?), the material (what is it made of?), and the final (what is it for?).

the worlds is untiring," as the Polish poet Zbigniew Herbert intones. This is more than finding "books in brooks and tongues in trees"; it is a longing for an authority at once benevolent and stern, inexorable yet elastic, at once rooted in the local and spanning the global. But standing "in front of the elements," the human voice realizes, with alarm or relief, that it will not "bring out stillness / from the grainy silence."[21]

Bibliography

Primary sources are listed in chronological order. The title by which the work is best known in the anglophone scholarly literature is given first, followed by English translation in brackets. No publishing information has been given for early works, which exist in multiple editions; it is given for primary sources published in the twentieth century. Secondary sources are listed alphabetically by the author's last name.

Primary Sources

Hesiod, *Works and Days* (8th c. B.C.E.)

Hippocratic Corpus, *On Airs, Waters, and Places* (5th c. B.C.E.)

Plato, *Gorgias* (4th c. B.C.E.)

Aristotle, *Physics* (4th c. B.C.E.)

Aristotle, *Nicomachean Ethics* (4th c. B.C.E.)

Virgil, *Georgics* (29 B.C.E.)

Pliny the Elder, *Natural History* (77)

Anon., *Physiologus* [*The Naturalist*] (2nd c.)

Galen, *Quod animi mores corporis temperamenta sequantua* [*That the Faculties of the Soul Follow the Temperaments of the Body*] (ca. 200)

Augustine of Hippo, *De civitate Dei* [*The City of God*] (413–20)

Justinian, *Digesta* [*Digest*] (533)

Peter Damian, *Liber Gomorrhianus* [*Book of Gomorrah*] (ca. 1050)

Alain de Lille, *De planctu naturae* [*The Complaint of Nature*] (ca. 1170)

Albertus Magnus, *De animalibus libri XXVI* [*On Animals*] (ca. 1250)

Thomas Aquinas, *Summa theologica* (ca. 1260)

Guillaume de Lorris and Jean de Meun, *Le roman de la rose* [*The Romance of the Rose*] (ca. 1230, 1270)

Gonzalo Fernández de Oviedo, *Historia General y Natural de las Indias* [*General and Natural History of the Indies*] (1535)

Michel de Montaigne, *Essais* [*Essays*] (1588)

Hugo Grotius, *De iure belli ac pacis* [*The Law of War and Peace*] (1625)

Thomas Browne, *Pseudodoxia Epidemica: Or, Enquiries into Very Many Received Tenents and Commonly Presumed Truths* (1646)

21. Zbigniew Herbert, "Mr. Cogito Considers the Difference between the Human Voice and the Voice of Nature" (1974), trans. from the Polish by John and Bogdana Carpenter, *Mr. Cogito* (Hopewell, N.J.: Ecco Press, 1993).

René Descartes, *Les principes de la philosophie* [*Principles of Philosophy*] (1647; 1st Latin ed.: 1644)

René Descartes, *Les passions de l'âme* [*Passions of the Soul*] (1650)

Thomas Hobbes, *Leviathan* (1651)

Samuel Pufendorf, *De officio hominis et civis iuxta naturalem* [*On the Duty of Man and the Citizen according to Natural Law*] (1673)

Robert Boyle, *A Free Inquiry into the Vulgarly Received Notion of Nature* (1686)

William Derham, *Physico-Theology* (1713)

Bernard Mandeville, *The Fable of the Bees* (1714)

Giambattista Vico, *Scienza nova* [*The New Science*] (1725)

Alexander Pope, *An Essay on Man* (1734)

David Hume, *Treatise of Human Nature* (1739)

Charles-Louis de Secondat, baron de Montesquieu, *De l'esprit des lois* [*The Spirit of the Laws*] (1748)

Julien Offray de La Mettrie, *L'Homme machine* [*Man a Machine*] (1748)

Georges-Louis Leclerc, comte de Buffon, *Histoire naturelle* [*Natural History*], (1749–1804)

Adam Smith, *The Theory of the Moral Sentiments* (1759)

Carl Linnaeus, *Dissertatio academica de politia naturae* [*On the Police of Nature*] (1760)

Jean-Jacques Roussseau, *Du Contrat social* (1762)

Jean-Jacques Rousseau, *Emile* (1762)

Paul Heinrich Dietrich, baron d'Holbach, *Le Système de la nature, ou les lois du monde physique et du monde moral* [*The System of Nature, or the Laws of the Physical World and of the Moral World*] (1770)

Denis Diderot, *Supplément au voyage de Bougainville* [*Supplement to the Voyage of Bougainville*] (1772)

Johann Friedrich Blumenbach, *Geschichte und Beschreibung der Knochen des menschlichen Körpers* [*History and Description of the Bones of the Human Body*] (1786)

Gilbert White, *The Natural History and Antiquities of Selbourne* (1789)

Constituent Assembly of France, *La Déclaration des droits de l'homme et du citoyen* [*Declaration of the Rights of Man and the Citizen*] (1789)

Immanuel Kant, *Kritik der reinen Vernunft* [*Critique of Pure Reason*] (1781)

Immanuel Kant, *Prolegomena zu einer jeden künftigen Metaphysik, die als Wissenschaft wird auftreten können* [*Prolegomena to Any Future Metaphysics*] (1783)

Mary Wollstonecraft, *A Vindication of the Rights of Women* (1792)

Thomas Malthus, *An Essay on the Principles of Population* (1798)

Friedrich Wilhelm Joseph von Schelling, *Erster Entwurf eines Systems der Naturphilosophie* [*First Draft of a System of Natural Philosophy*] (1799)

Pierre Jean Georges Cabanis, *Rapports du physique et du moral de l'homme* [*Relationships of the Physical and Moral in Man*] (1801)

William Paley, *Natural Theology* (1802)

Ralph Waldo Emerson, *Nature* (1837)

Alexander von Humboldt, *Kosmos* (1844)

Henry David Thoreau, *Walden* (1854)

Charles Darwin, *On the Origin of Species by Means of Natural Selection, or the Preservation of Favoured Races in the Struggle for Life* (1859)

George Perkins Marsh, *Man and Nature* (1864)

Francis Galton, *Hereditary Genius: An Inquiry into Its Law and Consequences* (1869)

John Stuart Mill, *The Subjection of Women* (1869)

John Stuart Mill, "Nature" (1874, but written between 1850 and 1858), in *Essays on Ethics, Religion and Society*

Oscar Wilde, *The Decay of Lying* (1889)

Charles Darwin, *The Descent of Man, and Selection in Relation to Sex* (1871)

Friedrich Engels, *Dialektik der Natur* [*Dialectics of Nature*] (1883)

Thomas Henry Huxley, *Evolution and Ethics, and Other Essays* (1893)

Herbert Spencer, *The Synthetic Philosophy* (completed 1896)

Ernst Haeckel, *Die Welträthsel. Gemeinverständliche Studien über monistische Philosophie* [*The Riddle of the Universe at the Close of the Nineteenth Century*] (1899)

Peter Kropotkin, *Mutual Aid: A Factor of Evolution* (Montreal: Black Rose Books, 1989 [1902])

G. E. Moore, *Principia Ethica* (Cambridge: Cambridge University Press, 1993 [1903])

Sigmund Freud, *Civilization and Its Discontents* (New York: Norton, 1989 [1930]) [original title: *Das Unbehagen in der Kultur*]

Paul Sears, *Deserts on the March* (Norman: University of Oklahoma Press, 1980 [1935])

Fairfield Osborn, *Our Plundered Planet* (Boston: Little, Brown, 1948)

Aldo Leopold, *A Sand County Almanac* (New York: Oxford University Press, 2001 [1949])

Rachel Carson, *Silent Spring* (Boston: Houghton Mifflin, 1994 [1962])

Paul Ehrlich, *The Population Bomb* (Rivercity, Mass.: Rivercity Press, 1975 [1968])

Barry Commoner, *The Closing Circle* (New York: Bantam Books, 1974 [1971])

E. O. Wilson, *Sociobiology: The New Synthesis* (Cambridge: Harvard University Press, Belknap Press, 2000 [1975]

Richard Dawkins, *The Selfish Gene* (Oxford: Oxford University Press, 1999 [1976])

Hans Jonas, *The Imperative of Responsibility: In Search of an Ethics for the Technological Age* [original title: *Das Prinzip Verantwortung: Versuch einer Ethik für die technologische Zivilisation*] (Chicago: University of Chicago Press, 1984 [1979])

Richard Alexander, *Darwinism and Human Affairs* (Seattle: University of Washington Press, 1979)

Carolyn Merchant, *The Death of Nature: Women, Ecology, and the Scientific Revolution* (London: Wildwood House, 1980)

Robert Frank, *Passions within Reason: The Strategic Role of the Emotions* (New York: Norton, 1988)

Judith Butler, *Gender Trouble: Feminism and the Subversion of Identity* (New York: Routledge, 1990)

Frans de Waal, *Good Natured: The Origins of Right and Wrong in Humans and Other Animals* (Cambridge: Harvard University Press, 1996)

Secondary Sources

Beer, Gillian, *Darwin's Plots: Evolutionary Narrative in Darwin, George Eliot, and Nineteenth-Century Fiction* (London: Routledge, 1983)

Caplan, Arthur L., ed., *The Sociobiology Debate: Readings on Ethical and Scientific Issues* (New York: Harper & Row, 1978)

Charlton, D. G., *New Images of the Natural in France: A Study in European Cultural History 1750–1800* (Cambridge: Cambridge University Press, 1984)

Chiffoleau, Jacques, "*Contra naturam:* Pour une approche casuistique et procédurale de la nature médiévale," *Micrologus* 4 (1996): 265–312

Coates, Peter, *Nature: Western Attitudes since Ancient Times* (Cambridge: Polity Press, 1998)

Collingwood, R. G., *The Idea of Nature* (New York: Oxford University Press, 1945)

Cronon, William, ed., *Uncommon Ground: Rethinking the Human Place in Nature* (New York: Norton, 1996)

Davis, Mike, *Ecology of Fear: Los Angeles and the Imagination of Disaster* (New York : Metropolitan Books, 1998)

Descola, Philippe, and Gísli Pálsson, eds., *Nature and Society: Anthropological Perspectives* (London: Routledge, 1996)

Economou, George, *The Goddess Natura in Medieval Literature* (Cambridge: Harvard University Press, 1972)

Ehrard, Jean, *L'idée de la nature en France dans la première moitié du XVIIIe siècle* (Paris: Albin Michel, 1994 [1963])

Farber, Paul Lawrence, *The Temptations of Evolutionary Ethics* (Berkeley: University of California Press, 1998 [1994])

Glacken, Clarence J., *Traces on the Rhodian Shore: Nature and Culture in Western Thought from Ancient Times to the End of the Eighteenth Century* (Berkeley: University of California Press, 1967)

Gottlieb, Robert, *Forcing the Spring: The Transformation of the American Environmental Movement* (Washington, D.C.: Island Press, 1993)

Gould, Stephen Jay, *The Mismeasure of Man* (New York: Norton, 1981)

Grgas, Stipe, and Svend Erik Larsen, eds., *The Construction of Nature: A Discursive Strategy in Modern European Thought* (Odense: Odense University Press, 1994)

Grove, Richard. H., *Green Imperialism: Colonial Expansion, Tropical Island Edens and the Origins of Environmentalism, 1600–1860* (Cambridge: Cambridge University Press, 1995)

Hacking, Ian, *The Social Construction of What?* (Cambridge: Harvard University Press, 1999)

Haraway, Donna, *Primate Visions: Gender, Race, and Nature in the World of Modern Science* (New York: Routledge, 1989)

Harding, Sandra, ed., *The "Racial" Economy of Science: Toward a Democratic Future* (Bloomington: Indiana University Press, 1993)

Harvey, David, *Justice, Nature, and the Geography of Difference* (Oxford: Blackwell, 1996)

Heinemann, Felix, *Nomos und Physis: Herkunft und Bedeutung einer Antithese im griechischen Denken des 5. Jahrhunderts* (Darmstadt: Wissenschaftliche Buchgesellschaft, 1965 [1945])

Henderson, John, *The Development and Decline of Chinese Cosmology* (New York: Columbia University Press, 1984; Neo-Confucian Studies, 11)

Hofstader, Richard, *Social Darwinism in American Thought: 1860–1915* (Boston: Beacon Press, 1992 [rev. ed. 1955]), with an introduction by Eric Foner

Hubbard, Ruth, *The Politics of Women's Biology* (New Brunswick: Rutgers University Press, 1990)

Jordan, Mark D., *The Invention of Sodomy in Christian Theology* (Chicago: University of Chicago Press, 1997)

Lewis, C. S., "Nature," in *Studies in Words* (Cambridge: Cambridge University Press, 1990 [1960])

Lovejoy, Arthur O., *The Great Chain of Being: A Study of the History of an Idea* (Cambridge: Harvard University Press, 1936)

Lovejoy, Arthur O., and George Boas, *Primitivism and Related Ideas in Antiquity* (Baltimore: Johns Hopkins University Press, 1935)

Lovejoy, Arthur O., "Nature as an Aesthetic Norm," in *Essays in the History of Ideas*, 69–78 (Baltimore: Johns Hopkins University Press, 1948)

MacCormack, Carol P., and Marylin Strathern, eds., *Nature, Culture and Gender* (Cambridge: Cambridge University Press, 1980)

Maienschein, Jane, and Michael Ruse, eds., *Biology and the Foundation of Ethics* (Cambridge: Cambridge University Press, 1999)

McCulloch, Florence, *Medieval Latin and French Bestiaries* (Chapel Hill: University of North Carolina Press, 1962)

Midgley, Mary, *The Ethical Primate: Humans, Freedom, and Morality* (London: Routledge, 1994)

Mitman, Gregg, *Reel Nature: America's Romance with Wildlife on Film* (Cambridge: Harvard University Press, 1999)

Moderson, Mechthild, *Natura als Goettin im Mittelalter: Ikonographische Studien zu Darstellungen der personifizierten Natur* (Berlin: Akademie Verlag, 1997)

Morris-Suzuki, Tessa, "Concepts of Nature and Technology in Pre-Industrial Japan," *East Asian History* 1 (1991): 81–97

Nasr, Seyyed Hossein, "The Meaning of Nature in Various Perspectives in Islam," in Nasr, *Islamic Life and Thought*, 96–101 (London: George Allen & Unwin, 1981)

Needham, Joseph, "Human Law and the Laws of Nature," in Needham, *The Grand Titration: Science and Society in East and West*, 299–331 (Toronto: University of Toronto, 1969)

Oelschlaeger, Max, *The Idea of "Wilderness": From Prehistory to the Age of Ecology* (New Haven: Yale University Press, 1991)

Richards, Robert J., *Darwin and the Emergence of Evolutionary Theories of Mind and Behavior* (Chicago: University of Chicago Press, 1987)

Roetz, Heiner, "Etymologische Vorbemerkung zum chinesischen Begriff der 'Natur,'" in Roetz, *Mensch und Natur im alten China*, 113–17 (Frankfurt am Main: Peter Lang, 1984)

Russett, Cynthia Eagle, *Sexual Science: The Victorian Construction of Womanhood* (Cambridge: Harvard University Press, 1989)

Schiebinger, Londa, *Nature's Body: Gender in the Making of Modern Science* (Boston: Beacon Press, 1993)

Sieferle, Rolf Peter, and Helga Breuninger, *Natur-Bilder: Wahrnehmungen von Natur und Umwelt in der Geschichte* (Frankfurt: Campus, 1999)

Singer, Peter, *The Expanding Circle: Ethics and Sociobiology* (New York: Farrar, Straus & Giroux, 1981)

Sivin, Nathan, "State, Cosmos, and Body in the Last Three Centuries B.C.," *Harvard Journal of Asiatic Studies* 55, no. 1 (1995): 5–37

Soper, Kate, *What Is Nature?* (Oxford: Blackwell, 1995)

Teich, Mikulás, Roy Porter, and Bo Gustafsson, eds., *Nature and Society in Historical Context* (Cambridge: Cambridge University Press, 1997).

Thomas, Keith, *Man and the Natural World: A History of the Modern Sensibility* (New York: Pantheon, 1983)

Tillyard, E. M. W., *The Elizabethan World Picture* (Harmondsworth: Penguin, 1984 [1943])

Tocanne, Bernard, *L'idée de nature en France dans la second moitié du XVIIe siècle: Contribution à l'histoire de la pensée classique* (Paris: Klincksieck, 1978)

Torrance, John, ed., *The Concept of Nature* (Oxford: Clarendon Press, 1992)

Williams, Raymond, "Ideas of Nature," in Williams, *Problems in Materialism and Culture: Selected Essays* (London: Verso, 1980)

———. "Nature," in *Keywords: A Vocabulary of Culture and Society,* rev. ed., 219–24 (London: Fontana, 1988 [1983])

Worster, Donald, *Dust Bowl: The Southern Plains in the 1930s* (New York: Oxford University Press, 1979)

Zimmermann, Albert, and Andreas Speer, eds., *Mensch und Natur im Mittelalter,* 2 vols. (Berlin: de Gruyter, 1991)

PART ONE

Values

Pure gold, exotic plants, noble deeds, illuminated manuscripts, unicorn horns, intricate tapestries, sacred relics, a good will, sublime landscapes, a Cezanne, heirlooms—value takes a myriad of forms. Its sources—rarity, labor, virtue, utility, divinity, beauty, sentiment—are equally varied. Value can be as concrete as the Hope Diamond and as abstract as duty. It encompasses the market and the museum, the sacred and the profane. There is a great mystery about value in modern social thought: how is value created? Another mystery is that of authority, this time a mystery of conservation: how is authority sustained? Both value and authority belong to the realm of the human—at least in the context of the modern West. To speak of the values of animals is to invent fables; to invoke the authority of the sun and moon is to wax metaphorical. What kind of value, then, is lodged in nature: what are its sources, its forms, its functions? In ancient Greece, the Renaissance, and modernity, terms of worth blurred economic, moral, and aesthetic categories. "Due measure," "abundance," "utility," and "pleasure"—such is the coin of nature's value. The essays in this section explore the subtle processes by which nature is made a repository for value and how that value comes to wield authority.

The best kind of authority works invisibly. If it must brandish weapons and admonish its subjects, it only advertises its weakness; the stablest order is unfelt and unquestioned. Values can replace law with an internalized authority, such as Adam Smith's "man within the breast" or Kant's "moral law within" that rules without coercion. To recognize a value as such is *ipso facto* to grant it a title to authority. But there are degrees in the authority exercised by values. Values may be shallowly or deeply held, irrefragable or contested. At one end of the spectrum are values so entrenched as to seem intuitively evident; at the other are values that must do battle with rivals. Nature can be invoked at both extremes. When Goethe praised the beauties of nature above those of classical art, he appealed to the immediacy of his experience (see Richards, essay 5), but when pedagogical reformers in nineteenth-century Prussia fought for a curriculum based on *Natur* rather than *Bildung,* they entered an arena of argument and counterargument (see Fuchs, essay 6). At any point along this continuum, of course, a number of meanings of nature may be at stake. Values are sometimes shored up by appeals to human nature: these lay claim to being built into the definition of the human—how we think, see, and feel. To deviate from these values is to raise questions about membership in the species, to invite the accusation of inhumanity. Detailed narratives about the world's physical resources can evoke a utopian vision of a golden age or set boundaries on human livelihood, beyond which lurks vengeful nature (see Slatkin, essay 1). And other values are grounded in nature as a universal order, a trump to the claims of the lesser orders of art, civilization, society.

The familiarity of such appeals to the values of nature tends to obscure the oddity of the strategy. Behind the question as to whether nature can really be saddled with the values at issue lies the prior question of why nature's imprimatur should certify values at all. Human nature can be seen as deeply marred by original sin or some other intrinsic flaw, and therefore as an obstacle to be overcome rather than as a beacon to be followed. The "nature" side of oppositions like "nature versus art" or "nature versus nurture" need not automatically be the superior one. Locke and his contemporaries praised the cultivated lands of European civilization over the wilderness of American nature; gardeners prize the tulips perfected by horticulture over the natural variety found in the Levant. Nature must be made valuable and values must be found in nature before nature's values can be converted into authority. The mechanisms of value construction are as important as the subsequent disposition of values in structures of authority. The essays in this section document a range of such mechanisms, ranging from the psychological to the visual to the rhetorical. Eighteenth-century naturalists redeemed objects once deemed contemptible by showering attention on them (see Daston, essay 4); Renaissance iconographers pictured nature as a lactating mother, a symbol of generous bounty (see Park, essay 2); contemporary environmentalists argue for nature's worth by enlisting economics and ethics (see Price, essay 7). As these examples suggest, both the value and the values of nature are established as often by appeal to the imagination and senses as to reason and argument.

Once in place, the values of nature ideally assert their authority by appearing not to. "Doing what comes naturally" in the modern context implies freedom from felt struggle. The glib phrase hides, however, a whole history of attempts to define what *should* come naturally. Nature's values sustain an authority that seeks to transcend its own coerciveness. Hence the importance of ideas of harmony and balance; hence also the complicity between first and second nature, between human nature and habit (see Allen, essay 3). As Bernard Mandeville pointed out, "ease" is taken to be an infallible sign of the natural, even though the "cunning" of socialization can produce that ease as effectively as nature itself. Key to the efficacy of the "cunning" of second nature is that it be unconscious, inaccessible to deliberation and therefore to choice. To act in accordance with the values of nature is not to be, like Hamlet, "sicklied o'er with the pale cast of thought." In these cases, the order of nature seems to be at once the strictest and gentlest of all orders: strict, because ineluctable; gentle, because spontaneous. Even those moralists who, like Kant, rejected nature as a source of value yearned for something like the effortless authority of nature: "the law I give myself." In the paradox of unauthoritarian authority lies the almost irresistible conflation of the descriptive and the normative that is so characteristic of nature's values. When these values have been thoroughly internalized, conduct becomes "normal" in all senses of the word: at once the usual and the desirable, where "is" coincides with "ought."

＊ I ＊

Measuring Authority, Authoritative Measures:
Hesiod's *Works and Days*

Laura M. Slatkin

μέτρα φυλάσσεσθαι· καιρὸς δ' 'ἐπὶ πᾶσιν ἄριστος.
[Observe due measure: and best in all things is the right time and right amount.]

HESIOD, *Works and Days* 694

"Ἥλιος οὐχ ὑπερβήσεται μέτρα· εἰ δὲ μή,
'Ερινύες μιν Δίκης ἐπίκουροι ἐξευρήσουσιν.
[The sun will not transgress his measures. If he does, the Furies, ministers of Justice, will find him out. Trans. C. H. Kahn.]

HERACLITUS, fr. 94D-K

If in Greek tragedy the Furies pursue human beings who violate the laws of kinship—of family exchanges properly conducted, taboos properly observed—in Heraclitus's discourse they pursue the potentially transgressive

The Moral Authority of Nature seminar at the Max Planck Institut für Wissenschaftsgeschichte was a memorably exhilarating model of *eris*-free collegial exchange and collaboration; my deep thanks to all the participants—especially to Lorraine Daston and Fernando Vidal—for their immeasurable intellectual generosity. The clarifying suggestions of Raine and Fernando, as well as those of Danielle Allen and Julia Thomas, continue to inform my thoughts on Hesiodic norms and values; I hope their comments will all find a place in my forthcoming study on archaic poetics from which the present essay draws. Many more debts will be acknowledged there to my Hellenist colleagues, among them Pat Easterling, Liz Irwin, Nicole Loraux, Gloria Pinney and Jamie Redfield, who have helped me to think through the issues touched on here, and above all, to Maureen McLane, for invaluable encouragement and illuminating, rigorous critique of all aspects of this project. This essay is dedicated to Nicole Loraux, cherished friend.

sun. Fragment 94 of this pre-Socratic philosopher (ca. 500 B.C.E.) offers an apparent paradox, not so much in its use of mythological personae to formulate a theory of cosmic structure (a practice common to all surviving sixth-century natural philosophy) as in its account of the relations among these figures. Throughout early Greek literature, the all-seeing and all-revealing sun bears witness to every action in both the human and the divine domains, functioning as the ultimate monitor of events that even the gods wish to conceal. Here, however, the sun's own celestial operations are themselves subject to the scrutiny of the shadowy, chthonic Furies, irascible informants whose realm lies deep within the earth.

Why then, in Heraclitus's book on the structure of the cosmos, are these dark and vengeful spirits imagined to oversee the movement of sun? And equally puzzling, why is Justice concerned with the passage of the day? At the center of this conjunction of radiance and obscurity, of sky and land, of the ephemeral journey of the sun (who, as Heraclitus tells us in another passage, is new every day) and the eternal, inflexible reckoning of the Furies, allies of Justice, is the element on which everything is staked: measure *(metra)*. Heraclitus's formulation posits terms, concepts, and problematic relations that will inform the rest of this essay, in which I propose to explore the relationship of nature to morality within the context of the philosopher's antecedents among the poets of early (preclassical) Greece. I hope to show that in early Greek thought and poetics, both values and norms—whether proposed as "cultural" or "natural" or "divine"— are derived through and embedded in a rich discourse and figurative complex of what Hesiod invokes as "due measure," which Heraclitus revisits in his fragment. For Greek thinkers of the archaic period (late eighth through early sixth centuries B.C.E.), right order—in the cosmos, between gods and men, among humans, between men and the earth—is not, as it were, *given* or transparently natural, but is rather in need of extensive poetic elaboration and explanation. For these writers, the processes and patterns of nature are read as a system ideally in equilibrium. Human beings are enjoined to model their behavior on, and to accord their actions with, the equilibrating logic of nature. That nature is seen to manifest such equilibrium, however violently its upheavals may appear at any time, is of course a feat of human cognition.[1] The values encoded in the notion of measure thus involve a transference from the nature that is their imagined source

---

1. For archaic thinkers' turn to nature as a cognitive model as well as a field for perception, see J. H. Finley Jr., *Four Stages of Greek Thought* (Stanford: Stanford University Press, 1965).

to the social and ethical order that is the explicit concern of such poets as Hesiod. To examine the reciprocity of nature and human values, I follow Heraclitus back to the tradition of didactic literature represented by the earliest extant Greek example of what is classified as a "wisdom" text: Hesiod's *Works and Days,*[2] a poem from the Greek mainland in the eighth century B.C.E., which stands as the first Greek text we possess that is overtly and explicitly moralizing.

In contrast to the reticence of Homeric epic, which is free of narrative judgment about, for example, the origins and conduct of the Trojan War, the *Works and Days* offers instruction on how to live an orderly and productive life. Hesiod's poem imagines a personal dispute as the occasion for a meditation on ethical behavior and on the role of justice in the construction of social order. In some eight hundred verses, the *Works and Days* takes the form of a set of instructions delivered by the poet to his brother Perses, who, in the division of their joint inheritance (so the poet tells us), has unfairly seized an unequal share. In making his illegitimate grab, the poem complains, Perses has been supported by corrupt leaders (kings) who also function, as was customary, as arbiters and judges; both Perses and the judges constitute the poem's purported addressees:

> [L]et us settle our dispute here with true judgement which is of Zeus and is perfect. For we had already divided our inheritance, but you seized the greater share and carried it off, greatly swelling the glory of our bribe-swallowing lords who love to judge such a cause as this.[3] (35–39)

In enjoining its audience(s), both within the poem and outside it,[4] to eschew the fecklessness of a Perses and the greed of those in power, the poem invokes cosmic patterns of the natural world within the context of defining human nature and the human condition.

2. Hesiod, *Works and Days,* in *Hesiod, the Homeric Hymns and Homerica,* trans. H. G. Evelyn-White (London: Heinemann, Loeb Classical Library, 1929). All translations of *Works and Days* are from this edition.

3. ἀλλ' αὖθι διακρινώμεθα νεῖκος
ἰθίῃσι δίκῃς, αἵ τ' ἐκ Διός εἰσιν ἄρισται.
ἤδη μὲν γὰρ κλῆρον ἐδασσάμεθ', ἀλλὰ τὰ πολλὰ
ἁρπάζων ἐφόρεις μέγα κυδαίνων βασιλῆας
δωροφάγους, οἳ τήνδε δίκην ἐθέλουσι δίκασσαι.

4. On the issue of the audiences (external and internal) of the *Works and Days,* as well as questions of persona and identity, see Mark Griffith, "Personality in Hesiod," *Classical Antiquity* 2, no. 1 (1983): 47–62.

Notably, there is no abstract term "Nature" to designate the natural world in the literature of this period in Greece; the word *phusis* never appears in Hesiod, nor ever in this sense in Homer.[5] What "nature" is and how it operates emerges from accounts of its specific, concrete individual elements, and from their presence in a range of tropes, including metaphor, personification, riddle, fable, and proverb. Nature is most often represented by the characteristics and seasonal processes of "the earth," through which the poem figures questions of morality and social order. Thus the poem tells us:

> Neither famine nor disaster ever haunt men who do true justice, but light-heartedly they tend the fields which are all their care. The earth bears them victual in plenty, and on the mountains the oak bears acorns upon the top and bees in the midst. Their woolly sheep are laden with fleeces; their women bear children who resemble their parents.[6] (230–35)

Significantly, the man who is identified as practicing "true justice" is not a judge but a farmer. Because humankind must toil to earn a livelihood from the earth, farming serves, in the *Works and Days,* as the governing trope for the human condition. The "Works" of the poem's title translates the (plural) noun *erga,* which in early Greek poetry specifically denotes agricultural work and also occurs with the meaning of "tilled fields." The *Works and Days,* however, does not aim to teach lessons about farming.[7] In its concern with justice and ethical behavior, the poem uses the farmer to think with because it is through farming that humans are most immersed in natural processes, and the farmer is the human type who most obviously must accord his behavior with the exigencies and contingencies of nature's patterns.

Although Hesiod refers to "true justice," justice itself is never defined in the literature of this period; rather, justice emerges in the passage above

---

5. Similarly, neither Hesiod nor Homer uses the term *kosmos* (κόσμος) as an abstraction meaning "unified world order," as do sixth-century natural philosophers.

6. οὐδέ ποτ᾽ ἰθυδίκῃσι μετ᾽ ἀνδράσι λιμὸς ὀπηδεῖ
οὐδ᾽ ἄτη, θαλίης δὲ μεμηλότα ἔργα νέμονται.
τοῖσι φέρει μὲν γαῖα πολὺν βίον, οὔρεσι δὲ δρῦς
ἄκρη μέν τε φέρει βαλάνους, μέσση δὲ μελίσσας·
εἰροπόκοι δ᾽ ὄιες μαλλοῖς καταβεβρίθασιν·
τίκτουσιν δὲ γυναῖκες ἐοικότα τέκνα γονεῦσιν·

7. See Stephanie Nelson, *God and the Land: The Metaphysics of Farming in Hesiod and Vergil* (Oxford: Oxford University Press, 1998).

as homologous with the order of nature that it both generates and imitates. Nature cooperates with, as well as rewards, the man who does "true justice." Here we see a crucial conceptual link: the order of justice and the order of nature reciprocally substantiate each other.

Hesiod's account of the just farmer rewarded by plenty resembles the poem's vision of life in the earliest phase of mortal existence, the Golden Age:

> First of all the immortal gods who dwell on Olympus made a golden race of mortal men who lived in the time of Cronos when he was reigning in heaven. And they lived like gods without sorrow of heart, remote and free from toil and grief: wretched age rested not on them; but with legs and arms never failing they took pleasure in feasting beyond the reach of all evils . . . and they had all good things; for the fruitful earth of its own accord bore them fruit in abundance and unstinting. They lived at ease and peace upon their lands with many good things, rich in flocks and loved by the blessed gods.[8] (109–20)

In what way can our collaboration with nature in that long-vanished past be reproduced through the efforts of the farmer, laboring in the here and now? What is the relationship between the world measured by days and structured by work, in which Hesiod's poem plants us, and that golden era of spontaneously flourishing nature and proximity to the divine, in which aging—the passage of time—was not a burden?

From the very outset, *Works and Days* establishes our relationship to nature by framing it temporally, locating it within a mythic history of the evolved human condition—or to put it another way, the poem represents the evolution of the human condition precisely as a function of our

---

8. χρύσεον μὲν πρώτιστα γένος μερόπων ἀνθρώπων
ἀθάνατοι ποίησαν Ὀλύμπια δώματ' ἔχοντες.
οἳ μὲν ἐπὶ Κρόνου ἦσαν, ὅτ' οὐρανῷ ἐμβασίλευεν·
ὥστε θεοὶ δ' ἔζωον ἀκηδέα θυμὸν ἔχοντες
νόσφιν ἄτερ τε πόνων καὶ ὀιζύος· οὐδέ τι δειλὸν
γῆρας ἐπῆν, αἰεὶ δὲ πόδας καὶ χεῖρας ὁμοῖοι
τέρποντ' ἐν θαλίῃσι κακῶν ἔκτοσθεν ἁπάντων·
θνῆσκον δ' ὥσθ' ὕπνῳ δεδμημένοι· ἐσθλὰ δὲ πάντα
τοῖσιν ἔην· καρπὸν δ' ἔφερε ζείδωρος ἄρουρα
αὐτομάτη πολλόν τε καὶ ἄφθονον· οἳ δ' ἐθελημοὶ
ἥσυχοι ἔργ' ἐνέμοντο σὺν ἐσθλοῖσιν πολέεσσιν.
ἀφνειοὶ μήλοισι, φίλοι μακάρεσσι θεοῖσιν.

changed relationship to "the earth" and its productions. And indeed it is this change that creates the very notion and experience of temporality and the dependence of human circumstances on it. The *Works and Days* presents both work and the day as governing conditions for our lives; yet both work and the day—our laboring condition and the temporality in and through which we live, work, sacrifice, reproduce, and die—require explanation. What is presented as given, then, must also be explained.

The Hesiodic tradition—encompassing both the *Works and Days* and the *Theogony*—is both cosmological and didactic: it accounts for the way things are (what is), how they came to be so, and what this cosmic arrangement requires of us. In the *Works and Days,* description and explanation shift almost imperceptibly into prescription; the "is" modulates into the "ought," the given into the enjoined. Although Hesiod presents his narratives—the mythic content of the *Works and Days*—and his injunctions as equally authoritative, this modulation between mythic explanation and normative exhortation is not seamless: the coordination of explanation and prescription constitutes, in fact, the work of the poem's measures.

For Hesiod, the discourse of measure mediates between the "is" and the "ought," coordinating notions of timeliness and seasonality as well as prescriptions for good conduct. The supreme ought of the *Works and Days* is expressed in the poet's exhortation to his brother Perses: "Work, foolish Perses! Work the work which the gods ordained for men" (397–98).[9] This imperative marks not only the route Perses should take but also the condition human beings have had thrust upon them. For as Hesiod makes plain, human beings did not always have to work. The authority of the poem's injunction to work thus requires an explanation, an accounting, not only of why Perses should work but indeed why we all must do so. It is not only that men should work; they must work, as human beings who are now subject to need. Hesiod's *Works and Days* thus conjoins a mode of explanation (why we must work) to a rhetoric of exhortation (work!).

For, as Hesiod tells us, it was not always so. From the description of life in the Golden Age, we learn that originally humankind feasted with the gods and lived as the gods do, without toil or sorrow or old age, sharing all good things that the earth shared with them. In the Golden Age, men lived in effortless harmony with the earth, in which the "ground" of their lives was spontaneous bounty. In that epoch, there was no need for work;

---

9. . . . ἐργάζευ, νήπιε Πέρση,
ἔργα, τά τ᾽ ἀνθρώποισι θεοὶ διετεκμήραντο.

scarcity was unknown, as was contention, "toil and grief," and aging. This glorious myth of origins should be read as the negative image of the present—for now:

> [T]he gods keep [the means of] life *(bios)* hidden from men. Otherwise in a day you would easily accomplish enough to have a living for a full year even without working; immediately you could put away your rudder over the smoke, and the work of the ox and patient mule would be over.[10]
> (42–46)

The circumstance that necessitates work also necessitates the explanation of both work and life: namely, that the gods have hidden from mortals their means of life; *bios* is the term that designates both livelihood and life itself. The *bios* of human beings—life and the means of life—is concealed from them, such that they must discover both the means of life and the *meaning* of it; they must learn where *bios* is located in the cosmic scheme—or, to put it more precisely, where it has been relocated. For at one time, the *Works and Days* recounts, men and gods began on the same terms:[11]

> I will sum you up another story well and skillfully—and you lay it up in your heart,—how the gods and mortal men sprang from one source.[12]
> (106–7)

Now, however, the defining condition of human existence is not only that men must age, suffer, and die, but that first they must struggle to achieve their *bios*. Men, moreover, must strive to understand the arrangement by which they and the gods now coexist; no longer is this arrangement self-evident.

10. κρύψαντες γὰρ ἔχουσι θεοὶ βίον ἀνθρώποισιν·
ῥηιδίως γάρ κεν καὶ ἐπ' ἤματι ἐργάσσαιο,
ὥστε σε κεἰς ἐνιαυτὸν ἔχειν καὶ ἀεργὸν ἐόντα·
αἶψά κε πηδάλιον μὲν ὑπὲρ καπνοῦ καταθεῖο,
ἔργα βοῶν δ' ἀπόλοιτο καὶ ἡμιόνων ταλαεργῶν.

11. At fr. 1, R. Merkelbach and M. L. West, *Fragmenta Hesiodea* (Oxford: Oxford University Press, 1967), men and gods are said originally to have taken their places together at shared feasts.

12. εἰ δ' ἐθέλεις, ἕτερόν τοι ἐγὼ λόγον ἐκκορυφώσω
εὖ καὶ ἐπισταμένως· σὺ δ' ἐνὶ φρεσὶ βάλλεο σῇσιν.
ὡς ὁμόθεν γεγάασι θεοὶ θνητοί τ' ἄνθρωποι.

Why have the gods hidden the *bios* from men? The answer begins with a story of bad measuring, of distorted equilibrium, which provokes a quarrel between Zeus and Prometheus. Hesiodic tradition narrates the story twice. It has been demonstrated that the account of Zeus and Prometheus in the *Works and Days* presupposes and alludes to the version of the story given in Hesiod's *Theogony;* it is from the combined accounts that we may assemble the entire narrative.[13] From the two poems we learn that the unity of men and gods was disrupted by an unequal division of a feast that was held at a place called Mekone:

> For when the gods and mortal men were making a division at Mekone, even then Prometheus eagerly cut up a great ox and set portions before them, trying to deceive the mind of Zeus. Before the rest he set flesh and innards rich with fat upon the hide, covering them with an ox paunch; but for Zeus he put the white bones dressed up with treacherous skill and covered with shining fat. Then the father of men and of gods said to him:
> "Son of Iapetus, most glorious of all lords, my dear, how unevenly you have divided the portions!"[14] (*Theogony* 535–44)

Prometheus's intervention on men's behalf, whereby he misrepresents the shares to their benefit, prompts Zeus to retaliate by withholding the power of fire from men, reserving it for the gods' use; whereupon Prometheus in turn steals it for men. As the *Works and Days* tells us:

> But Zeus angered in his heart hid it, because scheming Prometheus deceived him; so Zeus devised calamitous sorrows against men. He hid fire; but the clever son of Iapetus stole it again for men from Zeus the contriver in a hollow narthex, so that Zeus who delights in thunder did not notice it. But afterwards cloud-gathering Zeus said to him in anger:

13. See the valuable study by J.-P. Vernant, "The Myth of Prometheus in Hesiod," in *Myth and Society in Ancient Greece,* trans. J. Lloyd (New York: Zone Books, 1988), 183–201.

14. καὶ γὰρ ὅτ' ἐκρίνοντο θεοὶ θνητοί τ' ἄνθρωποι
Μηκώνῃ, τότ' ἔπειτα μέγαν βοῦν πρόφρονι θυμῷ
δασσάμενος προέθηκε, Διὸς νόον ἐξαπαφίσκων.
τοῖς μὲν γὰρ σάρκας τε καὶ ἔγκατα πίονα δημῷ
ἐν ῥινῷ κατέθηκε καλύψας γαστρὶ βοείῃ,
τῷ δ' αὖτ' ὀστέα λευκὰ βοὸς δολίῃ ἐπὶ τέχνῃ
εὐθετίσας κατέθηκε καλύψας ἀργέτι δημῷ.
δὴ τότε μιν προσέειπε πατὴρ ἀνδρῶν τε θεῶν τε·
'Ιαπετιονίδη, πάντων ἀριδείκετ' ἀνάκτων,
ὦ πέπον, ὡς ἑτεροζήλως διεδάσσαο μοίρας.

"Son of Iapetus, beyond all others in cunning, you rejoice that you have outsmarted me and stolen fire—a great bane for you yourself and for men who shall come after. But I will give men in exchange for fire [*anti puros*] an evil thing in which they may all take pleasure in their hearts while they embrace their own destruction."

So said the father of men and gods, and laughed out loud. And he bade renowned Hephaestus as speedily as possible to mix earth with water and to put in it the voice and strength of a human being, likening her face to the immortal goddesses, a beautiful, desirable maiden. . . . And he called this woman Pandora, because all those who live on Olympus gave her as a gift, a bane to men who eat bread. . . .

. . . For before this the tribes of men lived on earth remote and free from troubles and hard toil and painful sicknesses which bring upon men their deaths.[15] (47–63, 80–82, 90–92)

The autonomy of mortals as a group, their separateness, thus begins over an asymmetrical apportionment. Because of an unequal division, men must live divided from the gods; men must also begin their vexed relationship with Pandora and her kind. We may understand the Prometheus/Pandora story, in the Hesiodic version, as prompting Zeus to establish not only justice but a new mechanism for relations between gods and men, profoundly far-reaching both in itself and as an example for human relations.

This cosmic reorganization simultaneously moves men farther from the gods but closer to each other; they become each others' neighbors. Sepa-

15. ἐν κοίλῳ νάρθηκι λαθὼν Δία τερπικέραυνον.
τὸν δὲ χολωσάμενος προσέφη νεφεληγερέτα Ζεύς·
'Ιαπετιονίδη, πάντων πέρι μήδεα εἰδώς,
χαίρεις πῦρ κλέψας καὶ ἐμὰς φρένας ἠπεροπεύσας,
σοί τ' αὐτῷ μέγα πῆμα καὶ ἀνδράσιν ἐσσομένοισιν.
τοῖς δ' ἐγὼ ἀντὶ πυρὸς δώσω κακόν, ᾧ κεν ἅπαντες
τέρπωνται κατὰ θυμὸν ἑὸν κακὸν ἀμφαγαπῶντες.
ὣς ἔφατ'· ἐκ δ' ἐγέλασσε πατὴρ ἀνδρῶν τε θεῶν τε.
῞Ηφαιστον δ' ἐκέλευσε περικλυτὸν ὅττι τάχιστα
γαῖαν ὕδει φύρειν, ἐν δ' ἀνθρώπου θέμεν αὐδὴν
καὶ σθένος, ἀθανάτῃς δὲ θεῇς εἰς ὦπα ἐίσκειν
παρθενικῆς καλὸν εἶδος ἐπήρατον·
. . . ὀνόμηνε δὲ τήνδε γυναῖκα
Πανδώρην, ὅτι πάντες 'Ολύμπια δώματ' ἔχοντες
δῶρον ἐδώρησαν, πῆμ' ἀνδράσιν ἀλφηστῇσίν . . .
Πρὶν μὲν γὰρ ζώεσκον ἐπὶ χθονὶ φῦλ' ἀνθρώπων
νόσφιν ἄτερ τε κακῶν καὶ ἄτερ χαλεποῖο πόνοιο
νούσων τ' ἀργαλέων, αἵ τ' ἀνδράσι Κῆρας ἔδωκαν.

rated from the gods, they become (relative to life in the golden age) more dependent on the gods, but also on each other. All-encompassing consequences, needless to say, result from this situation, prominent among which is the development that, separated from the gods, men will now not only labor by themselves among themselves, but will, among themselves, attempt to rectify or overcome unequal apportionment, all the while perpetually recapitulating that original imbalance. Only through and over time—rather than within any given exchange—is equilibrium achieved.

Hesiod thus represents our mortal, laboring human condition not simply as a fall—an ontological change—but rather as a qualitative change in relations between gods, men, and the bounty of the earth. In the new mortal condition of separation from the gods, the previous model of commensality and sharing equally is replaced by a mode of transaction that will be the distinctive paradigm for relations at every level and in every arena of our existence: not collectivity and common property and sharing, Golden Age-style, nor theft, as practised by Prometheus, but exchange. Zeus's gift of Pandora "in exchange for fire" *(anti puros)* or, as the *Theogony* says, "in exchange for good,"[16] is nothing less than the originary gift exchange.

Because Zeus will now withhold natural fire—the thunderbolt— Pandora is given in exchange for fire. The emphasis on Pandora as a gift is reiterated a number of times throughout this passage; the *Works and Days* names her and etymologizes her name in a line that means not that the gods gave her a gift, but that they gave her as a gift.[17] In this exchange the gods get their unique and undivided status as immortals who receive sacrifices and live at ease without suffering or cares, and men get Pandora—and all the contents of her jar:

> For ere this the tribes of men lived on earth remote and free from ills and hard toil and heavy sicknesses which bring the Fates upon men; for in misery men grow old quickly. But the woman took off the great lid of the jar with her hands and scattered all these and her thought caused sorrow and mischief to men. Only Hope remained there in an unbreakable home within under the rim of the great jar, and did not fly out at the door; for ere that, the lid of the jar stopped her, by the will of Aegis-holding Zeus

16. *Theogony* 585. See the important article by Marylin Arthur, "Cultural Strategies in Hesiod's *Theogony*," *Arethusa* 15 vol. 1, no. 2 (1982): 63–82, who discusses reciprocity in the *Theogony* with implications for the *Works and Days*.

17. M. L. West, *Hesiod Works and Days* (Oxford: Oxford University Press, 1978), 166.

who gathers the clouds. But the rest, countless plagues, wander amongst men; for earth is full of evils, and the sea is full. Of themselves diseases come upon men continually by day and by night, bringing mischief to mortals silently.[18] (90–103)

Pandora is given to men "in exchange for fire"; now she and her kind will sear and scorch and consume men. As Zeus puts it, men will embrace their own ruin. In the *Theogony,* Pandora is called a "sheer deception:"[19] woman is an insidious subterfuge; a Trojan horse. But Pandora is also given "in exchange for fire" in another sense: in exchange for the pyre. Although men must die and be given to the funeral flames, through Pandora they will generate offspring and be provided with a means of overcoming that burning finality, that ultimate oblivion.[20] In this way the *Works and Days* explains that the quarrel between Zeus and Prometheus alters the terms on which the prerogatives of immortality will be accessible to men. Immortality for mortals will henceforth consist of fame through celebra-

18. Πρὶν μὲν γὰρ ζώεσκον ἐπὶ χθονὶ φῦλ' ἀνθρώπων
νόσφιν ἄτερ τε κακῶν καὶ ἄτερ χαλεποῖο πόνοιο
νούσων τ' ἀργαλέων, αἵ τ' ἀνδράσι Κῆρας ἔδωκαν.
[αἶψα γὰρ ἐν κακότητι βροτοὶ καταγηράσκουσιν.]
ἀλλὰ γυνὴ χείρεσσι πίθου μέγα πῶμ' ἀφελοῦσα
ἐσκέδασ'· ἀνθρώποισι δ' ἐμήσατο κήδεα λυγρά.
μούνη δ' αὐτόθι Ἐλπὶς ἐν ἀρρήκτοισι δόμοισιν
ἔνδον ἔμιμνε πίθου ὑπὸ χείλεσιν, οὐδὲ θύραζε
ἐξέπτη· πρόσθεν γὰρ ἐπέλλαβε πῶμα πίθοιο
[αἰγιόχου βουλῆισι Διὸς νεφεληγερέταο].
ἄλλα δὲ μυρία λυγρὰ κατ' ἀνθρώπους ἀλάληται·
πλείη μὲν γὰρ γαῖα κακῶν, πλείη δὲ θάλασσα·
νοῦσοι δ' ἀνθρώποισιν ἐφ' ἡμέρῃ, αἵ δ' ἐπὶ νυκτὶ
αὐτόματοι φοιτῶσι κακὰ θνητοῖσι φέρουσαι
σιγῇ, ἐπεὶ φωνὴν ἐξείλετο μητίετα Ζεύς.

19. See *Odyssey* viii. 494 where *dolos,* the *Theogony*'s term for Pandora, is applied to the Trojan horse.

20. See Arthur, "Cultural Strategies" note 16), 74, on Pandora as the principle of reproduction. As Froma Zeitlin points out to me, the *Works and Days* does not explicitly associate Pandora with immortality through reproduction; it is suggestive, though, that the *Catalogue of Women* links Pandora—as parent—with the ultimate progenitors of mankind, Deucalion and Pyrrha, and so with the prospect of future generations and the inextinguishability of humankind. According to a scholion at Ap. Rhod. 3.1086, "Hesiod says in the *Catalogue*" that Deucalion is the son of Prometheus and Pandora. In the text, the name of Pandora is badly corrupt (fr. 2 in Merkelbach and West, *Fragmenta Hesiodea* [note 11]), but fr. 4 maintains the connection, calling her the mother of Pyrrha.

tion in epic—of poetic immortalization—and of the capacity to reproduce their own kind, a capacity the immortal gods no longer possess.

Interestingly, how men were produced in the golden age in the first
place is not clear, but that does not mean that it was a mystery; in Hesiod's
account their existence simply doesn't require explanation. On the contrary, it is through the introduction of woman that one's progeny (or the
relation of progeny to paternity) can become a mystery, once the *bios* is
hidden from men. What has also required explanation, the poem attests,
is the appearance of the race of women and of human progeny—that is,
our contemporary situation as sexually dimorphic, reproductive, mortal
beings.[21]

As inaugurated by the gods' gift of Pandora, the dynamic of reciprocal
exchange becomes fundamental to the condition of a redefined human existence, and is tied to the struggle for survival. Human existence for the
first time now involves *need*—as well as the need for explanation. Pre-
cariously dependent on cosmic forces and natural phenomena, men are
obliged both to extrapolate patterns from nature and to impose them, in
turn, in the task of cultivation. No longer does the earth spontaneously
bear for mortals what they need in order to keep themselves alive. As a sign
of their difference from the gods rather than as a punishment,[22] human
beings are compelled to work: they must cultivate their food. In their
changed relationship with nature, they now must contend with cooking
and culture. The anthropologist Lévi-Strauss says of myths that sometimes
they do not seek to depict what is real, but to justify the shortcomings of
reality, since the extreme positions are only imagined in order to show that
they are untenable. From an anthropological perspective, the Mekone
story is perhaps a prototypical example of such a mythic thought experiment: why must we sacrifice? must we sacrifice? "The mythic narrative
founds sacrifice, whereby we transform our relation with nature (meat-

---

21. On women imagined as a race apart, see the definitive discussion in Nicole Loraux,
"Sur la race des femmes et quelques-unes de ses tribus," *Arethusa* 11, nos. 1–2 (1978): 43–88,
reprinted in *The Children of Athena: Athenian Ideas about Citizenship and the Division between the
Sexes,* trans. C. Levine (Princeton: Princeton University Press, 1993), 72–110.

22. It is important to observe that in the Greek tradition work is not meted out simply as
a punishment, as it is in ancient Near Eastern (including Mesopotamian and Hebrew) traditions. It is true that the need to work emerges in Hesiodic explanation as part of a series of
contentious exchanges between Zeus and Prometheus, and that, as part of this serial "payback"
for theft and bad apportioning, it can resemble punishment. Yet the archaic Greek tradition is
more ambiguous about how and whether to valorize aspects of the human condition (e.g.,
work, sex, death) than a reading of work-as-punishment will allow.

eaters, but cooked meat) into our relation with the gods, which turns out to be founded on our failure to take advantage of Zeus (*our* because Prometheus is our agent, and ancestor as well)." We commemorate this failure every time we eat meat, and suffer for it every time we labor, die, and have children.[23]

The Hesiodic tradition thus authorizes agricultural work as the right relation of men to the earth, sacrifice as the proper exchange relation between men and the gods, sexual reproduction as the constitutive relation between male and female, and economic transactions as the relations required among men. All these require men to take the measure of their actions.

The definition of the human condition as a laboring condition is installed, as we have seen, after a crisis of bad apportioning, mis-taking, and mis-measure. The primordial conflict between Zeus and Prometheus, between the greatest of the gods and the agent of men, is the quarrel to begin all quarrels, responsible for the irrevocable change from the golden age; it thus offers the prototype for the quarrel between Hesiod and Perses, which stands as a distant but direct descendent of that first conflict. The *Works and Days* begins with strife *(eris),* in a passage that serves as a programmatic introduction to the poem as a whole. The poet's grievance with his brother purports to be the occasion for the poem, but the address to his brother, enjoining him to *work,* is also, more broadly, an order that he desist from destructive conflict. What the poem introduces at the outset is not a particular instance of strife, but a reflection on the nature of strife:

> So there was not only one kind of Strife [*eris*], after all—but in fact on earth there are two. The one Strife, a man would praise once he knew her; but the other is blameworthy: and they are different at heart. For one fosters evil war and conflict, cruel as she is: no mortal cares for her; but of necessity, through the will of the immortals, mortals give honor to harsh Strife. But the other Strife dark Night bore first, and the son of Cronos who sits on high, dwelling in the aether, placed her in the roots of the earth: and she is much better for men; she rouses even the feckless man. For a person grows eager to work when he looks at his neighbor.[24] (11–22)

23. In J. M. Redfield's formulation (private communication). See J. M. Redfield, "The Sexes in Hesiod," in *Reinterpreting the Classics,* ed. C. A. Stray and R. A. Kaster, special issue of *Annals of Scholarship* 10, no. 1 (1994).

24. οὐκ ἄρα μοῦνον ἔην Ἐρίδων γένος, ἀλλ' ἐπὶ γαῖαν
εἰσὶ δύω· τὴν μέν κεν ἐπαινέσσειε νοήσας,

Hesiod tells us that there is not one strife, one *eris,* but two—namely, the one that produces wars and violence and the one that produces work. He makes it clear that in a kind of *mise en abîme,* the two strifes (or *erides*) are in contention with each other, bespeaking their primal character: strife was always already there.

Each *eris* herself needs to outdo the other; so, for example, Hesiod insists that it is the bad *eris* that is keeping Perses from activating her good counterpart. The personification of destructive Strife with a capital "S" (or *Eris* with a capital epsilon) appears in the *Iliad,* stirring up the contending armies on either side to a greater pitch of violence. But the realm of the beneficent strife *eris* is adversarial as well: she does not exist to promote (so to speak) self-motivation or auto-competition; there is no notion of bettering one's own record. There is, in fact, no "one" in this realm (any more than in that of her counterpart)—only two. Thus each participant in an *eris,* reciprocally, keeps the other one productive:

> For a man is eager to work when he looks at his neighbor, a rich man who hurries to plough and plant and put his house in good order; and neighbor vies with his neighbor as he hastens after wealth. This Strife is good for men.[25] (21–24)

In a sense, *strife* defines both the problem and the solution for the human condition.

Through its story of the two strifes, the *Works and Days* divides strife

---

ἢ δ' ἐπιμωμητή· διὰ δ' ἄνδιχα θυμὸν ἔχουσιν.
ἣ μὲν γὰρ πόλεμόν τε κακὸν καὶ δῆριν ὀφέλλει,
σχετλίη· οὔτις τήν γε φιλεῖ βροτός, ἀλλ' ὑπ' ἀνάγκης
ἀθανάτων βουλῇσιν Ἔριν τιμῶσι βαρεῖαν.
τὴν δ' ἑτέρην προτέρην μὲν ἐγείνατο Νὺξ ἐρεβεννή,
θῆκε δέ μιν Κρονίδης ὑψίζυγος, αἰθέρι ναίων,
γαίης ἐν ῥίζῃσι, καὶ ἀνδράσι πολλὸν ἀμείνω·
ἥτε καὶ ἀπάλαμόν περ ὁμῶς ἐπὶ ἔγειρεν.
εἰς ἕτερον γάρ τίς τε ἰδὼν ἔργοιο χατίζει
πλούσιον, ὃς σπεύδει μὲν ἀρώμεναι ἠδὲ φυτεύειν
οἶκόν τ' εὖ θέσθαι· ζηλοῖ δέ τε γείτονα γείτων
εἰς ἄφενος σπεύδοντ'· ἀγαθὴ δ' Ἔρις ἥδε βροτοῖσιν.
25. εἰς ἕτερον γάρ τίς τε δὼν ἔργοιο χατίζει
πλούσιον, ὃς σπεύδει μὲν ἀρώμεναι ἠδὲ φυτεύειν
οἶκόν τ' εὖ θέσθαι· ζηλοῖ δέ τε γείτονα γείτων
εἰς ἄφενος σπεύδοντ'· ἀγαθὴ δ' Ἔρις ἥδε βροτοῖσιν.

and theorizes its various actions; the division of strife also provides the opportunity for the poem to valorize one kind of conflict and to authorize some human activities (for example, work) and not others.[26] The *Iliad* juxtaposes them and renders the parallelism between them in a passage that describes a particularly grueling phase of the fighting between Greek and Trojan warriors, where an impasse in the battle is compared to a scene of agricultural activity, a dispute between neighbors over the common boundary of a field:[27]

> For neither were the powerful Lykians able to break through the wall of the Danaans and make a path to the ships, nor could the Danaan spearmen push the Lykians back from the wall, once they had reached it. But as two men holding measuring ropes in their hands quarrel over boundaries in a shared field, and in a narrow space contend [*erizeto* = have an *eris*] each for his equal share . . .[28] (*Iliad* 12.417–423)

Although a consideration of the competition between poetic genres, to which Hesiodic poetry alludes,[29] leads beyond the scope of this discussion, we may note that both the *Works and Days* and the *Iliad* generate their measures out of what they take to be a given, primordial crux: strife. And if strife is a generative matrix for early Greek poetry—its competing valorizations of warfare and work, its reflections on exchanges between gods,

---

26. Greek culture develops, we might say, a poetics of strife, out of which competing and conceptually interdependent genres emerge: one (Hesiodic) oriented toward proper conduct given the limits of the human condition and the regular measures of the cosmos, the other genre (Homeric epic) oriented toward exceptional conduct, which aims at transcending the very limits that the other genre has theorized.

27. This is one of a number of similes throughout the *Iliad* that liken the warriors' efforts on the battlefield to scenes in the wheatfield, or the vicinity thereof. See also, e.g., *Iliad* 11.67–72, as well as the striking passage at *Odyssey* 18.365–380, in which Odysseus equates the *eris* of work (using the phrase *eris ergoio*) with that of warfare, in proposing the terms of a contest between himself and Eurymachus.

28. οὔτε γὰρ ἴφθιμοι Λύκιοι Δαναῶν ἐδύναντο
τεῖχος ῥηξάμενοι θέσθαι παρὰ νηυσὶ κέλευθον,
οὔτέ ποτ' αἰχμηταὶ Δαναοὶ Λυκίους ἐδύναντο
τείχεος ἂψ ὤσασθαι, ἐπεὶ τὰ πρῶτα πέλασθεν.
ἀλλ' ὥς τ' ἀμφ' οὔροισι δύ' ἀνέρε δηριάασθον
μέτρ' ἐν χερσὶν ἔχοντες ἐπιξύνῳ ἐν ἀρούρῃ,
ὥ τ' ὀλίγῳ ἐνὶ χώρῳ ἐρίζητον περὶ ἴσης . . .

29. *Works and Days* 24–26. I offer more discussion of the generic nature of the *Works and Days* in a forthcoming study, *Figure, Measure, and Social Order in Early Greek Poetry*.

men, and the earth, its theory of the origin of two sexes and reproduction—it is not surprising that we see, accompanying the endless work of strife, the endless work of measure, since what is struggled over, what is labored for, what is contended is always, literally or metaphorically, a "share" or "portion."[30]

Because the earth no longer of its own accord unstintingly provides human beings their *bios* in all seasons, the ultimately pressing question for them becomes how much will it provide and when? Thus their existence also comes to be bound up with calculation and measurement, and with the exigencies of the calendar. If human existence were still as it had been in the golden age (pre-mortal, as it were) there would be no calendar and no measuring. There would be, in a sense, no difference between a day and a year.[31]

> For the gods keep [the means of] life *(bios)* hidden from men. Otherwise in a day you would easily accomplish enough to have a living for a full year even without working.[32] (45–47)

As it is, time and timing must be measured, and the seasons and days take on a distinctive character and significance. The farmer performs a precarious balancing act, adjusting to the forces of nature and the rhythm of the year. These adjustments are mirrored by the ceaseless imperative of reciprocal exchanges with his neighbors on the land.

If resources are no longer unlimited, they must be carefully weighed and protected; thus relations with others who have a claim on them be-

---

30. In a world where destiny expresses a notion of allotment—where the word "fate" *(moira)* is used to denote a serving of meat, a share of land, a portion of the spoils of war, as well as to refer to the final shape of a man's life—we understand that the largest framework in which human lives are viewed is that of division and apportionment. Throughout the Homeric poems, for instance, we observe that a basis in appropriate distribution and reciprocal exchange is explicitly invoked by formal procedures such as honorific feasting, distribution of booty, awards of prizes, return for specific services, ransom arrangements, and other transactions, where equitable division is the inflexible requirement. But this exigent imperative bespeaks a comprehensive view of social relations, in the *Iliad* as well as in the *Works and Days*, among human beings and between humans and gods.

31. See R. Hamilton, *The Architecture of Homeric Poetry* (Baltimore: Johns Hopkins University Press, 1989), 84.

32. κρύψαντες γὰρ ἔχουσι θεοὶ βίον ἀνθρώποισιν·
ῥηιδίως γάρ κεν καὶ ἐπ᾽ ἤματι ἐργάσσαιο . . .
ὥστε σε κεῖς ἐνιαυτὸν ἔχειν καὶ ἀεργὸν ἐόντα·

come fraught. Skill in conducting social and economic transactions thus becomes as crucial to survival as work itself:

> Invite your friend to a meal; but leave your enemy alone; and especially invite the one who lives near you: for if any trouble happens in the place, your neighbors come ungirt, but your kin stay to gird themselves. A bad neighbor is as great a disaster as a good one is a great blessing; whoever enjoys a good neighbor has a precious possession. Not even an ox would die if your neighbor weren't a bad one.[33] (342–48)

Hesiod's advice is to make sure that your neighbor needs you more than you need him:

> And so you will have plenty till you come to silvery springtime, and will not look wistfully to others, but another man will need your help.[34] (477–78)

Need becomes its own measurable resource, a measurable share: debt. This logic of measurable shares, the ground of Hesiodic economics, presumes finite resources, as Paul Millett has observed:

> Certainly Hesiod sees it as being in every man's interest to get for himself as much wealth as possible; but he also assumes that the stock of wealth— effectively the quantity of land—is finite and fixed. So what one man gains, another must necessarily lose, and there is no scope for an overall growth in prosperity. . . . And that is presumably why it is so important to work harder than your neighbor; it is a guarantee that his and not your *oikos*[35] will be the one to decline. . . . This negative view of wealth and prosperity as being feasible only at the expense of other people is apparently typical of peasant societies.[36]

33. τὸν φιλέοντ' ἐπὶ δαῖτα καλεῖν, τὸν δ' ἐχθρὸν ἐᾶσαι·
τὸν δὲ μάλιστα καλεῖν, ὅς τις σέθεν ἐγγύθι ναίει·
εἰ γάρ τοι καὶ χρῆμ' ἐγχώριον ἄλλο γένηται,
γείτονες ἄζωστοι ἔκιον, ζώσαντο δὲ πηοί.
πῆμα κακὸς γείτων, ὅσσον τ' ἀγαθὸς μέγ' ὄνειαρ.
ἔμμορέ τοι τιμῆς, ὅς τ' ἔμμορε γείτονος ἐσθλοῦ.
οὐδ' ἂν βοῦς ἀπόλοιτ', εἰ μὴ γείτων κακὸς εἴη.
34. εὐοχθέων δ' ἵξεαι πολιὸν ἔαρ, οὐδὲ πρὸς ἄλλους
αὐγάσεαι· σέο δ' ἄλλος ἀνὴρ κεχρημένος ἔσται.
35. *oikos* = "household," "holdings."
36. P. Millett, "Hesiod and His World," *Proceedings of the Cambridge Philological Society* 209 (1983): 84–115.

The line between cooperation and competition is thus constantly blurred and redrawn throughout the account of agricultural work:

> Give to one who gives, but do not give to one who does not give. A man gives to the giver, but no one gives to the non-giver.[37] (354–55)

If the destructive strife—the war-producing kind—generates negative reciprocity whereby social relations are frustrated and inhibited, the good strife promotes those sequences of exchanges through which in balanced reciprocity social relations are extended and made continuous; gift and countergift return each party to the other's position: each is by turns giver and receiver. As described in detail by Marcel Mauss[38] and other students of primitive exchange, maintaining a system of reciprocity involves the paradox of equilibrium built on imbalance in that each strives to give more than he has received. It is not simply that giving is *better than* receiving— it is a better *way to receive,* that is, a guarantee of future receiving. This dialectic of giving and receiving, the transmutation of giving into good receiving, presents a problem for Perses, who does not understand the relation between *giving* and *receiving* in an exchange system. Thus his options have become, according to Hesiod, either to seize illegitimately or to beg; neither furthers the desired economy.

> But I will give you no more nor give you further measure. Work, foolish Perses! Work the work that the gods ordained for men, so that in the grief of your heart, you with your children and wife do not seek your livelihood among your neighbors, who do not care.[39] (397–400)

The daily, and perennial, effort undertaken by the farmer—an effort that informs all his activities—to achieve a balance between too early and too late, too hot and too cold, too dry and too wet, between too little and too much, is replicated in the necessary effort to equalize exchanges be-

---

37. καὶ δόμεν ὅς κεν δῷ, καὶ μὴ δόμεν, ὅς κεν μὴ δῷ.
δώτῃ μέν τις ἔδωκεν, ἀδώτῃ δ' οὔτις ἔδωκεν.

38. M. Mauss, *The Gift: Forms and Functions of Exchange in Archaic Societies* (New York: Norton, 1967, trans. of *Essai sur le don* [1925]).

39. . . . ἐγώ δέ τοι οὐκ ἐπιδώσω
οὐδ' ἐπιμετρήσω· ἐργάζευ, νήπιε Πέρση,
ἔργα, τά τ' ἀνθρώποισι θεοὶ διετεκμήραντο,
μή ποτε σὺν παίδεσσι γυναικί τε θυμὸν ἀχεύων
ζητεύῃς βίοτον κατὰ γείτονας, ὃ δ' ἀμελῶσιν.

tween himself and his neighboring farmers over time. The task of proper calculation and the perception of temporality are what together enable a productive economy of strife:

[T]o the best of your ability, sacrifice to the immortal gods purely and cleanly, and in addition burn glorious thigh-pieces . . . that they may be propitious to you in heart and spirit, and so you may buy another's holding and not another yours.[40] (336–41)

In his prescriptions, Hesiod repeatedly articulates the essential character of reciprocal exchange, of gift and countergift, namely, that it is inherently and perpetually in disequilibrium and must be continually rebalanced over time, so that every exchange begets a further exchange:

Take fair measure from your neighbor and pay him back fairly with the same measure, or better, if you can; so that if you are in need afterwards, you may find him reliable.[41] (349–51)

We have seen that, after the crisis of bad apportioning—the scandal at Mekone and the theft of Prometheus—exchange has become an imperative: the system of reciprocity, of paybacks conducted through time, is now, we might say, what *is*. How one *ought* to behave, given these conditions, returns Hesiod to the problem of measure. The *Works and Days* prescribes adherence to due season and fair measure, both as literal mandates in daily activity and—beyond practicality—as tropes of mortal temporality, mutual (if agonistic) dependence, and the ethical ordering of society:

Observe due measure *(metra):* and best in all things is the right time and right amount.[42] (694)

40. κὰδ δύναμιν δ' ἔρδειν ἔρ' ἀθανάτοισι θεοῖσιν
ἁγνῶς καὶ καθαρῶς, ἐπὶ δ' ἀγλαὰ μηρία καίειν·
ἄλλοτε δὲ σπονδῇσι θύεσσί τε λάσκεσθαι,
ἠμὲν ὅτ' εὐνάζῃ καὶ ὅτ' ἂν φάος ἑρὸν ἔλθῃ,
ὥς κέ τοι ἵλαον κραδίην καὶ θυμὸν ἔχωσιν,
ὄφρ' ἄλλων ὠνῇ κλῆρον, μὴ τὸν τεὸν ἄλλος.
41. εὖ μὲν μετρεῖσθαι παρὰ γείτονος, εὖ δ' ἀποδοῦναι,
αὐτῷ τῷ μέτρῳ, καὶ λώιον, αἴ κε δύνηαι,
ὡς ἂν χρηίζων καὶ ἐς ὕστερον ἄρκιον εὕρῃς.
42. μέτρα φυλάσσεσθαι· καιρὸς δ' ἐπὶ πᾶσιν ἄριστος.

When Hesiod insists that he will no longer help Perses if the latter is in need—

> But I will give you no more nor give you further measure.[43] (397)

—Hesiod's "measure" may denote the actual material, the begrudged handouts, that he announces he will no longer give Perses. But elsewhere, as in the passage cited above, we see that "measure" functions as a more abstract, mobile counter in Hesiod's normative economy of good relations:

> Take fair measure from your neighbor and pay him back fairly with the same measure, or better, if you can; so that if you are in need afterwards, you may find him reliable.[44] (349–51)

Or, more pessimistically:

> Do not make a friend equal to a brother; but if you do, do not injure him first, and do not lie to please the tongue. But if he injures you first, either by saying or doing something offensive, remember to repay him double.[45] (707–11)

The notions of due measure and right season are thus *linked* in ways that at first seem to be quite literal—

> but let it be your care to arrange your work in due measure, that in the right season your barns may be full of grain.[46] (306–7)

—but that also function as ramifying tropes that figure a system in equilibrium. "Fair measure" then represents not a precise equivalent, but a just

---

43. οὐδ' ἐπιμετρήσω· ἐργάζευ, νήπιε Πέρση . . .
44. εὖ μὲν μετρεῖσθαι παρὰ γείτονος, εὖ δ' ἀποδοῦναι,
αὐτῷ τῷ μέτρῳ, καὶ λώιον, αἴ κε δύνηαι,
ὡς ἂν χρηίζων καὶ ἐς ὕστερον ἄρκιον εὕρῃς.
45. μηδὲ κασιγνήτῳ ἶσον ποιεῖσθαι ἑταῖρον·
εἰ δέ κε ποιήσῃς, μή μιν πρότερος κακὸν ἔρξῃς.
μηδὲ ψεύδεσθαι γλώσσης χάριν· εἰ δὲ σέ γ' ἄρχῃ
ἤ τι ἔπος εἰπὼν ἀποθύμιον ἠὲ καὶ ἔρξας,
δὶς τόσα τίνυσθαι μεμνημένος
46. σοὶ δ' ἔργα φίλ' ἔστω μέτρια κοσμεῖν,
ὥς κέ τοι ὡραίου βιότου πλήθωσι καλιαί.

amount that will continue the sequence of exchanges. And the "right season" takes account of those exchanges as part of a cycle, operating with and through the patterns of nature.

We see, then, that the Hesiodic etiology of work and exchange (the mythic narrative) and his prescriptions regarding proper conduct (the didactic discourse) both partake of the figurative system of the seasonable, the timely, the natural—that which is "in season" (horaios; hora = season). Life can be regulated according to calculable elements, says the Hesiodic tradition, and if you measure your actions and exchanges appropriately (erga . . . metria kosmein) you can recapitulate that order. Because of the predictability of at least some vital natural phenomena on which the life—the bios—of mortals depends, that which is "timely" (horaios) (for example, the appearance and disappearance of constellations throughout the year, their rising and setting, the sequence of the seasons) becomes a figure both for the ordered life and for a standard of appropriateness within it. Thus Hesiod's recourse to the imperative of due season:

But you, Perses, remember all works in their season.[47] (641–42)

"Due measure"—a figure for fair treatment and appropriate interactions—and "seasonability," a figure for order, first reinforce each other and then function as metonyms of one another; so that in a passage on right conduct and relations, inappropriate, improper behavior (like harming a suppliant or sleeping with your brother's wife) is called "unseasonable," "untimely" (parakairia):

Alike with whoever wrongs a suppliant or a guest, or does acts contrary to nature [unseasonable/untimely acts], climbing into his brother's wife's bed in covert lust, or who thoughtlessly injures orphaned children, or who abuses his old father at the grim threshold of old age and attacks him with harsh words, with this one truly Zeus himself is angry, and in the end imposes on him a harsh requital for his unjust acts.[48] (327–34)

47. τύνη δ', ὦ Πέρση, ἔργων μεμνημένος εἶναι
ὡραίων πάντων.
48. ἶσον δ' ὅς δ' ἱκέτην ὅς τε ξεῖνον κακὸν ἔρξῃ,
ὅς τε κασιγνήτοιο ἑοῦ ἀνὰ δέμνια βαίνῃ
κρυπταδίης εὐνῆς ἀλόχου, παρακαίρια ῥέζων,
ὅς τέ τευ ἀφραδίης ἀλιταίνεται ὀρφανὰ τέκνα,
ὅς τε γονῆα γέροντα κακῷ ἐπὶ γήραος οὐδῷ
νεικείῃ χαλεποῖσι καθαπτόμενος ἐπέεσσιν·

Hence the poem's insistence on acting "in season," on the importance of the calendar and of observing the proper timing for accomplishing work— this is the *Days* part of the *Works and Days:*

> But when Orion and Sirius are come into midheaven, and rosy-fingered Dawn sees Arcturus, then cut off all the grape-clusters, Perses, and bring them home. Show them to the sun ten days and ten nights: then cover them over for five, and on the sixth day draw off into vessels the gifts of joyful Dionysus. But when the Pleiades and Hyades and strong Orion begin to set, then remember to plough in season: and so the completed year will fitly pass beneath the earth.[49] (609–17)

Another perspective on the problematic of seasonality is offered by Hesiod's account of the history of humankind through time in the myth of the Five Ages, of which the Golden Age is the first:

> Thereafter, I wish that I were not among the men of the fifth generation, but either had died before or been born afterwards. For now truly is a race of iron, and men never rest from labor and sorrow by day, and from perishing by night; and the gods will give them sore cares. But, notwithstanding, even they shall have some good mixed with their evils. And Zeus will destroy this race of mortal men also when they come to have grey hair on their temples at birth.[50] (174–81)

τῷ δ' ἤ τοι Ζεὺς αὐτὸς ἀγαίεται, ἐς δὲ τελευτὴν
ἔργων ἀντ' ἀδίκων χαλεπὴν ἐπέθηκεν ἀμοιβήν.
49. ὦ Πέρση, τότε πάντας ἀποδρέπεν οἴκαδε βότρυς·
δεῖξαι δ' ἠελίῳ δέκα τ' ἤματα καὶ δέκα νύκτας,
πέντε δὲ συσκιάσαι, ἕκτῳ δ' εἰς ἄγγε' ἀφύσσαι
δῶρα Διωνύσου πολυγηθέος. αὐτὰρ ἐπὴν δὴ
Πληιάδες θ' Ὑάδες τε τό τε σθένος 'ωαρίωνος
δύνωσιν, τότ' ἔπειτ' ἀρότου μεμνημένος εἶναι
ὡραίου· πλειὼν δὲ κατὰ χθονὸς ἄρμενος εἰσιν.
50. μηκέτ' ἔπειτ' ὤφελλον ἐγὼ πέμπτοισι μετεῖναι
ἀνδράσιν, ἀλλ' ἢ πρόσθε θανεῖν ἢ ἔπειτα γενέσθαι.
νῦν γὰρ δὴ γένος ἐστὶ σιδήρεον· οὐδέ ποτ' ἦμαρ
παύονται καμάτου καὶ ὀιζύος, οὐδέ τι νύκτωρ
φθειρόμενοι. χαλεπὰς δὲ θεοὶ δώσουσι μερίμνας·
ἀλλ' ἔμπης καὶ τοῖσι μεμείξεται ἐσθλὰ κακοῖσιν.
Ζεὺς δ' ὀλέσει καὶ τοῦτο γένος μερόπων ἀνθρώπων,
εὖτ' ἂν γεινόμενοι πολιοκρόταφοι τελέθωσιν.

The sign of the last stage of corruption among mortals, when they have become so degenerate that Zeus will destroy them, is a stunning one: the mark of their corruption is that their *timing* is *out of sync*. Hesiod says that Zeus will destroy this age when babies are born with grey hair, that is, when the seasons and generations of man have collapsed all together and become confused. When newborns have the features of old men, the seasons of our lives are truly out of joint. So to observe the due sequence of things—to pay attention to the calendar—is not only to bring temporality, that inescapable fact of our lives, in some small way under control, but it is also to resist such moral chaos as is envisaged for the end of the fifth age, the age of iron—our own.

If that which is "in season" *(horaios)* emerges in Hesiod as a normative ground, the trope of due measure *par excellence,* we also know that it has required narrative explanation. Hesiod has, as we have seen, spent many measures explaining that time is, philosophically, a problem, not a given, for man; that the existence of the calendar measures, in fact, the distance men have come from their previous timeless ease; that the current measures of man, the conditions governing him—from temporality to mortality to work—are all developments requiring explanation.

Due measure and right season are invested with an ethical dimension, most fully realized as the basis for the operations of Justice.[51] In the poetic tradition transmitted by the *Works and Days,* Justice herself brings to the imagery of balance and fair measure—dealing "straightly" as opposed to askew—the dimension of appropriate temporality. For we discover that, in the cosmic genealogy narrated in Hesiod's *Theogony,* Justice herself is none other than one of the Seasons *(Horai).*[52] Thus Hesiodic etiology allows us to see what kind of norm justice is for archaic Greek thought: not just any norm, but an order; not just any order, but the cyclic order of the seasons, which defines time itself.[53]

---

51. *Works and Days* 256–62: "And there is virgin Justice, the daughter of Zeus, who is honored and reverenced among the gods who dwell on Olympus, and whenever anyone hurts her with lying slander, she sits beside her father, Zeus the son of Cronos, and tells him of men's wicked heart, until the people pay for the mad folly of their princes who, evilly minded, pervert judgement and give sentence crookedly. Keep watch against this, you princes, and make straight your judgements, you who devour bribes; put crooked judgements altogether from your thoughts."

52. *Theogony* 901–2.

53. I owe this formulation to the incisive editorial response of Raine Daston and Fernando Vidal.

The problem of measure thus informs justice as arbitration and calibration; it is brought equally to natural processes and to human affairs. Because days are more self-regulating than human works, it is not surprising that in the *Works and Days* Hesiod orients us to the problem of measuring human transactions:

> But you, Perses, lay up these things within your heart and listen now to Justice, ceasing altogether to think of violence. For the son of Cronos has ordained this law for men, that fishes and beasts and winged fowls should devour one another, for Justice is not in them; but to mankind he gave Justice which proves far the best.[54] (274–79)

Hesiod's identification of Justice and seasonality, and his coordination of the right time and the right amount in a metaphorical discourse of ethical behavior, help us to read Heraclitus's conjunction of the sun's measures, the avenging Furies and the Justice that administers their function. Heraclitus invokes the sun's measures—usually the very trope of regularity, of the predictable, of the obvious—*as a problem;* he does so, moreover, through a strikingly peculiar, counterfactual rhetoric, as through beginning the trope of the *adynaton*—the figure of impossibility (of the sort, "when rivers run back to their source . . . ," "when fish fly . . ."). Why is justice concerned with the passage of the day? Are (not) the sun's measures inviolable? The extended coordination of Justice with the sun's measures, or with—more broadly—that which is seasonable *(horaios),* must be seen as a complex, provocative wager: betting on the sun's regularity, troping norms out of nature, one gains in figurative power what one loses, perhaps, in ethical force. Heraclitus's imagined transgression, even if offered through a conditional rhetoric of the improbable, invites us to continue our explorations of the thought experiments conducted through poetry as well as philosophy.

As I hope to have shown for the *Works and Days,* the measures of early Greek poetry accomplish their cognitive and ethical work through com-

---

54. ὦ Πέρση, σὺ δὲ ταῦτα μετὰ φρεσὶ βάλλεο σῇσι,
καὶ νυ δίκης ἐπάκουε, βίης δ' ἐπιλήθεο πάμπαν.
τόνδε γὰρ ἀνθρώποισι νόμον διέταξε Κρονίων
ἰχθύσι μὲν καὶ θηρσὶ καὶ οἰωνοῖς πετεηνοῖς
ἐσθέμεν ἀλλήλους, ἐπεὶ οὐ δίκη ἐστὶ μετ' αὐτοῖς·
ἀνθρώποισι δ' ἔδωκε δίκην, ἣ πολλὸν ἀρίστη
γίγνεται·

plex tropes and figurations: even if "nature" as such does not quite yet exist as a category in archaic epic, nevertheless the processes and ordination of the natural world—its measures from the seasons to cosmic rhythms—everywhere inform Hesiodic explanations of and prescriptions for the social and ethical orders of man, and indeed, of suprahuman Justice. Any analysis of early Greek thought—whether construed as "prescientific," "mythological," "poetic," or in later periods "philosophical"—requires an extensive investigation of such figurative textures. The present reading, restricted as it is, aims to offer an invitation for future comparative readings across periods and cultures: through sustained collaborative analyses of the figurative bases of representations of cultural order (whether figured as "natural" or not), we can refine our understanding of human value-making and the role of figurally based ideologics in the constitution of communities.[55] We can also better approach the ways ancients, and indeed moderns, have chosen to represent themselves to themselves.

In the authoritative explanations and pronouncements of Hesiod, we have an example of archaic Greek culture thinking about itself, authorizing itself, taking its own measure. In early Greek poetry, man may not yet be what he later becomes for the fifth-century philosopher Protagoras, who called him the *measure of all things;* but as a worker, he is already what he must be first: the *measurer* of them.

55. For a powerful example of such a cultural reading, see G. Ferrari, *Figures of Speech: Men and Maidens in Ancient Greece* (Chicago: University of Chicago Press, 2002), and her earlier "Figures of Speech: The Picture of *Aidos*" in *Mētis* 5 (1990): 185–200.

2

Nature in Person: Medieval and Renaissance Allegories and Emblems

Katharine Park

In the context of medieval and Renaissance European literate culture, to write of nature's authority was already to engage in personification, for ideas of authority, as the word suggests, were closely tied to ideas of authorship. "Authority"—*auctoritas* in Latin—referred to the words of an author *(auctor):* one of the chain of especially revered and trusted writers or teachers, from the ancients to near contemporaries, whose texts established the framework within which questions in the learned disciplines were debated and explored.[1] By extension, authority was the defining characteristic of these writers: they might be mistaken on small points, but it was understood that if they appeared to contradict one another or to have missed the mark on important matters, the error lay with the interpretation placed on their words by careless or ignorant readers and listeners. Thus authority implied not only a high degree of credibility but an individual persona; to confer authority on an abstraction was to confer on it a face, a figure, and a voice.

Western Europeans had long written of nature in this way. A number of influential medieval texts and images personified the physical world and the principles that governed it as a female figure of great dignity—

I would like to thank Alisha Rankin for her help in researching this article.

1. On authority, authorship, and the role of *auctores* in medieval and Renaissance learning, see A. J. Minnis, *Medieval Theory of Authorship: Scholastic Literary Attitudes in the later Middle Ages,* 2nd ed. (Aldershot: Scolar Press, 1988), 10; Jole Agrimi and Chiara Crisciani, *Edocere medicos: Medicina scolastica nei secoli XIII–XV* (Naples: Guerini, 1988), chap. 3, esp. 81–104; G. Berndt, "Auctores-Auctoritas," in Robert Auty et al., eds., *Lexikon des Mittelalters,* 10 vols. in 15 (Munich: Artemis, 1977–99), vol. 1, cols. 1189–90.

majestic, clothed, and energetic—who was delegated by the Christian God to shape individual beings through the physical process of generation, to guide her creatures, and to maintain order in all such matters.[2] Just as this tradition had a beginning, in the years around 400 C.E., so too it had an end. Although isolated instances of this particular personification can be found in the later sixteenth century—Dame Nature, in Edmund Spenser's long allegorical poem *The Faerie Queene,* for example[3]—by that time it was becoming obsolete, supplanted by a new figure of Nature as a lactating woman, partly or completely naked, or as a woman endowed with many breasts. This new image was a humanist creation; it seems to have been invented in Naples, in the 1470s, the product of a collaboration between the Roman scholar Luciano Fosforo and the miniaturist Gaspare Romano, who had been charged with illustrating a manuscript copy of Pliny's *Historia Naturalis* (Natural history) for Cardinal John of Aragon. According to another contemporary humanist, Fosforo directed Gaspare to portray nature as "a seated woman, of wonderful beauty, dressed in the antique style, who holds a world [*mondo*] in front of her bosom and spills milk on it from her breasts"—an image that probably derived from the description of Physis as the "great wetnurse" in the tenth Orphic hymn, which had recently been translated by Marsilio Ficino.[4]

Although this particular manuscript is lost, it seems to have inspired a number of similar depictions. Three Italian books illustrated in the 1490s include related images: a printed copy of Pliny, with painted decoration;[5]

---

2. The fundamental study of this subject is Mechthild Modersohn, *Natura als Göttin im Mittelalter: Ikonographische Studien zu Darstellungen der personifizierten Natur* (Berlin: Akademie Verlag, 1997). See also Wolfgang Kemp's dissertation, *Natura: Ikonographische Studien zur Geschichte und Verbreitung einer Allegorie* (Tübingen: Eberhard-Karls-Universität, 1973), 9–14.

3. Edmund Spenser, *The Faerie Queene,* 7. This image persisted longer in texts than it did in images, and longer in Britain than on the Continent; see Elsa Berndt, *Dame Nature in der englischen Literatur bis herab zu Shakespeare* (Leipzig: Mayer & Müller, 1923).

4. Pietro Summonte, letter to Marcantonio Mihiel, transcribed in Fausto Nicolini, *L'Arte napoletana del Rinascimento e la lettera di P. Summonte a M. A. Mihiel* (Naples: R. Ricciardi, 1925), 165; see Kemp, *Natura* (note 2), 17–18. An English translation of the hymn appears in George Economou, *The Goddess Natura in Medieval Literature* (Cambridge: Harvard University Press, 1972), 40–41; Economou notes (pp. 55, 78) that Prudentius and Ambrose also referred to nature as "the wetnurse of humankind [*altrix* or *nutrix hominum*]." Ficino never published his translation, which he finished in the early 1460s; see Peter Dronke, "Bernard Silvestris, Natura, and Personification," *Journal of the Warburg and Courtauld Institutes* 43 (1980): 16–17.

5. Parma, Biblioteca Palatina , Inc. Pal. 1158, title page to Book II (1492); discussed and reproduced in Hermann Walter, "An Illustrated Incunable of Pliny's *Natural History* in the Biblioteca Palatina, Parma," *Journal of the Warburg and Courtauld Institutes* 53 (1990): 208–16,

a Latin manuscript translation of Themistius's commentary on Aristotle's *Physics;*[6] and a Greek manuscript copy of the *Physics* itself.[7] The first two show Nature clothed, with the exception of her lactating breast or breasts, as in Summonte's description, set against the backdrop of the heavenly spheres, while the third depicts a naked female figure, sitting in a purely terrestrial landscape and sprinkling the globe of the Earth with her milk.

In this essay I trace the trajectory by which the old image of Nature gave way to the new, as seen in illustrated books from the late fifteenth through the early seventeenth centuries, with special emphasis on the burgeoning literature of visual allegory and emblem. I explore this transformation for its changing meanings regarding the natural order and the moral authority attributed to nature. Although shifts in the understanding of the natural world did not necessarily cause the shift in personifications of nature, the two developments were linked in complicated ways. New ideas concerning the natural order, its meaning, and its relationship to both God and humans found expression in the new personification and contributed to its elaboration, while the new personification, with its much more precise set of associations with fertility and nurture, both shaped and limited the stories that artists and writers could tell about nature in the allegorical mode. The result was to reinforce a view of nature that tied it more strongly to the workings of matter, reconceiving it in turns of force and process and increasingly separating it from the realm of morals, voluntary action, and human will. In the course of this transformation, the natural order came to seem increasingly opaque, its physical structure and moral lessons still discernible, but only through an increasingly elaborate process of interpretation. Where medieval writers and artists personified nature as an articulate, speaking figure, their early modern counterparts saw her as both a valuable resource and an intrinsically enigmatic entity, whose teachings needed to be deciphered through human ingenuity and wit.

---

fig. 21a. The image illustrates the first section of Book II of the *Historia Naturalis,* which describes nature *(natura rerum)* as "embracing" the world or worlds.

6. Naples, Biblioteca dei Gerolomini, n. 221, fol. 2r; reproduced in Hermann Julius Hermann, "Miniaturhandschriften aus der Bibliothek des Herzogs Andrea Matteo III. Acquaviva," *Jahrbuch der kunsthistorischen Sammlungen des allerhöchsten Kaiserhauses in Wien* 19 (1898): 147–216, pl. XIII and 191–92. See Kemp, *Natura* (note 2), 76–77.

7. Vienna, Österreichische Nationalbibliothek, MS. phil. gr. 2, fol. 1r; reproduced in Hermann, "Miniaturhandschriften" (note 6), pl. V and 156–62, and in Modersohn, *Natura* (note 2), fig. 6. See Kemp, *Natura* (note 2), 71–73.

Nature Clothed

While ancient Greek and Roman writers occasionally personified nature—
Nature does nothing in vain, Nature abhors a void—they did so for the
most part only in passing, to render vivid a philosophical abstraction,
rather than as part of a sustained allegorical fiction.[8] One of the first
and most elaborate ancient examples of this latter use of personification
appeared in the late Roman writer Macrobius's *Commentarii in somnium
Scipionis* (Commentary on the dream of Cicero, ca. 400), where he discussed "fabulous narratives" *(narrationes fabulosas),* defined as stories that
rest "on a solid foundation of truth, which is treated in a fictitious style."[9]
According to Macrobius, philosophers such as Plato use these to speak of
the gods and the soul,

> not . . . merely to entertain, but because they realize that a frank, open ex-
> position of herself is distasteful to Nature, who, just as she has withheld an
> understanding of herself from the uncouth senses of men by enveloping
> herself in variegated garments, has also desired to have her secrets handled
> by more prudent individuals through fabulous narratives.[10]

In Macrobius's brief personification allegory,[11] Nature hides her secrets
from the unworthy, wrapping herself in "variegated garments" that conceal her body from those who rely for understanding only on their senses,
and revealing herself only to the prudent, whom he characterizes as
"eminent men of superior intelligence." Thus Macrobius's personification
of nature both embodied the practice of allegorical writing and referred to
the acts of decipherment that such writing required on the part of its more
discerning readers.

Twelfth-century Latin authors seized on this passage. While William
of Conches and Peter Abelard developed their own theories concerning
the *integumenta* (coverings) with which philosophers such as Plato wrapped

8. See in general Economou, *Goddess Natura* (note 4), 1–57; Dronke, "Bernard Silvester"
(note 4), 17–19.

9. Macrobius, *Commentary on the Dream of Scipio*, 1.2.17, trans. William Harris Stahl (New
York: Columbia University Press, 1953), 85.

10. Ibid., 86.

11. As the term suggests, personification allegories are allegorical fictions in which the
principal characters are personifications of abstract categories, such as *Le Roman de la rose,* by
Guillaume de Lorris and Jean de Meun, and John Bunyan's *Pilgrim's Progress.*

important truths concerning the order of the universe, contemporary poets began to compose their own far more extensive and elaborate allegorical fictions, in which they not only imitated Nature (and Plato) in this regard but included her as a central character as well.[12] Bernard Silvester's *Cosmographia* (ca. 1145) and Alan of Lille's *De Planctu Naturae* (Complaint of Nature, ca. 1160–72) and *Anticlaudianus* (ca. 1181–84) depict nature as an articulate and powerful, if flawed, figure, to whom God has delegated the creation and continuation of the physical world.[13] While one anonymous author composed a short poem (ca. 1200) in the form of a dream-vision in which Nature takes her revenge on a poet who had the temerity to spy on her nakedness[14]—a clever reworking of the story of Diana and Acteon—Bernard, Alan, and others of their contemporaries retained Macrobius's conception: Nature is a majestic, clothed figure, whose elaborately ornamented dress, described in great ecphrastic detail by Alan,[15] is a figure for the poet's own work of allegory and fabulation.

In the illustrations that accompany Alan's poems—no illustrated medieval manuscripts of Bernard's *Cosmographia* are known—Nature is represented in this vein, as a robed and queenly, often crowned figure. The images show her engaged in various activities, including speaking and molding or carving the perfect man to whose creation the *Anticlaudianus* is devoted, male-identified activities that give her a somewhat androgynous cast;[16] a drawing in one manuscript of *De Planctu Naturae* depicts her as lecturing or preaching at a podium, dressed in male clerical dress.[17] Na-

12. Peter Dronke, *Fabula: Explorations into the Uses of Myth in Medieval Platonism* (Leiden: E. J. Brill, 1974), chap. 1.

13. Bernard Silvester, *The Cosmographia of Bernardus Silvestris,* trans. Winthrop Wetherbee (New York: Columbia University Press, 1973); Alan of Lille, *Plaint of Nature,* trans. James J. Sheridan (Toronto: Pontifical Institute of Mediaeval Studies, 1980); Alan of Lille, *Anticlaudianus; or The Good and Perfect Man,* trans. James J. Sheridan (Toronto: Pontifical Institute of Medieval Studies, 1973). For an overview of Nature's role and character in these works, see Economou, *Goddess Natura* (note 4), chap. 3. Also foundational are the older discussions by Ernst Robert Curtius, "Zur Literarästhetik des Mittelalters II," *Zeitschrift für romanische Philologie* 58 (1938): 129–232, here 180–97; and *European Literature and the Latin Middle Ages* [1948], trans. Willard R. Trask (New York: Harper and Row, 1953), 106–27. Peter Dronke points out correctly that these authors are wrong in referring to Nature as a goddess; see his important article, "Bernard Silvestris" (note 4).

14. See F. J. E. Raby, "*Nuda Natura* and Twelfth-Century Cosmology," *Speculum* 43 (1968): 72–77.

15. Alan of Lille, *Plaint of Nature* (note 13), prose 1, 73–105.

16. Modersohn, *Natura* (note 2), 35–44 and figs. 9–15.

17. Ibid., fig. 10. On Nature's occasionally masculine or hermaphroditic character in Alan of Lille's work, see Larry Scanlon, "Unspeakable Pleasures: Alain de Lille, Sexual Regulation

FIGURE 2.1. Nature forging infants to replace the humans taken by Death. Jean de Meun, *Le Roman de la rose,* New York, Pierpont Morgan Library, MS. M.948, fol. 156r. By permission of the Pierpont Morgan Library.

ture has similar characteristics in the many illustrated manuscripts of the *Roman de la rose* (Romance of the rose), the thirteenth-century allegorical French poem begun by Guillaume de Lorris and continued, at much greater length, by Jean de Meun—an iconographical tradition that persisted, in manuscript and print, well into the first half of the sixteenth century. Figure 2.1, for example, from a manuscript produced ca. 1519 for King Francis I, shows her as a smith, hammering out new humans in her forge, to replace those whom Death, visible through the doorway, is taking away.[18]

Implicit in these medieval personifications, both verbal and visual, is the idea that Nature's authority over the created world was both physical and

and the Priesthood of Genius," *Romantic Review* 86 (1996): 213–42, esp. 231–33; Mark D. Jordan, *The Invention of Sodomy in the Middle Ages* (Chicago: University of Chicago Press, 1997), 71.

18. New York, Pierpont Morgan Library, MS. M.948, fol. 156r. Corresponding text in Jean de Meun, *Le Roman de la rose,* ll. 15891–19410, in Guillaume de Lorris and Jean de Meun, *The Romance of the Rose,* trans. Charles Dahlberg (Hanover, N.H.: University Press of New England, 1983), 270–320. On illustrations of Nature in the *Romance of the Rose,* see Modersohn, *Natura* (note 2), 73–158 and 198–244 (catalogue of illustrated MSS.), and associated figures, and F. W. Bourdillon, *The Early Editions of the Roman de la rose* (London: Chiswick Press, 1906). There is a facsimile of the Pierpont Morgan manuscript in *Le Roman de la rose pour François Ier . . . ,* 2 vols. (Lyon: Sillons du Temps, 1993), vol. 1.

moral. Because she formed the minerals, plants, animals, and people that populated it, she determined their physical properties, which resided in the matter and the forms peculiar to each species.[19] But she also took responsibility for their behavior, particularly their generative behavior, which was how she maintained the physical continuity of the world. In the case of humans, as Joan Cadden's and Helmut Puff's essays in this volume emphasize, her concern took the form of strenuous, if often unsuccessful, efforts to encourage "natural" (that is, reproductive and heterosexual) sex and discourage its "unnatural" (barren or sodomitical) counterpart.[20] Delegated by God to carry out the act of physical creation, she had remarkable power and autonomy. At the same time, however, this autonomy entailed vulnerability. The physical and moral orders she embodied were fragile ones, and both were endangered by sexual misconduct, which threatened to disrupt her ceaseless round of generation. In *De planctu Naturae,* Alan underscored Nature's vulnerability; human sexual license has torn and stained her dress.[21]

Despite her vulnerability, however, Nature was an active and articulate—even, in the *Roman de la rose,* garrulous—figure. In the medieval poems that describe her, she literally hammers out new beings; instructs philosophers on the varieties of natural phenomena; and lectures, berates, laments, and apologizes for the wayward humans who constantly contravene her will. No longer Macrobius's silent and enigmatic figure, she demands to be heard and understood. Her lessons are embodied in her clothing, like those of Macrobius's Nature, but she supplements these with ardent, expressive, and copious speech.

Nature Naked

The new Renaissance depiction of nature, as largely or completely naked and defined by the anatomical attribute of breasts and the physiological attribute of lactation, dramatically reconfigured Nature's character in me-

19. See esp. Alan of Lille, *Plaint* (note 13), prose 1, 78–105 (where the different species are depicted on Nature's garments).

20. See in general Hugh White, *Nature, Sex, and Goodness in a Medieval Literary Tradition* (Oxford: Oxford University Press, 2000), esp. chaps. 3–4. On the contradictions and ambiguities of medieval poetic treatments of this matter, see Jordan, *Invention of Sodomy* (note 17), 68–88 (on the *Complaint of Nature*); Simon Gaunt, "Bel Acueil and the Improper Allegory of the *Romance of the Rose,*" in *New Medieval Literatures* 2, ed. Rita Copeland, David Lawton, and Wendy Scase (Oxford: Clarendon Press, 1998), 65–93.

21. Alan of Lille, *Plaint of Nature* (note 13), 98.

dieval allegorical texts and images. The late fifteenth and early sixteenth centuries in fact saw the appearance of two separate but related visual traditions: Nature as a lactating woman (figs. 2.2, 2.3, and 2.4) and Nature as possessed of many breasts (fig. 2.6). The first tradition, the origins of which I have traced above, does not seem to have been based on ancient visual sources, though it drew on scattered textual references to Nature as wetnurse.[22] Its pictorial roots lay rather in the medieval tradition of representing a variety of (clothed) female figures—the goddess Terra or Tellus (Earth), as well as the personifications Ecclesia (the Church), Sapientia (Wisdom), Philosophia, Grammatica, and Caritas (Charity)—suckling animals or humans, often adult males.[23] Like these figures, it conjured up associations with Christian religious images of the nursing Virgin Mary.[24] At the same time, it also echoed a tradition of court spectacle in which decorative or table fountains were made in the shape of naked nymphs or other female figures with liquids jetting from their breasts.[25]

The Renaissance image of Nature as a lactating woman appeared intermittently in Italian prints, book illustrations, and paintings of the sixteenth century, where it became entangled with another iconographical tradition, that of Opis, wife of Saturn and mother of the gods, who was also

22. See note 4 above.

23. Dronke, "Bernard Silvestris" (note 4), 20–24 and figs. 1a and 2a; Liselotte Möller, "Nährmutter Weisheit: Eine Untersuchung über einen spätmittelalterlichen Bildtypus," *Deutsche Vierteljahrsschrift für Literaturwissenschaft* 24 (1950): 347–59; Caroline Walker Bynum, *Holy Feast and Holy Fast: The Religious Significance of Food to Medieval Women* (Berkeley: University of California Press, 1987), pls. 15 and 17.

24. On this iconographical tradition in its Italian context, see, e.g., Margaret Miles, "The Virgin's One Bare Breast: Female Nudity and Religious Meaning in Tuscan Early Renaissance Culture," in *The Female Body in Western Culture,* ed. Susan Rubin Suleiman (Cambridge: Harvard University Press, 1986), 193–208; Bynum, *Holy Feast and Holy Fast* (note 23), esp. 269–75 and pls. 13–24 passim; and Marilyn Yalom's suggestive, if superficial, *A History of the Breast* (New York: Ballantine, 1997), 38–48. On breastfeeding as a theme in Christian devotional literature and hagiography, see Caroline Walker Bynum, "Jesus as Mother and Abbot as Mother," in *Jesus as Mother: Studies in the Spirituality of the High Middle Ages* (Berkeley: University of California Press, 1982), 110–69; and Clarissa Atkinson, *The Oldest Vocation: Christian Motherhood in the Middle Ages* (Ithaca: Cornell University Press, 1991), esp. 57–61, 80, 121, and 142.

25. Kemp, *Natura* (note 2), 49–55. One late-fifteenth-century manuscript of the *Roman de la rose,* Oxford, MS Douce 195, shows the fountain of Narcissus represented in this way; Rosemond Tuve, *Allegorical Imagery: Some Mediaeval Books and their Posterity* (Princeton: Princeton University Press, 1966), fig. 99. For sixteenth-century examples of this tradition, see Bynum, *Holy Feast and Holy Fast* (note 23), 16; and Hendrick Goltzius's fountain project, the *Allegory of Plenty,* reproduced in Jean Starobinski, *Largesse,* trans. Jane Marie Todd (Chicago: University of Chicago Press, 1997), fig. 26.

represented with breasts spurting milk.[26] The identification of this figure
with Nature was cemented, however, with the publication in 1593 of Cesare Ripa's *Iconologia*. This influential Latin handbook of personifications,
which was translated into a number of European vernaculars in the seventeenth century, imagines Nature as a "naked woman, with her breasts full
of milk and holding a vulture in her hand." Noting that nature as a principle is divided into an active part (form) and a passive part (matter), Ripa
identified the former with Nature's breasts, "because form is that which
nourishes and sustains all created things, as woman with her breasts nourishes and sustains infants," and the latter with the vulture, "which, by its
appetite for form, moving and altering itself, destroys little by little all cor-
ruptible things."[27] The illustration of this personification (fig. 2.2) shows
a naked female figure with driblets of milk running from her breasts.[28]

The association made by sixteenth-century Italian humanists between
Nature and the ancient goddesses Tellus and Opis underpinned a second
visual tradition that depicted Nature as a female figure with many breasts.
The history of this particular personification is complicated, but it seems
to have its roots in a text from Macrobius's *Saturnalia*: "All religions worship Isis, who is either the earth or the nature of things under the sun. For
this reason, the body of the goddess is entirely covered with breasts, since
she sustains with her nourishing moisture [*altu*] the entirety of earth and
the nature of things."[29] On the basis of this description, Renaissance writers and painters began to portray Isis, Terra, and Nature on the model of
the many-breasted Diana of Ephesus, statues of which were brought to
light in the early sixteenth century and appeared in numerous contemporary accounts of ancient sculpture and topographical guides. By the 1520s,
Italian artists were using this figure as a personification of nature in their

26. Kemp, *Natura* (note 2), 23–25.

27. Cesare Ripa, *Iconologia overo descrittione dell'imagini universali* . . . (Rome: Heirs of Giovanni Gigliotti, 1593), *ad litteram* N. Ripa describes this image as based on a Roman medal of
the Emperor Hadrian, which Kemp claims is a fabrication (*Natura* [note 2], 22).

28. Ripa, *Iconologia* (Padua: Pasquinati, per P. P. Tozzi, 1618), *ad litteram* N. For a partial
modern edition, see Ripa, *Iconologia,* ed. Piero Buscaroli, 2 vols. (Turin: Fògola, 1986), esp. 2:
75. The 1618 edition was the first in which the entry for Nature was accompanied by an illustration. (The first illustrated edition of the *Iconologia* appeared in Rome in 1603.) For information concerning the images and texts in the 1593 and subsequent editions, see Yassu
Okayama, *The Ripa Index: Personifications and Their Attributes in Five Editions of the Iconologia*
(Doornspijk: Davaco, 1992).

29. Macrobius, *Saturnalia* 1.20.18–20, ed. Iacobus Willis, 2nd ed. (Stuttgart: B. G. Teub-
ner, 1994), 115.

N A T V R A.

FIGURE 2.2. *Nature.* Cesare Ripa, *La più che novissima iconologia di Cesare Ripa* . . . (Padua: Donato Pasquardi, 1630), 518. This woodcut was made from the same block as the image in the Paduan edition of 1618, the first in which Ripa's personification of nature was illustrated. By permission of the Houghton Library, Harvard University.

own paintings and statues, and it quickly spread to northern Europe as well.[30] Over the course of the sixteenth century, the personification of nature as endowed with many breasts became at least as influential as the image of her as naked and lactating, with which it was easily combined.

The art historian Wolfgang Kemp explains this dramatic shift in visual representations of Nature, from the medieval image of a dignified but often nondescript clothed female figure, without well-defined attributes, to the Renaissance figure of a naked and hyperbolically breasted or lactating woman, as part of a broader shift in artistic modes of personification that began in late-fifteenth-century Italy. Images of medieval personifications were commonly accompanied by captions or texts that identified their characters fully and sometimes gave them voice. The images of Nature in manuscripts of the *Roman de la rose* are a case in point; the painting of Nature in figure 2.1, for example, is explicated by a quatrain: "[This is] how

30. On the history of this image and its sixteenth- and seventeenth-century *fortuna,* see Theodora Jenny-Kapper, *Muttergöttin und Gottesmutter in Ephesos: Von Artemis zu Maria* (Zurich: Daimon, 1986); Kemp, *Natura* (note 2), passim, esp. 25–31; and, especially, the exhaustive study by Andreas Goesch, *Diana Ephesia: Ikonographische Studien zur Allegorie der Natur in der Kunst vom 16.–19. Jahrhundert* (Frankfurt am Main: Peter Lang, 1996), 13–168.

subtle Nature / Constantly forges sons or daughters, / So that the human race / Will never fail by her fault."[31] In contrast, Renaissance canons of artistic naturalism required the elimination of captions, banderoles, and inscriptions, while personifications increasingly appeared separated from literary texts in public spectacles such as processions and theatrical productions. According to Kemp, these two developments required the elaboration of more visually articulate personifications, quickly and easily differentiated from one another by visual signs alone.

This new type of personification was particularly well suited to the burgeoning literature of emblems, which combined striking and often arcane pictures with short and allusive texts.[32] Confronted by the need for novel, erudite, witty, and memorable images in order to impress their learned patrons, artists looked to contemporary scholars for references and ideas, a process exemplified in the collaboration between the illuminator Gasparo Romano and the humanist Luciano Fosforo, which I described above. The new, naked, lactating Nature that emerged from that collaboration served these purposes well, as a novel and striking allegorical conception informed by classical texts and, increasingly, classical images and esthetic norms.[33]

Nature's Fertility and the Natural Order

Although the new Renaissance personification of nature appears to have developed initially for reasons relating more to esthetics and the emergence of new artistic and literary forms than to new views of nature, it quickly became a device through which learned writers articulated their changing understandings of the natural order; indeed, it no doubt helped to shape those views. The figure had multiple associations. In the first place, Nature as a many-breasted or lactating woman was a powerful emblem of fertility and embodied an optimistic view of a benign and bountiful natural world. This new sense of nature stood in clear contrast to

31. New York, Pierpont Morgan Library, MS. M.948, fol. 156r: "Comment nature la subtille / Forge toujours ou fils ou fille, / Affin que l'humaine lignée / Par son deffault ne faille mye."

32. The literature on emblems is enormous. For an overview and introduction to some problems of interpretation, see Alistair Fowler, "The Emblem as Literary Genre," in *Deviceful Settings: The English Renaissance Emblem and its Contexts,* ed. Michael Bath and Daniel Russell (New York: AMS Press, 1999), 1–31; Giancarlo Innocenti, *L'Immagine significante: Studio sull'emblematica cinquecentesca* (Padua: Liviana, 1981), esp. chap. 1.

33. Kemp, *Natura* (note 2), 14–18.

the acute concern for natural scarcity in the works of ancient writers like Hesiod (see Laura Slatkin's essay in this volume), in which generation and corruption, birth and death barely balanced each other in what was at best a zero-sum game. In many respects, medieval writers shared this view. The poems of Alan of Lille and Jean de Meun emphasize Nature's frenetic efforts to outstrip death and keep the recalcitrant human species going, and Nature and Death are paired in a number of illustrations of the *Roman de la rose* and related texts, as in figure 2.1.[34]

In contrast to this bleak and strenuous scenario, sixteenth-century artists celebrated a regime of plenty, as the Renaissance image of automatic and apparently effortless lactation supplanted the demanding medieval discipline of reproductive sex that many humans found so difficult to maintain. This vibrant image of inexhaustible plenty seems to have had its roots in late medieval tropes of princely generosity, as figured in the table fountains tricked out as lactating nymphs that graced fifteenth- and early-sixteenth-century court banquets, but it increasingly seemed an appropriate expression of the prosperity of Europe's expanding commercial economy, fed in large part by the natural resources of the fertile New World. (Indeed, contemporary visual personifications of America showed her as a voluptuous naked woman.)[35] This association of Nature with fertility and beneficence was cemented in Ripa's *Iconologia,* where other images of lactating women included the personifications of Sostanza (Substance), Carità (Charity) and Benignità (Benignity).[36]

Unlike these other three personifications, however, Nature was described by Ripa as naked (fig. 2.2), which gave her figure additional resonance. Ripa treated nakedness as a floating signifier, characterizing a wide variety of female personifications. It connoted, among other things, deprivation (Povertà, Miseria), beauty (Bellezza, Chiarezza), and desire (Cupidità, Desiderio, Lussuria).[37] Nature's nakedness seems to have had yet

34. See Modersohn, *Natura* (note 2), 138–39, 184–88. Cf. Jean de Meun, *Romance of the Rose* (note 18), ll. 15891–16012, pp. 270–71.

35. Mario Klarer, "Woman and Arcadia: The Impact of Ancient Utopian Thought on the Early Image of America," *Journal of American Studies* 27 (1993): 4–7 and figs. 1 and 3; Christopher Columbus described the American earthly paradise as corresponding topographically to a "woman's nipple." On Europe's expanding commercial economy and the culture of consumption it supported in this period, see Lisa Jardine, *Wordly Goods: A New History of the Renaissance* (New York: Norton, 1996), esp. chaps. 1, 2, and 6.

36. All three were first illustrated beginning with the edition of 1603; see Okayama, *Ripa Index* (note 28), *ad verbum.*

37. Ibid.

another set of meanings, however, associated with Ripa's single largest group of naked female personifications, which exemplified abstractions related to Christian felicity, including Felicità Eterna, Gratia, Predestinatione, Resurrettione, and Sapienza (Wisdom).[38] More specifically, it seems to have represented her closeness to God's original plan for creation, unsullied by the deformations of human artifice, as represented by braided hair and ornamental dress. Although Ripa did not spell out these associations in his description of Nature, they appear clearly in his account of the related personification of Legge Naturale (Natural Law). He notes that this figure's flowing hair is "natural . . . and not braided by art"; "she is represented half naked, with her hair natural, because this law is simple as made by God."[39] By close analogy, Nature's nakedness, too, symbolized her direct relationship to God and her simplicity unadorned by art.

It was possible to conceptualize and explicate this new personification of Nature so as to emphasize her dignity and her cosmic power. The magnificent folding woodcut that illustrates Robert Fludd's *Utriusque cosmi maioris scilicet et minoris metaphysica, physica atque technica historia* (Metaphysical, physical and technical history of both the macrocosm and the microcosm, 1617) shows Nature, naked, with spurting breast and flowing hair, as prescribed by Ripa, with her feet on the earth and her haloed head in the celestial and angelic realms (fig. 2.3). She has lost her vulture—Fludd emphasized her fertility and creativity rather than her work of destruction and dissolution—but has acquired a different animal companion; her left hand holds a chain attached to the left hand of an ape, Art, who squats within the elemental spheres, while the chain attached to her right hand is held in turn by the hand of God.[40] In the accompanying explanation, Fludd laid out an extended allegory that described Nature as a cosmic principle, from whose breasts, marked in the woodcut by the symbols of the sun and the moon, flows "the radical warmth and moisture that dwells in all creatures made of the elements, from which their life and flourishing [*vegetatio*] springs." This "milk" takes the form of a "golden light" that is identified with the influences of stars and planets, "which are borne down to the earth, penetrating to its center. . . . The planets work together

38. Ibid.

39. Ripa, *Iconologia*, 1593 ed. (note 27), *ad litteram* N; this entry was first illustrated in the edition of 1624.

40. Robert Fludd, *Utriusque cosmi maioris scilicet et minoris metaphysica, physica atque technica historia* (Oppenheim: De Bry, 1617). For a detailed analysis of this image and its sources, see Kemp, *Natura* (note 2), 88–101.

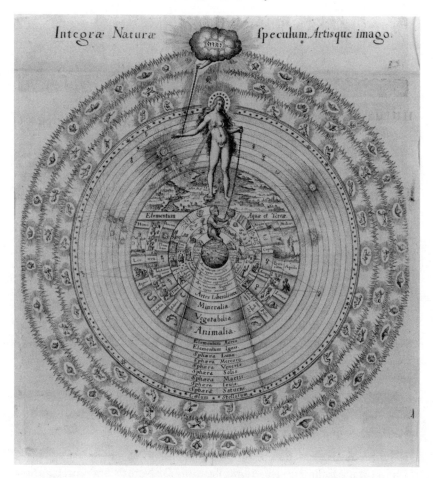

FIGURE 2.3. Attr. Matthias Merian, "Mirror of the Whole of Nature, and of Art." Robert Fludd, *Utriusque cosmi maioris scilicet et minoris metaphysica, physica atque technica historia*, 2 vols. (Oppenheim: de Bry, 1617), vol. 1, folding plate = [pp. 4–5]. Typ 620.17.399F, Department of Painting and Graphic Arts, Houghton Library, Harvard College Library.

through actions of this kind, by which they produce the various things that make up the animal, plant, and mineral realms."[41] Here Fludd was drawing on a well-established philosophical tradition, also invoked by Marsilio Ficino, that identified Nature with the world soul *(anima mundi),* one of the main hypostases in neo-Platonic metaphysics. According to Ficino, the world soul communicated with the corporeal universe through the cosmic

41. Fludd, *Utriusque cosmi maioris* (note 40), 8.

spirit *(spiritus mundi)*, a subtle and vivifying body that permeated it, which Fludd here identifies with the physiological principle of radical moisture and figures as Nature's warm and nourishing milk.[42]

Fludd's image associates Nature with Eve, who was also sometimes shown with an ape;[43] here, however, her halo and her nakedness identify her as Eve before the fall. Thus she stands for God's prelapsarian creation; by emphasizing the parallels between nature's relationship to God and art's relationship to nature, Fludd and his artist—probably Matthias Merian—showed her as the art of God. Where medieval poets and artists portrayed Nature as the creator of the physical world, in this woodcut, she becomes creation itself. Simultaneously, she loses much of her autonomy, as is clear from the chain that binds her wrist.[44] This set of associations served simultaneously to bind Nature to God and to reduce dramatically her independence and her attendant vulnerability to human corruption, both of which were salient characteristics of the medieval personifications. For medieval writers such as Alan of Lille, Nature and the moral order she embodied were largely defenseless against human perversity. Paradoxically, Nature's nakedness in the later sixteenth and early seventeenth century figured a kind of immunity against human disorder; it stood for prelapsarian purity, the natural order as God had intended it, unsullied by the often ill-conceived innovations, both moral and material, introduced into creation by human beings.

Nature's Moral Authority

Driving a wedge between the natural order and human artifice and culture had important and complicated implications for Nature as a moral authority. Increasingly, she was represented as *pre*moral in the sense understood by Renaissance writers, that is, unconcerned with human practices and conduct *(mores)*. Her contribution to human life was limited to conferring on individual humans the particular physical attributes and mental inclina-

42. On these concepts, see D. P. Walker, *Spiritual and Demonic Magic from Ficino to Campanella* [1958] (Notre Dame: University of Notre Dame Press, 1975), 12–13; Michael R. McVaugh, "The 'Humidum Radicale' in Thirteenth-Century Medicine," *Traditio* 30 (1974): 259–83.

43. Sabine Melchior-Bonnet, *The Mirror: A History*, trans. Katharine H. Jewett (New York: Routledge, 2001), 200.

44. On nature's loss of autonomy relative to God over the course of the seventeenth century, see Lorraine Daston and Katharine Park, *Wonders and the Order of Nature* (New York: Zone, 1998), 296–301.

tions with which they were born; she did not preside over or participate in the long process of socialization, education, and voluntary effort that shaped their character and moral worth.

An engraving by the Netherlandish printmaker Philip Galle, "Man Is Born Naked" (1563) clearly expresses this idea, representing Nature's role in the formation of human beings as prior to that of both the mother and the wetnurse. The first in a series of allegorical images entitled *The Misery of Human Life,* it depicts Nature as a majestic but primal figure; in addition to Diana of Ephesus, she conjures up the classical figure of Artemis, goddess of wildness, fertility, and childbirth.[45] Naked and many-breasted, like the animals that accompany her, she emerges from a forest and hands a limp and equally naked newborn to a clothed female figure, presumably its mother, whose bodice is open to reveal her own waiting breast. The mother and her attendant—her bared shoulder suggests that she is the wetnurse—are shown against a cultivated and inhabited landscape that contrasts dramatically with Nature's wood. It is the mother and the wetnurse, not Nature, who will nourish the infant and swaddle it in its first clothing, thus launching the process of civilization that differentiates humans from the rest of the natural world. The second image in the series ("Man Has to Learn Everything") shows the continuation of this process, as two women care for a group of children, some of whom are in the process of learning to walk.[46]

Galle's engravings reflect a strong strand in the postmedieval tradition of personifying nature that emphasized Nature's apparent ambivalence toward her human creatures. In this respect, she was seen as resembling an indifferent mother—often a stepmother or foster mother—or wetnurse. This theme emerged particularly strongly in northern European emblem books. Guillaume de la Perrière's *Morosophie* (Foolish wisdom, 1553), for example, explicitly identifies Nature as both mother and stepmother, and shows her as a Janus-faced figure who simultaneously nourishes infants and castigates the old.[47] Even in the former capacity, however, she exhibits none of the typical characteristics of a caring mother; rather than holding and nursing her children, she squirts them with breast milk as they lie on the ground at her feet. An emblem in Jean-Jacques Boissard's *Emblematum*

45. Reproduced in Adam von Bartsch, *The Illustrated Bartsch,* 165 vols. to date (New York: Abaris Books, 1978 –), vol. 56, *Netherlandish Artists: Philips Galle,* by Arno Dolders, fig. .072:1.
46. Reproduced in ibid., fig. .072:2.
47. Guillaume de la Perrière, *La morosophie* [1553] (Aldershot, Hants.: Scolar Press, 1993), emblem 42.

liber (Book of emblems, 1593) shows the same form of breastfeeding without intimacy (fig. 2.4), while the text emphasizes Nature's contribution to human life as largely physical:

> Nature has given to all human beings, without discrimination, the gift of living, but the honor of living well is accorded only by Virtue. Life is common not only to the good, but also to the evil. That which should truly be called life is conceded to His followers only by God, the Father of lights. Life devoid of virtue is not life, but rather must be valued as worse than death.[48]

The accompanying woodcut image underscores the point. On the left, Virtue, personified as a dignified woman, seated and fully clothed, reads to a group of attentive children, while Nature, on the right, naked from the waist up, expresses two streams of milk onto a more disorderly cluster of infants at her feet. This image, like Galle's, draws a clear distinction between nature and culture in the modern sense. Nature is nonverbal, like the infants she nourishes, and she supplies only physical sustenance; indeed, the impersonality of her actions distances her definitively from the human gesture of nursing as an affective and morally formative act.

Londa Schiebinger has analyzed eighteenth-century personifications of nature as many-breasted in the context of contemporary imperatives concerning motherhood and breastfeeding; she argues that, like Linnaeus's introduction of the term *mammalia* to refer to the group of animals traditionally known as quadrupeds, these images "helped legitimize the restructuring of European society by emphasizing how natural it was for females—both human and nonhuman—to suckle and rear their own children."[49] In the sixteenth century, however, although the benefits of maternal breastfeeding had long been a staple of humanist writing on the family, the new breast-centered personification of Nature did not yet serve to link the ideas of woman, nature, motherhood, and breastfeeding in any straightforward way.[50]

48. Jean-Jacques Boissard, *Emblematum liber* [1593] (Hildesheim: Olms, 1977), 18.

49. Londa Schiebinger, *Nature's Body: Gender in the Making of Modern Science* (Boston: Beacon Press, 1993), chap. 3; quotation on 74.

50. Most Western European elites employed wetnurses during this period. On this practice and the debates surrounding it, see, e.g., Christiane Klapisch-Zuber, "Blood Parents and Milk Parents: Wet Nursing in Florence, 1300–1530" [1980], in *Women, Family, and Ritual in Renaissance Italy*, trans. Lydia Cochrane (Chicago: University of Chicago Press, 1985), 132–64; Giuliana Calvi, *Il contratto morale: Madri e figli nella Toscana moderna* (Rome: Giuseppe

VITA VIRTVTIS EXPERS MORTE PEIOR.

FIGURE 2.4. "A Life Devoid of Virtue is Worse than Death." Jean-Jacques Boissard, *Emblematum liber* (Frankfurt am Main: Theodor de Bry, 1593), Emblem 11. Typ 520.96.225. Department of Painting and Graphic Arts, Houghton Library, Harvard College Library.

Unlike her medieval counterpart, who intermittently turned her hand to traditionally masculine activities such as blacksmithing and lecturing, Nature in the Renaissance was an unambiguously feminized figure, characterized by the anatomical feature of breasts and the physiological function of lactation. But it would be wrong to read these attributes as necessarily maternal, in a world where middle-class and patrician women rarely nursed their children. Furthermore, even in theory, Nature was hardly an unambiguous model for maternal breastfeeding, given her wasteful and impersonal style of nursing, which ran counter to the growing emphasis in sixteenth- and seventeenth-century Europe on the moral and emotional dimensions of motherhood.[51] If anything, she functioned more plausibly as a cautionary figure—an example of the shortcomings of the wetnurse or foster mother, although her bountiful supplies of milk make it difficult to identify her clearly with those more withholding figures. In this

Laterza, 1994), 107–9; Yalom, *History of the Breast* (note 24), 59–86; and in general Valerie Fildes, *Wet Nursing: A History from Antiquity to the Present* (Oxford: Basil Blackwell, 1988).

51. Calvi, *Contratto morale* (note 50), 63–64, 107–8, 160, and passim.

connection, it may make the most sense to read her as an expression of uncertainty about changing ideas of motherhood and the relationship between the several female functions of physical generation, material nurture, and moral and affective formation, which were shown as separable and separate in Galle's engraving of 1563.[52]

In general, however, it would be wrong to see this reconfiguring of Nature's role in the shaping of human beings as necessarily a demotion. Nature retained extraordinary importance and power, as the mistress of bodies, matter, and the principles that governed them, and she continued to be seen as an awesome cosmic force. But increasingly, her functions, unlike those of her medieval counterpart, were set apart from and opposed to the human activities of nurture, domestication, and moral education—whether by wetnurses, mothers, or teachers—and her functions were represented in purely physical terms. The power of Renaissance Nature was in and of the body, as represented by her nakedness, her breasts, and her inexhaustible lactation, rather than in intention, voice, or will. She provided value in the most basic sense of raw materials and resources, rather than values in the sense of ethical and religious norms. No longer the active shaper of natural phenomena and spokesperson for the moral order of God's creation, she became increasingly identified with its corporeal stuff: the inventory of God's creatures and the physical fabric of the natural world.

Nature's Body

Carolyn Merchant's important study *The Death of Nature* traces the transformation of European Christian ideas of nature through the seventeenth century, the period identified with the Scientific Revolution. According to Merchant, the Nature of medieval European writers had two aspects: that of "a nurturing mother: a kindly beneficent female who provided for the needs of mankind in an ordered, planned universe," and that of a wild and chaotic force. Early modern intellectuals, in contrast, replaced this "organically oriented mentality in which female principles played an important role" with a "mechanically oriented mentality that either elimi-

---

52. On the complicated relationship between these distinct functions, see Emilie L. Bergmann, "'Language and Mothers' Milk': Maternal Roles and the Nurturing Body in Early Modern Spanish Texts," in *Maternal Measures: Figuring Caregiving in the Early Modern Period,* ed. Naomi J. Miller and Naomi Yavneh, (Aldershot: Ashgate, 2000), 105–20.

nated or used female principles in an exploitative manner."[53] Merchant argued that this new view of nature found its first, and in some ways its most characteristic, expression in the natural philosophical writing of Francis Bacon, who promoted the human subjugation of the natural world, using metaphors and allegories organized around images of torture, enslavement, and rape.[54]

My own study of the changing personification of nature in this period suggests a somewhat different story. Although Nature in the Middle Ages was of course female, as required by the gender of the noun in both Latin and the European vernaculars, she had androgynous characteristics, as I have already mentioned. Furthermore, there was little emphasis either on her wildness—she functioned as a principle of order—or on her maternal functions; her generativity was almost always represented in terms of artisanal production (carving, coining, blacksmithing), rather than conception and birth. There is no developed medieval literary or visual tradition that identified Nature in the first instance with the role of mother, or even of wetnurse, or attributed maternal qualities to her in any but a passing way.[55] The reconstruction of Nature in terms of breasts, nursing, and lactation, which Merchant reads anachronistically as self-evidently maternal, although it drew on a few late antique references to nature as wetnurse, was in fact a Renaissance invention, a product of humanist imagination and scholarship in the late fifteenth and sixteenth centuries; indeed,

53. Carolyn Merchant, *The Death of Nature: Women, Ecology, and the Scientific Revolution,* 2nd ed. (New York: HarperCollins, 1990), 2.

54. Ibid., esp. chap. 7. Merchant associates these activities particularly with mining; for a suggestive iconographic treatment of this theme, see Horst Bredekamp, "Die Erde als Lebewesen," *Kritische Berichte* 9 (1981): 5–37.

55. As I have already mentioned, there are a few isolated texts and images that point in this direction; see, e.g., Economou, *Goddess Natura* (note 4), 18, 55, and 78. Modersohn has identified a single medieval image of Nature nursing—her nursling is in this case Scientia—on a twelfth-century English enamel casket in the Victoria and Albert Museum; see Modersohn, *Natura* (note 2), 47–49 and figs. 18a and 18b. But these representations are vastly outnumbered by other designations and descriptions of Nature. Merchant is not alone in her misapprehension. The power of the postmedieval personification of nature as nurse and mother is such that even specialists regularly read it back into the earlier period. See, e.g., Curtius, "Literarästhetik" (note 13), 180, where he subtitles a long section "Natura mater generationis," despite the fact that not one of the fourteen functions of Nature that he discusses in that section is explicitly maternal. Similarly, Dronke claims to have found references to nature as *mater* in the *Novum glossarium mediae latinitatis;* cf. vol. M–N (Copenhagen: Ejnar Munksgaard, 1959–69), col. 1090–91, where the only personifications I find are those of *artifex* and *opifex.*

this is the period to which almost all of Merchant's evidence for her "medieval" and "organic" theory of nature actually belongs.

Merchant is right to call attention to Bacon's choice of gendered metaphors to describe the natural philosopher's relationship with nature, although she, like some other feminist scholars, has overstated the element of violence in those metaphors, as well as the degree of consistency with which they are elaborated.[56] Bacon's personifications of nature tend to be both very brief and considerably rarer than one would gather from the critical literature on the topic; many apparent personifications are in fact artifacts of Bacon's nineteenth-century English translators, who consistently capitalized "Nature" and referred to it as "she," even when there was no warrant for this in the Latin original. At the same time, the gender implications of the passages in Bacon's work that do indeed personify nature are more ambiguous than Merchant suggests; indeed, Bacon's only extended allegorical discussion of nature personifies it as the male figure, Pan.[57] There is no doubt that in some passages Bacon exploited, to powerful effect, the identification of nature with the exposed female body, most notably in those passages where he described his new method as fo-

56. See the references to these critiques in Alan Soble, "In Defense of Bacon," *Philosophy of the Social Sciences* 25 (1995): 192–215; although Soble is quite correct in pointing up the inconsistency of Bacon's metaphors concerning nature, he understates the element of misogyny involved in the most explicit ones, which he qualifies as only "mildly sexist" (p. 212). For a more nuanced treatment of this topic, and one that mercifully avoids assuming that every reference to nature in Bacon involves personification, see Elizabeth Hanson, *Discovering the Subject in Renaissance England* (Cambridge: Cambridge University Press, 1998), chap. 5. More general discussions of the gender implications of medieval and early modern personifications in the context of ideas of natural knowledge appear in Londa Schiebinger, *The Mind Has No Sex? Women in the Origins of Modern Science* (Cambridge: Harvard University Press, 1989), 121–50; and Alcida Assmann, "Der Wissende und die Weisheit—Gedanken zu einem ungleichen Paar," in *Allegorien und Geschlechterdifferenz*, ed. Sigrid Schade, Monika Wagner, and Sigrid Weigel (Cologne: Böhlau, 1994), 11–25.

57. Francis Bacon, *De sapientia veterum*, 6, in Bacon, *The Works of Francis Bacon*, ed. and trans. James Spedding, Robert Leslie Ellis, and Douglas Denon Heath, 7 vols. (London: Longman, 1857–70), 6:635–41 (English translation in ibid., 6:707–14); Bacon, *De augmentis scientiarum*, 2, in Bacon, *Works*, 1:521–30 (English translation in ibid., 4:320–32). On Bacon's discussion of the myth of Pan, see Charles William Lemmi, *The Classic Deities in Bacon: A Study in Mythological Symbolism* (Baltimore: Johns Hopkins, 1933), 61–74; and, especially, Barbara Carman Garner, "Francis Bacon, Natalis Comes and the Mythological Tradition," *Journal of the Warburg and Courtauld Institutes* 33 (1970): 281–85. Andrea Alciato also used Pan to personify nature *(vis naturae)* in his influential *Emblemata* (Lyon: Mathieu Bonhomme, 1550), 106.

cused on exploring, penetrating, or dissecting Nature's viscera (visceribus), womb (utero), or bosom (sinu).[58] However, Merchant is wrong to imply that this figure of nature was recent, while the figure of nature as maternal was traditional: rather, both are versions of the same new, naked, and feminized Renaissance personification of nature. Bacon merely gave this new personification a particular twist, emphasizing the erotic and reproductive aspect of Nature's body, so that she became an appropriate object of courtship and seduction, and conferring on her as well a fertile uterus, which the natural philosopher could open and dissect.[59]

The new Renaissance figuration of nature as naked and female certainly authorized a more exploitative attitude toward the natural world, although the dominant metaphor was consumption (of nature's inexhaustible "milk" or bounty) rather than inquisition, dissection, or rape. It certainly supplied a powerful set of images in which such an attitude could be expressed. But my own evidence foregrounds as equally pervasive another aspect of the transformation of views of nature in this period: its increasing separation from the human world. In Renaissance allegories and emblems, Nature retained her power but lost her voice. Medieval writers had seen Nature as intentionally enigmatic; although she chose to hide the true lineaments of the natural order—her body—beneath her voluminous garments (which functioned also as a figure for poetic language), she revealed it to the learned in copious and forceful speech. In the sixteenth and seventeenth centuries, in contrast, Nature's body was exposed for all to see. But that body was itself opaque and difficult to interpret, like the alien, even grotesque, figure of Diana of Ephesus herself. Where Nature spoke

58. See, e.g., Bacon, *Instauratio magna,* in Bacon, *Works* (note 57), 1 : 137 *(visceribus); Novum organum,* 1.74 and 1.109, in ibid., 1 : 183 *(utero)* and 1 : 208 *(sinu)*. References to Nature's breast, in the ungendered sense of *sinus* (cf. John 1 : 18), can be found in Augustine and John Scotus Eriugenus, where they identify Nature's breast, usually qualified as "secret," "deep," or "hidden," as the place where the causes of natural phenomena are concealed; see Dronke, "Bernard Sylvestris" (note 4), 31, additional note (Augustine), and Edouard Jeauneau, *Quatre thèmes érigéniens* (Montreal: Institut d'Etudes Médiévales Albert-le-Grand; Paris: J. Vrin, 1978), 39–46 (Scotus).

59. Bacon did not invent the latter idea, which can be found in sixteenth-century medical and technical writers; see, for example, Katharine Park, "From the Secrets of Women to the Secrets of Nature," in Jane Donawerth and Adele Seeff, eds., *Crossing Boundaries: Attending to Early Modern Women* (Newark: University of Delaware Press; London: Associated University Presses, 2000), 29–47. On the courtship and seduction metaphors in Bacon, see Evelyn Fox Keller, "Baconian Science: The Arts of Mastery and Obedience," in *Reflections on Gender and Science* (New Haven: Yale University Press, 1985), esp. 36–37.

explicitly, voluminously, and directly to medieval writers such as Alan of
Lille in dreams and visions, she stood mutely before early modern naturalist inquirers or receded elusively from their grasp. One influential image, from Michael Maier's *Atalanta fugiens* (Atalanta fleeing, 1618), a book
of alchemical emblems, shows Nature walking at night before an al-
chemist, who, although handily equipped with the spectacles of experience, the lantern of erudition, and the walking stick of reason, can only

FIGURE 2.5. Matthias Merian, Al-
chemist following Nature's footprints.
Michael Maier, *Atalanta fugiens* (Op-
penheim: Johann Theodor de Bry,
1617), Emblem 42. Courtesy of the
Boston Medical Library in the Francis
A. Countway Library at the Harvard
Medical School.

FIGURE 2.6. Attr. Matthias Merian, Alchemists following Nature's footprints. *Musaeum
hermeticum* (Frankfurt: Sumptibus L. Jennisii, 1625), titlepage. Courtesy of the National Library
of Medicine.

just discern her footprints (fig. 2.5).[60] Although Nature is uncharacteristically clothed in this particular image, no doubt further to emphasize her inaccessibility, a slightly later version, attributed to the same artist, shows her in her more characteristic, many-breasted form (fig. 2.6).[61]

In this view, Nature was resistant to comprehension, not by intention, but by virtue of her very constitution. To be understood, her enigmatic body needed to be explicated and interpreted by learned naturalists, whose task was to explore the truths that she withheld. To discover those truths, it was no longer possible simply to listen to Nature, for she embodied rather than "knew" them. To this end, the natural phenomena that composed her body had to be studied—occasionally dissected—in painstaking and meticulous detail (see Lorraine Daston's essay in this volume). As Schiebinger has shown, the lessons Nature embodied continued to have a moral dimension; plant anatomy and animal physiology continued to reveal the "natural" superiority of males to females, together with ever more elaborate hierarchies of class and race.[62] But Nature herself no longer played an active part in propagating those basic truths, as she had in medieval allegories, where poets and artists depicted her as teaching explicitly and directly. Rather, they had to be extracted from her by learned men. In this sense, Nature retained her moral authority, but that authority was increasingly and more overtly mediated by the naturalists, theologians, and political philosophers who had appointed themselves to speak on her behalf.

60. Michael Maier, *Atalanta fugiens* (Oppenheim: Johann Theodor de Bry, 1618), emblem 42. This woodcut is the work of Matthias Merian, the artist to whom the woodcut in Fludd's *Utriusque cosmi maioris* (fig. 2.3) is also usually attributed.

61. *Musaeum hermeticum* (Frankfurt: Hermann van Sande, 1625), title page. The personification of nature and related concepts in medieval and Renaissance alchemical texts and images is a subject that demands further study.

62. Schiebinger, *Nature's Body* (note 49).

3

Burning *The Fable of the Bees:*
The Incendiary Authority of Nature

Danielle Allen

Long before Aristotle called not only human beings, but also bees, political creatures, these social insects had already assumed a firm place in the
Greek lexicon of poetic moralizations. Hesiod, in his didactic *Works and
Days,* exhorts his brother to industry by contrasting the laboring bee to
the drone: "Both gods and men are angry with a man who lives idle, for
in respect to his passions he is like the stingless drones who waste the labor
of the bees, eating without working; let it be your care to order your work
properly."[1] Like the human world, the beehive has social organization;
labor and value are divided and diversely apportioned. Indeed, Hesiod
elsewhere uses the bee/drone dichotomy to poeticize the human division
of labor along gender lines. In his narration of the creation of Pandora and
the birth of the race of women, he remarks:

> And as in thatched hives bees feed the drones whose nature is to do mis-
> chief—by day and throughout the day until the sun goes down the bees are
> busy and lay the white combs, while the drones stay at home in the cov-
> ered hives and reap the toil of others into their own bellies—even so Zeus
> who thunders on high made women to be an evil to mortal men, with a
> nature to do evil.[2]

Aristophanes (*Wasps* 1102–21) and Plato (*Republic* 552c–e) both expand
Hesiod's bee/drone contrast into fables about citizenship, stretching the

1. Hesiod, *Works and Days,* trans. Hugh G. Evelyn White (with some modifications)
(Cambridge: Harvard University Press, 1959 [1924]), 303–6.
2. Hesiod, *Theogony,* trans. Hugh G. Evelyn White (with some modifications) (Cambridge: Harvard University Press, 1959 [1924]), 590–601.

symbol to embrace all the classes a polity might contain and to convey the nature of political order *tout court*. But labor and sexuality remain central to their revisions too.[3]

One is tempted to close any history of the bee metaphor and its role in the moralizations of Western literature here, at its beginning, for uses of the trope continue in much the same vein for millennia.[4] In pre-Christian Latin texts, bees are as industrious as ever, but they are also loyal, capable of perfect concord, and chaste: Virgil thought they fetched their young from olive trees, or that they arose, without copulation, from the rotting flesh of oxen carcasses. In the Christian period the chastity of bees is especially important—indeed, issues of gender and sexuality are never far from the image's symbolic surface; but Christian writers also compared the nectar gathered by bees to divine grace and honey to god's great goodness.[5] In all periods, the bees' hive was used to exemplify perfect political order, whether that was taken to be monarchic (Virgil), communitarian

3. D. Allen, *The World of Prometheus: The Politics of Punishing in Democratic Athens* (Princeton: Princeton University Press, 2000), 131–33, 271. The topics of labor and sexuality seem to come up frequently in discussions of nature's moral authority. In this volume, see the essays by Slatkin, Puff, Cadden, Murphy, and Price.

4. J. W. Johnson ("That Neo-Classical Bee," *Journal of the History of Ideas* 22 [1961]: 262–66, here 262), argues for marked differences in the bee trope in different periods. But this is incorrect. The trope of the bee as moral and industrious persists throughout all periods, including the twentieth century. Richard Rolle (fourteenth century) wrote that the bee is "never idle and against them that will not work, and keeps her wings clean and bright" (H. M. Fraser, *History of Beekeeping in Britain* [London: Bee Research Association, 1958], 23). In the twentieth century, we find: "The working day begins with them at three or four o'clock in the morning, and is continued till dusk or even into the night, if the weather be warm and the moonlight good. It was stated by Godart, some two centuries ago, that some species appoint a trumpeter-bee to rouse the rest of the community; this statement was confirmed in the latter part of the nineteenth century by Hoffer. The doubt which was for so long entertained with regard to this trumpeter is perhaps due to the fact that most human beings receive their summons to leave their beds several hours after the humble-bees have been 'called,'" (O. H. Latter, *Bees and Wasps* [Cambridge: Cambridge University Press, 1913], 92). See also K. Frisch, *The Dancing Bees*, trans. D. Isle and N. Walker (1927; London: Methuen, 1966), 42; J. L. Gould and C. G. Gould, eds., *The Honey Bee* (New York: Scientific American Library, 1988), 19, 24, 28.

5. On chastity of bees, see E. Crane, *Bees and Beekeeping: Science, Practice, and World Resources* (Ithaca: Cornell University Press, 1990), 28; S. W. Mintz, "Sweet, Salt, and the Language of Love," *Modern Language Notes* 106 (1991) 852–60, here 857, n. 3; Fraser, *History* (note 4), 22; H. Friedmann, *A Bestiary for Saint Jerome: Animal Symbolism in European Religious Art* (Washington, D.C.: Smithsonian Institution Press, 1980), 153. For bees as blessed souls and honey as a symbol for divine words, see *Origines* of St. Isidore of Seville 560–636; *Hexameron* of St. Ambrose 340–97; E. A. W. Budge, ed. and trans., *The Book of the Bee: Solomon of Basra, The Syriac Text* (Oxford: Clarendon Press, 1886), 2.

(Christian writers), or egalitarian (some French revolutionaries).[6] The symbol of the bee as a model for the citizen, laborer, and monarch seems almost too trite—too consistent a poetic invocation of nature and too obvious in its implications—to be worth taking seriously in an examination of how various conceptions of nature, different as to time and place, contributed to the construction of moral authority in each.[7]

Yet one invocation of the trope is anomalous enough to open for us the whole tradition and its significance, and to expose the riches hidden beneath the easy pieties to those who wish to understand how nature comes to be valued in ways that sustain, in turn, its use in the construction of human values. In 1740 a text called *The Fable of the Bees*, written by a Dutch doctor residing in England named Bernard Mandeville, was burned in Paris by the public hangman on the appearance of its first French translation. Mandeville had begun his literary career as a translator of La Fontaine, and his age liked a good bee fable as much as any other. The enduring influence of Aristotle and Virgil had ensured the persistence into the seventeenth and eighteenth centuries of the bee as a symbol of industry and chastity, and early modern writers had themselves added to the bee's virtues, calling it gentle, clean, and master and (eventually) mistress of a "discret oeconomy," as well as "most just" (Charles Butler, *The Feminine Monarchy* [Oxford, 1609]). So what was odious enough about Mandeville's fable to get it burned?

The text, whose full title is *The Fable of the Bees; or Private Vices, Public Benefits*, had begun life in 1705 as a scarcely noticed 433-line octosyllabic rhyming poem called "The Grumbling Hive." Over the following eighteen years, Mandeville expanded the text dramatically, adding a preface, twenty prose "Remarks" to explicate passages in the poem, and dialogues and essays. This longer work frequently flaunts the paradox in its title and insistently maintains its "proposition set against the *doxa*, 'contre 'l'opinion commune,' as the *Dictionnaire Universel* (1727) put it."[8] Thus Mandeville

---

6. On the use of the bees in royal iconography, see J. Merrick, "Royal Bees: The Gender Politics of the Bee in Early Modern Europe," *Studies in Eighteenth-Century Culture*, ed. J. W. Yolton and L. E. Brown, 18 (1988): 7–38, here 11–13; on Christian iconography, see note 5 of this essay; on the iconography of the French revolution, see L. Hunt, "Hercules and the Radical Image of the French Revolution," *Representations* 1 (1983): 95–117.

7. In Daston's essay in this volume, it is precisely the supposed triviality of insects that makes them relevant to the question of nature's moral authority. On the consistency of moralizing uses of social insect images, see also the essay by Lustig.

8. E. J. Hundert, "Bernard Mandeville and the Enlightenment's Maxims of Modernity," *Journal of the History of Ideas* 56 (1995): 577–93, here 579.

writes: "I leave [the reader] with regret, and conclude with repeating the seeming paradox, the substance of which is advanced in the title page, that private vices by the dexterous management of a skillful politician may be turned to public benefits" (1:369 [428]).[9] By this paradox, Mandeville means simply that the drunkard's vice keeps in constant employment the barmaid, the tavern owner, the vintner, and the grower of grapes. For him, it is our "vices"—all the natural human appetites, as he defines the term, and all those acts "which man should commit to gratify any of his appetites" (1:48 [34])—that spur the labor that makes wealthy societies possible. Mandeville hoped to convince his audience of the benefits of vice and indulgence in luxury by discussing benefit in terms relevant to the polity, not to the individual. This switch in focus was incendiary.

The bees in Mandeville's "Grumbling Hive," where his idea about the public benefits to be derived from private vices was first laid out, grow rich as a community because they are individually despicable. In the very years that "The Grumbling Hive" was growing into *The Fable*, Isaac Watts wrote the famous "busy bee" poem later satirized by Lewis Carroll:

How doth the little busy Bee
Improve each shining Hour,
And gather Honey all the day
From ev'ry op'ning Flow'r![10]

How unlike these earnest bees are Mandeville's. Where Butler's and Watts's are charitable, clean, and single-mindedly industrious, Mandeville's cheat, lie, gamble, and indulge their basest appetites:

Vast numbers throng'd the fruitful Hive
Yet those vast Numbers made 'em thrive;
Millions endeavouring to supply
Each other's Lust and Vanity;
While other Millions were employ'd

9. All quotations from Mandeville in this essay are taken from B. Mandeville, *The Fable of the Bees; or Private Vices, Public Benefits*, 2 vols., ed. F. B. Kaye (1934; Oxford: Clarendon Press, 1966). The number before the colon is the volume number; the first number after the colon indicates the page in the Kaye edition; the number in brackets is Kaye's reference to the original pagination of the text.

10. Isaac Watts, *Divine Songs* (1720), quoted in Johnson, "That Neo-Classical Bee" (note 4), 262.

To see their Handy-works destroy'd;

. . . .

[And] Sharpers, Parasites, Pimps, Players,
Pick-pockets, Coiners, Quacks, and South-sayers. . . .

. . . .

These were call'd Knaves, but bar the Name
The grave Industrious were the same:
All Trades and Places knew some Cheat,
No Calling was without Deceit. (1:18−19 [3−4])

The English reading public reacted to these sardonic jingles much as
would the French nearly two decades later. In 1723, a presentment against
the book was made to the Middlesex Grand Jury on the grounds that it
was intended "to run down Religion and Virtue as prejudicial to Society,
and detrimental to the State; and to recommend Luxury, Avarice, Pride,
and all vices, as being necessary to Public Welfare."[11] John Dennis, one of
Mandeville's many eighteenth-century opponents, argued that "[t]he very
design and title of this book is a contradiction even to natural religion. . . .
So strange an assertion, contradicts the sense of mankind in all preceding
ages."[12] Dennis goes further: "A champion for vice and luxury, a serious,
a cool, a deliberate champion, that is a Creature intirely new, and has never
been heard of before in any Nation or any Age of the World" (pp. xvi–
xvii). "Man-devil," as he would come to be known, himself becomes, by
virtue of being the author of a perverse fable about bees, a prodigy, a creature whose very existence contradicts the laws of nature.[13]

And indeed, his inversion of the bee trope was prodigious, in the sense
of being without prior example (although Hobbes had objected to Aristotle's idea that bees are political; *Leviathan* chap. 17). But it is not the mere
inversion of the trope that inspires the hyperbolic reactions of Mandeville's
contemporaries. It is his use of the inversion to undercut an age-old
method of using nature to support constructions of moral and political authority. The bee image had traditionally been used to justify the concept

11. E. J. Hundert, ed., *The Fable of the Bees and Other Writings* (1732; Indianapolis: Hackett,
1997), xv.

12. J. Dennis, *Vice and luxury publick mischiefs: or remarks on a book intituled, The fable of
the bees, or private vices, public benefits* (London, printed for W. Mears, 1724), 12. "Mandeville's
paradoxical thesis, 'private vices, publick benefits,' was the most notorious maxim of the
eighteenth century" (Hundert, "Bernard Mandeville" [note 8], 578).

13. For "Man-devil," see Hundert, ed., *Fable* (note 11), xv.

of order *per se* by grounding it in phenomena antecedent to the human will and by insisting that the functional aims of particular social and political orders are themselves also external to the human will. This Mandeville ridiculed. I later return to this point but first must dissect Mandeville's critique of the moralizing use of nature images.

On the most superficial level, Mandeville uses the *Fable of the Bees* to show his contemporaries what has become transparently clear to us: in all the many lovely fables, the bees act as men do. Or as Mandeville puts it, "These Insects liv'd like Men, and all / Our Actions they perform'd in small" (1 : 18 [2]). Thus, Mandeville jokes about his bees: "we may / Justly conclude, they had some Play [gambling]; Unless a Regiment be shewn / Of Soldiers, that make use of none." Moralists who used the bee trope were not in fact learning from nature and then teaching human society, but rather projecting human norms onto nature in order to read them back out again, perhaps somewhat changed by their application to new material, perhaps not.

But, as Mandeville himself knew, his ironizing gestures on this point were not the real fillip to his opponents' fury. In face of the presentments, Mandeville wrote a vindication in which he volunteered, presciently, not only to recant, if someone could reasonably show that his book included "anything tending to immorality or the corruption of manners" (1 : 412 [477]), but even to "burn the book [him]self at any reasonable time and place . . . (if the hangman might be thought too good for the office)" (ibid.). Judge of his own sentence, Mandeville also indicted himself, confessing that he alone "would dare openly contradict, what by everybody else was thought *criminal* to doubt of," namely, "the fine notions they had received concerning the dignity of rational creatures" (1 : 45−46 [32−33], emphasis added). Mandeville wanted to deflate humanity's overestimation of itself and of human nature, for he recognized the term "nature" as one that had value for moralizers precisely because it could itself be used to assign value to particular practices and thereby influence normative moral prescription and obedience. But what was it about nature that made it useful for such valorization?

Two types of moral theory, both of which were inclined to naturalizing fallacies, were explicitly the targets of Mandeville's efforts to undercut the exaltation of human nature: on the one hand, the "rigorist" argument that human beings, as rational creatures, should use reason in order to conquer their appetites and thereby achieve virtue in accordance with their rational natures; and on the other, the "Shaftesburian" argument that human beings are virtuous by nature.[14] Against the rigorist argument, Mandeville argues that however much self-denial someone might seem to practice, those acts of self-denial in reality merely fulfill some other appetite: "It is not likely that anybody could have persuaded them [human beings] to dis-approve of their natural inclinations, or prefer the good of others to their own, if at the same time he had not shown them an equivalent to be en-joyed as a reward for the violence which by so doing they of necessity must commit upon themselves" (1:42 [28]).

Mandeville was not the first to reject the rigorist argument because of its internal contradiction. The other butt of his jokes, the third Earl of Shaftesbury, had already recognized that it made little sense to say that vir-tue *according* to nature requires people to *conquer* nature.[15] Shaftesbury had therefore argued that human beings are naturally virtuous and sociable; na-ture's is's and ought's are fully compatible, and there is no need for people to overcome nature in order to harmonize with it. Of this, Mandeville was simply scornful: "He [Shaftesbury] seems to require and expect goodness in his species, as we do a sweet taste in grapes and China oranges. . . . His notions . . . are a high compliment to humankind . . . inspiring us with the most noble sentiments concerning the dignity of our exalted nature. What a pity it is that they are not true" (1:323−24 [371−72]). Moralists who call the manners of human beings "natural virtues"—regardless of whether those manners are thought to be innate or are to be wrung from our baser nature on behalf of our better nature—lie about how many man-made incentives have been introduced to guide natural appetites into situations where those appetites are best fulfilled through the use of good manners (1:45−46 [32−33]); and they lie about the fact that the manners them-selves reflect human constructions of value.

To call good manners natural virtues, Mandeville suggests, is mere self-congratulatory hypocrisy, and this hypocrisy, not virtue, is the true target of his satire (cf. 1:349 [402−3], 2:108−9 [105−7]).[16] Such hypocrisy masks the importance of art and education (or what we would now call "cul-

14. On the rigorist definition of virtue and Mandeville's critique, see P. Harth, "The Satiric Purpose of the Fable of the Bees," *Eighteenth-Century Studies* 2 (1969): 321−40, here 336−40.

15. Treatments of nature, as a moral authority, frequently lead to such paradoxes: see Cadden's essay in this volume on how nature is the source of the vices that threaten natural virtues; see also the essays by Thomas and Murphy on efforts to "conquer" nature.

16. The point that Mandeville's satire is aimed at hypocrisy was first made by Harth ("Satiric Purpose" [note 14], 328) and has been very influential in Mandeville studies.

ture") to human life. Cleomenes, a character who speaks for Mandeville in one of the fable's dialogues, takes "uncommon pains to search into human nature, and [leave] no stone unturned to detect the pride and hypocrisy of it" (2:18 [xxi]), and is able "to demonstrate to you that the good qualities men compliment our nature and the whole species with are the result of art and education" (2:306 [365]). In the *Fable* Mandeville himself gives examples of forms of human behavior that are considered natural when in fact they should be seen as the products of education and authority. For example (and here we get our first hint that Mandeville's deconstruction of the bee trope will be intimately linked to questions of sex and gender):

> The multitude will hardly believe the excessive force of education, and in the difference of modesty between men and women ascribe that to nature, which is altogether owing to early instruction. Miss, is scarcely three years old, but she is spoken to every day to hide her leg . . . ; while Little Master at the same age is bid to take up his coats, and piss like a man. (1:71–73 [63])

Mandeville criticizes his contemporaries for failing to distinguish between nature and authority (or the "excessive force" behind and resulting from education), and so he thereby contributes to "the birth of the sociological imagination, which demystifies what appears given by recognizing it as, not natural, but social or cultural." [17]

Mandeville's choice to demystify the effects of human artifice by commenting on Miss and Little Master is by no means accidental. Mandeville is most frequently read as an economic theorist, a proto–Adam Smith. And as we shall see, his demystifications of "what is given" are ultimately oriented toward redirecting his contemporaries' understandings of labor. Adam Smith's "invisible hand" will naturalize once again the very idea that Mandeville is uncovering: that our political and social orders, especially the division of labor, are the products of human interaction; they are not made by nature or invisible forces, but by men and women. Gender was central to the division of labor in Hesiod's world—we have already seen how he

17. M. McKeon, "Historicizing Patriarchy: The Emergence of Gender Difference in England, 1660–1760," in *Eighteenth-Century Studies* 28 (1995): 295–322, here 303. On the relationship between nature and education in the nineteenth century, see the essay by Fuchs in this volume.

conjoined gender and labor in the bee image—and it was no less central in Mandeville's. He could not proceed with a demystification of divisions of labor and their surrounding social orders without a sensitivity to the different forms of "work" assigned to men and women.

But I return to the topic of labor later. Or rather, I will continue to approach it via the less direct, but more interesting, route of value. Mandeville's acts of demystification are in the first instance acts of devalorization, for he claims to replace the traditional, valorized idea of nature, particularly human nature, with his own value-free account of it. On Mandeville's view, there is only one natural fact relevant to understanding human behavior, and it is neither good nor bad; it simply is: pride inheres in all animals, even in cats who lick their faces (2:130–31 [135–36]). If we cease to assume that there is inherent good in nature, we will not be tempted to call particular human practices natural simply in order to charge those practices with the value we mistakenly attribute to nature. It will then be easier to distinguish what exists despite us from what comes to exist only because of us. According to Mandeville, the two human practices that do exist despite us arise from the natural fact of pride: *flattery* feeds our pride (1:42–46 [29–33], 1:51–55 [37–40]), and we engage in *imitation* in order to obtain that flattery.[18] "Thus sagacious moralists draw men like angels, in the hopes that the pride at least of some will put them upon copying after the beautiful originals which they are represented to be" (1:52 [38]). "Sagacious moralists," who confound nature and education, and encourage others to do so, are primarily engaged in a project of flattery, manipulating pride and drawing on the resources of human imitativeness. It is they, and neither nature nor an invisible hand, who establish social order.

Indeed, "sagacious" is not what Mandeville usually calls these moralists. He is much more likely to call them "cunning." He invokes this term regularly in his argument that "the first rudiments of morality" "were chiefly contrived that the ambitious might reap the more benefit from, and govern vast numbers of them with the greater ease and security" (1:46 [33]; cf. 1:44 [31], 1:50 [35], 1:52 [38]). As politicians had long ago discovered, Mandeville says, flattery is simply the best means of carrying out the "cunning" administration of human affairs: "[T]he nearer we search," he writes, "into human nature, the more we shall be convinced, that the

18. On flattery in Mandeville, see Hundert "Bernard Mandeville" (note 8), 589. On imitation, see A. M. Hjort, "Mandeville's Ambivalent Modernity," *Modern Language Notes* 106 (1991): 951–66, here 963.

moral virtues are the political offspring which flattery begot upon pride"
(1:52 [38]). The project of such cunning is precisely to use vanity to confound nature and culture. Thus Cleomenes argues:

> By diligently observing what excellencies and qualifications are really ac-
> quired in a well-accomplished man; and having done this impartially, we
> may be sure that the remainder of him is nature. It is for want of duly sepa-
> rating and keeping asunder these two things that men have uttered such ab-
> surdities on this subject, alleging as the causes of man's fitness for society,
> such qualifications as no man ever was endued with, that was not educated
> in a society, a civil establishment, of several hundred year's standing. *But the
> flatterers of our species keep this carefully from our view. Instead of separating what
> is acquired from what is natural, and distinguishing between them, they take no
> pains to unite and confound them together.* (2:301 [358–59]; emphasis added)

The epithet "natural" is one we use to flatter each other, but the flattery
must be kept from view if it is to do its work. The real point of flattery is
to erect structures of authority without seeming to. It is to make order—
whether moral or political—seem to issue, as it were, full-grown and
"perfite" from nature, as something that is not man-made. Hidden in the
human tendency to confuse nature and authority is, Mandeville suggests,
some secret about what authority needs to sustain itself. For some reason,
authority, in order to be stabilized, needs those over whom it has sway to
assume toward it such attitudes as they assume toward nature. But what
does Mandeville think is our habitual attitude toward nature?

Cunning does not depend ultimately on human vanity for its success,
but on forgetfulness, a by-product of the human disposition to imitate
(1:358–59 [414], 1:284 [336]). Thus Mandeville argues: "The fondness of
imitation makes them accustom themselves by degrees to the use of things
that were irksome, if not intolerable to them at first, till they know not
how to leave them, and are often very sorry for having inconsiderately
increased the necessaries of life without any necessity" (1:358–59 [414]).
After enough imitation, people can forget that they have learned certain
forms of behavior and were not born to them. More importantly, habit
transforms "the irksome" into "the easy." A form of behavior that began
as an onerous rule or as an obstacle to one's desires is transformed by habit
and forgetfulness until we no longer realize that we have learned how to
accommodate an obstacle. The importance of this idea becomes clear in a
long section of "The Third Dialogue between Horatio and Cleomenes"

where Mandeville juxtaposes the attitudes we assume toward man-made obstacles and those we assume toward natural obstacles:

> CLEO: In the pursuit of self-preservation men discover a restless endeavor to make themselves easie, which insensibly teaches them to avoid mischief on all emergencies. And when human creatures once submit to government, and are used to live under the straint of laws, it is incredible how many useful cautions, shifts, and stratagems they will learn to practice by experience and imitation from conversion together; . . . I am of opinion that men find out the use of their limbs by instinct, as much as brutes do the use of theirs; and that without knowing anything of geometry or arithmetic, even children may learn to perform actions that seem to bespeak great skill in mechanics, and a considerable depth of thought and ingenuity in the contrivance besides.
>
> HOR: What actions are they, which you judge this from?
>
> CLEO: The advantageous postures which they'll choose in resisting force, in pulling, pushing, or otherwise removing weight; from their slight and dexterity in throwing stones, and other projectiles, and the stupendous cunning made use of in leaping.
>
> HOR: What stupendous cunning, I pray?
>
> CLEO: When men would leap or jump a great way, you know, they take a run before they throw themselves off the ground. It is certain that by this means they jump further, and with greater force, than they could do otherwise. . . See a thousand boys, as well as men, jump and they'll all make use of this stratagem, but you won't find one of them, that does it knowingly for that reason. What I have said of this stratagem made use of in leaping, I desire you would apply to the doctrine of good manners, which is taught and practiced by millions who never thought on the origin of politeness, or so much as knew the real benefit it is of to society. (2:139−41 [146−48])

Cunning makes life "easie." It is a capacity to lessen the force with which all obstacles, whether man-made or natural, impinge on our lives. Cunning is, therefore, not only the ability to lead one's fellows into moral prescriptions; it is also the ability to help people inhabit those moral prescriptions and work within and around them, so that they can make themselves easy despite living within their constraints.

When sagacious moralists encourage us to understand moral rules as natural givens, they are actually encouraging us into a forgetfulness that mimics the sort of accepting attitude of easiness that we assume in respect to some aspects of nature, especially those we consider immutable.[19] It is not irksome to use a hot pad to handle a hot pan; it is not irksome to run before leaping; nor is it irksome to greet someone on the telephone before launching into our request of them. Just as people do not "knowingly" run before jumping, we do not "know" of what use manners are and what obstacles they allow us to circumvent. But whereas, in the case of nature, we simply ignore or unconsciously accommodate obstacles, in the case of authority and convention, we grow accustomed to them and forget them. In forgetting man-made obstacles, humanity loses the ability to see obstacles properly at all, and so to distinguish one type of obstacle (authority) from another (nature). The construction of authority, moral or political, depends on this confusion of education and nature, for the confusion marks the success that those producing authority have had at turning the onerous into the easy.

The cunning involved in the production of moral authority may be that of the ambitious politician; but it may also be the innocent pragmatism of those who learn to deal with obstacles in part by forgetting about them, or ceasing to think of them as obstacles. Whatever the case, moral and political authority resembles nature, according to Mandeville, precisely when it is most successful—when its constraints have been accommodated to the greatest degree possible through forgetfulness. The difficult question, then, is exactly how this *resemblance* between nature and *successful moral authority* works to make the construction of *moral and political authority in general* successful. Why do we rest easy in respect to some aspects of nature and why can that feeling of easiness be transferred to the realm of morality? How exactly can the idea of "nature" contribute to the process of forgetting? Here we return to the topic of nature's practical value to moralists.

The experience of forgetting that is relevant here should be more precisely defined as the experience of having "oughts," which one knows are obstacles, become "is's" in respect to which one feels easy. This transformation of "ought" to "is," which constitutes forgetting, can occur in two ways. First and foremost, the "oughts" underlying cultural norms are turned, through habituation, into the "is's" of actual practice. In contrast, the second method does not depend on imitation and habit, but only on moralists' success at linking their prescriptions to nature, whether through

19. Mandeville writes (1:3 [iii]): "Laws and government are to the political bodies of civil societies what the vital spirits and life itself are to the natural bodies of animated creatures."

argument and rhetoric or through poetic image, and at binding nature and culture to one another conceptually.[20] Mandeville's unusual sensitivity to this second method explains his ability to explode the bee trope as dramatically as he does. He has discerned that what is at stake in attaching virtues and values such as temperance and chastity to the image of the beehive is not the virtues themselves but rather the very idea of order, the process of constructing categories and delimiting of differences, that grounds the virtues.

Any prescription, or normative rule, implies a whole world or normative *order* that supplies the context in which the rule makes sense.[21] The successful production of moral and political authority eases into a position of cultural orthodoxy some interrelated and interactive (though not necessarily consistent) set of rules that constitute a world of relations, boundaries, and social roles. In other words, authority (as opposed to force or the threat of force) consists in making some particular ordering of the world feel easy for those who live within it. The principle task in ensuring that people can or will rest easy with authority is, therefore, to make sure that the idea of order itself (and not merely some particular order) is compatible with an experience of easiness. For order *per se* is imposition; order rules some things out, and in this sense must be irksome unless those who submit to that order are somehow able to accommodate the imposition and make themselves easy with it. What would our experience of the world be like if order *tout court* were irksome to us? Order *per se* must be justified when political and moral authority is constructed, so that we can be at ease in respect to it.

Mandeville's decision to launch his critique of the moralizing use of nature by inverting the bee image reveals that the trope had long provided a clear example of the use of nature to justify order and to make order itself a forgettable or negligible concept. From its earliest appearances, the figure of the beehive allowed those who invoked it to step, in their imagination, outside of the situatedness of human interaction for the sake of contemplating "the whole" of human society: its integrity and the integration of all parts and actions with respect to a single, final end. Thus, when Aeneas

20. McKeon, "Historicizing Patriarchy" (note 17), 302: "In fact, the determinant authority of the natural in modern culture derives precisely from its unprecedented separability from the sociocultural, which henceforth always stands at risk of a naturalizing takeover."

21. Here I am following both J. L. Austin (*How to Do Things with Words* [Cambridge: Harvard University Press, 1975]) and Martin Heidegger (*Being and Time,* trans. J. Macquarrie and E. Robinson [San Francisco: HarperCollins, 1962]).

arrives at a hill outside Carthage in the *Aeneid,* Virgil uses a bee metaphor
to capture Aeneas's experience of seeing a whole society at once, something that we can in reality never do:

> They climb the next ascent, and, looking down,
> Now at a nearer distance view the town.
> The prince with wonder sees the stately tow'rs,
>
> . . . .
>
> The gates and streets; and hears, from ev'ry part,
> The noise and busy concourse of the mart.
>
> . . . .
>
> To ply their labor: some extend the wall;
> Some build the citadel; the brawny throng
> Or dig, or push unwieldy stones along.
>
> . . . .
>
> Some laws ordain; and some attend the choice
> Of holy senates, and elect by voice.
> Here some design a mole, while others there
> Lay deep foundations for a theater;
>
> . . . .
>
> Such is their toil, and such their busy pains,
> As exercise the bees in flow'ry plains,
> When winter past, and summer scarce begun,
> Invites them forth to labor in the sun;
> Some lead their youth abroad, while some condense
> Their liquid store, and some in cells dispense;
> Some at the gate stand ready to receive
> The golden burthen, and their friends relieve;
> All with united force, combine to drive
> The lazy drones from the laborious hive.
> (*Aeneid* 1. 418, trans. John Dryden, 1697)

Meditations on the beehive provide an Archimedean point from which to
consider human society. Looking down on a town or surveying a beehive,
one has a sense of seeing all the actions in society at once as well as the end
product to which they contribute. But there can be no belief in "the
whole" and no interest in it without some expectation that all the diverse
and individuated members of human society are somehow related to one
another. Those relations make "the whole," which can therefore be imag-

ined only if one can also explain the interrelatedness of different people, the mutual dependence of those in socially differentiated positions on one another. The beehive, with its explicit division of labor among monarch, worker bees, and lazy drones provides the tools for thinking "the whole," by offering a model of social differentiation. Virgil focuses on this in describing the different tasks that are carried out by various classes in the city: some build, some make laws, some found the theater. Order consists precisely of this functional differentiation.

Mandeville himself uses the bee trope to flag his interest in part–whole and relational thinking.[22] In "The Grumbling Hive," he seeks to understand the "grandeur and worldly happiness of the whole" (1:7 [viii]), and argues that the point of politics is to integrate the diverse components of society: "This was the State's Craft, that maintain'd / The Whole of which each Part complain'd: This, as in Musick Harmony, / Made Jarrings in the main agree" (1:24 [10]). One might adduce many other examples of the emphasis on part–whole thinking and functional differentiation in the usages of the bee image in all periods;[23] suffice it to say that the greatest consistency in the use of the image is precisely this facilitation of thought about order.[24] These features of the bee image are what have been useful to moralists, not the diverse forms of moral behavior—such as chastity, temperance, moderation, and public-spiritedness—that have been hung upon the image's underlying conceptual structure.

The move of moralists, then, to cast human prescriptions in natural

22. "The real thrust of his [Mandeville's] line of reasoning does, however, become apparent once we understand that his true commitments lie with the holist terms, the language of individual virtue having been mobilized for the sake of rhetorical effect" (Hjort, "Ambivalent Modernity" [note 18], 957). Mandeville writes: "I hope the reader knows that by society I understand a body politic in which man either subdued by superior force, or by persuasion drawn from his savage state, is become a disciplined creature, that can find his own ends in laboring for others, and where under one head or other form of government each member is rendered subservient to the whole and all of them by cunning management are made to act as one" (1:347 [400]).

23. See, e.g., S. Hartlib, *The Reformed Commonwealth of Bees* (London: G. Calvert, 1655), 43 (no. 17, "A copy of a letter written by Mr. William Mew, Minister at Eastlington in Glocestershire to Mr. Nathanil Angelo, Fellow of Eaton Colledg"). Hartlib's *Reformed Commonwealth of Bees* is a collection of letters and notes on bees, only some of which are by Hartlib himself.

24. In the twentieth century, S. W. Itzkoff uses the image of the hive to assume precisely "a society where everything is in place. Our schools and libraries, our roads and shopping malls . . . were all there at our birth" (Itzkoff, *The Road to Equality: Evolution and Social Reality* [Westport, Conn.: Praeger, 1992], 16).

form, for example in the image of the beehive, is an effort to suggest that we get the idea of order itself from nature, from outside ourselves. We turn to nature for an idea of the whole and, more importantly, for an idea of an integrated whole. And when sagacious moralists find images in the natural world to embody both order *per se* and also the particular principles of order that their commandments imply, they vivify or ensoul those ordering principles, making them real and "visible" in the world. Once people "see" certain ordering principles operating in the world and beyond human control, those ordering principles, although in fact invented by the human imagination, themselves become "is's" (that is, they exist in the world), and not merely "oughts." No longer, for example, are we exercised by the question of how we ought to divide, categorize, and distinguish living creatures; at ease, we simply report the categories, divisions, and distinctions (for instance, of species and genus) that "do" exist among them. One late-seventeenth-century text called the *Congress of Bees: or, political remarks on the bees swarming at St. James's suppos'd to be wrote by Sir John Mandevil* (London, 1728) reports that for the Romans, swarms of bees were "looked upon as certain sign of the utmost Ease and Plenty and that the House to whom the Gods sent them, was sure of a lasting Tranquillity." The author then reads a swarm of bees that has in his own time settled at St. James' Palace as "certainly a Sign of Great Plenty and perfect Harmony" (5). "Plenty" is the fixed term. The bees signify this. But plenty also brings with it perfect harmony, tranquillity, peace, and ease. Principles of order, cast in forms external to the human world, leave us feeling easy, and therefore render the prescriptions based on them more effective.

Mandeville himself noticed how the bee image was used to connect order to ease and ironized it. As we have seen, he was explicitly aware that the bee trope was used for the part–whole and relational thinking involved in expressing ideas about order. He was also quite convinced that politics consists of establishing order. Like the many bee moralists, he too needed images of order to ground his political thinking. But he insists, in the sixth dialogue between Horatio and Cleomenes, on images that reinforce the fact that order arises from the human mind, and with that order only an at best uneasy ease:

CLEO: Yet I know nothing to which the laws and established economy of a well-ordered city may be more justly compared, than the knitting-frame. The machine, at first view, is intricate and unintelligible. Yet the effects of it are exact and beautiful, and in what is produced by it, there is a surpris-

ing regularity. But the beauty and exactness in the manufacture are principally, if not altogether, owing to the happiness of the invention, the con-
trivance of the engine . . . .

HOR: Though your comparison be low, I must own that it very well illustrates your meaning.

CLEO: Whilst you spoke, I have thought of another, which is better. It is
common now to have clocks that are made to play several tunes with great
exactness. There is something analogous to this in the government of a
flourishing city, that has lasted uninterrupted for several ages. There is no
part of the wholesome regulations, belonging to it, even the most trifling
and minute, about which great pains and consideration have not been employed, as well as length of time. . . . But then once that the laws and ordinances are brought to as much perfection, as art and human wisdom can
carry them, the whole machine may be made to play of itself, with as little
skill as is required to wind up the clock. (2:322–23 [386–88])

Mandeville's images of the knitting machine and the clock pointedly reproduce the conceptual frame for thinking about order that was present in
the bee trope by emphasizing part–whole relationships, differentiation,
and division of labor. But then Mandeville adds an ironizing envoi. As
with the bee images, order will lead to ease, or at least it will *almost* lead to
ease. Maintaining the order of the social world will take "as little skill as is
required to wind up the clock" (again, one cannot help but think of the
"invisible hand"). But it will inevitably take some labor. Mandeville does
not, in the end, obscure this. Indeed, his images bring nothing if not labor
to the fore. The knitting machine belongs to the world of manufacturing;
the clock was becoming, as Mandeville wrote, a machine for ordering
work.[25] Mandeville's images reinforce his interest in divisions of labor, an
interest that the bee images themselves had also always, for that matter,
evinced. The bee images, however, had naturalized divisions of labor,
something that Mandeville's figures stop just short of. In the gap between
the symbolic work being done respectively by the bees and by the knitting
machine and clock, labor's association with perfect and unmitigated ease
is lost.

Mandeville thus suggests that moralists' inclination to use nature images, and specifically the bee image, betrays a tradition of efforts to cover

25. See M. Postone, *Time, Labor, and Social Domination: A Reinterpretation of Marx's Critical
Theory* (Cambridge: Cambridge University Press, 1993).

up the force of necessity inherent in the social orders structuring labor. The stakes of constituting labor's orders are such that labor's irksomeness must be hidden. It can be hidden only by cunning arts that draw on nature, such as are used to turn human laborers into bees. This symbol's unusually great power to preserve the irksomeness of social order depends, ultimately, on the fact that humans desire honey, just like bees. On this point, our desires are not analogous to theirs, but identical. When the symbol is read as a parable about the relation between social order and desire, it suddenly seems to apply to our lives not metaphorically, but literally; hence its grip on the imagination. We are not merely modeling ourselves on bees, but competing with them; the symbol taps the psychological power of envy and emulation that Mandeville thought was crucial to the construction of social order. The symbol establishes human desire as the reason for social order, while also obscuring that fact, with important consequences.

The earliest images we have of humans relations to bees are cave paintings in Spain (6000 B.C.E.), South Africa (Stone Age), and Zimbabwe (age unknown) of human figures reaching into beehives, presumably to steal honey for which they have brought along containers; bees fly out and away from the hives.[26] The thieves seem to be oriented toward the same goal as the bees whom they displace, and the images suggest that the tradition of understanding human action as oriented toward ends is old: the honey gatherers are stretching themselves toward a goal, honey, the purpose of which is to provide for sustenance. By analogy to the thieves who displace them, the bees themselves become goal-oriented actors, but with goals that are antecedent to any choices of theirs.[27] As late as the early modern period, apiologists still believed that the bees gathered honey, a gift from god, fully formed from flowers. As Charles Butler put it:

> The greatest plenty of the purest Nectar cometh from above: which Almighty God doth miraculously distill out of the Aier . . . and hath ordained the Oak, among all the trees of the wood, to receive and keep the same upon her smooth and solid leaves until either the Bees tongue or the sun's heat have drawn it away. [This] is a honeydew. . . . I would rather

---

26. Gould and Gould, eds., *Honey Bee* (note 4), 2.

27. The analogy can be found everywhere. Here's one instance: "The new problem was to store enough honey during the summer to 'burn' through the winter. It is the colony's drive to store honey that makes commercial beekeeping possible and profitable" (ibid., 24).

judge this to be the very quintessence of all the sweetness of the earth . . .
this most sweet and soveraign Nectar.[28]

Honey is something perfect; the only real sweetener in most of the
world for most of history, most sweet and sovereign, it comes from God.[29]
Its gold color also suggests wealth. Thus, for the eighteenth-century bee-
man Dr. Joseph Warder of Croydon, a hive is like "a rich City, strong by
Nature made And every House with finest Treasure filled."[30] But since
honey is also divine, it does not suggest mammon to Warder but "the
chiefest good, future happiness of our immortal soul" (161). "Let us then,"
he says, "imitate the industrious Bee, who goes from Flower to Flower for
honey, and *labours not in vain*. So every leaf of our Bible is full of Honey,
full of Grace, full of Love" (162). The good "sought" by the bees is not
merely external to their willing but also perfect; at this point, they "labour
not in vain." Once the goal-oriented bees are set in relation to externalized and perfect goals, the analogy between the thieves and the bees can
be turned around; humans in pursuit of honey are made into creatures
who also pursue goods that are both external and antecedent to their own
wills.[31] The result, for them as for the bees, is the easiness that inheres in
"labouring not in vain."

Mandeville's ironizing of the bee image thus subtly discloses that moralists who invoke the bee trope do so to avoid treating labor as structured
by human desire. He might easily have pointed to his near contemporary

28. C. Butler, *The feminine Monarchy* (Oxford, 1634 [1609]), 110–11. Spellings have been
modernized.

29. On the relative value of honey and sugar, see Crane, *Bees and Beekeeping* (note 5),
28, 424.

30. J. Warder, *The true amazons, or The monarchy of bees. Being a new discovery and improvement of those wonderful creatures . . . with directions plain and easy, how to manage them . . . so that
with laying out but four or five pounds, in three or four years, if the summers are kind, you may get 30
or 40 pounds per annum*, 9th ed. (London: Baldwin, 1765 [1712]), 42. Warder's text was one of
the most popular bee books of the eighteenth century.

31. The tendency to associate the bee's functional perfection with the aspirations of humanity to perfection continues even now. Gould and Gould (*Honey Bee* [note 4], x) write:
"Honey bees are at the top of their part of the evolutionary tree, whereas humans are the most
highly evolved species on our branch. To look at honey bees, then, is to see one of the two
most elegant solutions to the challenges of life on our planet. More interesting, perhaps than
the many differences are the countless eery parallels—convergent evolutionary answers to
similar problems." Also: "In the European religious art, the image of man diving into beehive
persists and is used to symbolize the wholehearted seeker after the Good (Friedmann, *Bestiary*
[note 5], 152).

Samuel Hartlib, whose *Reformed Commonwealth of Bees* sets the beehive up as a model that even Adam and Eve should have followed:

> God meant Adam and Eve to exercise his Industry as well about the discovery of the fruitfulnesses of perfect nature. *Adam could have by dressing increased the pleasantness of the place . . . and made [it] compleatly answerable to the perfection of his own imagination.* For although there was nothing imperfect in Nature before the curse, *yet all the imaginable perfections, which the seminal properties of the Earth contained, were not actually existent in the first instant.*[32]

Adam's charge is to make his environment answerable to the perfections of his own imagination through labor. Yet everything that is imaginable is already contained in the "seminal properties of the Earth." Adam's imagination need go only where something has gone before. His attempts to labor and to establish an industrious order in the garden are therefore represented as a process of discovering goals that are both external and antecedent to him. In order to labor, in order to begin to create order, the laborer must believe, Hartlib's image suggests, that order waits to be made, and that we are not responsible for the shape it will take. The trope of the bee allows those who use it to step outside the chain of human actions and interactions in which they are situated and to imagine some single, final end—a world that is "answerable to the perfections of our own imagination"—while simultaneously allowing them to disavow the excessive force of their imagining and the necessitousness involved in how humans organize labor.

Mandeville, in contrast to moralists like Hartlib, refuses to accept that any but human desires give rise to our ends or to the order underlying labor. It is not from "the seminal properties of the Earth" that social order arises, for Mandeville, but only from the desire for wealth. As Annette Hjort puts it: "Mandeville's concept of 'public welfare,' is purely economic, being uniquely a matter of the production of material wealth. . . . Wealth, then, would be the ultimate end in itself, the 'transcendental term,' underlying social unity, the fundamental basis of all social bonds."[33] Hjort's phrase "transcendental term" is extremely revealing, for wealth is not, of course, a good that transcends the human will. The value consti-

---

32. Hartlib, *Reformed Commonwealth of Bees* (note 23), 43–46 (no. 18), "A Copy of Mr. Hartlibs Letter to that worthy Minister at Eastlington Mr. William Mew." Emphasis added.

33. Hjort, "Ambivalent Modernity" (note 18), 958.

tuting it, and constituting money especially, has always been among the forms of value most easily recognized as a cultural construct. Aristotle reports the view of his contemporaries that "money is often thought to be nonsense and entirely a convention and by nature nothing" (*Politics* 1257b12, I.3.16). Mandeville accepts the idea of order, believes it necessary for political thinking, and believes that authority itself is sustained when the idea of order is treated in ways that make it easily accommodated. But he insists that we understand order, and the authority it sustains, as being grounded in nothing other than human appetites. Most importantly, order arises from our desire for the values that we ourselves create.

Mandeville's redescription of social order and demystification of its sources is ultimately a vehicle for a new account of value, which for him arises from the labor made possible through order.[34] Here it is crucial that the bee trope had always been used to figure not merely order, but also value and hierarchy (that is, the application of value to order). In Hesiod's bee image, mere acknowledgment of the distinction between bees and drones turns immediately into a question of which creature deserves the greater distinction. In Virgil's image, too, the laboring bee is the picture of perfection, superior to the imperfect and lazy drone, and the bees are lauded for throwing the drones out of the hive. These images suggest that nature itself has values. Mandeville not only aimed to devalorize the ideas of nature and of human nature, but he was also trying to undercut just such an idea of natural value. When he sets "private vices" on the same footing as "public benefits," as in his controversial title, he is not so much revaluing "vice," traditionally the low term in a comparison with "benefit"; rather he is undoing the whole structure of naturalized value on which the opposition of "vice" to "benefit" rested.

Another look at his knitting machine and clock will help make this point clearer, for these two images are intimately connected with Mandeville's project to undercut the idea of natural value. Indeed, Cleomenes introduces the knitting image as part of a study in comparative value. He moves toward the image by suggesting that he and Horatio should analogize politics to weaving, and says: "There shall be no greater difficulty in governing a large city, than (pardon the lowness of the simile) there is in weaving of stockings." Horatio agrees that the simile is low, "Very low indeed" (2:322 [386]). The art of politics was traditionally a male art,

---

34. Hjort considers that "a perspective in which no moral categories whatsoever were applied to the global level at which a purely emergent order is described" would constitute "a genuine and complete break with tradition" (ibid., 964).

highly esteemed for its association with the intellect, prudence, and virtues like courage. Weaving, in contrast, was women's work associated with the body and the home. In comparing high to low, Mandeville openly rejects belle lettristic rules of propriety by which high language should be used for high subjects. Again, this is a rejection of natural value, where value is believed to inhere in things themselves. But more is at stake here than just a rejection of his age's pieties. Mandeville is taking us to the core of the age-old bee image. Just as his images equate weaving to the art of politics, his "Grumbling Hive" sets the stay-at-home drones on the same footing as the voyaging bees. Mandeville's focus on the bee image and his decision to situate that image in the context of a contrast between weaving and politics rest on the twin assumptions that the tradition of elevating male labor above female is paradigmatic of the construction of natural value, and that the bee image gives us access to none other than that paradigm. The contrast between women's stay-at-home work of weaving and men's far-reaching political work is, in fact, an ancient one. Both Aristophanes *(Lysistrata)* and Plato *(Statesman* 305e2–311c6) make much of it, and this relative evaluation of male and female work is indeed the backdrop to Hesiod's misogynistic comparison of men as bees and women as drones. But it was no less central to thought about bees in the seventeenth and eighteenth centuries, when the gender of the worker bees, leader, and drone was itself a burning scientific issue.

Hesiod wrote in an historical context where observers of insects were uncertain about the sex of the three kinds of bees in the hive: the single large bee that all the rest follow when the hive swarms; the tens of thousands of small "worker bees" which fly back and forth between hive and flower; and the few hundred mid-size bees that leave the hive only rarely and, in contrast to the other two types, have no sting. Four centuries later Aristotle decided that nature does not arm the female, and that the large bee and the workers, since they have stings, must therefore be male. His triad of king, worker-soldier, and feminine/effeminate drones held until 1586, when a Spanish scientist, Luys Méndez de Torres, identified the single large bee as the female egg-layer in the colony, thus turning the age-old king into the now proverbial queen bee. His arguments did not themselves have any influence in England,[35] but two books written in the first decade of the seventeenth century also argued for a queen: Charles Butler's *The feminine Monarchy* (1609) and John Lovett's *The ordering of bees*

---

35. In Thomas Moffett's *Theater of Insects,* in Edward Topsell, *The History of Four-Footed Beasts and Serpents and Insects* (London, 1658), 3:891, the drones are considered "eunuchs."

(written between 1600 and 1609, although it was not published until 1634).[36] The discovery of the queen led to a recategorization of the workers as female and the drones as male. And with this trio of discoveries came a flood of anxieties about the relative value of male and female work.

Butler accepted the idea that the female is capable of rule—he argues snidely that Aristotle should have known this since "Aristotle's master was Plato, Plato's was Socrates, and Socrates' was Xanthippe Queen of Shrews"—but he nonetheless worried about the possible effects his discovery would have on naturalized hierarchies of value and on the principle that "the male is more worthy":

> For albeit generally among all creatur's the mal's, as more worthy, doo master de females, yet in these, the females have the preeminence. . . . But let no nimble tongued Sophisters gather a false conclusions . . . that they by example of these may arrogate to themselves the like superiority: for *Ex particulare non est syllogizare;* and he that made them to command their males, commanded them to be commanded. . . . ; And this they may note by the way; that albeit the females in this kind have the sovereignty, yet have the males the louder voice: as it is in other living things, Doves, Ousels, Thrushes, etc. The males being known by their sounding and shrill notes from the silent females. Yea, the wives themselves will not suffer that Hen to live which presumes to crow as The Cock dooeth; nature teaching that silence and soft nois become that sex. (64)

When Butler tells potential "nimble-tongued" sophistical women that they cannot legitimate their own authority by referring to the bees—*Ex particulare non est syllogizare*—he is arguing for a practice of analogizing between human society and nature that is specifically unidirectional: such analogizing should *confirm* human values *in nature,* not get them *from nature.* At the end of the above paragraph, Butler employs precisely such reasoning in determining what "nature teaches." In order to argue that women should remember to remain more silent than men, he offers several examples from the animal kingdom of male creatures that have louder voices than females, but then he turns to a human example to draw the evaluative moral from the story: "Yea, the wives themselves will not suffer that

36. Pliny mentions but dismisses the possibility that the workers are female. Fraser (*History* [note 4], 17) reports that a medieval charm against swarming referred to the bees as "victor dames" and Anglo Saxon has the word "beo-mother," which is also found in Old High German; but Aristotle and Aristotelianism seem to have trumped such folk knowledge.

Hen to live which presumes to crow as The Cock dooeth; nature teaching that silence and soft nois become that sex." Butler first externalizes a principle of order by situating it within the natural world, and then evaluates it, in the context of the natural world, according to principles drawn from the human world. He uses this process of double-externalization to stabilize both a particular principle of order and the human values used to assess it. In the process, he also constitutes nature's stability and its values.

More importantly, Butler develops these anxious rhetorical techniques in response to a seeming reallocation of tasks to workers. And he was not alone. In *The true amazons, or The monarchy of bees* (1712), Joseph Warder writes about the drones:

> I ought to speak on behalf of this poor, abused, and despised creature; . . . I differ about the noble creature which I shall no longer call a Drone but the Male-Bee, since he is very industrious in the Work which Nature hath designed him for, which is not only Procreation but his great Usefulness in sitting upon and hatching the eggs, and by his great Heat doth keep warm the Brood. . . . And here, by the way, let me caution . . . against an unhappy Mistake . . . of killing the Male-Bee or Drone . . . by which they hinder their breed . . . to the great Damage, if not utter Destruction of the hive of Bees; for they had better kill six working bees, than one of these great Bees. (19, 23)

Warder explicitly establishes a calculus of worth that upends the traditional relationship of drone to worker bees. No longer is the drone worthless compared to the bee; now one is valued at the cost of six workers. But to effect this change of value, Warder first has to reorder the image: he changes the name of the drone and assigns it new labors of appropriately high productive value. Once the drone is reassigned to new tasks, the value of his labor converts into the natural value of the drone. Productive value, once transferred to the bodies of laborers, seems to inhere in nature. The symbol of the beehive naturalizes not only order, but also value, and does so for the sake of stabilizing a system of labor. Both Butler and Warder work hard to preserve the gendered social order, already present in Hesiod, that depends on an attribution of relative value not just to the products of different laborers but also to the personae and bodies of the laborers themselves.

Mandeville's knitting machine thus brings to the surface what had long lain beneath the bee image: the centrality of labor to the naturalizing project. The image, unpacked, provides the following account of naturalization: the human world produces order for the sake of productivity, and so order is necessarily geared toward structuring labor; labor, however, produces value, and this value reflects back on the order that generated the practices of laboring in the first place; the reflected value is distributed throughout the order's categories and classes accordingly as the labor was distributed; as a result, the value produced by labor is eventually seen to inhere not in the labor itself nor in its products but in the order enabling the labor and in the bodies that were ordered for that labor; when we externalize order, so as to rest easy with the idea of it, the values attached to the order are externalized with it and are accordingly lodged in nature; the values externalized in nature therefore support specific productive orders. On this account, we should expect to find, at the root of our most powerful naturalizations, an attempt to preserve some particular system of labor. Even Little Miss's lessons in hiding her leg must somehow contribute to the economic order of Mandeville's day. His willful and inflammatory act of disclosure makes the point that the eighteenth-century naturalizing project aimed not only to stabilize orders of labor but also to obscure the fact that they needed stabilizing.

It is especially important that Little Miss in particular not come to believe that her modesty has economic implications.[37] Mandeville is perhaps drawn to examples lifted from the realm of sex and gender precisely because it is in that context that the process of naturalizing labor has been most effective. His use of the bee and knitting images suggests that so much effort has gone into naturalizing bimorphic sex because of its status as a foundation for one of the most basic divisions of labor.[38] If one were to follow Mandeville's treatment of the naturalization of order and value, gender might be understood as the valorized version of what was originally a set of instructions for laboring stabilized by the externalization of sexual order in nature. The easiness with which so many people accept bimorphic sexual ordering proves Mandeville right in thinking that by external-

---

37. Mandeville writes: "'But why should pride be more encouraged in Women than in Men?' 'For the same reason that it is encouraged in soldiers. . . .' 'To keep both to their respective duties'" (2:124 [125]).

38. The bees, of course, serve as an image with which to transfer the model of order in the family to the polity and vice versa. The bees are both a microcosmic kingdom of the household and also a macrocosmic household of the kingdom (Merrick, "Royal Bees" [note 6], 10–11). On the relation between social differentiation and labor in the case of gender in the eighteenth century, see McKeon, "Historicizing Patriarchy" (note 17), 298. See also J. De Vries, "The Industrial Revolution and the Industrious Revolution," *Journal of Economic History* 54 (1995): 240–70, here 255–61.

izing order into nature we ensure its stability. That easiness also explains why the values that then attach to that order, the values that constitute gender, are themselves so remarkably stable.

— — —

Mandeville's efforts to lay bare the mechanics of naturalization are thus a powerful attempt at destabilization. The thrust of his argument works to unsettle not only the idea of order and the values of his contemporaries, but also the very idea of nature itself. Mandeville's knitting machine image repudiates the traditional evaluations of male and female labor and unravels the long-fixed conceptual ties between categories of people, their work, and the productive values that result from their work. With the image, Mandeville reclaimed value itself from nature. But the machine itself was more potent yet. It itself embodied a principle of nature's transformation. Sex, gender, and labor could be reorganized in respect to each other. Mandeville does not develop the knitting machine image until the 1728 edition of his text. It is as if he had decided to be more explicit about what was so offensive as to get his book before a jury in 1723. Mandeville's knitting machine reminds us that nature itself may prove to be less stable than we, for the sake of sustaining moral authority, treat it as being. The connection between any given order and labor can be disrupted, and with that disruption should come a shift in nature's values. Nature itself is no more stable than the social orders whose externalizations constitute the order we know as nature. In juxtaposing the beehive to the knitting machine, Mandeville suggests that nature's value, its stability, is itself something we have made through our efforts, on behalf of labor, to stabilize order.[39] In reclaiming value from nature, Mandeville also made nature as changeable as our values, and drove his devilish *Fable* on, from the Middlesex court to the Parisian fire.

39. This whole paragraph depends heavily on the work and ideas of Judith Butler, especially *Gender Trouble: Feminism and the Subversion of Identity* (New York: Routledge, 1990).

4

Attention and the Values of Nature in the Enlightenment

Lorraine Daston

Introduction: There Be Gods Even Here

The most oft-quoted mottoes of the Enlightenment did not just commit what philosopher G. E. Moore was later to call the "naturalistic fallacy,"[1] they insisted upon it: "Whatever IS, is RIGHT."[2] This essay is about the ways in which value in nature was created in eighteenth-century natural history, with an emphasis on practices rather than theses: how certain regimens of experience (rather than proofs and arguments) established nature's values in an age that looked to nature as its guide in every realm, from the fine arts to weights and measures. Four aspects of how nature served as a source of value are in play here. First, how specific domains of the natural were redeemed as worthy objects of scientific study and personal dedication: because insects were deemed trivial or even disgusting, the efforts to elevate their status within natural history were attempts to turn dross into gold, to create value out of the least promising materials. Hence my focus on Enlightenment entomology. Second, the mechanisms by which this value in specific naturalia was not merely asserted, but made into a felt, even a self-evident reality: the disciplines of attention practiced by the naturalists beatified even the most inauspicious objects. Third, the kinds of value

My thanks to Matthias Doerries, Amy Johnson, and Andrew Mendelsohn for their comments on an earlier draft of this essay. Unless otherwise noted, all translations are my own. For archival material, the original text is given in the footnotes.

1. G. E. Moore, *Principia Ethica* (1903; Cambridge: Cambridge University Press, 1976), 37–58.

2. Alexander Pope, *An Essay on Man, Being the First Book of Ethic Epistles to Henry St. John, L. Bolingbroke* (London: John Wright for Lawton Gilliver, 1734), I.286, p. 22.

created by attention: the naturalists wove together theological, aesthetic, moral, and economic strands of valuation in their observations and descriptions, often deploying the single word "utility" for this knot of the good, the beautiful, and the useful. Fourth and finally, how the valorization of parts of nature underwrote the valorization of nature as a whole: through the tropes of allegory and pesonification so rife in Enlightenment poetry and prose, the naturalists' paeans to the intestines of worms and the stamens and pistils of flowers became the body and substance of Nature writ large, who famously spoke with such authority in eighteenth-century works on politics, ethics, religion, and economics. "Great Nature spoke; observant Men obey'd."[3]

My aim in this essay is to uncover some of the sources and workings of the Enlightenment value and authority of nature in forms of cultivated experience, rather than in discursive legitimation. More specifically, I argue that valorization was built into highly elaborated modes of attention, observation, and description applied to natural objects. Which objects to study, and how to study them? The choices made by naturalists already expressed a hierarchy of values about naturalia, the discipline of the mind and senses, and the proper investment of time, labor, and capital. These choices often clashed with prevailing value codes, as the sharp criticism of contemporaries bears witness. Values also saturated, in both obvious and subtle ways, how the naturalists observed and described their objects of inquiry: "there be gods even here."[4] I am particularly concerned with how the appreciation of the beauty and the function of certain features of organic nature—the underside of a leaf, the viscera of a worm, the tongue of a bee—merged in an immediate and pleasurable perception. In the painstaking observations of the naturalists, the argument from design became less an argument from evidence than an experience of self-evidence.

Natural theology was the framework within which the Enlightenment naturalists carried out their investigations, and it remains the framework of my analysis. However, natural theology was a house of many mansions, and my path through it diverges from that of most earlier treatments in two respects. First, I am less concerned with natural theology as a form of argument than as a mode of cultivated experience. Hence my principal sources are the reports of naturalists rather than the treatises of philosophers and theologians—Jan Swammerdam and René Antoine Ferchault de

3. Ibid., III.200, p. 48.
4. See the oft-quoted passage from Aristotle, *On the Parts of Animals*, 645a5–23.

Réaumur rather than Joseph Butler and William Paley. Second, I take the metaphor of God-the-artisan seriously, comparing eighteenth-century descriptions of divine workmanship in natural history with those of human workmanship in political economy. The habit of dissecting organisms and tools into component parts, the keen eye for optimal solutions to mechanical problems, the penchant for copious and minute description, even the conventions of keying text to illustrations broken down into functional parts—these practices united the work of the naturalists and the political economists. It is not coincidental that prominent naturalists like Réaumur were also actively involved in projects to describe and rationalize handiwork. The image of perfected craftsmanship common to Enlightenment natural history and political economy blended the values of utility and beauty into a single judgment of what Adam Smith called "fitness."[5] In the work of Enlightenment naturalists such as Charles Bonnet and Adam Schirach, the normative aspects of nature melted together, the useful into the beautiful, the moral into the sublime, the sacred into the pleasing.

Serious upon Trifles

In 1676 playwright Thomas Shadwell amused London audiences with his comedy *The Virtuoso,* in which the title character, Sir Nicholas Gimcrack, has squandered two thousand pounds of his nieces' money to finance experiments ranging from the microscopic observation of mites in cheese, to blood transfusions from a sheep to a madman, to reading the Bible by the light of a luminescent leg of pork—all parodies of experiments actually performed by members of the Royal Society of London. At one point, a visitor to Sir Nicholas's laboratory suggests that such preoccupations are unworthy of a gentleman: "What does it concern a man to know the nature of an ant?" To which another visitor replies: "O it concerns a virtuoso mightily; so it be knowledge, 'tis no matter of what."[6] Shadwell's play was only one of several late-seventeenth-century satires attacking new-

5. Adam Smith, *Theory of Moral Sentiments,* ed. D.D. Raphael and A.L. Macfie (1759; Oxford: Oxford University Press, 1976), 179–80. See also Neil De Marchi, "Adam Smith's Accommodation of 'Altogether Endless' Desires," in Maxine Berg and Helen Clifford, eds., *Consumers and Luxury: Consumer Culture in Europe 1650–1850* (Manchester: Manchester University Press, 1999), 1–17.

6. Thomas Shadwell, *The Virtuoso,* ed. Marjorie Hope Nicolson and David Stuart Rodes (1676; London: Edward Arnold, 1966), III.iii, p. 69.

style natural philosophers for their unseemly preoccupation with trivial and base objects (insects were a favorite object of scorn) at the expense of their worldly and religious duties. French moralist Jean de la Bruyère mocked the passions of a butterfly collector: "Another man loves insects; . . . You've chosen the wrong time to visit him; you find him sunk in deep despair; he is in the blackest and bitterest of moods, and his whole family are the victims of it; for he has suffered an irreparable loss. Go up and look at what he's showing you on his finger, a lifeless object that has just breathed its last: it is a caterpillar, and what a caterpillar!"[7] As Danielle Allen describes in her essay in this volume, Enlightenment Europeans were heirs to a venerable tradition of idealized visions of beehives as model polities. However, this allegorical admiration for insect societies was a far cry from complete absorption in the minute details of particular insects, or of insect anatomy. Even John Locke, himself a Fellow of the Royal Society and self-declared admirer of the new experimental philosophy pursued there, warned that young gentlemen ought to be taught only so much natural philosophy as to "as it were begin an acquaintance, but not to dwell there."[8]

The ridicule heaped on the preoccupations of the new-style naturalists was directed not so much at natural philosophy *per se* as at the disproportion between the time, resources, and passion consumed and their objects. It was the intensity and misdirection of attention that moralists and satirists deplored. As the English essayist Joseph Addison remarked:

> However, since the world abounds in the noblest fields of speculation, it is, methinks, the mark of a little genius to be wholly conversant among insects, reptiles, animalcules, and those trifling rarities that furnish out the apartment of a virtuoso. . . . [Such pursuits] make us serious upon trifles, by which means they expose philosophy to the ridicule of the witty and the contempt of the ignorant. In short, the studies of this nature should be the

7. Jean de la Bruyère, "Of Fashion," *Characters,* trans. Jean Stewart (1688; Harmondsworth: Penguin, 1970), 254; *Les caractères de Théophraste traduit du grec avec les caractères ou les moeurs de ce siècle,* ed. Robert Pignarre (Paris: Garnier-Flammarion, 1965), 338.

8. John Locke, *Some Thoughts Concerning Education* in *English Philosophers of the Seventeenth and Eighteenth Centuries* (1623; New York: Collier, 1910), 83. On further Restoration satires of the virtuosi and the new experimental philosophy, see Walter Houghton, "The English Virtuoso in the Seventeenth Century," *Journal of the History of Ideas* 3 (1941): 51–73, 190–219; R. H. Syfret, "Some Early Reactions to the Royal Society," *Notes and Records of the Royal Society* 7 (1950): 207–58; Marjorie Hope Nicolson, "Virtuoso," in Philip P. Wiener, ed., *Dictionary of the History of Ideas,* 5 vols. (New York: Charles Scribner's Sons, 1973), 4:486–90.

diversions, relaxations, and amusements, not the care, business, and concern of life.[9]

The virtuosi of the early seventeenth century had been amateurs in the root sense of the word, pursuing their interests in artworks, antiquities, mathematics, and natural rarities for the love of it, as an avocation and diversion. Their successors, however, took up these studies in deadly earnest. The very disinterestedness that had made the hobbies of the amateur, whether Roman coins or seashells or the solution of problems in number theory, admirable became a cause for reproach when attached to the too-singleminded study of those very same objects. English writer Mary Astell reproached the virtuosi for the uselessness of their knowledge, so ill-matched to their indefatigable efforts to procure it: "What Discoveries do we owe to their Labours? It is only the Discovery of some few unheeded Varieties of Plants, Shells, or Insects, unheeded only because useless."[10]

Too much attention paid to the wrong objects spoiled one for polite society as well as for the sober duties imposed by family, church, and state. There was a pedantry of things as well as words, and the naturalists passionate for microscopes or insects bored their interlocutors by speaking of nothing else. The incompatibility of ostentatious learning with refinement and polish was a recurring theme in late seventeenth-century conversation manuals;[11] similar reproaches were hurled in the direction of naturalists too enthralled by their pet objects of research to accommodate themselves to shared topics of conversation, as courtesy required. As a result, complained the poet Samuel Butler, the virtuoso was a boor in company: "He seldom converses but with Men of his own Tendency, and wheresoever he comes treats all Men as such, for as Country-Gentlemen use to talk of their Dogs to those that hate Hunting, because they love it themselves; so will he of his Arts and Sciences to those that neither know, nor care to know any Thing of them."[12] The pathology of misdirected attention rendered its victims oblivious to the social cues of age, rank, sex, vocation, and education that skilled conversationalists effortlessly registered and to which

9. Joseph Addison, *The Tatler*, No. 216 (26 August 1710).

10. Mary Astell, *An Essay in the Defense of the Female Sex,* 2nd ed. [1696], quoted in Houghton, "The English Virtuoso" (note 8), 53.

11. Dominique Bouhours, *Les Entretiens d'Arise et d'Eugène* [1669], in Jacqueline Hellegouarc'h, ed., *L'Art de la conversation,* preface Marc Fumaroli (Paris: Dunod, 1997), 35.

12. Samuel Butler, *Characters,* in R. Thyer, ed., *The Genuine Remains in Verse and Prose of Mr. Samuel Butler,* 2 vols. (London: J. & R. Tonson, 1759), 2:185.

they adapted their themes and manner accordingly.[13] Pope summed up the reproaches pithily: "O! would the Sons of Men once think their Eyes / And Reason giv'n them but to study *Flies?*"[14]

Creature Love

Naturalists thus accused of channeling their energies and emotions toward unworthy objects defended themselves by elevating their insects or phosphors or microscopic mites to the dignity of divine handiwork. The natural philosopher Robert Boyle, butt of some of Shadwell's best jokes, was himself worried that the attention lavished by his fellow naturalists on such objects bordered on idolatry: it was wrong to admire "corporeal things, how noble and precious soever they be, as stars and gems, the contentment that accompanies our wonder, is allayed by a kind of secret reproach grounded on that very wonder; since it argues a great imperfection in our understandings, to be posed by things, that are but creatures, as well as we, and, which is worse, of a nature very much inferior to ours."[15] Ultimately, only God was a fitting object for such rapt attention. Throughout the eighteenth century, natural theology—the worship of God through the study of His works—supplied the motivation and rationale for an expenditure of attention that contemporaries perceived as uncomfortably close to religious reverence. Entomologists were particularly fervent in their declarations that divine providence could be discerned in the design of a fly's wing or the industry of the beehive—in part to defend themselves against charges of triviality, but also in part to redeem even the most lowly objects as repositories of divine artistry and benevolence. The Dutch naturalist Swammerdam insisted that the tiny ant showed God's exquisite handiwork to as good or better effect than the largest animals;[16] his French colleague Réaumur defended his six volumes of observations on insects as a work of religious devotion, albeit with a touch of defensiveness (probably in response to the barb of his lifelong rival Georges Leclerc de Buffon

13. Peter Burke, *The Art of Conversation* (Ithaca: Cornell University Press, 1993), 102.

14. Alexander Pope, *The Dunciad: As It Was Found in the Year 1741* (London: T. Cooper, 1742), IV.429–30, p. 28.

15. Robert Boyle, *Of the High Veneration Man's Intellect Owes to God* [1685], in Thomas Birch, ed., *The Works of the Honourable Robert Boyle,* 6 vols. (1772; Hildesheim: Georg Olms, 1965–66), 5:153.

16. Jan Swammerdam, *Histoire générale des insectes* (Utrecht: Guillaume de Walcheren, 1682), 3–4.

that insects should occupy no greater place in the naturalist's head than they do in the scheme of nature): "Must we blush to place even among our occupations observations and investigations that have for their object the works in which the Supreme Being seems to have been pleased to shut up so many marvels, and to vary them so greatly?"[17]

Yet despite the piety intended to ennoble the meanest objects, the discipline of attention cultivated by the naturalists differed in several key respects from that of religious meditation and prayer. Although the techniques of meditation practiced in early modern Europe often made a concrete, prosaic object—"Upon the Sight of Flies" or "Upon the Sight of a Drunken Man"—their point of departure, they aimed to transport contemplatives to a more abstract, spiritual realm beyond that of the senses.[18] Post-Reformation works of meditation, both Protestant and Catholic, insisted on a purely interior realm of religious vision distinct from the carnal and the sensual.[19] In contrast, the observations of the naturalists, including those of the most fervent natural theologians, focused unswervingly on a particular object—a cocoon, a leaf, a queen bee—for hours and sometimes days at a time. Genevan naturalist Charles Bonnet dedicated every waking hour from 5:30 A.M. to 11:00 P.M. for twenty-one days to the observation of a single aphid (*mon puceron*, later *ma pucerone* after it bore offspring) in order to determine whether the species could reproduce parthenogenetically. Although Bonnet justified his researches on aphids on the grounds that they served to demonstrate "the adorable Wisdom" of God, he became ever more emotionally attached to the objects themselves, speaking of the "delights [*délices*]" of observation. When he lost sight of his aphid one June morning, he was disconsolate: "One easily judges that I was not insensible to this loss. I witnessed the birth of this aphid: I followed her constantly for more than a month; and it would

17. René Antoine Ferchault de Réaumur, *Mémoires pour servir à l'histoire des insectes,* 6 vols. (Paris: Imprimerie Royale, 1734–42), 1:4. Réaumur here uses the word "occupation" in the consuming sense of "vocation." On the range of Réaumur's scientific work, see Centre International de Synthèse, *La Vie et l'oeuvre de Réaumur* (Paris: Presses Universitaires de France, 1962).

18. See, e.g., Frank Livingstone Huntley, *Bishop Hall and Protestant Meditation in Seventeenth-Century England: A Study with Texts of the Art of Divine Meditation (1606) and Occasional Meditations (1633)* (Binghamton, N.Y.: Center for Medieval and Early Renaissance Studies, 1981).

19. This was the view expressed even in illustrated Jesuit works: François Lecercle, "Image et médiation: Sur quelques recueils de méditation illustrés de la fin du XVIe siècle," in Cahiers V. L. Saunier, *La Méditation en prose à la Renaissance* (Paris: Ecole Normale Supérieure, 1990), 44–57.

have given me pleasure to observe her until her death."[20] After Réaumur had closely observed the desperate efforts of some worker bees to revive a half-drowned queen bee, he was so moved by their "good intentions" that he gave them some honey.[21] Originally motivated by piety, unwavering attention directed to humble objects became an end in itself, infusing them with aesthetic and sentimental value.

A distinctive brand of natural theology arose from this displacement of sentiment—not from God to Nature, the latter conceived either as an all-encompassing whole or as a personified goddess, but to highly specific natural objects. Linnaeus thought that God had arranged for the human spirit to take pleasure in the variety of natural particulars so that the ardor to examine and admire His works would be inexhaustible, a constant stimulus to fastidious inquiry in natural history.[22] Bonnet believed that his detailed account of how aphids reproduce would inspire readers with "the grandest ideas of the SUPREME BEING who is the author of them [insects]," at the same time teaching them "the art of observation."[23] Adam Gottlob Schirach, Lutheran minister at Klein-Bautzen in Saxony and enthusiastic student of bees, filled his *Melitto-Theologia* (1767) with exacting observations of bee anatomy and honey collection and defended his activities as a beekeeper as fully compatible with pastoral duties: "I want only thereby to show that one can very well combine inquiry into Scripture and in the Bible of Nature; indeed that they must be connected to one another."[24] The detailed observations and arguments to be found in this genre of natural theology contrast with the more sweeping accounts to be found, for example, in the Abbé Pluche's *Spectacle de la nature* (1732), which were content to describe only "the exterior decoration of the world" without plumbing hidden causes.[25]

The naturalists themselves were not, however, unaware of tensions between their natural theological precepts and their natural historical practices. In principle, each minute observation of a moulting aphid, each

20. Charles Bonnet, *Traité d'insectologie, ou Observations sur les pucerons* [1745], in *Oeuvres de l'histoire naturelle et de philosophie,* 18 vols. (Neuchatel: Samuel Fauche, 1779–83), 1:36.

21. Réaumur, *Mémoires* (note 17), 5:260.

22. Carl von Linné [Linnaeus], "Curiositas naturalis," in *L'Equilibre de la nature,* trans. Bernard Jasmin, intro. and notes Camille Limoges (1748; Paris: Vrin, 1972), 136.

23. Bonnet, *Traité d'insectologie,* in *Oeuvres* (note 20), 1:xxxi.

24. Adam Gottlob Schirach, *Mellito-Theologia: Die Verherrlichung des glorwürdigen Schöpfers aus der wundervollen Biene* (Dresden: Waltherischen Hof-Buchdruckerey, 1767), xiv.

25. Noel Antoine La Pluche, *Le spectacle de la nature,* 2nd ed., 8 vols. (Paris: Veuve Estienne and Jean Dessaint, 1732), 1:x.

dissection of a worker bee, each experiment with a cocoon-weaving cater-pillar was the bottom rung on a ladder that ascended to God's wondrous providence. In practice, though, the observations became an end in them-selves, mesmerizing and delighting the naturalist—and at the same time detaching him from all other concerns, including salvation. Swammerdam, who experienced a religious conversion under the spiritual guidance of the Flemish mystic Antoinette Bourignon, seems to have suffered from peri-odic fits of remorse that his studies of insects were a distraction from reli-gious duties. Perhaps he, like Boyle, sensed that intense and prolonged acts of attention, combined with awed admiration for their objects, too closely approximated the sentiments of reverence due to God alone.[26]

Bonnet frankly introduced his major work on natural theology as a "great whole, of which the smallest [part] would absorb the Naturalist who would wish to make it the unique object of his researches," and therefore begged not to be judged by the standards that would apply to a work of natural history *per se,* suggesting that natural history and natural theology had already parted ways on the issue of painstakingly observed detail.[27] Conversely, Schirach worried that it might be possible to love bees too much, and prayed to God to

> extinguish from my heart all my hidden, exaggerated creature-love, that I
> might not thereby displease Thee. It is true, that whoever has once clearly
> perceived the charm of the beauties in nature; whoever has once sampled
> the commodious, harmonious, the delightful [in them], is so to speak car-
> ried away to deny all other delights, and as a thinking being to seek his plea-
> sures therein.[28]

Natural theology alone was no longer the sole or even the principal ground for the naturalists' absorption in their objects, nor for the feats of attention that rendered those objects so hypnotizing to initiates.

Mastery of the disciplines of attention increasingly defined who was a naturalist worthy of the name. When Bonnet was called on to sort out the puzzling observations of a society of beekeepers in the Lausitz on the al-leged reproduction of worker bees, he praised the enthusiasm of his cor-

26. Hermann Boerhaave, "Life of John Swammerdam," in Jan Swammerdam, *The Book of Nature; or, The History of Insects,* trans. Thomas Flloyd (1737–38; London: C. G. Seyfert, 1758), ix.

27. Bonnet, *Contemplation de la nature* [1764], in *Oeuvres* (note 20), 7:xvii.

28. Schirach, *Melitto-Theologie* (note 24), 204.

respondents, but nonetheless distinguished these *amateurs* from *des Observateurs ou des Naturalistes de profession*—by which he meant not naturalists who earned their living from their science, but those qualified to make exact and exacting observations and to describe them in painstaking detail.[29] To be attentive in the naturalists' sense meant to be inattentive to almost everything else, for observation could be a round-the-clock job. Whereas the annals of seventeenth-century natural philosophy and natural history were peppered with accounts of frustrated observers interrupted by visits or occupied with guests at the crucial moment, the accounts of the eighteenth-century naturalists suggest that it was the guests who were neglected in the interests of an urgent observation. The English astronomer Edmond Halley was chagrined when a dinner party kept him from observing an aurora borealis; Réaumur in contrast left instructions to be called the moment the bees in the hive he'd been studying began to swarm.[30] When he moved out of central Paris in order to have more room for his beehives and natural history collections, Réaumur noted that his new address would be inconvenient for visitors but well-suited to his researches.[31] In gardens, private studies, open fields, and country promenades, the naturalists sought out their quarry with increasingly undivided attention.

The Cult of Attention

Bonnet's *naturaliste de profession* was seldom salaried for work in natural history, and was usually an autodidact. What distinguished genuine naturalists in the eyes of their peers was not professional status, but rather the practice of heroic observation, described as at once a talent, a discipline, and a method. None sufficed without the others. Attentive observation was firmly distinguished from mere seeing, and even from remarking upon. In his letters to Réaumur, the young Bonnet dismissed his own earlier observations on caterpillars as what "mediocre attention could

29. Bonnet, "Deuxième mémoire sur les abeilles," in *Oeuvres* (note 20), 10:171n.

30. Edmond Halley, "An Account of the late surprizing Appearance of the *Lights* seen in the *Air,* on the sixth of *March* last; with an Attempt to explain the principal *Phaenomena* thereof," *Philosophical Transactions of the Royal Society of London* 29 (1714–16): 406–28; Réaumur, *Mémoires* (note 17), 5:249.

31. Letter to Jean-François Séguier, Paris, 25 April 1743, in Académie des Belles-Lettres, Sciences et Arts de La Rochelle, *Lettres inédites de Réaumur* (La Rochelle: Veuve Mareschal & Martin, 1886), 15.

perceive," since he "was not yet well instructed in all the precautions required for the exactitude of an observation."[32] To observe an object attentively meant first and foremost to observe it distinctly, which the naturalists defined as a kind of mental as well as visual dissection. Microscopes and magnifying glasses were standard tools of the trade, and some of the most vaunted feats of observation in nineteenth-century history, such as Swammerdam's remarkable account of the ovaries of the queen bee, were anatomies of minutiae: the intestines of a caterpillar or the subcutaneous membranes of plants.

Naturalists took apart their objects with the mind and eye as well as the scalpel: Réaumur spoke of the need to view every operation in the beehive "sharply and distinctly";[33] Bonnet thought the mental act of abstraction was at root nothing more than a form of attention that focused on one aspect of an object to the exclusion of others.[34] Hence *l'art d'observer* was as essential to metaphysics as to physics.[35] Through masterworks of attentive observation one could scale the heights of genius in natural history as well as in celestial mechanics. When Bonnet was first confronted with observations suggesting that worker bees could sometimes lay eggs and even metamorphose into queens, he justified his incredulity by an appeal to the lynx-eyed virtuosity of the greatest observers of the age: "Nothing in the world is better confirmed by the repeated researches of SWAMMERDAM, of MARALDI, of REAUMUR than the absolute sterility of worker bees. How would it be possible for the ovaries of these bees to have escaped the great Anatomist of Holland, he who so well described and represented the ovaries of the Queen-bee?"[36]

32. Bonnet to Réaumur, 3 November 1738, Dossier Charles Bonnet, Archives de l'Académie des Sciences, Paris. Original: "Elles [observations] serviront à vous prouver que je ne vois guères que ce qu'une médiocre attention peut apercevoir; . . . je n'étois pas encore bien instruit de toutes les précautions requises pour l'exactitude d'une observation."

33. Réaumur, *Mémoires* (note 17), 5:223.

34. Bonnet, *Philalethe ou Essai d'une méthode pour établir quelques vérités de philosophie rationelle* (comp. 1767–68), in *Oeuvres* (note 20), 18:237.

35. Bonnet, "Préface," in *Oeuvres* (note 20), 1:x.

36. Bonnet, "1re Lettre à Monsieur Wilhelmi, Au sujet de la découverte de M. SCHIRACH, sur les Abeilles," 10 November 1768, in *Oeuvres* (note 20), 10:99; compare 10:163. For the eighteenth-century controversy over Schirach's observations, see Renato G. Mazzolini and Shirley A. Roe, eds., *Science against Unbelievers: The Correspondence of Bonnet and Needham, 1760–1780* (Oxford: Voltaire Foundation, 1986), 157–69. Recent observations confirm that worker bees do occasionally deposit eggs and can undergo ovary development when the colony loses its queen. R. E. Page Jr. and E. H. Erikson Jr., "Reproduction by Worker Honey Bees (*Apis mellifera* L.)," *Behavioral Ecology and Sociobiology* 23 (1988): 117–28. However, dissections reveal that only one in about ten thousand workers has a fully developed egg in her

What finally persuaded Bonnet to take the startling new observations seriously was the care with which they had been executed and, above all, the detail with which they were reported in subsequent accounts.[37] Eighteenth-century naturalists developed many ingenious instruments to observe the otherwise invisible and the hidden, ranging from magnifying lenses to glass-fronted, flattened beehives.[38] But the *sine qua non* of the sterling observation was fastidious attention to detail in word and deed. And since deeds were ultimately reduced to words in the naturalists' reports, it was above all the techniques of description that certified attentiveness. They are long, articulate in the root sense of the word, and stuffed with adjectives. Here is Bonnet on a caterpillar he found in October 1740:

> It was of a middling size, half-hairy, with 16 legs, of which the membranous [ones] have only a half-crown of hooks. The base of the color on the bottom of the body is a very pale violet, on which are cast three yellow rays, which extend from the second ring to about the eleventh [the description continues for about a page]. . . Yellow spots are strewn on the sides. The head is violet-colored.[39]

And here on the colors and sheens of aphids:

> The greatest diversity that one observes among different species of Aphids is in the color: there are green, yellow, brown, black, white [ones]. Some have a matte color; that of the others has a sort of lustre; but often that lustre is due to a small worm that the aphid nourishes in its interior, and which kills it. Finally, some species are prettily spotted, sometimes brown and white, sometimes green, black, or other colors.[40]

The tools of description were the journal, the table, and, to a lesser extent, the illustration.[41] Somewhat alarmingly, considering the length of his

---

ovary, which explains why Swammerdam et al. would not have found any even after many dissections. Francis L. W. Ratnicks, "Egg-Laying, Egg-Removal, and Ovary Development by Workers in Queenright Honey Bee Colonies," *Behavioral Ecology and Sociobiology* 32 (1993): 191–98.

37. Bonnet, "Ier Mémoire sur les Abeilles, où l'on rend compte de la découverte de Mr. Schirach," in *Oeuvres* (note 20), 10:131–32.

38. Réaumur, *Mémoires* (note 17), 5:220ff.

39. Bonnet, *Observations diverses sur les insectes,* in *Oeuvres* (note 20), 2:212–13.

40. Bonnet, *Traité d'insectologie,* in *Oeuvres* (note 20), 1:6.

41. Swammerdam was unusual among naturalists of this period in making his own drawings. Réaumur depended heavily on the skills of his artist Mlle Dumoutier de Marsilly, whom

printed descriptions, Bonnet told his readers that these were only excerpts from his far more copious journal.[42] Early in his career as a naturalist he had formed the habit of writing down everything that he saw.[43] Like the journal, the tables structured repeated observations chronologically—in the case of his aphid observations, Bonnet noted day and hour of each birth as exactly as he could, marking cases where he was not actually present with an asterisk.[44] Both methods of recording strung details together one after another, like beads on a chain; their narrative structure, like that of the journal and the table, was that of temporal sequence, intertwined with the daily rhythms of the observer (fig. 4.1).

It was not so much the content as the texture of these descriptions that made a naturalist's reputation. Modern naturalists have found it difficult to make taxonomic identifications on the basis of Bonnet's descriptions, despite (or perhaps because of) their length and specificity.[45] Descriptions bristled with details, each as sharply etched and distinct as the lettered parts of the accompanying figures. The object as a whole shattered into a mosaic of details, and even the tiniest insect organ loomed monstrously large. The literary effect achieved by these descriptions uncannily mimicked in words the visual experience of magnification, in which the part overwhelms the whole, and the small grows in significance. As Diderot remarked of his attempts to describe a very small painting by Hubert Robert at the Salon of 1767: "There's an infallible way of making the person listening to you take a greenfly for an elephant; all you have to do is describe in extreme detail the anatomy of this living atom." Inevitably, the example Diderot adduced for this kind of field-filling description came from the natural history of insects: "Monsieur de Réaumur didn't suspect

he praised warmly in his will and made his heir. Maurice Caullery, *Les Papiers laissés par de Réaumur et le tome VII des Mémoires pour servir à l'histoire des insectes* (Paris: Paul Lechevallier, 1929), 8–9. Bonnet was dissatisfied with his Parisian engraver and contended that the descriptions were more exact than the figures. Bonnet, *Traité d'insectologie*, in *Oeuvres* (note 20), 1:xxiv.

42. Bonnet, *Traité d'insectologie*, in *Oeuvres* (note 20), 1:21.

43. Bonnet to Réaumur, 3 November 1738, Dossier Charles Bonnet, Archives de l'Académie des Sciences, Paris.

44. Bonnet, *Traité d'insectologie*, in *Oeuvres* (note 20), 1:27. Most of the starred observations are between 5:00 and 6:00 A.M., presumably the first observations of the day, suggesting that the birth had occurred sometime during the night.

45. Jean Wüet, "Bonnet face aux insectes," in Marino Buscaglia, René Sigrist, Jacques Trembley, and Jean Wüest, eds., *Charles Bonnet: Savant et philosophe (1720–1793)* (Geneva: Editions Passé Présent, 1994), 149–62, 153.

TABLE des jours & heures auxquels sont nés les Pucerons qu'enfanta depuis le premier Juin jusqu'au vingt-un inclusivement, celui qui depuis sa naissance avoit été tenu dans une parfaite solitude.

Jours de Juin.	Nombre des Pucer. nés dans chaque j.	Nombre des Pucerons nés chaque matin, & les heures de leur naissance.		Nombre des Pucer nés chaque après midi, & les heures de leur naissance.	
1.	2 puc.	. . . . . .	0 p.	à 7 h. $\frac{1}{2}$	1 p.
				9.	1 p.
2.	10 puc.	à 5 h.	2 p.*	à 12 h. $\frac{1}{2}$	1 p.
		6	1 p.	1 $\frac{1}{2}$	1 p.
		6 $\frac{1}{2}$	1 p.	6 $\frac{1}{2}$	1 p.
		7 $\frac{1}{2}$	1 p.		
		8 $\frac{1}{2}$	1 p.		
		8 $\frac{3}{4}$	1 p.		
3.	7 puc.	à 10 h.	1 p.	à 3 h.	1 p.
		11	1 p.	4	1 p.*
				4 $\frac{3}{4}$	1 p.
				6	1 p.
				9	1 p.
4.	10 puc.	à 5 h.	3 p.*	à 12 h. $\frac{3}{4}$	1 p.
		6	1 p.	1 $\frac{1}{4}$	1 p.
		6 $\frac{3}{4}$	1 p.	6	1 p.
				9	2 p.*

FIGURE 4.1. Table of Observations. Charles Bonnet, *Traité d'insectologie*, in *Oeuvres d'histoire naturelle et de philosophie* (Neuchatel: Samuel Fauche, 1779–1783), vol. 1. Courtesy of Bibliothek der Freien Universität, Berlin.

this, but get someone to read you a few pages of his *Treatise on Insects* and you'll discover the same absurd effect as in my descriptions."[46]

Despite the specificity of these descriptions, they were usually composites of multiple observations under varying conditions, for the naturalists insisted on numerous repetitions.[47] The point was not always or even habitually to replicate controversial observations, but rather to redirect the necessarily narrow focus of attention to other facets of the same object or phenomenon. As the Genevan pastor Jean Senebier explained in his treatise on observation, which was larded with examples taken from the works

46. Denis Diderot, "Salon of 1767," in *Selected Writings on Art and Literature*, ed. Geoffrey Bremner (Harmondsworth: Penguin, 1994), 280; Diderot, *Oeuvres complètes*, ed. J. Assézat and Maurice Tourneux, 11 vols. (Paris: Garnier, 1875–77), 11:233.

47. See, e.g., Bonnet, "IIIe Mémoire sur les Abeilles, Où l'on donne un précis des observations faites sur ces Mouches, par M. RIEM," in *Oeuvres* (note 20), 10:190; Réaumur to Jean-Baptiste Ludot, 1 March 1751, in *Lettres inédites* (note 31), 124; see also, more generally, Jean Senebier, *L'Art d'observer* (Geneva: Chez Cl. Philibert & Bart. Chirol, 1775), 188 passim.

of Bonnet, Réaumur, and other naturalists, one could shift one's attention from one part of the object to another with each repetition, "reserving for another time the circumstantial examination of those one neglects," like a spotlight systematically scanning a surface, point by point.[48] Once again, observation cultivated an intense, laser-like attention. Horace Bénédict de Saussure, Bonnet's nephew and protégé, reassured Albrecht von Haller that he had observed many plants, "in all the ways and in all the positions that I could imagine, with different Microscopes, some simple, others compound; on all sorts of days: and when my Microscope was found to be strong enough, when the object was sufficiently illuminated, and my attention was redoubled, I constantly saw that membrane."[49] Hence it is more accurate to speak of a regimen of observations in eighteenth-century natural history, rather than individual sightings or aperçus.

"Regimen" is meant here in the double sense of a program and a discipline. The program might consist of a series of experiments, such as Réaumur's introduction of foreign queens (painted to distinguish them from the native queen) into beehives, or the intensive scrutiny of a single phenomenon, such as Bonnet's study of aphid generation or Saussure's investigation of plant membranes. The program of vigilant observation imposed a strict discipline on the observer that was scarcely compatible with any other activity. Réaumur, for example, counted the number of bees leaving the hive in one day, arriving at a sum of over eighty-four thousand departures in fourteen hours (approximately one hundred per minute).[50] Bonnet's sequestered study of his *pucerone* had its precedent in Swammerdam's arduous researches on bees, which "began at six in the morning, when the sun afforded him light enough to survey such minute objects," and continued long into the night, when he recorded his observations.[51] Both naturalists paid for these strenuous efforts with ruined health, including badly weakened eyesight.[52] Senebier thought the risks to natural-

48. Senebier, *L'Art d'observer* (note 47), 192. On Senebier's relationship to Bonnet, see Carole Huta, "Bonnet-Senebier: Histoire d'une relation," in Buscaglia et al., eds., *Charles Bonnet* (note 45), 211–24.

49. Horace Bénédict de Saussure, *Observations sur l'écorce des feuilles et des pétales* (Geneva: n.p., 1762), xviii (dedicatory epistle to Albrecht von Haller).

50. Réaumur, *Mémoires* (note 17), 5:433; see also Jean Torlais, "Réaumur et l'histoire des abeilles," *Revue d'Histoire des Sciences* 11 (1958): 51–67.

51. Boerhaave, "Life" (note 26), ix.

52. Ibid., ix–xii; Jean Trembley, *Mémoire pour servir à l'histoire de la vie et les ouvrages de M. Charles Bonnet* (Bern: Société Typographique, 1794), 26–27 ; see also Raymond Savioz, ed., *Mémoires autobiographiques de Charles Bonnet de Genève* (Paris: J. Vrin, 1948).

ists (especially of premature loss of sight) so great that he exhorted them to avoid long bouts of exacting observations; moreover, "when the senses are fatigued, they become unfaithful."[53] But he agreed with the prevailing view among naturalists that observations demanded stamina and courage. François Burnens, servant and collaborator of the blind naturalist François Huber, often observed bees for twenty-four hours at a stretch, "without permitting himself any distraction, taking neither food nor rest," and on one occasion examined every single bee in a hive by hand, refusing to stun them first with a cold bath lest certain "small differences" in form be obliterated, "examining them with attention without fearing their rage . . . and even when he was stung, he continued his examination with the most perfect tranquility."[54] The meticulous patience and manual delicacy of great naturalists was proverbial: magnifying glass in hand, staring for hours at the tongue of a bee or deftly stripping the skin of a caterpillar.

Yet despite the rigorous demands that attentive observation made on mind and body, the naturalists testified unanimously to its pleasures. Describing his taxing and dangerous ascent of an alpine glacier, André Deluc claimed that the diversions of observing new objects staved off vertigo along treacherous mountain passes, and reported that "the attention with which I carried out my [thermometric] observation" assuaged the pain of a foot injury.[55] Réaumur forgot the broiling summer heat while happily watching bees in a glassed-in hive (fig. 4.2).[56] Bees replaced poetry in Schirach's affections, as he came to distinguish the insects from "my usual objects" and turned to them with "a special attention."[57] Saussure rhapsodized over the beauty of the surface of a flower petal viewed under the microscope;[58] Swammerdam delighted in the "beautiful appearance" of a dissected caterpillar, "especially as the pulmonary tubes were at the same time observed to glitter like pearls"[59] (fig. 4.3.). Linnaeus found in natural history a foretaste of "heavenly delight, a constant joy of the spirit, the beginning of perfect consolation and the highest summit of human felic-

53. Senebier, L'Art d'observer (note 47), 213.

54. François Huber, Nouvelles observations sur les abeilles, adressées à M. Charles Bonnet (Geneva: Chez Barde, Manget & Compagnie, 1792), 10.

55. (André Deluc and Jean Dentand), Relation de différents voyages dans les Alpes du Faucigny (Maestricht: Chez J. E. Dufour & Ph. Roux, 1776), 29, 38–39.

56. Réaumur, Mémoires (note 17), 5:251.

57. Schirach, Melitto-Theologia (note 24), xi.

58. Saussure, Observations (note 49), 92.

59. Swammerdam, Book of Nature (note 26), Part II, 53.

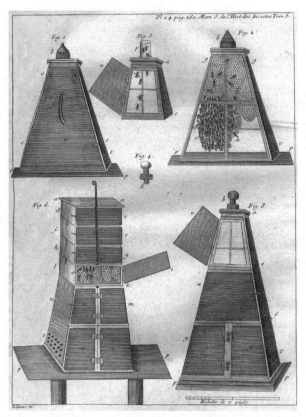

FIGURE 4.2. Observation hive. René Antoine Ferchault de Réaumur, *Mémoires pour servir à l'histoire des insectes* (Paris: Imprimerie Royale, 1734–1742), vol. 5. Courtesy of the Library of the Max Planck Institute for the History of Science.

ity."[60] These pleasures partook of the aesthetic and the sensuous, but they also partook, quite explicitly, of the absorption and pinpoint focus of attention.

The faculty of attention and its distinctive pleasures were objects of considerable theorizing in the late seventeenth and eighteenth centuries.[61]

60. Linnaeus, *L'Equilibre* (note 22), 143.

61. For a general bibliography, see Lemon L. Uhl, *Attention. A Historical Summary of the Discussions concerning the Subject* (Baltimore: Johns Hopkins Press, 1890); on the place of absorption in eighteenth-century French art criticism, see Michael Fried, *Absorption and Theatricality: Painting and Beholder in the Age of Diderot* (Berkeley: University of California Press, 1980); and on links between forms of attention and modernism in art, Jonathan Crary, *Suspensions of Perception: Attention, Spectacle, and Modern Culture* (Cambridge: MIT Press, 1999); in literature, Roger Chartier, "Richardson, Diderot et la lectrice impatiente," *Modern Language Notes* 114 (1999): 647–66, and Adela Pinch, *Strange Fits of Passion: Epistemologies of Emotion, Hume to Austen* (Stanford: Stanford University Press, 1996), esp. 152–63; in pedagogy, Christa Kerstig, "Die Genese der Pädagogik im 18. Jahrhundert. Campes 'Allgemeine Revision' im Kontext der neuzeitlichen Wissenschaft," Ph.D. diss., Freie Universität Berlin, 1992; and in

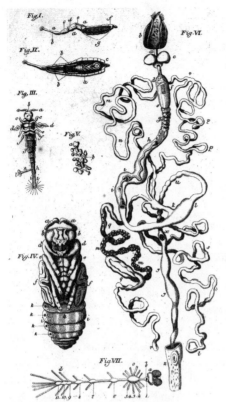

FIGURE 4.3. Anatomized caterpillar. Jan Swammerdam, *Histoire générale des insectes* (Utrecht: Guillaume de Walcheren, 1682). Courtesy of the Houghton Library, Harvard University.

Key to the enjoyments of attention was a cultivated oblivion to one's surroundings and, as critics complained, to all other ordinary obligations. According to contemporary physiology, attention diverted nervous fluid from all other fibers of the brain so that no other external impression could register upon the soul in that rapt state. The saturated fibers held the soul in contented captivity, mingling intellectual concentration with sensory and emotional transports. The philosopher and theologian Nicolas Malebranche had conceded that the passions and senses, although in excess detrimental to attention, may be useful, even necesary, to excite and sustain it.[62] Bonnet posited that attention was proportional to the pleasure ex-

medicine, Michael Hagner, "Psychophysiologie und Selbsterfahrung: Metamorphosen des Schwindels und der Aufmerksamkeit im 19.Jahrhundert," in Aleida Assmann and Jan Assmann, eds., *Aufmerksamkeiten* (Munich: Wilhelm Fink Verlag, 2001), 241–64.

62. Nicolas Malebranche, *Recherche de la vérité* [1674–75] (Paris: Galerie de la Sorbonne, 1991), VI.i.3, pp. 735–39.

cited by the object.[63] But the experiences of observation narrated by the naturalists suggested that the equation could be reversed: attention also created pleasure, even when directed to objects initially deemed trivial or disgusting.[64]

Attention infused its objects with affect: the naturalists came to regard their bees and aphids and even insects extracted from horses' dung with wonder and affection. Huber mused on the tendency of such sentiments to foster anthropomorphism, once again returning to the power of attention to dignify its objects:

> In general, naturalists who have long observed animals, and especially those who have chosen insects for the favorite objects of their studies, have too easily ascribed to them our sentiments, our passions and even our views. Entranced by the desire to admire, shocked perhaps as well by the disdain with which one speaks of insects, they believe themselves under the obligation to justify the use of time consecrated to them and they have embellished various traits of the industry of these little animals with all the colors furnished by an exalted imagination.[65]

In fact, the naturalists, including the entomologists, were considerably more cautious about anthropomorphizing their animals than Huber's fears would suggest. Réaumur, for example, debunked the Aesopian view that ants prudently store up food for the winter, and warned against projecting human sentiments onto bees.[66] Bonnet rejected the Cartesian view of animals as machines and attributed a certain "character" to each species, but he also argued that animals differ profoundly from humans in possessing only sensations without signs, and therefore refused to impute motives to them.[67] Although their works were still peppered with anthropomorphisms (particularly concerning political organization and gender roles among social insects),[68] the sentiments they entertained for their objects

63. Bonnet, *Essai analytique sur les facultés de l'ame* [1760], in *Oeuvres* (note 20), 13:128.

64. Senebier, *L'Art d'observer* (note 47), 215.

65. Huber, *Nouvelles observations* (note 54), 320.

66. Réaumur, *Histoire des fourmis*, intro. E. L. Bouvier, notes Charles Pérez (Paris: Paul Lechevalier, 1928), 31–34; Réaumur, *Mémoires* (note 17), 5:258, 266–67.

67. Bonnet, *Contemplation de la nature*, 2 vols. (Amsterdam: Chez Marc-Michel Rey, 1764), 2:99, 235–36.

68. Edwin Mueller-Graupa, "Der Hochzeitsflug der Bienenkönigen: Die Geschichte eines biologischen Problems von Aristoteles bis Maeterlinck," *Sudhoffs Archiv für Geschichte der Medizin und der Naturwissenschaften* 31 (1938): 350–64; Jean-Marc Drouin, "L'Image des sociétés d'insectes à l'époque de la Révolution," *Revue de Synthèse* 113 (1992): 333–46; Jeffrey

of observation did not depend on transforming insects into tiny humans. Rather, the naturalists extended the faculty of sympathy, so in vogue among Enlightenment moral theorists, beyond the realm of the human. They did so self-consciously, aware of the apparent disproportion between emotion and object. When Bonnet became alarmed about the health of his closely watched aphid, he realized that his concern would strike his readers as odd: "Perhaps I will be accused of puerility, if I recount the worries that my aphid caused me during the last moulting."[69] Evidently the anticipated raised eyebrows were an insufficient check: sentiment suffused observation, and hence crept willy-nilly into description. Attention had, to use the Freudian term, cathected its objects. Abigail Lustig's essay in this volume documents the tenacity of this link between attention and affection in nineteenth- and twentieth-century entomology.

Conclusion: The Fitness of Things

The value that the naturalists poured into the parts of nature they studied so attentively was simultaneously moral, aesthetic, and economic. The eighteenth-century word for this compound of norms was "utility," and it pervaded Enlightenment meditations on the good, the beautiful, and the useful—perhaps no other shibboleth was so often invoked, in so many varied contexts.[70] Utility also permeated natural history, not only in the justifications of natural history as edifying or (as in the case of apiculture) profitable, but also in the habits of observation and the pleasures of attention that marked the *naturaliste de profession*. These judgments of utility, as applied to the cocoon of a caterpillar or cell of a honeycomb, were strikingly similar to the judgments of utility applied to the fabrication of pins or the workmanship of locksmiths in Enlightenment descriptions of the

Merrick, "Royal Bees: The Gender Politics of the Beehive in Early Modern Europe," *Studies in Eighteenth-Century Culture* 18 (1988): 7–37. The lore of bees in particular is vast, beginning with Book IV of Virgil's *Georgics*. For an introduction, see Robert Delort, *Les Animaux ont une histoire* (Paris: Editions du Seuil, 1984), section "Abeilles"; Remy Chauvin, ed., *Traité de biologie de l'abeille*, 5 vols. (Paris: Masson, 1968), vol. 5, *Histoire, ethnographie et folklore*. On the importance of social insects as models for human polities, see the essays by Allen and Lustig in this volume.

69. Bonnet, *Traité d'insectologie*, in *Oeuvres* (note 20), 1:22.

70. On the history and nuances of the term "utility" and its cognates in medieval Latin and modern European vernaculars, see O. Höffe, "Nutzen, Nützlichkeit," in Joachim Ritter and Karlfried Gründer, eds., *Historisches Wörterbuch der Philosophie*, 10 vols. (Darmstadt: Wissenschaftliche Buchgesellschaft, 1971–98), 6:992–1008. See also the essay by Price in this volume.

arts and crafts, most famously that of the *Encyclopédie*. Naturalists were not only interested in the final cause or purpose of the underside of a leaf or honeybee drones; they sought to understand what Bonnet called the "organic Economy," in which the "arrangement and play of different parts of organized bodies" explained operations like growth and generation.[71] The choice of the word "economy," used here in its eighteenth-century sense of an intricate system of interrelated, functional parts, was not accidental. The patterns of observing and describing in Enlightenment natural history and political economy resembled one another strongly, and indeed several prominent naturalists—Réaumur, Duhamel du Monceau, Saussure—also wrote on matters of political economy.[72] As Danielle Allen in her essay in this volume points out apropos of Bernard Mandeville's analogies to both beehives and knitting machines, a fascination with part–whole relationships and the division of labor bridges the natural and the artificial in eighteenth-century political economy. In both political economy and natural history, observers analyzed objects into interlocking parts and traced the fit of form to function with an eagle eye for "fitness."

What was meant by fitness can best be gleaned from specific examples. Réaumur reported with evident satisfaction that mathematical demonstrations proved the pyramidal base of honeycomb cells to be a more economical use of wax than a flat base would have been. Even the slight asymmetries of the hexagonal cell, studied minutely by Réaumur, would, he was sure, eventually make economic sense: "I don't know if this disposition [of angles in the hexagon] also saves wax, but it is indubitable that it tends to render the work more perfect, that it has some utility that will be admired, as soon as it is made known."[73] Describing the spinning of a cocoon, Bonnet admired the funnel-shaped structures woven like fishermen's nets to keep out predators, and the efficiency of the caterpillar: "Our worker showed itself to be as diligent as it was industrious: in less than three-quarters of an hour, the new funnel was already quite recognizable."[74] The skill and ingenuity of human workers prompted praise in

71. Bonnet, *Contemplation de la nature* [1764], rev. ed. in *Oeuvres* (note 20), 7:291–92.

72. For Réaumur and Duhamel du Monceau, see Académie Royale des Sciences, *Description des arts et métiers*, 26 vols., Archives de l'Académie des Sciences, Paris; for Saussure's involvement in the Genevan Société pour l'avancement des arts (est. 1776), see Jean Senebier, *Mémoire historique sur la vie et les écrits de Horace Bénédict de Saussure* [An VIII/1801], in Horace-Bénédict de Saussure, *Voyages dans les Alpes*, 4 vols. (1779–96; Geneva: Editions Slatkine, 1978), 1:20.

73. Réaumur, *Mémoires* (note 17), 5:390–91.

74. Bonnet, *Observations diverses sur les insectes*, in *Oeuvres* (note 20), 2:212, 235.

similar terms. Just as the naturalists deplored public scorn of insects, Du-hamel du Monceau complained of the general public indifference to the craft of locksmiths, both subjects so well-suited to nourish a "reasonable curiosity."[75]

Duhamel and Réaumur collaborated with the engineer Jean-Rodolphe Perronet to write an elaborate description of pin-making for the Acadé-mie Royale des Sciences that formed the basis of Perronet's *Encyclopédie* ar-ticle on the subject, which was in turn the likely source of Adam Smith's famous description of the division of labor in the opening chapters of *The Wealth of Nations* (1776). Although the cheapness of pins was noted, the di-vision of labor by person did not figure prominently in the earlier French articles. It was instead the speed, efficiency, and delicacy of the operations (sixteen steps in the Académie description, eighteen in the *Encyclopédie*) that commanded the admiration of the authors, just as the deft zig-zags of the cocoon-weaving caterpillar had enchanted Bonnet.[76] The rhetoric of *multum in parvo* so prominent in the natural history of insects also orna-mented the account of pin-making: "The pin is of all mechanical works the smallest, the most common, the least precious, and however one of those that perhaps requires the most combinations: from which it results that art like nature displays its prodigies in small objects."[77]

Shared practices of observing, analyzing, describing, explaining, figur-ing, and, above all, evaluating united the naturalists and the political econ-omists. Réaumur made the analogy explicit in a draft introduction to the Académie Royale des Sciences project for a description of the arts and trades (with a revealing slip of the pen): "The history of the arts that we have undertaken to describe is not that of their progress, their decadence; it is that of their practices that are currently in use; we have called this kind of description history of arts, as we name natural history {that of the prac-

75. Henri-Louis Duhamel du Monceau, "Art du Serrier" [1767], in Académie Royale des Sciences, *Description* (note 72), 4:159.

76. *Art de l'Epinglier*, Par M. de Réaumur. Avec des additions de M. Duhamel du Mon-ceau, & des Remarques extraites des Mémoires de M. Perronet, Inspecteur Général des Ponts & Chausées, 3. (This was a part of the Académie des Sciences *Description*, printed separately but not published. The exemplar I consulted was at the Niedersächsische Staats- und Univer-sitätsbibliothek Göttingen, and was possibly originally sent to the Akademie der Wissen-schaften zu Göttingen.) On the *Description*, see Arthur H. Cole and George B. Watts, *The Handicrafts of France as Recorded in the Descriptions des Arts et Métiers, 1761–1788* (Boston: Parker Library, 1952).

77. "Epingle *(Art. Méchaniq.)*," in Jean d'Alembert and Denis Diderot, eds., *Encyclopédie, ou Dictionnaire des sciences, des arts et des métiers* [1751–80], 17 vols. (Stuttgart/Bad Cannstatt: Friedrich Fromm, 1988), 5:804–8, here 5:804.

tices} [struck through in MS.] the description of the productions of nature, such as animals, soils, stones, minerals etc."[78] Réaumur and his colleagues anatomized locks and worms, divided the blurringly fast operations of pin makers and worker bees into steps, counted the bees in a hive and the production of pins in a day, explored how the bee's tongue was as exquisitely fitted to the gathering of nectar and pollen as key fitted into lock, parsed and numbered their engraved figures of the viscera of insects and machines into functional parts, and exclaimed over the delicacy, intricacy, and efficiency of the workmanship.[79] Not only were organisms works of divine craftsmanship; they were also workers: the bees manufactured wax and honey, the underside of the leaf drew in nourishment, the caterpillar wove its labyrinthine cocoon. Moral (diligence), aesthetic (symmetry), and economic (efficiency) criteria of fine workmanship merged in these equally fine-grained observations and descriptions. Bonnet's caterpillar, like Perronet's pin maker, was at once industrious, adroit, and speedy. Naturalists and political economists were connoisseurs of fitness in contrivances, whether the craftsmanship in question was human or divine.

Apparent violations of fitness provoked displeasure and consternation. When caterpillar nests appeared to Bonnet to be arranged randomly, he complained that this impression was "quite contrary to the pleasure that I have when I can observe some new industry in insects," and immediately set about searching for a hidden purpose in the hodgepodge.[80] Seventeenth-century naturalists like Robert Plot had not hesitated to attribute the beauty and variety of flowers to God's regal desire to ornament His creation.[81] But later authors found such flouting of function intoler-

78. René Antoine Ferchault de Réaumur, "Préface inédite pour la Description des Métiers," fMSTyp.432.1(1), Houghton Library, Harvard University. Original: "L'histoire des arts que nous avons entrepris décrire, n'est point celle de leurs progres, de leurs decadences, c'est celle des pratiques qui sont actuellement en usage, nous avons appeller cette espece de description histoire des arts, comme on nomme histoire naturelle {celle des pratiques} la description des productions de la nature, comme des animaux, terres, pierres, mineraux & c." On the manuscript sources for the "Description," see Martine Jaoul and Madeleine Pinault, "La Collection 'Description des Arts et Métiers': Etude des sources inédites de la Houghton Library Université Harvard," *Ethnologie française* 12 (1982): 335–60.

79. On the visual analysis of work and tools into parts, see William H. Sewell Jr., "Visions of Labor: Illustrations of the Mechanical Arts before, in, and after Diderot's *Encyclopédie*," in Steven Laurence Kaplan and Cynthia J. Koepp, eds., *Work in France* (Ithaca: Cornell University Press, 1986), 258–86, here 268–70.

80. Bonnet to Réaumur, 27 July 1739, Dossier Charles Bonnet, Archives de l'Académie des Sciences. Original: "Cette pensée etoit pourtant bien contraire au plaisir que j'ai quand je peux observer chés les Insectes quelque nouvelle industrie."

81. Robert Plot, *The Natural History of Oxfordshire* (Oxford: Theater, 1677), 121.

able. Julius Bernhard von Rohr, author of a natural theology of plants, firmly rejected the notion that blossoms "were merely for decoration"; their function was to nourish fruit.[82] It was not so much that utility had replaced beauty as a criterion of excellence in workmanship, as that utility had become beautiful. Everything in nature, Linnaeus claimed, had been created for a determinate end—or as Pope put it, "All Nature is but Art, unknown to thee, / All Chance, Direction, which thou canst not see."[83]

But what kind of art? The Enlightenment naturalists who imbued nature with value thought, felt, and above all observed in a world different from that of either their Renaissance predecessors or even their natural theologian contemporaries. The multifaceted standard of utility may have resembled Aristotelian final causes, and it was certainly buttressed by natural theological doctrines of God as artisan, nature as art. But the naturalists' accolade of utility went far beyond these general notions of purpose and artistry. Fitness articulated the ways in which the smallest parts of organisms—or machines—meshed together in a smoothly functioning whole, and revealed itself only to the most acute and persevering observers. Judgments of utility were intrinsically analytical, anatomizing natural and artificial objects into minute, interlocking parts—figurative and literal clockwork of the finest stamp. Ingenuity was more admirable than variety; efficiency more pleasant than magnificence. As David Hume exclaimed in his reflections on "Why Utility Pleases": "What praise, even of an inanimate form, if the regularity and elegance of its parts destroy not its fitness for any useful purpose!"[84]

Nature miniscule merged with Nature majuscule in the figure of personification or prosopopeia, so ubiquitous in Enlightenment poetry and prose. Although scorned by Romantic poets, prosopopeia—putting a face to an abstraction—inspired neoclassical critics and electrified Enlightenment readers: how else to render a universal norm vivid and sensually intelligible, as sensationalist psychology demanded?[85] Erasmus Darwin defended his lavish use of personification in his verse exposition of the Linnaean system on the grounds that it was among the "arts of bringing

82. Julius Bernhard von Rohr, *Phyto-Theologia* (Frankfurt: Michael Blochberger, 1740), 104–5.

83. Pope, *Essay on Man* (note 2), I.281–82, p. 21.

84. David Hume, *An Inquiry Concerning the Principles of Morals,* ed. Charles W. Hendel (1751; Indianapolis: Bobbs-Merrill, 1957), V.i, p. 41.

85. Earl R. Wasserman, "The Inherent Values of Eighteenth-Century Personification," *Publications of the Modern Language Association* 65 (1950): 435–63. On earlier visual traditions of personified Nature, see the essay by Park in this volume.

objects before the eye, or of expressing sentiments in the language of vision," the sense that rendered ideas clearest. In contrast to personifications in art, prosopopeia in poetry was "generally indistinct, and therefore does not strike us so forceably as to make us attend to its improbability."[86] Darwin's determination to yoke poetry to scientific explanations (in footnotes) may have been unusual, but his views on the centrality of personification to poetry and the necessity of drawing on the natural history descriptions were widely shared. In *Les Saisons* (1769) the French poet and *philosophe* Jean François de Saint-Lambert insisted not only that "[f]or the poet Nature [always capitalized] is the only model, the only subject, the only mistress," but also that the poetry of Nature must be descriptive, incorporating the myriad details revealed by "Physics, Astronomy, Chemistry, Natural History, etc." Since Nature was the one true standard in political economy, religion, and morals, Saint-Lambert argued that the poet instructed humanity in virtue and happiness by illuminating this standard, still more so by making it sensually irresistable. In contrast to the richly embroidered gown of medieval allegorical representations of nature or the naked fecundity of Renaissance personifications described in Katharine Park's essay in this volume, Enlightenment Nature was to be elevated by scenes of sublimity, embellished with descriptions of its "beauties and riches," made interesting in relation to "truths of physics and morals"— drawing upon the fine-grained details of the attentive naturalists' descriptions.[87] Nature personified was no longer an allegory to be deciphered, but a mirror of minute observations.

It is noteworthy that poetry which strikes modern sensibilities as an indigestible and often ludicrous blend of lofty abstraction and opulent concreteness should have, by all accounts, enthralled eighteenth-century readers.[88] It is also worth remarking that when nature's authority was invoked

---

86. Erasmus Darwin, *The Botanic Garden, Containing the Loves of the Plants, a Poem with Philosophical Notes,* Part II (1789; Oxford: Woodstock Books, 1991), 41–44.

87. Jean François de Saint-Lambert, *Les Saisons, Poëme* (Amsterdam [Paris]: n.p., 1769), "Discours préliminaire," xi–xii, xvi. On eighteenth-century French nature poetry, see Edouard Guitton, *Jacques Delille (1738–1813) et le poème de la nature en France de 1750 à 1820* (Thèse: Université de Paris IV, 1972), especially 201–88; on French views of nature more generally, Jean Ehrard, *L'Idée de la nature en France dans la première moitié du XVIIIe siècle* (Paris: Albin Micel, 1994).

88. Joseph Priestley, for example, wrote in 1774 of how personification puts the mind "under a temporary deception, the personification is neither made nor helped out by the speaker, but obtrudes itself upon him; and while the illusion continues, the passions are as strongly affected, as if the object of them really had the power of thought." Quoted in Wasserman, "Inherent Values" (note 85), 442.

FIGURE 4.4. Worship at the Temple of Nature. Erasmus Darwin, *The Temple of Nature; Or, the Origin of Society* (London: J. Johnson, 1803), frontispiece. Courtesy of the Library of the Max Planck Institute for the History of Science.

in the eighteenth century—in philosophical treatises, Revolutionary pageants, political proclamations, didactic poems—nature at that moment almost inevitably became Nature (or even NATURE), that is, a personified being who exhorted and admonished, and who was made visible to the mind's eye by exquisite details of the kind observed and described by the naturalists. When Erasmus Darwin chose to present to the imagination "beautiful and sublime images of the operations of nature" in the form of a long poem with philosophical notes, his mezzotint frontispiece portrayed the Eleusinian mysteries interpreted as the worship of a personified nature, who combined the many breasts of Diana of Ephesus with the classical dress of a Greek goddess[89] (fig. 4.4.) Darwin's poetry fused the authority of an ancient deity with lush descriptions of Enlightenment natural history: "SHRIN'D in the midst majestic NATURE stands, / Extends o'er earth

89. Erasmus Darwin, preface to *The Temple of Nature; Or, the Origin of Society* (London: J. Johnson, 1803), n.p.

and sea her hundred hands; / Tower upon tower her beamy forehead crests, / And births unnumber'd milk her hundred breasts." [90]

To put the point simply, no doubt too simply: for Enlightenment sensibilities, authority required personification, and personification in turn required description. This description in turn drew heavily upon the practices of seeing and writing cultivated by the naturalists. This was the path that led from the values with which the tiniest and most trivial parts of nature came to be infused by dedicated observers to the values that thundered in the name of Nature in countless treatises and proclamations. This was a mechanism that vested moral authority in nature not through a stratigraphy of legitimation, in which one abstraction is layered upon the next, until the foundational bedrock of Nature is reached. Rather, it derived authority from habit, which in turn created new forms of experience. In eighteenth-century habits of attention and imagination, forged in the practices of observing nature and reading poetry, nature and NATURE, value and authority, came together with the force of direct perception. The enchained argument *from* design became the immediate intuition *of* design.

90. Ibid., I.129–32, p. 12.

5

The Erotic Authority of Nature: Science, Art, and the Female during Goethe's Italian Journey

Robert J. Richards

> Often I have composed poetry while in her arms, and have softly beat out
> the measure of hexameters, fingering along her spine.
>
> GOETHE, *Römische Elegien* V

In a late reminiscence, Goethe recalled that during his close association with the poet Friedrich Schiller, he was constantly defending the "rights of nature" against his friend's "gospel of freedom."[1] Goethe's characterization of his own view was artfully ironic, alluding as it did to the French Revolution's proclamation of the "Rights of Man." His remark implied that values lay within nature, values that had authority comparable to those ascribed to human beings by the architects of the Revolution. During the time Goethe made his defense, he also faced another revolution, one in which Schiller was a partisan—that of Kant in the intellectual sphere. Both upheavals had undermined the autonomy of nature, replacing its authority with that of human will and understanding.

Previous essays in this volume record the shifting fortunes of nature that led to this stage of jeopardy. In the early classical period, the notion of nature as a unified entity had not yet arisen. Animate and inanimate objects

This essay was originally composed as a contribution to the conference on "The Moral Authority of Nature" held at the Max-Planck Institut für Wissenschaftsgeschichte in 1999–2000. It was subsequently integrated into my book *The Romantic Conception of Life: Science and Philosophy in the Age of Goethe* (Chicago: University of Chicago Press, 2002).

1. Johann Wolfgang von Goethe, "Einwirkung der neueren Philosophie," *Zur Morphologie* 1.2, in *Sämtliche Werke nach Epochen seines Schaffens* (Münchner Ausgabe), ed. Karl Richter et al., 21 vols. (Munich: Hanser Verlag, 1985–98), 12:97.

had natures—characteristic modes of action—but there was as yet, according to Laura Slatkin, no articulate concept of nature as a whole standing over against human beings. In her essay, Katharine Park describes a long period of transition during which nature became personified in the form of a didactic female, a figure imaginatively based, it would seem, on the system of a natural philosophy *(philosophia naturalis)* that stood in contrast to the revealed wisdom of God. Nature in this guise yet derived her authority and nurturing capacity from that higher, divine power. During the seventeenth century, as Danielle Allen shows, writers like Mandeville began to suspect that nature might be a chimera, a fictive creature that disguised humanity's own hidden desires and inclinations. These doubts grew over the course of the following century—with the likes of Hume accelerating the skepticism—until finally, in the two revolutions that so troubled Goethe, nature was completely stripped of her authority.

Goethe had been confirmed in his defense of the rights of nature during his travels to Italy in 1786–88. After his return to Germany, he set out to develop a science that would recognize nature's autonomy and authority. In this reconfiguration, though, nature would come to exhibit features distinctively altered from those of her earlier incarnation. Goethean morphology would not reinstate nature as emissary of an aloof, divine power. Nature would no longer stand apart from human beings, designed for their instruction, but would encompass the authority of both divinity and humanity: fecund, creative nature would replace God; and man would find himself an intrinsic part of nature and able to exercise, in the role of the artist, her same creative power. This transition in the conception of nature would ultimately lead to the kind of evolutionary theory that Goethe himself would introduce and Darwin would later cultivate.[2] The hinge of this great transition was, for Goethe, the experience of his southern travels, when so much depended on an Italian girl.

Art and science, nature, and women obsessed Goethe from adolescence to old age. From the surface of his concerns to the deeper recesses of his personality, these consuming interests formed many links, both rational and passional. Nature would have a dominating authority in his life, and both art and science would be the principal modes by which he would seek to reveal her commands. On the theoretical side, he adopted quite early on the Spinozistic principle *Deus sive natura:* God and nature are one.

---

2. I discuss Goethe's evolutionary theory and its relation to Darwin's in *The Romantic Conception of Life: Science and Philosophy in the Age of Goethe* (Chicago: University of Chicago Press, 2002), chap. 11.

This philosophical perspective encouraged him to seek, through empirical research, those *adequate* ideas about which Spinoza spoke—ideas that would reveal essential structures in nature, those very *Urtypen* used by nature in the creation of life.

On the passional side, Goethe's poetry expressed the joys and sorrows of an ardent embrace of nature, whether in the figure of a golden grove or a golden girl. In his autobiography *Dichtung und Wahrheit* (Poetry and truth), he poetically represented his approach to both nature and women in similar fashion: at first from a distance, with fertile imagination, then more immediately and intimately, and finally in the fabrications of memory. His many intense relationships with women—what Schiller referred to as "the *Weiberliebe* that plagues him"[3]—became the passional means by which he explored the aesthetic and the universal in nature.

The erotic authority that nature exercised over Goethe rose to insistent consequence during his first travels through Italy. He was at that time beginning to formulate the conception of the *Urpflanze,* the ideal plant that would exhibit the fundamental structure of all plants. But the archetypal plant would be more than that; it would also be a creative life force. It was, I believe, Goethe's sensuous encounters with Italy—its brilliant days, its lush landscapes, its ancient art—and with one woman in particular that enabled him to give substance and shape to his more abstract morphological notions. This essay aims, in short, to show how Goethe's experiences in Italy turned him into a romantic biologist.

The Weimar Councilor and the Frustrated Lover

From November 1775 to September 1786, Goethe served in Weimar as a member of the Privy Council at the pleasure of Carl August, duke of Saxe-Weimar-Eisenach. During this period, Goethe achieved considerable power both as friend and advisor to his prince. He rose to the presidency of the council, which required immersion in the various civil concerns of the duchy; and he kept the court entertained with plays, poetry, and songs. He also formed close friendships that extended beyond these official duties and cultural responsibilities. Two individuals in particular shaped his passional life and intellectual outlook, Charlotte von Stein and Johann Gottfried Herder—she touched him deeply and gently swayed his intellectual appetites toward the recognition of certain ideal structures in living

---

3. Friedrich Schiller to Christian Gottfried Körner, 1 November 1790, in *Schillers Werke (Nationalausgabe)*, ed. Julius Petersen, 42 vols. (Weimar: Böhlaus Nachfolger, 1943–), 26:55.

reality; and he brought the poet to appreciate Spinoza, who provided a theory of such ideal structures.

Herder served as administrator of ecclesiastic affairs in the duchy and had come to Weimar at the behest of Goethe. During the early 1780s, they conferred often on Herder's book project, *Ideen zur Philosophie der Geschichte der Menschheit* (Ideas toward a philosophy of the history of mankind, 1784–91). The theoretical considerations of this multivolume work, interlaced with assumptions drawn from Spinoza, left important residues in the recesses of Goethe's consciousness for the later emergence of his theories of the primal plant and the primal animal, the transcendent forms of all organic life. Herder conceived the earth as undergoing a vast series of developments that exhibited an underlying plan, a unity of organization running throughout nature:

> Now it is undeniable that given all the differences of living creatures on the earth, generally a certain uniformity of structure and a *principal form* [*Hauptform*] seem more or less to govern, a form that mutates into multiple varieties. The similar bone structure of land animals is obvious: head, rear, hands, and feet are generally the chief parts; indeed, the principal limbs themselves are formed according to one prototype and vary in almost infinite ways.[4]

Herder assumed that organisms instantiated a general form that was realized phenomenally in a multitude of instances. In constructing his theory, he relied largely on Spinoza's conception of "adequate ideas"—ideas corresponding in essence to physical entities. Goethe himself had briefly examined Spinoza's *Ethica* during his student years. He undertook serious study of the philosopher, however, only in Weimar, during the winter of 1784–85, when he and von Stein together read the "holy" Spinoza. Goethe had long since abandoned orthodox ties to Christianity, but certainly welcomed the sentiment that Herder inscribed in a Latin copy of the *Ethica,* which he gave to the pair at Christmas: "Let Spinoza always be for you the holy Christ."[5] Spinoza's conception that God and nature were one became a leitmotiv for Goethe's aesthetic and scientific approaches to nature. The nature that he studied would have resident ideas, or archetypes,

4. Johann Gottfried Herder, *Ideen zur Philosophie der Geschichte der Menschheit,* in *Werke,* ed. Martin Bollacher, 10 vols. (Frankfurt am Main: Deutscher Klassiker Verlag, 1989–98), 6:75 (I, ii, 4).

5. See Goethe, *Sämtliche Werke* (note 1), 2.2:875.

that not only represented the essential structures of particular natural beings—hence foundations for aesthetic and scientific evaluation—but that
also functioned to produce the very objects they represented.

During his Italian travels, this Spinozistic conception would begin to
merge for Goethe with ideas drawn from Johann Joachim Winckelmann
the great critic and historian of ancient art. Winckelmann believed that
classical sculpture in particular expressed certain ideals of beauty that had
to be comprehended in order to appreciate and produce significant work
of art. Only after the experience of classical art could one, according to
this view, begin to discover beauty in nature. Goethe initially accepted this
Winckelmannian propadeutic, and would come to identify Spinozistic
archetypes as the ground for beauty both in works of art and in works of
nature. Yet he would finally, even while in Italy, begin to reverse Winckelmann's priority, namely, it would be the immediate experience of natural beauty that would allow him to appreciate classical art and to penetrate to the archetypal reality behind its expression. I believe Goethe was
led to comprehend the value of immediate experience of nature primarily (though certainly not exclusively) through his experience of several
women—his real and fancied relationships with them—climaxing in his
sexual fulfillment with a woman in Rome. They revealed to him an archetypal reality behind their individual embodiment, a standard of beauty and
a force of nature. This experience infused reality into his emerging notions
of the *Urpflanze,* which would also be an archetype of beauty and a force
of nature. Goethe's engagement with various women began prior to his sojourn in Italy, most poignantly in the case of the one woman who realized
for him an ideal yet remained as distant as a Platonic form—Charlotte von
Stein (fig. 5.1).

She was born Charlotte Albertine Ernestine von Schardt and raised
by a pious mother to those accomplishments—reading, sums, music, and
dance—befitting a daughter of the minor nobility. When the twenty-six-
year-old Goethe first met her, she was thirty-three, had been married
eleven years, and had born seven children, four girls who died immediately
after birth and three boys who survived. Just prior to meeting Goethe, her
physician, Johann Georg Zimmermann, described this most interesting
woman to a friend:

> The Lady-in-Waiting, Mistress of the Stables, and Baroness von Stein of
> Weimar. She has large black eyes of the greatest beauty. Her voice is soft
> and low. Every man, at first glance, notices in her face earnestness, tender-
> ness, sweetness, sympathy, and deeply rooted, exquisite sensitivity. Her

FIGURE 5.1. Charlotte von Stein. Self-portrait in pencil (1790).

courtly manners, which she has perfected, have been ennobled with a very fine simplicity. She is rather pious, which moves her soul with a quiet enthusiasm. . . . She is thirty-one years old, has quite a few children and weak nerves. Her cheeks are very red, her hair completely black, and her skin, like her eyes, is Italianate. Her figure is slim, and her entire being is elegant in its simplicity.[6]

When von Stein learned that the famous author of *Werther* might be coming to live at Weimar, she wrote Zimmermann to ask what he knew about the man. Her friend, who had not yet met Goethe, related what he had heard from a "woman of the world." "Goethe," he reported, "is the most handsome, liveliest, most original, fieriest, stormiest, softest, most seductive, and for the heart of a woman, the most dangerous man that she had ever seen in her life."[7] And for von Stein, he would prove to be all of that.

6. Johann Georg Zimmermann to Johann Kaspar Lavater, 1774, quoted in Wilhelm Bode, *Charlotte von Stein,* 3rd ed. (Berlin: Mittler & Sohn, 1917), 63.

7. Johann Georg Zimmermann to Charlotte von Stein, 19 January 1775, as quoted in ibid., 69–70.

The relationship between Goethe and von Stein grew to be quite complex during the decade of their closeness (1776–86). She was obviously alluring, coquettish, sensitive, sexually attractive, if not a great beauty, and had a lively intellect, at least one that might appreciate Spinoza. Goethe swooped into her dull existence like a frenzied swan from Olympus. He was maddeningly impetuous and foolish with her, trampling on her sensitivities and shocking her with his ribaldness and vulgarities. His genius was matched by the size of his ego. When she would remonstrate about his behavior, which she frequently did, he would hang-doggedly become apologetic, only shortly to do something else that upset her. Yet he loved her for her efforts at taming him; and, in time, she—and the responsibilities of the court—did rein in his actions. Goethe's initial hyperbolic behavior serves as an index, partly at least, of his frustrated desire for her. Typical of his early letters to her is one from spring 1776:

> Why should I plague you! Lovely creature! Why do I deceive myself and plague you, and so on—We can be nothing to one another and are too much to one another.—Believe me when I speak as clear as crystal to you; you are so close to me in all things.—But since I see things only as they are, that makes me crazy. Good night angel and good morning. I do not wish to see you again—Except—You know everything—I have my heart—Anything I could say is quite stupid.—I will look at you just as a man watches the stars—think about that.[8]

Despite her occasional lapse into peevishness, she was strongly attracted to him. He gave her instructions in English, read to her his poems and works in progress; and together they discussed everything from Spinoza to the latest gossip at court. She was his confidant, support, and, for the early years, his passion.

During their long relationship, Charlotte von Stein remained mistress of Goethe's soul, though very probably not of his body. The Weimar court, a small, ingrown and gossiping community, knew of his devotion to her, but seems never to have suspected a physical relationship of the sort their duke enjoyed with a number of women, save his wife. When Schiller visited the court in summer of 1787, a year after Goethe had departed for

8. Johann Wolfgang von Goethe to Frau Charlotte von Stein, 1 May 1776, in *Goethes Briefe* (Hamburger Ausgabe), ed. Robert Mandelkow, 4 vols., 3rd ed. (Munich: C. H. Beck, 1988), 1:213.

Italy, he heard no hint of scandal. He wrote his friend Christian Gottfried Körner a note that suggests he perhaps did expect to hear some: "This woman possesses perhaps over a thousand letters from Goethe, and he has written her every week from Italy. They say that their relationship is completely pure and blameless." [9] Based on the ardent outpourings of his missives, Goethe undoubtedly wished the relationship to have become less pure. She stoked his passions, but restrained by her piety, by her reputation, and by thoughts of her children (hardly, it seems, by thoughts of her husband), she would immediately pull back when he came too close. He deeply valued her not only as a sexual being but even more specially as a friend; and he certainly would not have wished to destroy that friendship by a reckless attempt at seduction. One can believe that his physical desire for her, though fluctuating, never completely died, until, that is, his journey to Italy.

If the pitch of their relationship, that love in all its complexity, should be measured by the poetry she inspired, then one must judge their love the most significant of his life. With her in mind, he penned a myriad of verses in that apparently effortless way that made Schiller think of him as a sheer wonder of nature. In a letter that accompanied one of his most evocative poems, Goethe made the connection between this particular woman, nature, and the poetic expression of both: "That I am always envisioning the phenomena of nature and my love for you, you will see from the enclosed." [10] In the poem *An den Mond* (To the moon), which he tucked into his letter, the poet compares the soft, haunting light of the moon to his lover's watchful eye, which holds his fiery heart like a ghost is held on a river bank.

> Your gaze softly spreads
> Over my field
> Like the eye of my beloved
> Gently watching over my fate.
> You know it to be so voluble,
> This heart on fire,
> You hold it like a ghost
> Bound by the river. [11]

9. Schiller to Körner, 12–13 August 1787, in *Schillers Werke* (note 3), 24:131.

10. Goethe to von Stein, 11 August 1777, *Goethes Briefe* (note 8), 1:559.

11. Goethe, "An den Mond," in *Sämtliche Werke* (note 1), 2.1:34. The first "you" is in the familiar form *(du)*, the second "you" *(ihr)* is plural, and seems to mean both the moon and the lover. Fritz von Stein, Charlotte's son and a virtual ward of Goethe, later suggested that

In these remarkable lines, the poet's burning heart and the ghost—perhaps it is a lonely spirit that arises from the earth—melt into one, as does the moon's light and the lover's look. For Goethe, woman and nature flow into a single being.

By the summer of 1786, Goethe had become exhausted and frustrated. His court and administrative duties occupied enormous amounts of time. His love for Charlotte von Stein had settled into a comfortable companionship, though without prospect of his passion's being requited; he undoubtedly feared his virginal state might become permanent. His literary life was nigh on virginal as well, with his many compositional efforts during this Weimar period lying scattered into various unfinished piles. Perhaps as a means to convince himself that his talent had not completely withered, he had contracted, in summer 1786, to have a small edition of his collected works brought out, including drafts and occasional pieces. Much later he told Eckermann that "in the first ten years of my Weimar ministerial and court life, I accomplished virtually nothing."[12] His scientific and philosophic excitement flickered and threatened also to die away. And he was getting older, approaching his thirty-seventh birthday. There was, Goethe believed, but one way out. He had cultivated in his heart since childhood images of a land his father once visited, where disappointment and Teutonic gloom might vanish under a sun-drenched sky; it was a land becoming enshrined in German consciousness as a sensuous retreat where the inhibitions of the north could be cast away as easily as a heavy woolen cloak on a warm summer's day. In Goethe's unfinished *Wilhelm Meisters Sendung* (Wilhelm Meister's mission), on which he labored at this time, the enigmatic and androgynous Mignon, an Italian adolescent of mysterious origin, sings a beautiful song of longing:

Do you know the land where the lemon trees flower,
Where in verdant groves the golden oranges tower?
There a softer breeze from the deep blue heaven blows,
The myrtle still and the lovely bay in repose.
Do you know it?

---

the ghost of the poem referred to a young woman who cast herself into the river Ilm over an unrequited love.

12. Goethe to Eckermann, 3 May 1827, in Johann Peter Eckermann, *Gespräche mit Goethe in den letzten Jahren seines Lebens,* 3rd ed. (Berlin: Aufbau-Verlag, 1987), 539. He added "despair drove me to Italy."

There, there
Would I go with you, O my master fair.[13]

Goethe formed secret plans to escape to that land where the lemon trees flowered—to Italy. On 24 July 1786, he set out for Carlsbad, the social and therapeutic resort to which the court—including von Stein and Herder—had already decamped. He remained there for the celebration of his birthday on 28 August. But then at three o'clock on the morning of 3 September, after posting notes to the duke and his immediate friends, he slipped away from the company and boarded a coach taking him to that southern country, where he hoped for a rebirth.

Goethe's Italian Journey: Love, Art, and Morphology

The rebirth that Goethe sought in Italy would begin with his own artistic conception of the journey: he traveled with pen in hand, recording his impressions, thoughts, and reactions in diaries and letters that he prepared for von Stein and his circle of friends in Weimar. He had planned to use this written material to compose a volume, which materialized, however, only during his later years.[14] The *Italienische Reise* recreates the period that gave his life a new beginning and a new meaning. This rebirth and the artistry of its telling are signaled by the motto Goethe chose for the book: "Auch ich in Arcadien"—"I, too, am in Arcadia." He appears to have taken this resonant epigram from a painting by Guercino (Giovanni Francesco Barbieri, 1591–66), whose works he viewed in Cento (17 October 1786). The painting portrays two shepherds in an Arcadian setting. They are gazing at a skull atop a rock slab, perhaps a tomb, which is inscribed with the legend "Et in Arcadia ego"—death, too, can be found in paradise.[15] In no

13. Goethe, *Wilhelm Meisters Sendung,* in *Sämtliche Werke* (note 1), 2.2:170. In the published version, *Wilhelm Meisters Lehrjahre,* "master" *(Gebieter)* becomes "lover" *(Geliebter).*

14. The first two parts of Goethe's *Italienische Reise* came out in 1816 and 1817, and the less redacted third part not until 1829.

15. In a famous essay on the epigram, Erwin Panofsky maintained that Goethe did not know the provenance of the phrase and that in his use of the motto "the idea of death has been entirely eliminated." See Erwin Panofsky, *"Et in Arcadia ego:* Poussin and the Elegiac Tradition," in his *Meaning in the Visual Arts* (Chicago: University of Chicago Press, 1982), 319. I believe Panofsky to be wrong on both counts. Goethe would use a similar epigram for the third installment of his autobiography, *Campagne in Frankreich,* the story of his part in the German attempt to quash the French Revolution. The motto of that latter work—"Auch ich in Campagne"—suggests he was there amidst a great deal of death. The epigram stood on the first two installments of the *Italienische Reise* (1816, 1817), and conveys, perhaps, not only the

less dramatic terms, Goethe would portray his entrance into the Arcadia of Italy, where he would experience a death, that of his former self, and a rebirth. At the beginning of his trip, he wrote von Stein that

> the rebirth that is transforming me from the inside to the outside continues apace. I thought I would, indeed, learn something here, but that I would have to return to primary school, that I would have to unlearn so much—well, I didn't count on that.[16]

And on the denouement of the journey, he explained to Carl August the rationale for this reeducation and its scope:

> The chief reason for my journey was: to heal myself from the physical-moral illness from which I suffered in Germany and which made me useless; and, as well, so that I might still the burning thirst I had for true art. . . . When I first arrived in Rome, I soon realized that I really understood nothing of art and that I had admired and enjoyed only the pedestrian view of nature in works of art. Here, however, another nature, a wider field of art arose before me.[17]

He might have added, as he did in other letters to his patron, that a wider field of the female also opened before him. The *Italienische Reise* and other material from the time speak insistently about the major features of his rebirth and, indeed, the great themes of his life: art, nature, and women. What must be argued, though, is the intimately affective and causally transforming relations instantiating these features. The changes, the *Bildung* he experienced, though retrospectively poured by Goethe into the two years of his journey, spilled over into the next decade and a half of his life, when his new self developed in the cultural milieu of his friendship with Schiller and his interactions with the Romantic group at Jena. But the changes do have their source in his Italian journey.

The Roman Community   The initial goal and stabilizing anchor of Goethe's travels in Italy was Rome, the city about which his father, who had himself visited there as a young man, had spun so many magical tales.

---

idea of Goethe's death and rebirth but also the wistful recognition that a long-departed self had once visited paradise.

16. Goethe to von Stein, 20 December 1786, in *Goethes Briefe* (note 8), 2:33.

17. Goethe to Carl August, 25 January 1788, in ibid. (note 8), 78.

Shortly after reaching his destination, on 1 November 1786, Goethe made the acquaintance of the German community of artists, with whom he quickly dropped his protective pseudonym of Herr Möller. The community, though initially in awe of the great poet in their midst, adopted him as a friend and provided an intimate circle he always remembered with great affection. The company included the art critic Johann Heinrich Meyer, a pupil of Winckelmann and someone with whom Goethe would have a lasting relationship; the artist Angelica Kauffmann, who had worked with Joshua Reynolds and who developed a fancy for Goethe, later painting his portrait (fig. 5.2); and Johann Friedrich Reiffenstein, another pupil of Winckelmann and by reason of age and wisdom the padrone of the group. The two members of the community that most influenced Goethe were the painter Johann Heinrich Wilhelm Tischbein, who would leave memorable portraits of his new friend, and the writer Carl Philipp Moritz, whose psychological novel *Anton Reiser* evoked a powerful response in Goethe, as it continues to do in readers today. Tischbein would be Goethe's roommate and guide in Italy, and the one who most educated his eye for the practical aspects of artistic technique. The painter's own considerable abilities seem to have eventually convinced Goethe, by painful comparison, of his own lack of real talent in the medium. Shortly after Goethe met Moritz, this melancholic fellow suffered a badly broken arm. With great solicitude, Goethe visited his new friend, sometimes twice daily, during the six weeks of the convalescence.[18] They became constant companions during the poet's time in Italy, and Goethe came to so respect Moritz's critical sense that he included an essay on aesthetics by his friend in the *Italienische Reise*.

"Without love, . . . Rome would not be Rome" (Römische Elegien I)　　As Goethe entered the city of Rome, he expected the very stones to cry out the genius of their past. But he also listened most attentively for the whisper of a beckoning female voice. Charlotte von Stein had not responded physically to his overtures, and so he abandoned the northern chill for a warmer climate that promised more. Hardly had he stepped onto Italian soil before he began to be on the watch. In Vincenza he noticed some "quite pretty creatures, especially the sort with black curly hair," and in

---

18. Carl Phillip Moritz to J. H. Campe, 20 January 1787, in *Goethe in vertraulichen Briefen seiner Zeitgenossen,* eds. Wilhelm Bode and Regine Otto, 3 vols., 2nd ed. (Berlin: Aufbau-Verlag, 1982), 1:326.

FIGURE 5.2. Johann Wolfgang von Goethe. Portrait by Angelika Kauffmann (1787). Goethe remarked of this painting, "He is indeed a handsome fellow, but no hint of me."

Venice "some beautiful faces and figures."[19] During his second stay in Rome, he fell for a striking Milanese girl, whom he tutored in English; but he underwent a Werther-like moment when he learned she was already betrothed.[20] In Naples he would become entranced by Emma Lyon, the mistress of the English ambassador Sir William Hamilton.[21] Hamilton was fifty-seven at the time, and she was twenty-two, "very beautiful and well built." She entertained Hamilton's guests by striking a series of "attitudes"—appearing draped in a shawl or in Greek costume, now in this

19. Goethe, *Italienische Reise*, 22 and 29 September 1786, in *Sämtliche Werke* (note 1), 15:67, 80.

20. Ibid., October 1787, in *Sämtliche Werke* (note 1), 506–11.

21. Hamilton had excavated statuary dedicated to Priapus in Isneria, about fifty miles outside of Naples. His friend Richard Payne Knight described this collection of ancient erotica in his *Account of the Remains of the Worship of Priapus* (London: Spilsbury, 1786). Goethe was familiar with Knight's work and likely examined the collection itself when he viewed Hamilton's "secret art treasury." See Goethe, *Italienische Reise*, 27 May 1787, in *Sämtliche Werke* (note 1), 15:400. Goethe's poem cycle *Römische Elegien* (see below) originally had as introduction and envoi two poems in which the garden god Priapus brandishes his tool; Goethe himself suppressed their publication.

pose, now in that. Her performances seemed to Goethe like an ancient dream, one that Tischbein attempted to capture on canvas.[22] From Goethe's descriptions one can well understand how this woman could entice Hamilton, who loved pretty things, into marrying her and then later seduce Lord Horatio Nelson into a scandalous affair.

And, of course, there were the artists' models. They were generally "lovely and happy to be seen and to be enjoyed." Yet, as Goethe wrote Carl August, "the French disease made this paradise uncertain."[23] By the middle of his second Roman stay, he had grown quite disconsolate. He wrote the duke another letter, lamenting the condition in which Priapus had left him:

> The sweet small god has relegated me to a difficult corner of the world. The public girls of pleasure are unsafe, as everywhere. The zitellen [the unmarried women] are more chaste than anywhere—they won't let themselves be touched and ask immediately, if one does something of that sort with them: e che concluderemo [and what is the understanding]? Then either one must marry them or have them married, and when they get a man, then the mass is sung. Indeed, one can almost say that all the married women stand available for the one who will take care of their families. These, then, are the lousy conditions, and one can sample only those who are as unsafe as the public creatures. What concerns the heart does not belong to the terminology of the present chancellery of love.[24]

Goethe no doubt showed proper caution, even against the swelling tide of powerful desire. This was not the sort of caution that Carl August himself practiced, however, and he strongly recommended Goethe follow suit. Remarkably, Goethe appears to have taken the advice to heart, or at least fortuitous occasion accomplished what friendly council had prescribed. On 16 February 1788, he wrote Carl August:

> Your good advice, transmitted 22 January directly to Rome, seems to have worked, for I can already mention several delightful excursions. It is certain that you, as a doctor *longe experientissimus,* are perfectly correct, that an appropriate movement of this kind refreshes the mind and provides a wonderful equilibrium for the body. I have had this experience more than once

22. Goethe, *Italienische Reise,* 16 March 1787, in *Sämtliche Werke* (note 1), 15:258.
23. Goethe to Carl August, 3 February 1787, in *Goethes Briefe* (note 8), 2:49.
24. Goethe to Carl August, 29 December 1787, in ibid. (note 8), 2:5.

in my life, and also have felt discomfort when I have deviated from the broad road to the narrow path of chastity and safety.[25]

The adolescent boast of the last sentence undoubtedly suggested to the duke, as it does to us, that Goethe only recently acquired more than theoretical knowledge of that delightful motion.

One of the greater mysteries of Goethean scholarship has been the identity of his Italian mistress, assuming, of course, that his representation to Carl August had substance. In the *Römische Elegien*,[26] the poem cycle that he began either during his last weeks in Italy or shortly after his return to Weimar, the poet sings of his affair with a young Italian woman named Faustine:

Then one day she appeared to me, a tawny-skinned girl [bräunliches Mädchen]
Whose dark, luxuriant hair tumbled down from her brow.[27]

Blessedly, she knows nothing of Lotte and Werther. They nightly honor the god Amor "devilishly, vigorously, and seriously."[28] The object of this affection has not been identified with any certainty, though, from the evidence of the poems, she seems to have been a young widow of modest circumstance with a small child, an individual safe enough from the French disease.[29] Yet the persona of the poems seems to mask a second woman as well, Christiane Vulpius, whom Goethe met back in Weimar shortly after his return and who quickly became his mistress and later officially his wife.

25. Goethe to Carl August, 16 February 1788, in *Briefwechsel des Herzogs-Grossherzogs Carl August mit Goethe,* ed. Has Wahl, 2 vols. (Berlin: Ernst Siefgried Mittler & Sohn, 1915), 2:117.

26. The *Römische Elegien* were originally published as a twenty-poem song cycle in Schiller's *Horen* in 1795, though completed, it seems, by 1790. Goethe had initially composed a cycle of some twenty-four poems, bearing a draft title of *Erotica Romana.* Four of the original poems were suppressed, two of the Priapus poems by Goethe himself and two others at Schiller's urging because of their highly erotic character. The elegies and their variations are printed in Goethe, *Sämtliche Werke* (note 1), 3.2:8–82. I return to the poems below.

27. Goethe, *Römische Elegien,* no. 4, in *Sämtliche Werke* (note 1), 3.2:45.

28. Ibid.

29. Nicholas Boyle tentatively identified her as Faustina Antonini, an innkeeper's daughter. See N. Boyle, *Goethe, the Poet and the Age,* 2 vols. to date (Cambridge: Cambridge University Press, 1991–), 1:506. This identification has been disputed by Roberto Zapperi in his *Das Inkognito: Goethes ganz andere Existenz in Rom,* trans. Ingeborg Walter (Munich: Beck, 1999), 206. After considerable detective work (201–38), however, Zapperi is in no doubt of the real existence of Goethe's Roman affair.

During the period of his reworking of the poems, he rejoiced in love and sexual pleasure with this new mistress, and her scent lingers over their final composition. In this way, Goethe's art transformed the artist: he became in Weimar the lover the poems describe; and to further this metamorphosis, he refused to staunch speculation about the authenticity of the events depicted in the poems. The cycle yet celebrates Rome and the experiences of the poet in that city—and it is, to that extent, biographically revealing, even if the original experiences have been transmuted through a resexualized imagination.[30] But even more importantly, these poems also express Goethe's deeply felt convictions about the relationship of the female ideal to art and science—a complex matter that I discuss in greater detail below.

The Artist in Rome   Goethe traveled to Rome with a satchel full of manuscripts, undoubtedly expecting the new milieu to liberate an inspiration stunted in colder climes. He also had the task of making these manuscripts publishable for the edition of his collected works. Before leaving Weimar, he had a decent draft of a prose version of his play *Iphigenie auf Tauris;* but Italy warmed him to the more classical format of a verse rendition, which he completed in the middle of December 1786. The next year he finished another play previously set aside, his *Egmont,* a tale that celebrates moments of happiness even in the darkening shadow of mortality. Italian musical heritage, as well as popular songs heard on every street corner, provided the atmosphere, if not the needed genius, for rewriting two operettas he had earlier composed, *Erwin und Elmira* and *Claudina von Villa Bella.* He worked on his early draft of *Faust,* and completed a few new scenes that gave the drama a different orientation; but his vision did not quite embrace the whole, and *Faust* remained a fragment. During the last months of his sojourn, however, he all but finished the play *Torquato Tasso.* Italy obviously sharpened a pen that had become dull from scratching out official documents and frustrated love letters.

When he first went to Italy, Goethe brought along his sketchpads, and always attempted to quicken his eye to the art that could be found in museums and even beneath his feet in that ancient land. Tischbein and two other artist friends, Johann Heinrich Meyer and Christoph Heinrich

---

30. While seeking biographical clues in works of art is risky, there is no reason not to make prudential judgments. As R. G. Collingwood insisted, the epistemic value of any document—whether letter, book, or recorded conversation—must be judged on the basis of evidence. Imaginative literature should be no different. Poetry, especially Goethe's, can be as revealing, sometimes more revealing, than, say, polite letters that disguise deeper sentiments.

FIGURE 5.3. Anatomy sketch by
Goethe, done in Rome.

Kniep, provided constant instruction not only in art history but in the
practice of sketching and painting (fig. 5.3). Toward the end of his journey, though, Goethe reluctantly came to the conclusion that he had only
a modest talent for painting. He simply was not able to accomplish what
Tischbein and Kniep could in a few deft strokes. Yet his efforts at acquiring artistic skill revealed to him how such practical experience proved a
necessary condition for the intuition of ideal structures that might lie more
deeply below surface appearances.

Goethe studied the paintings and statuary of the ancients with Winckelmann's *Geschichte der Kunst des Altertums* (History of ancient art, 1764) as
his vade mecum. Winckelmann disdained those critics who constructed
their histories and judgments, not from the meticulous examination of
the works themselves, but from journals and books. This insistence on the
immediate experience of objects appealed to Goethe, though Winckelmann's considerations of what constituted ideal beauty and how it should
be realized—for example, ideal proportions of different bodily parts, representations of dress, and so on—left Goethe rather confused and uncertain.[31] As a result he became initially convinced of a sharp boundary be-

31. See Johann Joachim Winckelmann, *Geschichte der Kunst des Altertums* (1764; Darmstadt:
Wissenschaftliche Buchgesellschaft, 1972), 168–88.

tween the realm of art and the realm of nature. Just after receiving a new translation of Winckelmann's book in early December, he wrote Duchess Louise (wife of Carl August) to explain his abrupt departure from Weimar. In the course of his account of the artistic riches of Rome, he observed

> that nature is more congenial and easier to view and evaluate than art. The lowliest product of nature has a sphere of perfection within itself and I need only have eyes to see in order to discover these relationships; I am convinced that a whole, true existence is enclosed within this small circle. A work of art, on the other hand, has its perfection outside of itself. The "best" of it is in the idea of the artist, which he seldom or never achieves, and follows from certain assumed laws, which are derived from the character of art and of craft. But these latter are not so easy to understand and decipher as the laws of living nature. In works of art, there is tradition; the works of nature, however, are always like words immediately spoken by God.[32]

Winckelmann had insisted that artistic genius harbored an ineffable ideal of beauty, which could never be given adequate expression. A given work of art might only dimly reflect that beauty, refracting it through historically determined styles of realization. Greek art—sculpture in particular—did, however, achieve the most luminous instantiations of ideal beauty. And this is why ancient art offered models for the cultivation of taste and aesthetic judgment. The great critic contended that only after coming to appreciate ideal beauty as exemplified in the work of the Greeks could one begin aesthetically to judge nature according to the proper standard.[33] Goethe initially endorsed these Winckelmannian notions, though his letter to Louise suggests they were not his spontaneous convictions. He continued to immerse himself in the art of the ancients, attempting to discover there the ideals of beauty that his guide so evocatively described. And as he confronted the art of different centuries—from the ancient period to the Renaissance and modern times—he came quickly to appreciate what Winckelmann taught about the historical character of artistic expression. "No judgment in this field is possible," Goethe wrote in January 1787, "unless one has made it in light of historical development."[34]

---

32. Goethe to Duchess Louise, 12–23 December 1786, in *Goethes Briefe* (note 8), 2:31.

33. Johann Joachim Winckelmann, *Gedanken zur Nachahmung der Griechischen Werke in der Mahlerey und Bildhauerkunst,* 2nd ed. (1756; Stuttgart: Reclam, 1995), 15.

34. Goethe, *Italienische Reise,* 28 January 1787, in *Sämtliche Werke* (note 1), 15:200.

Thus the relativity of artistic styles had to be distinguished from the essentiality of nature, whose eternal structures could be read off the surface of their appearances—at least this was Goethe's opinion at the beginning of his journey. Only after he had reflected more deeply on the character of nature, especially as the result of his experiences in Sicilian parks and Roman arms, would his notions of nature and art undergo yet further development. By the time he returned to Weimar, Goethe would come to see as well the essential in art and the historical in nature, both as confirmed by an Italian girl.

The Morphological Conception of Nature and Art   On Ash Wednesday, 21 February 1787, just after the madness of the Roman Carnival had subsided, Goethe wrote Frau von Stein one of his typically hesitating letters, professing both love ("all the fibers of my being depend upon you") and enervation ("today I am confused and rather weak").[35] He was getting ready to depart for Naples and Sicily, and his preparations reminded him of what he had left in Germany. He was drawn irresistibly south, where grew the lemon trees whose remembered fragrance had produced great longing in Mignon. Those lands, as well, promised adventure—he had learned that ancient Vesuvius had begun spewing fire.

He arrived in Naples on 25 February, and a week later set out to climb the fabled volcano—as his great admirer Ernst Haeckel, in conscious imitation, would do three-quarters of a century later. In his first attempt to reach the new crater, the sulfurous fumes drove him back. He tried again four days afterward, this time with a reluctant Tischbein in tow. A small eruption showered them with ashes and rocks, and one large stone came close enough to produce that frisson of imminent death, which Goethe on this and other occasions found so exhilarating. He would return once more two weeks later to watch the lava flow from the new site.[36]

The dramatic experience of wildly beautiful nature subtly suggested to Goethe the poverty of the merely artistic imagination, Winckelmann's assertions notwithstanding. The calmer scenes of beauty made no less a striking impact. A few days after his last experience with Vesuvius, he was returning to Naples in a small coach driven by a young lad when they came upon an elevated view of the city, in all its magnificence, with the harbor in the distance. The boy let out a yell, which startled Goethe, who turned

35. Goethe to von Stein, 21 February 1787, in *Goethes Briefe* (note 8), 2:50–51.
36. See Goethe, *Italienische Reise*, 2, 6, 20 March 1787, in *Sämtliche Werke* (note 1), 15:229–30, 235–40, 265–67.

in irritation to the rough Neapolitan youngster. But the boy, pointing his finger toward the city, exclaimed, "Sir, forgive me. This is my native land!" Goethe confessed, "tears came to the eyes of this poor northerner."[37]

Goethe and his new friend Kniep left Naples by boat for Palermo on 29 March 1787. The sailing took five days, for most of which Goethe lay seasick and delirious in his cabin.[38] In the early morning of 1 April, a gale tossed the ship about, creating not a little fear in this man of German soil. However, at dawn the sky cleared, and at noon the coast of Sicily came into view. They landed in Palermo on 2 April, and would remain in the city for two weeks, with ventures into the countryside. Just before leaving the town, a set of events occurred that gave impetus to Goethe's developing understanding of the relations among art, nature, and the female.

On 16 April 1787, Goethe went to the public gardens in the city to relax with a copy of the *Odyssey*, which he was reading as preparation for the composition of a play tentatively titled *Nausikaa*. In Homer's epic, Odysseus, naked and filthy, washes up on the shore of an island and immediately falls asleep. He is awakened by the play of a most beautiful girl, Nausikaa, daughter of King Alcinous. Once cleaned and properly dressed, the now magnificent Odysseus awakens keen desire in Nausikaa. She takes him to the garden of Alcinous, while she goes forth to prepare the court, where he will sing of his many adventures.[39] Goethe, who regarded the garden of Palermo more lovely than the imagined garden of Alcinous, returned the next day to continue his meditation on his proposed play, which ultimately never got beyond a few lines. But as he sat down to ruminate, as he recalled, "another spirit seized me, which had already been tailing me during these last few days." He gazed around the garden, and inquired of himself:

> Whether I might not find the *Urpflanze* within this mass of plants? Something like that must exist! How else would I recognize that this structure or that was a plant, if they were not all formed according to a model?[40]

Goethe thus moved in imagination from contemplating the lovely Nausikaa and the glorious Odysseus in the garden of Alcinous to the striking

37. Ibid., 23 March 1787, in *Sämtliche Werke* (note 1), 15:275.

38. A story later told to Zacharias Werner by Kniep, from Werner's *Tagebuch* (7 May 1810), in *Goethes Gespräche,* ed. Flodoard von Biedermann, enlarged by Wolfgang Herwig, 5 vols. (Munich: Deutscher Taschenbuch Verlag, 1998), 1:421.

39. Homer, *Odyssey* 6.

40. Goethe, *Italienische Reise,* 17 April 1787, in *Sämtliche Werke* (note 1), 15:327.

notion that in the real garden of Palermo he might discover a comparably beautiful and magnificent form.

Though Goethe believed he might actually find the *Urpflanze* in Sicily, his conception of the entity remained labile, as indicated by a hypothesis he formed at this time. Regarding the organic parts constituting the *Urpflanze,* he supposed that "all is leaf, and through this simplicity the greatest multiplicity is possible." The leaf, he thus conjectured, might be transformable into all the other parts. In this sense, "a leaf that only sucks fluid under the earth we call the root; a leaf that spreads out from those fluids, we call a bulb, an onion for instance; a leaf that stretches out, we call the stem" and so on for all the other parts of the plant.[41] But the leaf, as Goethe here mused, must be understood symbolically: it represented a unitary dynamic force beneath the multiple transformations to which it gave rise—indeed, a short time later he would refer to "the leaf in its most transcendental sense."[42] Just so, the *Urpflanze* itself might be conceived as the dynamic, unifying idea that lay behind the great variety of living plants. Much later, when he published the *Italienische Reise* in 1816, he wrote Christian Nees von Esenbeck that his friend would be amused at the diary of his journey because "I sought at that time the *Urpflanze,* unaware that I sought the idea, the concept whereby we could develop it for ourselves."[43] But that awareness had already reached the penumbra of Goethe's consciousness, as his remarks on the leaf suggest. Spinoza had already set him on the path that would lead to Kant and then to the Romantics.

A theme of this essay has been the deep emotional and aesthetic connection between Goethe's experience of female forms—in literature and life—and his ideal biological structures. The eternal feminine and the eternal plant were for Goethe both ideals of beauty and models for the comprehension of their many empirical instantiations—illustrated in the former instance by the many women Goethe conjured up in his autobiographical writings and sang of in his poetry. Another tale from the *Italienische Reise* strongly evinces this conjunction. On 25 March 1787, just

41. These lines come from brief notes Goethe made just after his time in the gardens of Palermo. See Johann Wolfgang von Goethe, *Zur Morphologie: von den Anfängen bis 1795, Ergänzungen und Erläuterungen, Goethe Die Schriften zur Naturwissenschaft,* ed. Dorothea Kuhn, 21 vols. (Weimar: Harmann Böhlaus Nachfolger, 1977), 2.9a:58.

42. The phrase comes from unpublished notes that stem from late 1788. See Goethe, "Einleitung," in *Sämtliche Werke* (note 1), 3.2:306.

43. Goethe to Christian Nees von Esenbeck (middle of August 1816), in *Goethes Werke* (Weimar Ausgabe), 142 vols. (Weimar: Böhlau, 1887–1919), IV, 27:144.

a few days before Goethe left for Palermo, Kniep invited him to the roof of his apartment to enjoy the magnificent view of the bay of Naples and the coast of Sorrento. As they were engrossed in the scene, up from the trap door leading to the roof a beautiful head suddenly emerged, that of Kniep's mistress. To Goethe she seemed like the angel of the Annunciation: "But this angel really had a beautiful figure, a pretty little face, and a fine natural comportment. I was delighted to see my new friend so happy under this wonderful sky and in view of the most beautiful region of the world."[44] Immediately after he narrated this story, Goethe went on to say that he took a walk along that striking seacoast, where he had a flash of insight concerning his botanical ideas. The passage reads: "I have come to terms with the *Urpflanze;* only I fear that no one will wish to recognize the rest of the plant world therein."[45] Beautiful nature, a beautiful woman, and the primal plant.

I have drawn these conjunctions out of Goethe's *Italienische Reise,* a book he composed of letters and diaries over a quarter of a century after the events written about took place. When he cast a melancholic gaze back over this time in Italy, he recalled it as that period in his life when he discovered "what it really was to be a man"; and by comparison to his existence in Rome, he judged that he had "never again been happy."[46] Whether the connections that I have followed, then, were the actual fusions of ideas and motivational associations made during the period or whether they were imposed later by a reminiscent imagination is, of course, impossible to know, since few of the original letters and diaries have survived. Yet, other contemporary sources are available, namely, new scenes for *Faust* and his poem cycle *Römische Elegien.* He likely began both of these during his last weeks in Italy and further worked on them shortly after his return to Germany, while memory and longing were still green.

"I have softly beat out the measure of hexameters, fingering along her spine"
Fifteen years prior to his Italian journey, Goethe had composed a draft of *Faust,* which he had only desultorily reexamined in the intervening years. But during his time in Italy, he was determined to finish it, since he wanted to include it in the Göschen edition of his collected works. He added several scenes, but was simply unable to bring the work to completion, at least to his own satisfaction. It would be published in the collected works as

44. Goethe, *Italienische Reise,* 25 March 1787, in *Sämtliche Werke* (note 1), 15:276.
45. Ibid., 25 March 1787, in *Sämtliche Werke* (note 1), 15:277.
46. Goethe to Eckermann, 9 October 1828, in *Gespräche mit Goethe* (note 12), 248–49.

Faust, a Fragment. The most important of the scenes added during his Italian sojourn speaks to the theme of this essay. In "A Witch's Kitchen," Mephistopheles brings Faust to a witch who will concoct a magic elixir to make him young again.[47] While in the kitchen, Faust gazes into an enchanted mirror and sees

> The image of a woman so very fair!
> Can a woman be that exquisite?
> Should I see in this body lying there
> The essence of a heavenly visit?
> Is there anything on earth to compare?

After Faust quaffs the magic potion and wishes once more to look at the enticing image, Mephistopheles exclaims:

> No! No! You shall soon see the model rise
> Of every woman bodily before your eyes.

And then adds as an aside:

> With this drink you will see in flesh and nail
> Soon lovely Helen in every female[48]

In the Italian redraft of *Faust,* the force of the story has shifted from that of the seduction of a girl—Gretchen—to the quest for an ideal of beauty and love, but one discoverable only in immediate experience. Goethe suggests here that experience must drive down to the ideal, to the active force that lies in the depths of reality, a force that must be comprehended in order to construe its empirical appearances. Helen—the very form of beauty itself—is present in every female, something that Faust must comprehend, must come to see, even if too late in the case of Gretchen. The fundamental model, the force, that furnishes the rationale for Faust's striving, is that of beauty, of Helen, of the very form of the female.

Goethe's *Römische Elegien* expresses a similar theme, though with greater lyrical intensity. Of the twenty published poems that make up the cycle,

---

47. The inserted scene remains, even in the completed play, incongruous, since at the beginning of the drama Faust is a young man and hardly requires rejuvenation—it was the forty-year-old Goethe who, rereading his old play, felt the need for the elixir that Italy provided.

48. Goethe, *Faust, ein Fragment,* in *Sämtliche Werke* (note 1), 3.1:548, 553–54.

the fifth achieves an elegance and beauty not otherwise matched. In this poem, Goethe blends images, feelings, and ideas within alternating hexameter and pentameter lines that evoke the poetry of the Roman elegists. Yet the poem is distinctively Goethe's own: it erotically synthesizes experiences of classical sculpture and of a real woman, showing them to be constituted by the same eternal form:

> Happy, I find myself inspired in this classical setting;
> The ancient world and the present speak so clearly and evocatively to me.
> Here I follow the advice to page through the works of the ancients,
> With busy hands and daily with renewed joy.
> Ah, but throughout the nights, Amor occupies me with other matters.
> And if I wind up only half a scholar, I am yet doubly happy.
> But do I not provide my own instruction, when I inspect the form
> Of her lovely breasts, and guide my hands down her thighs?
> Then I understand the marble aright for the first time: I think and compare,
> And see with feeling eye, and feel with seeing hand.
> Though my beloved steals from me a few hours of the day,
> She grants me in recompense hours of the night.
> We don't spend all the time kissing, but have intelligent conversation;
> When sleep overcomes her, I lie by her side and think over many things.
> Often I have composed poetry while in her arms, and have softly beat out
> The measure of hexameters, fingering along her spine.
> In her lovely slumber, she breathes out, and I inspire
> Her warm breath, which penetrates deep into my heart.
> Amor trims the lamp and remembers the time
> When he performed the same service for his three poets.[49]

Even through inadequate translation, perhaps some impression of the whole has yet been preserved. A conceptual rendition of the poem further strips it of its aesthetic meaning. While recognizing that crucial liability, let me attempt one reading that coheres with much else that I have concluded thus far about Goethe's aesthetics and metaphysics. In the first part of the poem, the poet claims to understand great, classical art only when he can embrace its living embodiment. The white, hard marble of the statue speaks to him only after he has experienced the brown, pliant flesh of the girl ("I inspect the form / Of her lovely breasts, and guide my hands down

---

49. Goethe, *Römische Elegien*, in *Sämtliche Werke* (note 1), 3.2:47. The poets referred to in the last line are the Roman elegists Propertius, Catullus, and Tibullus.

FIGURE 5.4. Christiane Vulpius, with a copy of Tischbein's *Goethe in the Campagna* over the table. Sketch by J. H. Lips (1791).

her thighs"). As a result, the visual aspects of each have taken on a new depth of tactility and, reciprocally, haptic awareness of their form has been imbued with visual qualities ("see with feeling eye, and feel with seeing hand")—thus he might "page through the works of the ancients" with eyes manually instructed. The first part of the poem, then, suggests that the sculptured marble and the living girl embody the same *Urform* of the female, a form necessary for both the actual creation of each—by the artist and by nature—and for the aesthetic comprehension of each. The second part of the poem relates how this now-understood dynamical form transmutes into another artistic instantiation, this time in poetry. In an unforgettable image, the poet lies in the arms of his lover, inspiring her living spirit and beating out the hexameters of a poem he is composing—the very poem we are reading?—by counting along her vertebrae. The poet is actually following the natural form of his lover in order to impress that same form on his words. The erotic power of nature thus transmutes into an artistic force that realizes beauty in another medium. Touch and vision, reason and sense combine in love to produce an aesthetic and intuitive understanding of the unity grounding nature and art. Well, there are obvious dangers in conceptually disassembling what exists only in poetic fusion.

Journey's End

The conclusion of the Italian journey really marked a new beginning for Goethe as he returned to Weimar. This new beginning had several interrelated components. With Winckelmann, he maintained that the artist worked with an ineffable conception of beauty, whose expression would be conditioned by historical circumstance. Classical artists—whether Phidias or Homer—yet achieved a realization of the archetype of beauty in a more objective way than modern artists. Upon returning from Sicily (17 May 1787), Goethe made this point in a letter to Herder: "Let me express my thoughts briefly. They [ancient writers] represent real existence, we usually represent its effects. They portray what is frightful, we portray frightfully; they the pleasant, we pleasantly, and so on. As a result everything we produce is overdone, mannered, without real grace, a mess."[50] Goethe then immediately confessed to his friend that he perhaps did not initially appreciate what Homer had wrought; but now, on his return from Sicily—his head reeling with images drawn from immediate experience of rocky coasts and sand-strewn bays—"the Odyssey for the first time has become for me a living word."[51] Just so the marbles of Polycleitos and Myron required the real experience of the lover's caress to appreciate what they had achieved. Only under the erotic authority of the living female could the forms buried in the marble come alive.

In the *Italienische Reise*, Goethe associated his remarks to Herder about the ancients' objectifying ability with a passage from a letter to von Stein and the Weimar group. The letter, actually written two weeks after the one to Herder, related what he had learned about the *Urpflanze*:

> Tell Herder that I am very near to the secret of the generation of plants and their organization. Under these skies, one can make the most beautiful observations. Tell him that I have very clearly and doubtlessly uncovered the principal point where the kernel [Keim] is located, and that I am in sight, on the whole, of everything else and that only a few points must yet be determined. The *Urpflanze* will be the most wonderful creation on the earth; nature herself will envy me. With this model and its key, one can, as a consequence, discover an infinity of plants—that is, even those that do not yet exist, because they could exist. It will not be some sketchy or fictive shadow or appearance, but will have an inner truth and necessity. The same law [Gesetz] will be applicable to all other living things.[52]

In his *Italienische Reise*, Goethe juxtaposed his observations about the objective idea used by superior artists with this passage, in which he de-

50. Goethe, *Italienische Reise*, 17 May 1787, in *Sämtliche Werke* (note 1), 15:393.
51. Ibid.
52. Goethe to von Stein, 8 June 1787, in *Goethes Briefe* (note 8), 2:60; also partly in *Italienische Reise*, in *Sämtliche Werke* (note 1), 15:394.

scribed the objective idea—or law—used by nature. The connection between the artist's idea and nature's idea was, via this conjunction, only implicit. But in another letter to von Stein and his Weimar circle, written not long after the one quoted above, he made the connection between the artist and nature quit explicit: "These great works of art are comparable to the great works of nature; they are created by men according to true and natural laws. Everything arbitrary, imaginary collapses. Here is necessity, here is God."[53] The notion that the artist operates according to the same laws as nature received a comparable expression in an essay that Goethe jotted down in his travel diary at this time (and published shortly after his return). In *Einfache Nachahmung der Natur, Manier, Styl* (Simple imitation of nature, manner, style, 1789), he distinguished artists of modest ability, who would imitate nature very precisely in their simple compositions, from those of greater talent, who would discover within themselves a language, a manner, by which to express more complex subjects. But the truly great artist would dialectically incorporate both of these stages and move beyond them. He or she would study the varying phenomenal aspects of natural objects, penetrating to their essential features, and thus be able to express in an artistic medium what nature expressed in her phenomenal medium.[54] Goethe's assertions in this essay reversed the priority that Winckelmann had given to the imitation of ancient art. The great critic had proclaimed that immersion in ancient works would develop the kind of taste necessary to appreciate beauty in nature.[55] Goethe proposed that authority flowed in a different direction: namely, that perception of nature revealed a kind of beauty and reality that would illuminate the work of ancient authors. Despite this alternative emphasis, the stamp of Winckelmann on Goethe's thought is unmistakable. Goethe reached his conclusion about the artist working according to the same laws as nature through his reading of Winckelmann, but it was a reading sifted through his own experiences in Italy—and added to this mix, a tincture of Spinoza. This conviction would leaven Goethe's biological science during the rest of his life. He would find a comparable conception, somewhat differently formulated, in Kant's third *Critique,* which he would take up not long after his return to Germany. And later, under Schiller's tutelage, he would come to admit that nature, as experienced, has its humanly constructed features,

53. Goethe, *Italienische Reise,* in *Sämtliche Werke* (note 1), 15:478. The original of this letter has not survived.

54. Johann Wolfgang von Goethe, *Einfache Nachahmung der Natur, Manier, Styl,* in *Sämtliche Werke* (note 1), 3.2:186–19.

55. Winckelmann, *Gedanken zur Nachahmung der Griechischen Werke* (note 33), 15.

but also to recognize with the Romantics that there was a deep aspect of human nature that connected it with external nature, making the interchange between man and the world harmonious. He epitomized this notion in his posthumously published maxim: "There is an unknown, law-like something in the object that corresponds to an unknown, law-like something in the subject."[56]

I have argued that nature in the form of the female had a command over Goethe, an erotic authority that directed him to a deeper scientific understanding of nature writ large, as well as to a comprehension of that nature found in great works of art. In making this argument, I have done violence to the *Römische Elegien* by stripping those poems of their aesthetic beauty. Perhaps, though, W. H. Auden offered the only kind of comment suitable to the medium—another poem. In "Good-by to the Mezzogiorno," he recognized Goethe (despite their different sexual tastes) as the very emblem of the poet:

> Goethe,
> Tapping homeric hexameters
> On the shoulder-blade of a roman girl, is
> (I wish it were someone else) the figure
> Of all our stamp.

Two years after Goethe had returned to Weimar, he attempted a second journey to Italy; he felt the need to recapture that originating experience. After about six weeks, however, he admitted he could not make the past come alive again. He reluctantly headed back home in melancholy disappointment, but never thereafter ceased to recall the time of his rebirth. Auden's poem ends with a remark about Italy that might reflect Goethe's own lasting feelings about the land that so changed his life:

> Though one cannot always
> Remember exactly why one has been happy,
> There is no forgetting that one was.[57]

56. Johann Wolfgang von Goethe, *Maximen und Reflexionem,* no. 1344, in *Sämtliche Werke* (note 1), 17:942.

57. W. H. Auden, "Good-bye to the Mezzogiorno," in *Selected Poems,* ed. Edward Mendelson (New York: Vintage Books, 1989), 239–42.

6

Nature and *Bildung:* Pedagogical Naturalism in Nineteenth-Century Germany

Eckhardt Fuchs

In the beginning was Rousseau. His pedagogical anthropology, so it seems, marked the turning point to a new "natural pedagogy." His vision was called a "nature's gospel of education" (Goethe) and inspired natural education in Germany from the philanthropists during the late Enlightenment to the reform movement of "new education" at the beginning of the twentieth century.[1] At first glance, the existence of a natural pedagogy in Germany might be surprising since no idea is so closely associated with German culture as the neohumanistic concept of *Bildung* that dominated the educational discourse in Germany throughout the nineteenth century. Although the term *Bildung* bears no single definition but has different meanings depending on the respective discursive fields, it is shaped by a specific characteristic, that is, the absence of nature. In the early nineteenth century, *Bildung* replaced the Enlightenment's focus on natural education, utility, and welfare, a process that led to an exclusion of nature and science from pedagogical theory and educational instruction.[2]

However, despite the discursive and institutional hegemony of neo-humanistic philosophers, educationalists, and school officials in the nine-

1. A critique of the image of Rousseau in historical perspective can be found in Jürgen Oelkers, "Die pädagogische Erfindung des Kindes," paper delivered at the University of Konstanz, http://www.paed.unizh.ch/ap/Konstanz.rtf (30 August 2002).

2. Within the pedagogical discourse in the nineteenth century, the term *Bildsamkeit,* coined by Johann Friedrich Herbart in 1835, played an important role. Nevertheless, in this essay I focus on *Bildung* since it was the concept referred to by pedagogical naturalism. On *Bildsamkeit,* see *Lexikon der Pädagogik der Gegenwart,* vol. 1 (Freiburg: Herder, 1930), entry "Bildsamkeit."

teenth century, a subversive discourse aiming at incorporating the "natural" and the sciences into educational theory and practice never ceased to exist. I will call this discourse "pedagogical naturalism." The common theme of pedagogical naturalism was child education and school reform, but one has to keep in mind that it encompasses a variety of meanings. The entries in pedagogical encyclopedias and dictionaries reflect these differences. The most general definition can be found in the *Real-Encyclopädie des Erziehungs- und Unterrichtswesens nach katholischen Prinzipien* of 1864: "Pedagogical naturalism therefore consists in that one wants to make all education depend on nature."[3] Twenty years later we can find a more subtle definition. From then on, pedagogical naturalism was divided into two main branches: first, as education according to nature *(Naturgemäßheit)*, the principle of which is to follow (objective) nature; second, as the concept of "natural tendencies," which meant the development of the (subjective) nature of the individual.[4]

Taking up this division, in this essay I suggest dividing pedagogical naturalism into two main strands: first, there is the romantic strand with its concept of the natural as the original, naturally grown form of life and education;[5] second, the rational strand that aimed at an empirical foundation of natural education.[6] I show how the proponents of these two strands referred to nature and science in order to reform educational practice and establish a scientific pedagogy, and also how they interacted with each

3. Hermann Rolfus and Adolph Pfister, eds., *Real-Encyclopädie des Erziehungs- und Unterrichtswesens nach katholischen Prinzipien*, vol. 3 (Mainz: Florian Kupferberg, 1865), entry "Naturalismus," 398.

4. Gustav. A. Lindner, ed., *Encyklopädisches Handbuch der Erziehungskunde mit besonderer Berücksichtigung des Volkschulwesens* (Vienna: A. Pichler's Witwe & Sohn, 1884), entry "Naturalismus in der Erziehung," 541–44. See also Joseph Loos, ed., *Enzyklopädisches Handbuch der Erziehungskunde,* vol. 2 (Vienna: A. Pichler's Witwe & Sohn, 1908), entry "Naturgemäßheit des Unterrichts," 102–3.

5. Dieter Birnbacher, "'Natur' als Maßstab menschlichen Handelns," *Zeitschrift für philosophische Forschung* 45 (1991): 60–76.

6. Buyse called these two different strands "la pédagogie expériencée" and "la pédagogie expérimentale." See Raymond Buyse, *L'Expérimentation en pédagogie* (Brussels: Lamertin, 1935). For Joppich pedagogical naturalism was divided into a technical–rational and a romantic–irrational group. His terms, however, have a different meaning than mine. See Gerhard Joppich, "Die Theorie des pädagogischen Naturalismus," *Zeitschrift für Pädagogik* 2 (1956): 154–72. A multitude of different types of pedagogical naturalism—as suggested in some dictionaries—does not appear useful if one looks for the unity and commonalities of this movement. For examples, see Winfried Böhm, *Wörterbuch der Pädagogik* (Stuttgart: Alfred Kröner, 1994), entry "Naturalismus, pädagogischer," 493; Hermann Schwartz, *Pädagogisches Lexikon,* vol. 3 (Bielefeld: Velhagen & Klasing, 1939), entry "Natur und Erziehung," 854–66.

other. The philanthropic movement of the late Enlightenment, which still combined the romantic and the rational strands of pedagogical naturalism, provides my starting point. A section on the debate over naturalism in the 1840s in which pedagogical naturalism was attacked by defenders of religious education is followed by the genesis of rational pedagogical naturalism in the form of scientific pedagogy. I then continue with the romantic strand that had its highlights in the reform movement for a "new education" in the early twentieth century and a particular reform project, the Waldorf School founded by Rudolf Steiner, as an example of how nature and the natural were transformed into the mystification of the child and Nature. In my conclusion I address the relationship between pedagogical naturalism and the concept of *Bildung*.

I argue that the discourse of pedagogical naturalism was by no means uniform. Different values resulted from the specific notions of nature educators used, and these values were the basis for their educational practice. The romantic notion of nature that combined (objective) Nature with the (subjective) nature of the human being contained an immanent moral, social, and—partly—aesthetic value for the science and art of education. For them Nature had an emancipatory connotation since living, acting, and educating according to nature meant doing so according to a reasonable *(vernünftige)* nature that was the opposite of an unreasonable society. For the rational pedagogical naturalists, nature writ large did not have a value *per se* that would imply a certain authority for education. For them it was pedagogical science that was seen as the mediator between nature and pedagogy. Only scientific results about human nature bore authority since they were seen as objective and truthful and not based on the subjective experience of the educator. Although their psychological-experimental approach to education did not challenge the concept of *Bildung,* both romantic and rational naturalists were contested by the neohumanists, who did not assign any value to nature and science.[7]

7. I will not deal with the discussion about the introduction of natural sciences in the school curriculum and the debate between "humanists" and "realists" in terms of the reform of the *Gymnasium* since the midcentury that indicated this crisis and dissolution. See Werner Kutschmann, *Naturwissenschaft und Bildung. Der Streit der "Zwei Kulturen"* (Stuttgart: Klett-Cotta, 1999); Andreas Daum, *Wissenschaftspopularisierung im 19. Jahrhundert. Bürgerliche Kultur, naturwissenschaftliche Bildung und die deutsche Öffentlichkeit 1848–1914* (Munich: Oldenbourg, 1998), chap. 2; Ulrich Herrmann, "Pädagogisches Denken und Anfänge der Reformpädagogik," in Christa Berg, ed., *Handbuch der deutschen Bildungsgeschichte,* vol. 4, *1870–1918: Von der Reichsgründung bis zum Ende des Ersten Weltkrieges* (Munich: C. H. Beck, 1991), 147–78, esp. 150ff.; Walter Schöler, *Geschichte des naturwissenschaftlichen Unterrichts im 17. bis 19. Jahrhun-*

Pedagogical Naturalism and the Philanthropists:
The Combined Authority of Nature and Science

The philanthropists of the late eighteenth century, such as Johann Bern-hard Basedow, Joachim Heinrich Campe, Ernst Christian Trapp, and Christian Gotthilf Salzmann, shared the enlightened ideal of the creation of a *new* man.[8] Educators and school reformers believed themselves to be living in a "pedagogical century," identified with the idea of perfectibility and moral improvement by the power of education.[9] The emancipatory claim of the Enlightenment to civic society was reflected in child educa-tion and instruction and became signified by the new term "natural edu-cation." The philanthropists were deeply influenced by Rousseau's peda-gogical anthropology. With his understanding of the child's nature and individuality, Rousseau broke radically with the traditional feudal concept of child rearing. In his view, education had to let children develop ac-cording to their own natural tendencies *(natürliche Anlagen)*.[10] He propa-gated a "negative education," which limited the task of the educator to protect the child from external influences. Whereas during infancy, fur-thermore, when children are allowed to persist in their natural condition (nature teaches the child), during childhood the educator becomes the advocate of nature. The educator observes the laws of the child's nature and then promotes them with indirect actions. This natural education was based on the premise that human nature is originally good. Rousseau asked

dert. Erziehungstheoretische Grundlegung und schulgeschichtliche Entwicklung (Berlin: de Gruyter, 1970); Otto Brueggemann, *Naturwissenschaft und Bildung: Die Anerkennung des Bildungswertes der Naturwissenschaften in Vergangenheit und Gegenwart* (Heidelberg: Quelle & Meyer, 1967), 41 ff.

8. See Christa Kersting, "Wissenschaft vom Menschen und Aufklärungspädagogik in Deutschland," in Fritz-Peter Hager, ed., *Bildung, Pädagogik und Wissenschaft in Aufklärungsphi-losophie und Aufklärungszeit* (Bochum: Winkler, 1997), 77–107, esp. 85–92; Paul Mitzenheim, "Philantropine als pädagogische Provinz: Historische Vorläufer der Reformpädagogik," in Michael Seyfarth-Stubenrauch and Ehrenhard Skiera, eds., *Reformpädagogik und Schulreform in Europa: Grundlagen, Geschichte, Aktualität*, vol. 1, *Historisch-systematische Grundlagen* (Balt-mannsweiler: Schneider-Verlag Hohengehren, 1996), 23–35; Heikki Lempa, *Bildung der Triebe. Der deutsche Philanthropismus (1768–1788)* (Turku: Turun Yliopisto, 1993).

9. Ullrich Herrmann, "Perfektibilität und Bildung: Funktion und Leistung von Kontin-genzformeln der Anthropologie, Kulturkritik und Fortschrittsorientierung in den reflexiven Selbstbegründungen der Pädagogik des 18. Jahrhunderts," in Dietrich Hoffmann, Alfred Langewand, and Christian Niemeyer, eds., *Begründungsformen der Pädagogik in der 'Moderne'* (Weinheim: Deutscher Studien Verlag, 1994): 79–100.

10. However, Rousseau's concept that was part of a general discourse on natural education and child medicine at the time was not received without critique by his contemporaries. See Oelkers, "Die pädagogische Erfindung des Kindes."

how we can know if what we call man's "nature" is not already the result of historical development and therefore not "pure" anymore. In his *Émile* and especially in the *Discourse on Inequality* he acknowledged this paradox. The man of nature becomes a being without sociality, rationality, and history. For Rousseau, the natural man cannot be human yet, while the human being is no longer natural. In his anthropology, humanity is the product of historical development, which causes the distinction between man as natural being and man as social being, the distinction between nature and history.[11]

Taking up Rousseau's ideas, the philanthropists fought for reforming the inadequate school system according to the needs of the bourgeois society and tried to create a systematic foundation of pedagogy. In 1774 the most famous philanthropist, Johann Bernhard Basedow, opened his demonstration school, the *Philanthropinum,* in Dessau. Campe, Trapp, and Salzmann worked at this experimental school at times. At the *Philanthropinum,* playing as learning, visual instruction, activity in nature, and a commune of teachers and students were promoted. Field trips and excursions provided a basis for observing nature, learning technical processes, and improving physical fitness. The curriculum included natural sciences, gymnastics, and handicrafts. Under the slogan "Nature! School! Life!" this experienced-based concept of learning aimed at improving the character of the children for the common good. Basedow developed a didactic that correlated the different psychological phases of child development and used observation and empirical experiment for instruction.

Basedow and the other philanthropists were mainly concerned with school reform and educational practice. It was Ernst Christian Trapp, however, who became the first educationalist who tried to establish a systematic and scientific foundation of pedagogy. In 1780, a year after he had become the first professor of pedagogy (and philosophy) in Germany at the

---

11. Jean-Jacques Rousseau, *Emil oder über die Erziehung,* trans. Karl Große (Leipzig: Otto Wigand, 1851); Rousseau, *Diskurs über die Ungleichheit / Discours sur l'inégalité,* Kritische Ausgabe des integralen Textes, ed. Heinrich Meier (Paderborn: Schöningh, 1984). See Peter Tremp, *Rousseaus Émile als Experiment der Natur und Wunder der Erziehung. Ein Beitrag zur Geschichte der Glorifizierung von Kindheit* (Opladen: Leske & Budrich, 2000); Frithjof Grell, *Der Rousseau der Reformpädagogen: Studien zur pädagogischen Rousseaurezeption* (Würzburg: Ergon Verlag, 1996), 257; Christoph Lüth, "Der Unterricht über die Natur in Rousseaus Theorie einer natürlichen Erziehung," in Hager, *Bildung, Pädagogik und Wissenschaft* (note 8), 109–28; Paul Mitzenheim, "Zur Auffassung Rousseaus über den Begriff der Natur und ihre Bedeutung für die Weiterentwicklung des pädagogischen Denkens," in *Philosophie und Natur: Beiträge zur Naturphilosophie der deutschen Klassik* (Weimar: Böhlau, 1985), 116–23.

University of Halle, Trapp published his *Versuch einer Pädagogik*. In this compendium of philanthropic pedagogy, he stated that the knowledge of human nature was the precondition for education. Just as for Goethe it was clear, as Robert Richards demonstrates in his essay in this volume, that the "dynamic idea" of an *Urpflanze* lay behind the variety of all plants, philanthropists believed in the idea of a general "human nature" that had its varieties in different individuals. Pedagogical principles can only be derived from the knowledge of this human nature. Only this knowledge guarantees the appropriate education for the individual's blissfulness and public welfare. The nature of the individual, therefore, bears a utilitarian value. The more the individual is educated according to his natural tendencies, the more useful is the role he plays in society. Trapp stated that such a scientific pedagogy could not be developed from a theological or philosophical point of view, but had to be founded on induction and empirical observation of the child. For him, anthropology and experimental psychology offered the methods for establishing this new natural educational science.[12]

Since for the philanthropists anthropology and psychology did not provide sufficient knowledge about human nature and the natural genesis of the child, they took over the research about the human and developed empirical and experimental research programs: from investigating the question how nature shapes man without human influence to finding systematic anthropological-sociological methods of observing children. In his sixteen-volume encyclopedia from 1785, *Allgemeine Revision des gesamten Schul- und Erziehungswesen,* Joachim Heinrich Campe produced the most complete pedagogical handbook of the Enlightenment. Here he suggested systematically observing infants and recording these observations in baby diaries in order to study the physical and mental development of the child and to apply the knowledge to educational theory. As eighteenth-century naturalists established a regimen of experience, as Lorraine Daston shows in her essay in this volume, and created nature's value by the "cult of attention," educators tried to investigate human nature by the same method and for the same purpose. Their descriptions covered practically every minute of the child's development and were used to provide educational rules that would promote human development in the best interests of the individual and of society.

12. Ernst Christian Trapp, *Versuch einer Pädagogik* (Berlin: Friedrich Nicolai, 1780). A reprint was published by Ulrich Herrmann (Paderborn: Schöningh, 1977).

For the philanthropists, the authority of nature results from two assumptions. The first is the divine origin of nature. Nature is the mirror of God's will. The Creator has endowed nature with an overall harmony to give the human being pedagogical guidelines. Pedagogy is part of nature and has to be discovered. All education has to follow this natural order. Nature and norm are united: the *lex naturae* coincides with the *lex divina*. Second, if, as Rousseau claimed, social conditions did not correspond with the nature of man, education could not find its norms in society but through the understanding of (human) nature.[13] Pedagogical action can be derived from the teleological principle of nature. The realization of the moral authority of nature requires the scientific study of the nature of the child, especially child psychology. In order to pursue a natural education, the educator has to understand human nature. Thus a natural education and science complemented each other.

The *Naturalism* Debate: Nature versus Religion

Although the ideas of the philanthropists were soon forgotten after the turn of the century, the concept of pedagogical naturalism found its followers in the nineteenth century. Already in 1810, the most important educator of this time, the Swiss Johann Heinrich Pestalozzi, proclaimed in his book *Wie Gertrud ihre Kinde lehrt* that education had to follow the natural order and was subordinated to the psychological laws of human nature. In 1774, the young Pestalozzi had observed the early development of his son and kept a diary in an attempt to apply Rousseau's ideas.[14]

However, it took until the 1840s before a controversy over natural education shaped the German public discourse on pedagogy.[15] The quarrel between the famous liberal educationalist and student of Pestalozzi, Friedrich Adolph Wilhelm Diesterweg, and the proponents of the neopietist Elberfeld revival movement *(Erneuerungsbewegung)* stood at the center of this debate, which had already begun in the late 1810s when Diesterweg

13. See Siegfried Jaeger, "The Origin of the Diary Method in Developmental Psychology," in Georg Eckert, Wolfgang G. Bringmann, and Lothar Sprung, eds., *Contributions to a History of Developmental Psychology: International William T. Preyer Symposium* (Berlin: Mouton Publishers, 1985), 63–74, here 66.

14. See Helmut Heiland, "Wegbereiter der Reformpädagogik: Rousseau, Pestalozzi, Fröbel—Die Grundgedanken und ihre Rezeption durch die Reformpädagogik," in Seyfarth-Stubenrauch and Skiera, eds., *Reformpädagogik* (note 8), 36–57, esp. 46 ff.

15. The following is based on Grell, *Rousseau* (note 11), 121 ff.

was assistant headmaster at the Latin school in Elberfeld. For Diesterweg, education was characterized by natural development. Everything on Earth follows natural laws; only human beings are able to break these laws, but this causes obstacles in the natural development of man and society. Natural education has to overcome these obstacles, not by teaching but by showing the "natural" development of nature that requires corresponding natural actions. The human being belongs to nature, physically to the outer Nature, mentally to the inner nature. The development of both parts follows specific laws. As part of Nature the human being can only develop according to natural laws—which are again linked to human nature. For Diesterweg, the idea that human development could follow other laws than the ones that determine human nature was unthinkable, for everything was natural.[16] Therefore, the educator appears only as nature's helping hand. For Diesterweg, pedagogical naturalism referred to reason; acting according to Nature *(naturgemäß)* was the reasonable, and the artificial *(Unnatürliche)* or antinatural was the unreasonable. Education therefore had to bring about reason by corresponding natural action. The normative authority of nature was absolute in Diesterweg's view because all actions of Nature were useful and good. The ethical imperative for the educator is thus founded in this assumption: to educate according to Nature *(erziehe naturgemäß)*. This principle had to be applied in practice—in education, in school, and in teaching. The educator acts as Nature's servant. Consequently, he cannot act against the child's nature.[17]

The proponents of the religious and mystical system of the Elberfeld movement condemned Diesterweg's method as a sinful throwback to the political and anthropological theories of Rousseau. It was the mineralogist and historian Karl von Raumer who became Diesterweg's main opponent. He offered the Elberfeld movement the arguments and authority to attack Diesterweg from a religious and politically conservative and restorative position. His *Geschichte der Pädagogik vom Wiederaufblühen klassischer Studien bis auf unsere Zeit,* published in 1843, espoused romantic, theological, and antiliberal views. For Raumer the restoration of a utopian ideal of a me-

16. Friedrich A. W. Diesterweg, "Der Naturalismus [1845]," in *Friedrich Adolph Wilhelm Diesterweg, Sämtliche Werke,* vol. 6 (Berlin: Volk & Wissen, 1963), 441–50, esp. 44ff. See also Diesterweg, "J. J. Rousseau [1844]," in ibid., 302–31; "J. J. Rousseau (gegen Professor Raumer in Erlangen) [1847]," in ibid., vol. 7 (1964), 160–89. On Diesterweg, see Horst F. Rupp, *Fr. W. Diesterweg. Pädagogik und Politik* (Göttingen: Muster-Schmidt, 1989).

17. Diesterweg, "Über das oberste Prinzip der Erziehung," in *Sämtliche Werke* (note 16), vol. 2 (1957), 21–31; Diesterweg, "Über Natur- und Kulturgemäßheit in dem Unterricht [1832]," in ibid., 445–54.

dieval society completely subordinated to God was the only way to master the political, economic, and social changes that had occurred since the early nineteenth century and to oppose the threat to a traditional religious life posed by the rationalization, secularization, and dechristianization of society. It is therefore not surprising that he traced back the "negative anthropological political and pedagogical ideas" to their origins in the eighteenth century. These ideas, according to Raumer, had led to a natural reform pedagogy that replaced the religious foundation of education by belief in Nature, the nature of human beings, and natural development. In showing the long tradition of this naturalistic movement, he devalued Rousseau, blaming him as the main source of the principle "naturam sequi."[18] Natural education to Raumer, therefore, meant the abandonment of God and His supernatural realm of mercy.

Raumer's critique and rejection of Diesterweg—embedded in his theological-religious worldview—denounced a pedagogical naturalism that denied a supernatural world outside of nature and, therefore, the possibility of a divine power to interfere with the laws of nature and the fate of man. Nevertheless, the debate over naturalism between Raumer and Diesterweg signaled the end of religious influence on pedagogical discourse. It was the new opposition of nature, sciences, and *Bildung* that would soon dominate the educational debates.

Rational Pedagogical Naturalism:
The Authority of Science over Nature

As we have seen above, the beginnings of a scientific pedagogy in Germany go back to the eighteenth century. Whereas the philanthropists Trapp and Campe tried to introduce experience and experiment into pedagogical theory, they did not succeed in establishing methods to investigate empirically the "natural tendencies" of the child. But in the middle of the nineteenth century the situation changed. The development and success of the sciences led to belief in the universally valid objectivity of their methods and trust in their emancipatory potential. Observation and experiment seemed also applicable to the science of man, and the impact of positivism and evolutionism on pedagogical thinking was immense. In the process of the scientization of pedagogy, the authority of nature for natural education

---

18. Karl von Raumer, *Geschichte der Pädagogik vom Wiederaufblühen klassischer Studien bis auf unsere Zeit,* 4 vols. (Stuttgart: Liesching, 1844–54), here 4:8.

was replaced by the authority of science. This change influenced the mechanism of value construction. Whereas for the romantic pedagogical naturalists the moral and utilitarian value of nature was so self-evident that it did not need any deeper investigation, for the representatives of rational pedagogical naturalism moral values could not be derived from Nature as such. Scientific pedagogy had to provide material teaching conditions that corresponded to the natural (physical) development of the child. It was not the aim of pedagogical science to establish new norms.

The contribution of science to understanding the child and its nature can be seen in two domains: experimental pedagogy and pedagogical psychology.

(i) Around the middle of the nineteenth century, medical scientists entered the pedagogical discourse. Their observations about overfatigue and school hygiene, for instance, led to debates about school reforms in order to improve the health and hygiene of students.[19] The discourse on pedagogically relevant matters was thus not initiated by educationalists, who only began to participate in the 1880s. But whereas medical surveillance and service aimed at securing the natural physical development of the child, it could not help to understand its mind. From the 1880s on, experimental pedagogy and pedagogical psychology became the main fields of child research. Early developmental psychology using the diary method, which Campe had suggested for child observation, was followed by the method of questionnaires and the application of experimental methods. Still, there was not much systematic information and data available about child development in the nineteenth century. Educationalists articulated the need for such data obtained scientifically and over a long period of time to understand child development and hence to create a scientific pedagogy.

It was Wilhelm Preyer who laid the basis for a psychological and physiological investigation of the child based on Darwin's evolutionism in the 1880s.[20] By the turn of the century, anthropometry, craniometry, and

---

19. See Annette Miriam Stroß, "Zwischen Emphase, Kritik und Methodenbewußtsein: Schulhygiene, Medizin und wissenschaftliche Pädagogik im Deutschen Kaiserreich," in Peter Drewek and Christoph Lüth, eds., *History of Educational Studies (Paedagogical Historica, Supplementary Series vol. 3)* (Gent: CSHP, 1998), 561–78; Jürgen Oelkers, "Physiologie, Pädagogik und Schulreform im 19. Jahrhundert," in Philipp Sarasin and Jacob Tanner, eds., *Physiologie und industrielle Gesellschaft: Studien zur Verwissenschaftlichung des Körpers im 19. und 20. Jahrhundert* (Frankfurt am Main: Suhrkamp, 1998), 245–83.

20. Wilhelm Preyer, *Naturforschung und Schule* (Stuttgart: W. Spemann, 1887), 4. Preyer became famous with the publication of *Die Seele des Kindes: Betrachtungen über die geistige Entwickelung des Menschen in den ersten Lebensjahren* (Leipzig: Th. Grieben's Verlag [L. Fernau],

cephalometry had become part of the experimental research on children as the result of evolutionary theories and eugenics. Experimental and test psychology as conducted by Gustav Theodor Fechner and Wilhelm Wundt created the framework for introducing experimental methods into pedagogy.[21]

Wilhelm August Lay and Ernst Meumann were the founders of experimental pedagogy in Germany around the turn of the century. Both wanted to establish a new pedagogical science based on the model and results of the natural sciences.[22] Although in terms of methods Meumann saw psychology as the mother of empirical pedagogy, the latter was not just applied psychology or the auxiliary discipline of any other science. Its main goal was to find empirically the real conditions of the mental and physical development of the child. Meumann wanted to replace a deductive derivation of pedagogical norms and speculation about a general human nature. For him the development of the child was not only the result of its natural tendencies independently from human influence, but also of an artificial transfer of the child's psychic-physical habit into that of an adult.[23] Both had to be explored by systematic observations, psychological experiments, and, as Wilhelm Lay notes, statistics.[24] Lay differentiated between three factors that have an impact on education: individual, natural, and social factors. Pedagogy was, therefore, divided into individual, natural, and social pedagogy. The natural factors, such as geography, climate, water, air, light, warmth, fauna, and flora, influence the physical and mental conditions of the child. A child, he concluded, is not a grownup in a smaller dimension, as traditional pedagogy claimed; it is, therefore, not sufficient to explore adults and apply these observations to children.[25] For Lay, nature

---

1882). A reprint with a long introduction to Preyer's life and work was edited by Georg Eckardt (Berlin: Deutscher Verlag der Wissenschaften, 1989).

21. Walter Herzog, "Psychologische Wissenschaft und pädagogische Reform: Die experimentelle Psychologie als Basis einer neuen Pädagogik?" in Jürgen Oelkers and Fritz Osterwalder, eds., *Die neue Erziehung. Beiträge zur Internationalität der Reformpädagogik* (Frankfurt am Main: Peter Lang, 1999), 265–303. On German psychology, see Mitchell G. Ash and Ulfried Geuter, eds., *Geschichte der deutschen Psychologie im 20. Jahrhundert. Ein Überblick* (Opladen: Westdeutscher Verlag, 1985).

22. Ernst Meumann, *Vorlesungen zur Einführung in die experimentelle Pädagogik und ihre psychologischen Grundlagen,* vol. 1 [1907] (Leipzig: Wilhelm Engelmann, 1916), v.

23. Ernst Meumann, "Entstehung und Ziele der experimentellen Pädagogik," *Die deutsche Schule* 5 (1901): 284.

24. Meumann, *Vorlesungen* (note 22), 10 f.

25. Wilhelm A. Lay, *Experimentelle Pädagogik mit besonderer Berücksichtigung auf die Erziehung durch die Tat* (Leipzig: B. G.Teubner, 1908), 8, 33 ff.

had to be studied scientifically and the appropriate place for that was the pedagogical lab.

(ii) The first attempts to create a modern child psychology can also be found by the end of the nineteenth century. Influenced by the idea of biological evolution, its representatives, such as Wilhelm Preyer, claimed that the individual passes through different stages in its development that are the same for all individuals but that have different results. This idea replaced the traditional assumption of natural tendencies in each human being. Ferdinand Kemsies, founder of the *Zeitschrift für pädagogische Psychologie,* argued that the educational problem could only be solved when the nomothetic connection between education and the soul of the child is explored. He saw the basis of this applied psychology in the laws of child development. For him, the shortcomings of school instruction are due to the lack of empirical data. Education was reduced to the knowledge of the inner life of the child.[26]

Scientific pedagogy was institutionalized within two decades. In 1899 the *Allgemeine deutsche Verein für Kinderforschung* was founded in Jena, in the same year the *Verein für Kinderpsychologie,* which published the *Zeitschrift für pädagogische Psychologie* in Berlin. In the years before World War I, institutes and laboratories of experimental pedagogy and psychology, congresses, and associations were founded. Rational naturalism, nevertheless, never gained a dominant position within academic pedagogy at the universities.[27] The leading school in pedagogy in the last quarter of the nineteenth century based their educational theory on Johann Friedrich Herbart's principles of pedagogy as *Geisteswissenschaft.* Stressing the separation of the moral and the physical, the main representatives of this Herbart school,

26. Ferdinand Kemsies, "Fragen und Aufgaben der pädagogischen Psychologie," *Zeitschrift für pädagogische Psychologie* 1 (1899): 1–20, here 2. See also Marc Depaepe, *Zum Wohl des Kindes? Pädologie, pädagogische Psychologie und experimentelle Pädagogik in Europa und den USA, 1890–1940* (Weinheim: Deutscher Studien Verlag, 1993), 321 ff.

27. Depaepe, *Zum Wohle des Kindes* (note 26), 68 ff. See also Heinz-Elmar Tenorth, "Versäumte Chancen: Zur Rezeption und Gestalt der empirischen Erziehungswissenschaft der Jahrhundertwende," in Peter Zedler and Eckhard König, eds., *Rekonstruktion pädagogischer Wissenschaftsgeschichte: Fallstudien, Ansätze, Perspektiven* (Weinheim: Deutscher Studien Verlag, 1989), 317–43. For a contemporary account, see Ferdinand Kemsies, "Die Entwicklung der Pädagogischen Psychologie im 19. Jahrhundert," *Zeitschrift für Pädagogische Psychologie, Pathologie und Hygiene* 4 (1902): 197–211, 342–55, 473–84; Ernst Meumann, "Die gegenwärtige Lage der Pädagogik," *Zeitschrift für Pädagogische Psychologie* 11 (1910): 193–206; Hermann Götz, "Zur Geschichte der Kinderpsychologie und der experimentellen Pädagogik," *Zeitschrift für Pädagogische Psychologie* 19 (1918): 257–68.

Tuiskon Ziller and Wilhelm Rein, rejected the influence of medical science on educational theory and, likewise, a scientific pedagogy that was based on the natural sciences. The aim of education was directed toward morality *(Sittlichkeit)* and toward the formation of conviction *(Gesinnungs-bildung)*. The point of departure for their educational theory was the teacher and instruction, but not the child and its development.[28]

Both the experimental pedagogists and the pedagogical psychologists aimed at providing empirical data for a school reform that provided such conditions that were adequate to the natural needs of the children. Preyer, for instance, criticized the mode of instruction in schools that ignored the physiological conditions and demands of the students and that led to over-exertion of the brain and many other health problems.[29] Preyer's efforts for school reform—he became the spokesman of the Allgemeiner Deutscher Verein für Schulreform that was founded in 1889—did not include a rejection of the general educational system and the ideal of *Bildung*. He was not a cultural critic like Rousseau and the philanthropists, and neither were the other rational naturalists. The scientific study of child development was restricted to methods derived from evolutionary theory and experimental biology.[30] For the rational naturalists, nature had lost its direct value for education. Now it was only the authority of sciences that provided the key to understanding the child. Their interest was not centered on child education according to nature but on scientific study that saw the child as a pure object of investigation. Their "objective" scientific works remained embedded in the concept of *Bildung*. Rational pedagogical naturalists failed to establish new norms for education but provided only a vast amount of data that hardly found its way into pedagogical theory.[31]

28. Wilhelm Rein, *Pädagogik in systematischer Darstellung,* vol. 1, *Die Lehre vom Bildungswesen* (Langensalza: Beyer, 1902). Herbart saw the foundations of scientific pedagogy in a planned formation of abilities. His "mathematical" but nonexperimental psychology became outdated by the experimental psychology. See Johann Friedrich Herbart, *Die Psychologie als Wissenschaft—neu gegründet auf Erfahrung, Metaphysik und Mathematik* (Königsberg: n.p., 1824–25). See Jürgen Oelkers, "Das Ende des Herbartianismus: Überlegungen zu einem Fallbeispiel der pädagogischen Wissenschaftsgeschichte," in Zedler and König, eds, *Rekonstruktionen* (note 27), 77–116; Rotraud Coriand and Michael Winkler, eds., *Der Herbartianismus—die vergessene Wissenschaftsgeschichte* (Weinheim: Deutscher Studien Verlag, 1998).

29. Preyer, *Naturforschung und Schule* (note 20), 8.

30. Georg Eckard, "Preyer's Road to Child Psychology," in Eckard, Bringmann, and Sprung, eds., *Contributions* (note 13), 177–85, esp. 183, 185.

31. This lack of establishing educational norms is discussed by Rudolf Lehmann, "Pädagogik und Biologie," *Zeitschrift für pädagogische Psychologie* 14 (1913): 497–504.

Romantic Pedagogical Naturalism:
The Authority of the Nature of the Child

Whereas the rational representatives of pedagogical naturalism focused on science rather than on nature, the romantic naturalists emphasized nature with their demand for an education that focused on the "nature" of the child. The latter considered *Bildung* as artificial and nature as authentic. The reform pedagogy of the end of the century would take up this opposition with its call for a child-centered education.[32] The rational naturalists, with their intentions to establish a new pedagogy based on the psychological and physiological needs of the children, created the preconditions for the reform pedagogy. It is interesting to note that many romantic naturalists were natural philosophers or scientists, or at least had a scientific or medical education. They all shared the romantic ideal of the child, based on two assumptions: first was the idea that the child embodies the natural and innocent Edenic human who lives in harmony with the divine order. The history of mankind is reflected in each child's development. Second, with the birth of a child, the divine returns to the world and enables the adult to go back to his origin. The child is not only the minor who needs education, but he is at the same time the model for the adult. Education of the child means therefore also the re- or posteducation of the adult.[33]

The most influential romantic naturalists and proponents of child-centered education were two women: the Italian Maria Montessori and

32. I cannot elaborate on the controversial debate on reform pedagogy *(Reformpädagogik)* that has been taking place among German historians of education since the late 1980s. The proponent of a new, revisionistic approach is Jürgen Oelkers. I share most of his interpretation that—in opposition to the traditional accounts by Herman Nohl and Hermann Röhrs—doubts the novelty of *Reformpädagogik* between 1890 and 1933, but rather puts it into a long tradition of reform pedagogy since the eighteenth century. Among the huge pile of literature, see Hermann Nohl, "Die pädagogische Bewegung in Deutschland," in Herman Nohl and Ludwig Pallat, eds., *Handbuch der Pädagogik,* vol. 1 (Langensalza: Beltz, 1933), 302–74; Hermann Röhrs, *Die Reformpädagogik: Ursprung und Verlauf in Europa* (Hannover: Hermann Schroedel Verlag, 1980); Jürgen Oelkers, *Reformpädagogik: Eine kritische Dogmengeschichte* (Weinheim: Juventa Verlag, 1989); Oelkers, "Break and Continuity: Observations on the Modernization Effects and Traditionalization in International Reform Pedagogy," *Paedagogica Historica* 31 (1995): 675–713.

33. Heiner Ullrich, "Ursprungsdenken vom Kinde aus—Über die widersprüchliche Modernität des reformpädagogischen Grundmotivs," in Tobias Rülcker and Jürgen Oelkers, eds., *Politische Reformpädagogik* (Frankfurt am Main: Peter Lang, 1998), 241–59; 249f. On the romantic image of the child, see Yvonne-Patricia Alefeld, *Göttliche Kinder: Die Kindheitsideologie in der Romantik* (Paderborn: Schöningh, 1996); Dieter Richter, *Das fremde Kind: Zur Entstehung des Kindheitsbildes des bürgerlichen Zeitalters* (Frankfurt am Main: S. Fischer, 1987).

the Swedish feminist Ellen Key. In the first chapter of her highly influential book *The Montessori Method,* published in 1912, Montessori dealt with the concept of "new education" as it related to modern science. She claimed that, despite the results of medicine, physiological and experimental psychology, and morphological anthropology, "scientific pedagogy has never yet been definitely constructed nor defined." Rather, up until the time at which she wrote, it had been "the mere intuition or suggestion of a science." The main problem Montessori saw in previous experimental research on the child was the confusion of experimental study with education, since it was expected that once knowledge of the individual was discovered, the art of education would then develop naturally. So-called scientific pedagogy was, therefore, really pedagogical anthropology.

Although pedagogical experimentalists, such as Meumann, tried to incorporate their scientific paradigm into the schools, rational educational naturalism had led to a break between educational theory and practice, that is, between education as a science and as an art. Whereas for the romantic representatives of pedagogical naturalism theory and practice were combined, in rational pedagogical naturalism the art of education, that is instructing and teaching, was separated from empirical research. Therefore, the latter could not deliver any values for the art of education. This, however, was the main concern of the romantic naturalists: for them the art of (natural) education was closely linked to empirical observation and pedagogical experiments.

It is not surprising that Montessori called for a "genuine fusing of these modern tendencies, in practice and thought; such a fusion as shall bring scientists directly into the important field of the school and at the same time raise teachers from the inferior intellectual level to which they are limited to-day." In praising the scientists, she prefers their "spirit" to their "mechanical skills." For education this means that the students must be made "worshippers and interpreters of the spirit of nature." The interest in natural phenomena makes it possible to receive revelation. This requires the appropriate education of the teacher in the "self-sacrificing spirit of the scientist" and with the love of God, but also an essential reform of the school: "The school must permit the *free, natural manifestations* of the *child* if in the school scientific pedagogy is to be born." This is only possible by accepting experiments in pedagogy that are based on empirical data derived from close observation of the child.[34]

34. Maria Montessori, *The Montessori Method: Scientific Pedagogy as Applied to Child Education in "The Children's Houses." With Additions and Revisions by the Author,* trans. Anne E. George, intro. Henry W. Holmes (New York: Stokes, 1912). See also Paul Oswald, "Päda-

Ellen Key also referred to the relation between sciences and child-centered education. Key, who proclaimed in her famous book the "century of the child," considered Herbert Spencer's book *Education: Intellectual, Moral, and Physical* as "the most noteworthy book on education in the last century." For her, Spencer's pedagogical theory was indebted to Rousseau. And her esteem of psychological pedagogy also embraced especially German thinkers, both pedagogists such as Basedow, Pestalozzi, Salzmann, Froebel, and Herbart but also philosophers such as Lessing, Herder, Goethe, and Kant who "took the side of natural training." In setting up a line of continuous tradition from Rousseau to experimental pedagogy, Key suggested that all these educationalists and thinkers are united by the same idea, namely, that education "should only develop the real individual nature of the child." It was the concept of "natural education" that linked eighteenth-century reformers with the child-centered education of the reform pedagogy in the beginning of the twentieth century. According to Key, "even men of modern times still follow in education the old rule of medicine, that evil must be driven out by evil, instead of the new method, the system of allowing nature quietly and slowly to help itself, taking care only that the surrounding conditions help the work of nature. This is education."[35]

Key's admiration of contemporary child psychology was based on the expectation "that when through empirical investigation we begin to get acquainted with the real nature of children, the school and the home will be freed from absurd notions about the character and needs of the child, those absurd notions which now cause painful cases of physical and psychical maltreatment, still called by conscientious and thinking human beings in schools and in homes, education." Key acknowledged that the speculative and aesthetic definition of the child's nature by the early proponents of natural education was insufficient. It had to be empirically defined and tested. On the occasion of the opening of the Jean Jacques Rousseau Institute in Geneva in 1912, she underlined the influence of experimental psychology for this task.[36]

---

gogik als Wissenschaft nach der Auffassung Maria Montessoris," *Vierteljahresschrift für wissenschaftliche Pädagogik* 46 (1970): 135–46.

35. Ellen Key, *Das Jahrhundert des Kindes* (Berlin: S. Fischer, 1905). The book was first published in Stockholm in 1900. On Key, see Reinhard Dräbing, *Der Traum vom "Jahrhundert des Kindes:" Geistige Grundlagen, soziale Implikationen und reformpädagogische Relevanz der Erziehungslehre Ellen Keys* (Frankfurt am Main: Peter Lang, 1989).

36. Ellen Key, "Ein internationales Institut für die Erziehungswissenschaften," *Das monistische Jahrhundert* 6 (1912):468–74, 495–502, here: 470.

As the philanthropists a century before, both Key and Montessori sought to combine the rational and the romantic strands of pedagogical naturalism. This was by no means only empty rhetoric. The unifying idea behind the two forms of pedagogical naturalism was the radical call for a child-centered education; the point of reference was nature.[37] This demand was tied to similar assumptions by Rousseau but was differently met. But whereas Rousseau's idea of nature was dissociated from its empirical basis, the romantic naturalists aimed at linking nature with reality. They focused on the individuality of the child, whereas in Rousseau's anthropology the child was seen as the particular of the general. For the romantic naturalists, the nature of the child was its individuality, not just its natural tendencies.

For both Key and Montessori education had to respect nature but also help to improve and utilize the child's predisposition. But how is this possible if life is determined by biological conditions? A central aspect of Montessori's "cosmic theory" was evolutionary theory. For her, evolution was biological growth that meant an increase in volume but also alteration in form. Biological growth, therefore, constituted the ontogenetic evolution, the development of the *individual*. This development occurs in different stages that vary—within a general framework—from one individual to another. The "cosmic mission" of the human being is to reform the outside world, the "supra-nature," in order to make it more human.[38] Key also referred to evolutionism:

> Nature herself, it is true, repeats the main types constantly. But she also constantly makes small deviations. In this way different species, even of the

37. It should be mentioned here that within reform pedagogy we can find another link to nature in addition to the aspect of the nature of the child. Embedded in the broad life movement the creation of *Landerziehungsheime* (private rural schools) symbolized a retreat from the big city and its pernicious moral and cultural influences, its harmful effects on nature, and other negative experiences of modernity. Their proponents, such as Hermann Lietz and Berthold Otto, were no pedagogical theorists, but they created the conditions making it possible to realize the concepts of "natural education." The character of the romantic naturalists—the cultural critique—could not become clearer than by propagating a natural, ascetic lifestyle. Besides this lifestyle and the quiet of the rural setting, nature as landscape was one of the pedagogical qualities. See Ralf Koerrenz, *Hermann Lietz. Grenzgänger zwischen Theologie und Pädagogik: Eine Biographie* (Frankfurt am Main: Peter Lang, 1989).

38. Maria Montessori, "Kosmische Erziehung," in *Kleine Schriften,* vol. 1, ed. Paul Oswald and Günter Schulz-Benesch (Freiburg: Herder, 1988). Harald Ludwig offers a good summary in "Die Montessori-Schule," in Seyfarth-Stubenrauch and Skiera, eds., *Reformpädagogik* (note 8), 237–52.

human race, have come into existence. But man himself does not yet see the significance of this natural law in his own higher development. He wants the feelings, thoughts, and judgments already stamped with approval to be reproduced by each new generation. So we get no new individuals, but only more or less prudent, stupid, amiable, or bad-tempered examples of the genus man.[39]

Nevertheless, the romantic naturalists so badly in need of a new theory of the child could not meet this challenge. They combined their vision of the child with evolutionary theory but did not apply this evolutionism to education because this would have made the romantic idea of child's nature and natural tendencies obsolete. There was no direct link between reform pedagogy and experimental pedagogy either, no transfer from empirical data and psychological theories into the classroom. Like Rousseau, the proponents of child-centered education eventually founded their educational practice on intuition that had no scientific basis. As Oelkers notes, reformers hesitated to use experimental pedagogy, because its results could question their intuition about the nature of the child and natural education and threaten their dogma, which was structured on moral and aesthetic grounds.[40]

The idea of the "nature of the child" was rhetorical and decisive for the discourse. It had to be understood as realistic, but there was no way to describe this reality empirically. For Montessori, the forces acting in nature and child were created by God. Creation, nevertheless, was a continuous process that was not accessible by rational thinking. It is not surprising that the empirically unprovable, secret nature of the child was transformed into a myth of the child. Not a scientific theory of the "nature of child" but the child's divine spirit, its holiness on the one hand, its suffering in the "hell" of school on the other, was the point of reference for child education and the basis for the mystification of the child. The child became the essence of religious inner contemplation.[41]

39. Key, *Jahrhundert* (note 35), 123.
40. Oelkers, "Break and Continuity" (note 32), 687; Oelkers, "Die Natur des Kindes: Theorieprobleme der Reformpädagogik," *Neue Sammlung* 28 (1988): 476. See also Andreas Flitner, *Reform der Erziehung. Impulse des 20. Jahrhunderts* (Munich: Piper 1992); Ullrich, "Ursprungsdenken" (note 33).
41. See Oelkers, "Die Natur des Kindes" (note 40), 482f.; Oelkers, *Reformpädagogik* (note 32), 76–79. On the relationship between religion and the image of the child, see Jan Weisser, *Das heilige Kind: Über einige Beziehungen zwischen Religionskritik, materialistischer Wissenschaft und Reformpädagogik im 19. und zu Beginn des 20. Jahrhunderts* (Würzburg: Ergon Verlag, 1995).

Waldorf Pedagogy and the Mystification of the Child: The Lost Authority of Nature

The anthroposophical pedagogy created by Rudolf Steiner reflected this mystification of the child most clearly. Steiner, who suffered from the effects of a rationalized and technologized world of mass culture, found the answer to his romantic question about the nature of the human being, namely, how the inner experience of the spirit could be mediated with the knowledge of nature, in Goethe's natural philosophy, especially in his theory of metamorphosis. Steiner studied Goethe very intensively and edited his scientific works for fifteen years.[42] In his book *Die Erziehung des Kindes vom Gesichtspunkte der Geisteswissenschaft,* published in 1907, Steiner laid the foundation of his educational theory. Drawing on Goethe's phenomenological view of nature about the unity of spirit and matter, Ernst Haeckel's philosophy of nature, and Fichte's romantic philosophy, Steiner created a mystic-occult worldview that became the foundation of his anthroposophical *Geisteswissenschaft,* in which he combined science with gnostic and Hindu religion.[43] Steiner assumed that there existed not only the physical world but also an invisible and spiritual world beyond the limits of the human senses and rational knowledge. His anthroposophical science opened the doors to this supernatural, supersensual world. Only through meditative exercise can the human attain spirituality and unite with the greater spirit of the cosmos. The basic laws of this occult and spiritual world are reincarnation, karma, and the relationship between macro- and microcosm. The human being embodies the world—the microcosm where all forces and ideas of the different natural stages manifest; the world is the human being enlarged.

Steiner, who had taught disabled and working class children, was able to realize his pedagogical concept when he founded the first Free Waldorf School in Stuttgart in 1919. In his *Allgemeine Menschenkunde als Grundlage der Pädagogik,* a summary of his ideas and manual for the teachers at the

42. In 1912 the *Anthroposophische Gesellschaft* was founded in Cologne. Steiner designed the building for this society, the *Goetheanum,* in Dornach near Basel. See Walter Kugler, *Rudolf Steiner und die Anthroposophie* (Köln: DuMont, 1979).

43. On Steiner, see Charlotte Rudolph, *Waldorf-Erziehung: Wege zur Versteinerung* (Darmstadt: Luchterhand, 1987), 178 ff. See also Heinrich Ullrich, *Waldorfpädagogik und okkulte Weltanschauung: Eine bildungsphilosophische und geistesgeschichtliche Auseinandersetzung mit der Anthropologie Rudolf Steiners* (Weinheim: Juventa Verlag, 1986); Peter Bierl, *Wurzelrassen, Erzengel und Volksgeister: Die Anthroposophie Rudolf Steiners und die Waldorfpädagogik* (Hamburg: Konkret Literatur Verlag, 1999).

Waldorf School, Steiner applied his anthroposophy to education. The child is embedded in "cosmic unity" where the human being and nature are deeply interwoven. Personal development is seen as a process of metamorphosis in which all mental and intellectual cosmic forces unfold in four successive stages. The child passes through these stages of anthropogenesis in its individual development. The four essential parts of the human being—physical body, ethereal body, astral body, and self-body—correspond to the four stages of cognition—sensual-material, imaginative, inspiratory, and intuitive. The succession of these parts in a cosmic rhythm of seven years is central for the education of the child because they correspond to specific learning styles that guarantee the spiritualization of the understanding of the world. All pedagogical actions are structured in cosmic orders; the school life follows specific "rhythms." The teacher has, therefore, to fulfill four tasks according to the stages of child development: he has to be a gardener, an artist, a therapist, and a seer. The instruction on nature depends on the age of the student: image-based learning for the six- to nine-year-old children who still live in magic-animalistic unity with nature; the physiognomic study of nature on the basis of anthroposophic all-unity thinking from the third grade on; and natural sciences from the seventh grade on. The schools are formed as small worlds related to the cosmic structure of the "big" world. Even the school buildings with their absence of right angles, their proportion, acoustics, colors, cardinal points, and design correspond to spiritual education.[44]

Whereas Steiner formulated his educational concepts in a naturalistic way, that is, the educational principles result automatically from human nature, he did not found his ideas on child psychology but on his cosmic spiritualistic anthroposophy. Cosmological and human history, and the spiritual-mental nature of the child and its physical function are inseparably linked. It was not Steiner's mystic pedagogical concept of anthroposophic education but its practical realization in the Waldorf School that

44. See Heiner Ullrich, "Rudolf Steiner und die Waldorfschule," in Seyfarth-Stubenrauch and Skiera, eds., *Reformpädagogik* (note 8), 253–67; Ullrich, "Vom Außenseiter zum Anführer der Reformpädagogischen Bewegung: Betrachtungen über die veränderte Stellung der Pädagogik Rudolf STEINERS in der internationalen Bewegung für eine Neue Erziehung," *Vierteljahresschrift für wissenschaftliche Pädagogik* 3 (1995): 284–97. On the practice of Waldorf schools, see also Christoph Lindenberg, *Waldorfschulen: Angstfrei lernen, selbstbewußt handeln. Praxis eines verkannten Schulmodells* (Reinbek bei Hamburg: Rowohlt Taschenbuch Verlag, 1975). Information about the curriculum can be found in E. A. Karl Stockmeyer, *Rudolf Steiners Lehrplan für die Waldorfschulen* (Stuttgart: Pädagogische Forschungsstelle beim Bund der Freien Waldofschulen, 1976).

justifies calling him a representative of reform pedagogy. This practice was centered on the child.

Steiner's concept of child education breaks with the assumptions of both strands of pedagogical naturalism. It is neither based on sciences nor on the romantic notion of the nature of the child. For him, the nature of the child is intuited from the anthroposophical mysticism and, therefore, beyond rational explanation. The aim of education is not the natural development of the child but the spiritualization of its soul and mind. Nature is used only as a means to achieve an end: spirituality and cosmic unity. The supernatural replaces the natural.

Bildung, Culture, and Politics: The Absence of Nature

The concept of nature and natural education did not affect the overall idea of *Bildung* in the nineteenth century. [45] Already in the early nineteenth century this neohumanistic educational idea replaced the anthropological-psychological concept of education and devalued the philanthropic tradition.[46] Whereas the philanthropists advocated civic education in terms of usefulness and utility, the neohumanists emphasized the education of the individual into a fully developed human being, to its individuality, its perfectness, and inner *Bildung*. Education now meant nonpurposiveness, inwardness, and scholarliness. In the public discourse on *Bildung,* education, instruction, and learning, the term "nature" and the ideas of natural education slowly vanished. Not nature but (Greek) culture was the key reference. Ancient Greece was seen as the cultural model for the ideal way of being a human and becoming the "perfect man." Indebted to Johann Joachim Winckelmann and Friedrich August Wolf, (Greek) philology became the main focus of education; imitating the aesthetics of the ancients

45. See Joachim Ritter, ed., *Historisches Wörterbuch der Philosophie,* vol. 1 (Darmstadt: Wissenschaftliche Buchgesellschaft, 1971), entry "Bildung," 921–37; Rudolf Vierhaus, "Bildung," in Otto Brunner, Werner Conze, and Reinhart Koselleck, eds., *Geschichtliche Grundbegriffe: Historisches Lexikon zur politisch-sozialen Sprache in Deutschland,* vol 1, (Stuttgart: Klett-Cotta, 1972), 508–51; Karl-Ernst Jeissmann, "Zur Bedeutung der 'Bildung' im 19. Jahrhundert," in Karl-Ernst Jeissmann and Peter Lundgreen, eds., *Handbuch der deutschen Bildungsgeschichte,* vol. 3, *1800–1870: Von der Neugründung Deutschlands bis zur Gründung des Deutschen Reiches* (Munich: C. H.Beck, 1987), 1–21; Georg Bollerbeck, *Bildung und Kultur; Glanz und Elend eines deutschen Deutungsmusters* (Frankfurt am Main: Suhrkamp, 1996).

46. See Friedrich Immanuel Niethammer, *Der Streit des Philanthropismus und Humanismus in der Theorie des Erziehungs-Unterrichts unsrer Zeit* (Jena: Frommann, 1808).

was preferred over nature and natural aesthetics. *Bildung* was seen as imitation, education of the spirit and of the inner self. In short, it was self-cultivation.[47]

In the course of educational reconstruction in Germany since the beginning of the nineteenth century, education and nature became alienated. Wilhelm von Humboldt's turn to subjectivity, that is, the derivation of *Bildung* from the needs of the subject, reduced nature to the material for the realization of these needs. Pure knowledge *(reines Wissen)* was acquired for its own sake and not because of its usefulness and practical application. The unity of nature and man as well as of the history of nature and the history of man was dissolved; the outside world was diminished from being a partner of the human being to a mere ingredient of his self-education. For Hegel, man as a being of spirit and reason was not by nature what he ought to be, but he needed to be educated to become reasonable *(vernünftig)*.[48] The "nature" of the human was now related to culture, a stage that was only achievable by the equal education of reason, heart, and taste, as Wilhelm T. Krug put it in 1827. For him *Bildung* was "only the beginning of culture."[49] The concept of *Bildung* reversed the argument of the naturalists: it is not nature that provides norms for the human being and therefore shapes human society, but instead it is culture that creates values, which define human nature and the position of the human being within nature. Culture did not mean educating according to nature but subduing and controlling it.[50]

Like pedagogical naturalism, the idea of *Bildung* was by no means static throughout the nineteenth century. But it soon became the ideological icon of the elite and mainly Protestant bourgeois class. Passing the *Gymnasium* was a privilege granted by state examination. Being educated *(gebildet)* was seen as a status symbol. Since *Bildung* was mainly a concept for the German middle and higher classes, the latter opposed the efforts of school

47. Jürgen Oelkers, "Das Konzept der Bildung in Deutschland im 18. Jahrhundert," in Jürgen Oelkers, Fritz Osterwalder, and Heinz Rhyn, eds., *Bildung, Öffentlichkeit und Demokratie* (Weinheim: Beltz, 1998), 45–70, esp. 58. See also Suzanne L. Marchand, *Down from Olympus: Archeology and Philhellenism in Germany, 1750–1970* (Princeton: Princeton University Press, 1976), esp. 24 ff.

48. See Vierhaus, "Bildung" (note 45), 534 f.

49. Wilhelm Traugott Krug, ed., *Allgemeines Handwörterbuch der philosophischen Wissenschaften nebst ihrer Literatur und Geschichte,* vol. 1 (Leipzig: F. A. Brockhaus, 1832), entry "Bildung," 358–60.

50. See Ernst Hering, "Pädagogischer Naturalismus: Ein Beitrag zur jüngsten Geschichte der Pädagogik," *Vierteljahresschrift für philosophische Pädagogik* 1 (1917–18): 282–88.

reformers to establish a general education for all social classes. This opposition meant a departure from the democratic principles of freedom and equality that were part of the original neohumanistic concept of *Bildung*. Humboldt's demand for a general human education in a unified school system remained a utopian dream. [51]

There is no question that the fundamental changes in the educational system such as the introduction of compulsory education, the effort to achieve general literacy, the beginning secularization of schools and curricula, and the training of teachers at special teachers colleges that occurred over the nineteenth century revolutionized the educational system—on the elementary level by the centralization of public schools and on the level of higher education by the establishment of high schools *(Gymnasium)*. These reforms were conducted by the state. As progressive as these state reforms were, the school system was by no means democratically structured.[52] In trying to train politically loyal teachers, the Prussian government fought pedagogical liberalism. After the revolution of 1848–49, educational decrees were passed that reestablished the traditional religious supervision over the public schools and restricted the subjects that dealt with natural studies. Although these regulations only existed until 1872, they indicated the still strong influence of the churches over the elementary school system after the middle of the century. The schools were subjected to social selection, status assignment, and security of power. Wilhelm II's speech at the Prussian School Conference in 1890 made the nationalistic and ideological aim of school instruction and education obvious. Whereas nature was replaced by culture at the beginning of the century, culture was replaced by politics at its end. Education was national and aimed at obedience, discipline, piety, authority of the teacher, and subordination of the student.[53] Neither moral and utilitarian values derived

51. The idea of a universal education as the natural determination of man, as was still stated by Krug, was never realized. Until the end of the nineteenth century, the natural ("realistic") sciences did not belong to the canon of subjects that were taught at schools but were subordinated under "arts" or natural history. See Krug, *Allgemeines Handwörterbuch,* 359. For more on the debate between the proponents of the incorporation of science into the concept of *Bildung* (the so-called *Realisten*) and the defenders of the neohumanistic ideal of education (the *Humanisten*) that shaped the educational discourse on secondary education in the second half of the nineteenth century, see Daum, *Wissenschaftpopularisierung* (note 7), chap. 2.

52. See Frank-Michael Kuhlemann, "Tradition und Innovation: Zum Wandel des niederen Bildungssektors in Preußen 1790–1918," in *Jahrbuch für Historische Bildungsforschung* 1 (1993): 41–67.

53. See Dorle Klika, *Erziehung und Sozialisation im Bürgertum des wilhelminischen Kaiserreichs: Eine pädagogisch-biographische Untersuchung zur Sozialgeschichte* (Frankfurt am Main: Peter Lang,

from Nature (romantic naturalists) nor moral values based on (Greek) culture (rational naturalists and neohumanists), but political and social values of nationhood and patriotism dominated in education at the turn of the nineteenth to the twentieth century.

Conclusion: Pedagogical Naturalism and *Bildung*

Contrasting pedagogical naturalism with *Bildung* requires considering the two spheres of pedagogical practice and educational theory. First, naturalism aimed at improving the physical conditions at schools according to the natural needs of the students. This holds true for both its strands but was the main concern for the rational naturalists. Second, just as proponents of *Bildung* disregarded the concept of natural education, naturalists refused the concept of *Bildung*. This, however, was limited to the romantic naturalists who were rejected and excluded from the official pedagogical discourse.

The representatives of rational pedagogical naturalism developed their theories without questioning the social status quo and the structure of the educational system. They neither intended to reform the society nor did they try to develop new educational norms. For the rational naturalists, the authority of science replaced the authority of nature. Since science was conducted as an objective, purely empirical enterprise it could not, however, produce educational norms. Even if some rational naturalists, such as Preyer, were engaged in school reform, and the efforts of medical doctors and child psychologists led to improvements in the classroom, they never questioned the overall structure and ideology of the educational system. Their reform ideas were integrated into the state reforms that followed the modernization process of German society. Their aim was the establishment of a scientific pedagogy or scientifically based education, but did not include having natural education as the theoretical foundation of pedagogy.

The educational aim of romantic naturalists was to overcome the theory, practice, and politics of German *Bildung*. They sought to reform the "unnatural" concept of education—alienated from nature and reality—which found its practical expression in an undemocratic, authoritative, and highly ideological school system. The romantic naturalists challenged the authority of culture with what they understood to be the moral and

---

1990). For more general information, see Katharina Rutschky, *Schwarze Pädagogik: Quellen zur Naturgeschichte der bürgerlichen Erziehung* (Frankfurt am Main: Ullstein, 1977).

utilitarian values of nature. Although they tried to reform society, for example Key in the woman's movement and Montessori in the "New Education Fellowship," they knew that the education of the "new" or "natural" human being could not be fulfilled within the structure of traditional education. Whereas the theoretical impact of natural education remained rather limited, its practical realization became the actual subversive discourse. The enclaves the reformers founded—from Emile's country home, the Dessau *Philantropinum,* and Pestalozzi's *Neuhof* to Montessori's "Children's Houses," to Lietz's Country Reformatory in Ilsenburg, and Steiner's Waldorf School—were a retreat from the world of culture. It seems that the authority of nature could only be followed when isolated from mainstream culture. These "pedagogical islands" as alternative types of learning and education were untouched by the school reforms under the auspices of the state and from the educational institutions.

Even if the romantic naturalists were indebted to Rousseau's natural education, there was no direct reception of his abstract and universal idea of natural tendencies. But his concept of "natural education" became their general educational aim. The notion of a Nature that was designed by the Creator in perfect harmony and that acted as a model for education was taken over from eighteenth-century philanthropists. This assumption—linked with evolutionary theory—offered the basis for an education according to nature *(Naturgemäßheit).* But whereas the philanthropists tried to combine cultural critique, social reform, natural education, and an empirical, scientific pedagogy, such a comprehensive program could not be realized in the nineteenth century. Although the romantic naturalists tried to found their child-centered education on scientific knowledge, they never really overcame the romantic notion of the child and intuition in educational practice. They were not able to give their idea of the nature of the child an empirical foundation. It was based on intuition and experience and therefore prevented the usage of pedagogical scientific knowledge. For the romantic naturalists, experiment only meant the practical test of their new pedagogical ideas. The pedagogical naturalists were not able to escape the trap of the use of nature as a seemingly objective phenomenon that itself does not need any further foundation. The idea that the answers to questions about what was ethically right or wrong was inscribed into the objective, reasonable Nature enabled the naturalists to draw values directly from it. Since such a notion of nature was descriptive *and* normative, it seemed to bridge the gap between the *is* and the *ought.* The moral philosopher George E. Moore in his *Principia Ethica* (1903) called this the "naturalistic fallacy."[54] According to Moore, normative statements cannot be logically derived from descriptive statements about nature. This "naturalistic fallacy" of the pedagogical naturalists explains that the call for an education according to nature could also justify—as in the case of Sophie in Rousseau's *Emile*—social gender differences and, therefore, the "natural" hierarchical subordination in human society by using the same arguments.

The religious background of romantic naturalists resembled the main-stream pedagogical discourse that was indebted to Christian ethics.[55] It was not the *real* but the *ideal* nature of the child that became the focus of the romantic naturalists.[56] The idealized image of the good and innocent child led to an antimodernist mystification of child and childhood beyond ex-perience that escaped the rationality of science. As the title of a book, *The Gospel of Natural Education,* indicated, antimodern cultural critique inter-twined with the religious beliefs in divine nature.[57] The ideal of the divine creation of nature and the mystification of the child suggested a religious foundation of nature and natural education. The call for nature implied the creation of a substitute religion.

Romantic pedagogical naturalism was sharply criticized and rejected.[58] On the one hand, this counterdiscourse of mainstream pedagogy was di-rected against evolutionary ideas in education. The defenders of tradi-tional education reversed the argument of the naturalists: nature occurred here as the "bad" model for mankind. All human atrocity had its example in nature and its laws. Therefore, education could not be the blind imita-tor of nature. Civilization was nothing other than a continuous war against nature. Its moral stage was seen as the complete opposite of the natural stage. Nature, with its egoistic pitting of the individual against everyone else could not be imitated.[59] Whereas for the naturalists reason and nature coincided, for the mainstream educators reason and nature clashed.

54. George E. Moore, *Principia Ethica* (Cambridge: Cambridge University Press, 1971). See Birnbacher, "'Natur' als Maßstab" (note 5), 64.

55. Jürgen Oelkers, "Wissenschaft und Wirklichkeit: Der pädagogische Diskurs in Deutschland am Ende des 19. Jahrhunderts," in Hoffmann, Langewand, and Niemeyer, eds., *Begründungsformen* (note 9), 101–24.

56. Oelkers, *Reformpädagogik* (note 32), 81.

57. Ewald Haufe, *Das Evangelium der natürlichen Erziehung* (Leipzig: K. G. Th. Scheffer, 1904).

58. For a general critique, see Georg Lunk, *Kritik des pädagogischen Naturalismus im Sinne einer Orientierung vom Kinde aus* (Leipzig: Klinkhardt, 1927).

59. Lindner, ed., *Encyklopädisches Handbuch* (note 4), 542.

On the other hand, it was the cultural critique that was refused by mainstream educationalists. The romantic naturalists can be linked to the broad life reform movement in Germany, an antimodernist protest movement opposing the effects of the industrial mass society with its unlimited exploitation and destruction of nature as well as increasing urbanization.[60] The emphasis on the nature of the child in contrast to the threatening tendencies of a technology-orientated industrial society shifted concepts, such as nature and growth, to the center of a discourse that was directed against all unnatural phenomena of modernity, such as tradition, reason, and intellectualism. The nature of the child represented a refuge from civilization. Unlimited growth and free development of the child were the catchwords of a liberal reform pedagogy with nature at its core. Natural education had to create the conditions for a natural development of the human being and, therefore, for the humanization of life and society. The myth of nature combined with a romantic conception of the child became the points of orientation in a time of radical social change.[61]

60. See Wolfgang R. Krabbe, *Gesellschaftsveränderung durch Lebensreform: Strukturmerkmale einer sozialreformerischen Bewegung im Deutschland der Industrialisierungsperiode* (Göttingen: Vandenhoeck & Ruprecht, 1974).

61. See Armin Bernhard, *Demokratische Reformpädagogik und die Vision von der neuen Erziehung: sozialgeschichtliche und bildungstheoretische Analysen zur Entschiedenen Schulreform* (Frankfurt am Main: Peter Lang, 1999), 60ff.

7

Economics, Ecology, and the Value of Nature

Matt Price

[L]abour makes the far greatest part of the value of things, we enjoy in this World: And the ground which produces the materials, is scarce to be reckon'd in, as any, or at most, but a very small part of it; So little, that even amongst us, Land that is left wholly to Nature, that hath no improvement of Pasturage, Tillage, or Planting, is called, as indeed it is, *wast;* and we shall find the benefit of it amount to little more than nothing.

JOHN LOCKE, *Second Treatise on Government*

What is nature worth? Most of the essays in this section chronicle parts of the long struggle to interpret the moral or spiritual values hidden in Nature's works—in the labors of bees, the turning of the seasons, the shape of a leaf. The present essay, by contrast, explores what I would argue is a specifically twentieth (and now twenty-first) century conundrum: how to assess the value of nature and natural goods, and to weigh that value against other goods in a moral calculus.[1]

The contemporary obsession with nature's value reopens a question that John Locke thought he had solved in the seventeenth century. Indeed, modern economics begins with Locke's all-but-categorical denial of the value of nature's works. The labor theory of value required it: Locke needed to show that human activity was the true source of all value, thereby grounding his theory of property, his liberal version of the social contract,

---

1. Lorraine Daston's contribution to this volume may seem to provide direct counterevidence to this claim, insofar as it treats of an effort to invest value in the minutiae of natural history. I would argue, however, that the process she so convincingly describes is quite distinct from the precise *calculations* of value that are the subject matter here.

and his arguments on political authority. If the lands untouched by human toil, and their value, had to be sacrificed on the altar of property, that was hardly controversial in an era when "wilderness" was a term of abuse. But ever since the hedonic theory of utilitarianism captured political economy from the dismal scientists, economists have rejected toil in favor of pleasure, and spaces once called *waste*lands are now more often named *wet*lands. When the word "nature" itself most often refers to the collectivity of lands and creatures least disturbed by human activity, Locke's assertion provokes the question: what is nature *really* worth? If the value of nature is not nothing, then how should it be measured—in what kind of units, against what other sorts of values? By what authority is the "true" value of nature to be asserted above all other contending claims? Such questions are more than idle contemplation; for a variety of reasons, they have risen in the past three decades to become some of the most urgent issues confronting the world political-economic order. Actors as far removed from each other as World Bank economists, grassroots activists, municipal governments, and individual citizens are increasingly called on to make decisions that pit human interests against those of nature. These decisions depend crucially on competing *valuations* of nature or natural goods. Whether the venue is the streets of Seattle, the forests of the Amazon, or the conference rooms of Kyoto, practical struggles over the value of nature are almost always informed by theoretical approaches developed at least partly in the academy. The question of nature's value organizes a *topos* of intellectual inquiry that brings together economists, ethicists, and natural scientists in a pragmatic, urgent and often vitriolic struggle over the future of economic development.

This essay lays out brief genealogies and central arguments for what I take to be the primary positions currently available within discussions of nature's value. Together these three frameworks—neoclassical economics, ecological economics, and environmental ethics—constitute a sort of space of possibility; their elements are the building blocks for almost all ascriptions of value to nature that can be made in academic discussions. By mapping the terrain on which valuations of nature can take place, I explore both theories of value and the place of nature's authority within them. I am particularly interested in the power of "nature," in all its manifold forms, to challenge hegemonic conceptions of value. So, in the first instance at least, the question is not "What do economics and ethics tell us about nature?" but "What does the question of nature tell us about economics and ethics?"

Neoclassical Economics, Externalities, and Contingent Valuation

In the late nineteenth century, a much-attenuated labor theory of value finally dissolved entirely under the pressure of utilitarianism, at least within the mainstream of the emerging economics profession. By abandoning labor, the leaders of the "marginalist revolution" were able to erect a rigorous mathematical discipline on a foundation of pure desire.[2] Their descendants have endured over a century of revilement, retrenchment, consolidation, and expansion to build up an overwhelmingly dominant position within contemporary economics, usually called "neoclassicism." Thus when the problem of *nature's* value reemerged in the latter part of the twentieth century, it was perhaps unsurprising that certain neoclassicist economists came up with their own solutions to it (after all, unlike Locke, they were not committed *a priori* to a theory of value that denied any to nature). The method they endorsed, known as "contingent valuation" (CV), remains the dominant procedure for assessing the value of nature, and therefore the most important antagonist for dissenting positions. Because CV rests on the marginalist theory of value, and because CV's opponents often aim their criticisms directly at that value theory, I explore it briefly here before returning to the question of nature's value.

Marginalist value theory is often, and aptly, called "the subjective theory." Most versions of the labor theory (and here we caricature a two-hundred-year tradition) had been theories of *objective* value: the labor expended in the production of a good was argued to be a measure of the value inhering in the good itself.[3] This objectivism was no doubt tied to a pre-utilitarian version of the virtue of utility—one in which, as John Stuart Mill said in a critique of Adam Smith, "use is opposed to pleasure."[4]

2. On the history of marginalism, see R. D. Collison Black, A. W. Coats, and Craufurd D. W. Goodwin, eds., *The Marginal Revolution in Economics. Interpretation and Evaluation* (Durham: Duke University Press, 1973).

3. There is, of course, substantial variation among labor theories of value. The word "objective" can be somewhat misleading in this context. Many labor theorists (Marx, for example) understood an object's value as a relational property—one which only held for that object within a specific social arrangement or historical context. Such relational properties can be termed "objective," but this sense is obviously weaker than the term's conventional usage. On this point, see John O'Neill, "The Varieties of Intrinsic Value," *Monist* 75, no. 2 (1985): 119–37.

4. *Principles of Political Economy with Some of Their Applications to Social Philosophy* (1848; London: Routledge, 1996), III.ii.4, p. 466. Lorraine Daston's contribution to this volume suggests how tenuous the opposition between utility and pleasure could be in the eighteenth century.

Where the useful can be distinguished from the frivolous by objective criteria, the true or natural value of a commodity might also be distinguished from its temporary, fluctuating market value. But utilitarianism eschews such judgments: for Mill and Bentham, utility was happiness, unburdened of puritanical connotations. The usefulness of a good, then, was "its capacity to satisfy a desire, or serve a purpose,"[5] and a good's value was a measure of the amount of pleasure it could provide to an individual. Of course, such pleasures can vary from person to person—my friend Keefe has no interest in chocolate, while I find myself fatally attracted—and so the values of things are all, under utilitarianism, *subjective*. Moreover, none of us has any direct access to the amount of pleasure a good brings to any other individual; utilities are *private* and *psychological*.[6]

The marginalists took this hedonic theory of value to a logical conclusion, by noting what was later called "diminishing marginal utility" of most goods. The more chocolate I have, the less I crave my next Hershey bar, and thus the less I value it (though when it comes to chocolate and other addictions, this formulation is dubious).[7] This quite simple move, combined with mathematization, allowed the marginalists to derive all sorts of neat results and solve a number of outstanding problems. Crucially important, they derived the law of supply and demand in a rigorous (if somewhat artificial) fashion. An early disciple and mathematical wizard, Vilfredo Pareto, was then able to show that under certain restrictive (and questionable) assumptions, a perfect free market would maximize a society's utility—the Holy Grail of bourgeois political economy since Adam Smith, and a result that played a significant role in the development of CV. Rather than go into the mechanics of these mathematical feats, I just want to point out that they rely on two subsidiary features of the subjective theory. First, the utilitarian universe is more or less pre-Socratic: behind all the various goods that circulate in the economy—coal and tea and steam

5. Ibid., III.i.2, p. 456.

6. For a Wittgensteinian critique of the philosophy of mind underpinning this tenet of utilitarianism, see Mark Peacock, "Interpersonal Comparison of Utility: Some Lessons from Wittgenstein," *Review of Political Economy* 8, no. 3 (1996): 279–90.

7. Marginalism intensifies the "subjective" character of utilitarian value theory by making "capacity" claims more difficult. That is, Mill could still claim that the value of a thing derived from its intrinsic *capacity* to satisfy a desire—and so resides in the thing itself. But if, of two identical Hershey bars, the *first* I consume gives me great pleasure, and the *second* gives me very little, then the claim that all Hershey bars have a certain capacity to give me pleasure just sounds more implausible. (Nonetheless, such a claim can of course be made, given enough qualifying clauses.)

engines and gin—lies but a single substance, pleasure.[8] Second, because the value of all these physically distinct goods are of the same kind, it is possible to order these values, these pleasures, and compare their magnitudes. Thus it is possible to make statements of the kind "For me, in my present circumstances, one bar of chocolate is worth ten apples" even though apples and chocolates are physically quite different from each other. And since this is true for all commodities, we can take one of them and, under the right conditions and with a sufficiently sophisticated system of hedging, use it as a measure of utility. Moreover, we have available to us the perfect candidate: a commodity that has, in itself, no particular use, that is entirely anonymous and independent of physical form—namely, money. One's willingness to pay for an item then becomes an excellent measure of the item's marginal utility, its utility relative to a given economic situation. This makes value *quantifiable*—in a strong sense, that is, scalar and cardinal[9]—and therefore useful as a variable in mathematical models.

A cardinal, subjective, hedonic theory of value is the most central assumption of the neoclassical theory, and underwrites the imperialistic ambitions of that discipline, including its attempts to assess the value of nature. Even when economists try to set brakes on this impulse, they tend to be unsuccessful, as in Alfred Pigou's distinction between economic and noneconomic welfare.[10] As John Foster perceptively points out in a recent critique,

> [A]ctually the distinction as [Pigou] sets it up is unable in principle to prevent economics from extending its purview to embrace all political life. For there are simply no choices which we are called on to make as sharers in a modern polity where monetary considerations do not bear at least indirectly on the goods envisaged. . . . [R]easons which can be *weighed* against reasons expressed in terms of the measuring-rod of money are themselves necessarily brought thereby into indirect relation with that measure.[11]

8. The claim is often made that under marginalism, value results from the combination of utility and scarcity. This formulation has certain explicatory virtues: the less chocolate I have, the more I value each bar. It is, however, technically imprecise, as marginalism actually bundles scarcity into utility itself (one Hershey bar has more than half the utility of two because of its relative scarcity)—this is the whole point of marginal utility! Thus English-speaking marginalists often imported Leon Walras's term *rareté* to replace both "utility" and "scarcity."

9. Later economists grew skeptical of cardinal ordering systems, replacing them with ordinal systems, where one can say which option is better, but not by how much.

10. Alfred C. Pigou, *The Economics of Welfare* (London: Macmillan, 1920), 14.

11. John Foster, "Introduction: Environmental Value and the Scope of Economics," in John Foster, ed., *Valuing Nature? Economics, Ethics and Environment* (London: Routledge, 1997), 14.

So there is a certain pressure within economics to expand further and further away from narrow considerations of markets. The neoclassical conception of the value of nature (to which I now return) is worked out precisely through such expansion, concerning what are called, after Alfred Pigou, "externalities."

Externalities are costs or benefits—utilities or disutilities—that result from economic processes, but are not "captured" by the market; often externalities have to do with effects of economic activity on "nature." For instance, a large number of industrial processes use river water as both an input and an output for manufacturing purposes; water is taken in as a coolant or a wash, and then ejected back into the river along with unwanted by-products of the production process.[12] The pollution in which this results inflicts certain costs on the surrounding community—for clean-up, say—but these costs are not borne by the producers themselves, and so do not enter into their economic calculations. The market has "failed," insofar as it fails to incorporate all utilities and disutilities, and so the market on its own will not be able to maximize social utility as it would under the ideal conditions of neoclassical theory. Pigou suggested regulatory mechanisms (taxes, subsidies, and so forth) to correct the market failure. In the 1960s, Ronald Coase came up with a much more radical suggestion: that markets be established for externalities.[13] Coase argued that market failures occur only when there are goods to which no one has property rights; if property rights are assigned to such goods (in this case, the river water) then utility-maximizing agents will correct the failure. This is called "internalization," because the externalities are brought back into the purview of market forces.

But both Pigou's and Coase's approaches to externalities raise the question of nature's value. What is clean water *worth?* Even if the optimal prices are to be set, in the fashion Coase suggests, by market forces, some initial value needs to be assessed. One straightforward method is to try and estimate the narrowly economic impacts of the externality. For instance, pollutants dumped into a river may need to be filtered out by downstream communities that use the river for drinking water; the illnesses that the pollution can be expected to cause will have certain computable economic consequences; and so on. There are a variety of methods for computing

12. For an account of this general feature of production, see Robert Costanza et al., *An Introduction to Ecological Economics* (Boca Raton, Fla.: St. Lucie Press, 1997), app. A.

13. Ronald H. Coase, "The Problem of Social Cost," *Journal of Law and Economics* 3 (1960): 1–44.

these economic effects, which together are referred to as "cost–benefit analysis." Yet clearly this narrow estimation leaves out some disutilities; and so the cost imposed on the polluter will not *fully* reflect the impact of the pollution. After all, what about the displeasure experienced by swimmers and sunbathers when confronted by the noxious fumes rising up off the water? Or the disappointment of recreational fishermen when all they can bring up is their own empty beer cans? What about the displeasure *I* feel when I read on the Internet about the pollution of a river five thousand miles from my apartment in Berlin? For that matter, what about the *Schadenfreude* that my misanthropic neighbors feel when they hear the same news—do these enter into calculations of the value of clean water?

Such questions are the bread and butter of contingent valuation analysis. As one helpful website defines it:

> The contingent valuation method involves directly asking people, in a survey, how much they would be willing to pay for specific environmental services. In some cases, people are asked for the amount of compensation they would be willing to accept to give up specific environmental services. It is called "contingent" valuation, because people are asked to state their willingness to pay, contingent on a specific hypothetical scenario and description of the environmental service.[14]

So, for instance, since taxes on pollution will most likely result in higher product costs, a CV poll might ask how much extra consumers would be willing to pay for chemical products in order to reduce plant emissions. In fact, most CV studies involve much more complex scenarios and employ more sophisticated sets of questions. The field is sufficiently robust to support papers on methodological questions, and a large array of approaches to solving them.[15] What I wish to emphasize, though, is that all versions

14. Denis M. King and Marissa Mazzota, "Contingent Valuation Method," *Ecosystem Valuation,* http://www.ecosystemvaluation.org/contingent_valuation.htm [17 August 2000].

15. See, e.g., Matthew J. Kotchen and Stephen D. Reiling, "Environmental Attitudes, Motivations, and Contingent Valuation of Nonuse Values: A Case Study Involving Endangered Species," *Ecological Economics* 32 (2000): 93–107; Judy Clark et al., "'I Struggled With This Money Business': Respondents' Perspectives on Contingent Valuation," *Ecological Economics* 33 (2000), 45–63. A standard reference work is D. W. Pearce and R. K. Turner, *Economics of Natural Resources and the Environment* (New York: Harvester, 1990). For critical discussions of contingent valuation methods, see Foster, *Valuing Nature?* (note 11); Wilfred Beckerman and Joanna Pasek, "Plural Values and Environmental Valuation," *Environmental Values* 6 (1997): 65–86, both also discussed below.

The two most significant issues for our discussion are the distinction between "Willing-

of CV accept the utilitarian theory of value, with its hedonism, its monism, and its formalism. Nature's value is solely nature's capacity to produce pleasure in us, and that value can inherently be substituted for by other forms of pleasure (for example, monetary payoffs). Under utilitarianism, the term "authority of nature" is close to incoherent; if anything, the laws of the market replace natural law as the arbiter of morality. A contemporary G. E. Moore might well accuse neoclassicists of a "marketistic fallacy": whatever the (perfected) market does, is right. This position does not deny that the utility functions of some individuals might be affected by their private, individual sense of morality; but as far as moral and economic calculations go, all that matters in such a case is the pleasure or displeasure that this sense of morality brings when confronted with some hypothetical state of affairs. "That human preferences should count and be 'sovereign' is the fundamental value judgement in cost–benefit analysis," writes one prominent practitioner.[16] Nature itself has no independent contribution to make to value; and this is what makes the so-called ecological economists start foaming at the mouth.

Ecological Economics

Cost–benefit analysis is the best-institutionalized method for integrating environmental concerns into economic analysis, and contingent valuation is the version of it that is most often employed to assess market-independent utilities. In the last thirty years, CV has come to be widely practiced in governmental decisionmaking in both Europe and the United States, as well as in nongovernmental organizations and by academic economists. Their very omnipresence is a testament to the impact that environmentalism has made in policy circles since the 1970s. And indeed, many of the proponents of CV are themselves moderate environmentalists who have struggled with their colleagues in economics over the legitimacy of their methods.[17] Yet they still have to contend with searing criticism

---

ness to Pay" and "Willingness to Accept" amounts, and the distinction between "use" and "non-use" values. Both of these are discussed extensively in the literature, and it is often suggested that the two issues are linked. However, neither affects the very general points this essay makes about contingent valuation.

16. David Pearce, "Cost–Benefit Analysis and Environmental Policy," *Oxford Review of Economic Policy* 14, no. 4 (1998): 84–100.

17. The main complaint orthodox economists level against contingent valuation is its reliance on "stated preferences," that is, assertions of "Willingness to Pay" or "Willingness to Accept," as against the "revealed preferences" of consumer behavior.

from the left. Since the early 1980s, and especially in the last ten years, CV has suffered bitter attacks by environmentalists. In part this is related to the Reagan administration's use of such methods in its dismantling of the environmental regulation system; but the criticisms run much deeper than that. Ultimately they go to the heart of neoclassical economics, the hedonic theory of value. The rest of this essay explores the two main lines of attack levied against contingent valuation, the first of which goes by the name "ecological economics."

We can perhaps best begin in 1972—on the eve of Watergate, at the tail end of the Vietnam war, in the early days of disco. Earth Day was two years in the past, and the American environmental movement was undergoing massive changes in both its establishment and radical wings. It was in this year that Herman Daly, halfway through a vituperative attack on orthodox economics, wrote the following:

> Unfortunately, economists long ago forgot about physical dimensions and concentrated their attention on value. Value is measured in money. Money, as a unit of account, has no physical dimension. A sum on deposit at Chase Manhattan Bank can grow forever at 5 percent! Income and wealth are value concepts; they too are measured in money; why cannot they too grow forever at 5 percent? Money fetishism triumphs completely! The concrete reality being measured is reduced to identity with the abstract unit of measure. The physical dimensions of wealth are "annihilated" by the Almighty Dollar! But in fact wealth always has a physical dimension.[18]

At the time Daly was an obscure associate professor of economics at Louisiana State University, who had in the last few years published several equally obscure papers on the rather poorly regarded notion of a steady-state economy. It might well have surprised his audience to learn that Daly would find himself, over the next thirty years, at the center of a movement to unseat neoclassical calculation from the throne of economic science; and that the arguments put forth in this very paragraph would form the heart of it.[19]

18. Herman Daly, "In Defense of a Steady-State Economy," *American Journal of Agricultural Economics* 54 (1972): 945–54.

19. For one thing, there were much more likely candidates to lead a revolt in economic theory. In the years around 1970, the Cambridge-trained Indian economist Amartya Sen published a series of books mounting precisely measured attacks on the illusions of economic theory, most importantly *Collective Choice and Social Welfare* (London: Holden-Day, 1970). Sen's work drew heavily on that of Kenneth Arrow, one of the towering figures in postwar economic theory; for a time, it seemed that game and set theory, not ecology, might be the tools

The central thesis of ecological economics is this: the neoclassical tradition is *contra naturam*. Its theory of value leads it down paths that contradict natural law, denying the reality of the material world and fundamentally misunderstanding the nature of life. So the world needs a successor science, one that can assume command of the vast apparatus of social decisionmaking neoclassicism currently controls, and lead us all down the road to sustainability. Daly articulated this challenge to neoclassicism both explicitly and polemically in a series of publications in the early 1970s, pirating basic concepts from heterodox economists (especially Kenneth Boulding and Nicholas Georgescu-Roegen) and prophets of doom such as Paul Ehrlich. During the 1970s Daly was one part of the diverse challenge to economic orthodoxy to which E. F. Schumacher (small is beautiful) and Garrett Hardin (tragedy of the commons) also belonged. Like these somewhat better-known figures, Daly harried his economist colleagues largely from the sidelines, publishing mostly semitechnical and popular works, often in a millenarian cast. But in the 1980s, this situation changed.

In 1980, Louisiana State hired a young ecologist named Robert Costanza; together, Costanza and Daly would erect the institutional foundation of ecological economics. Costanza had just published an article in *Science* based on his Ph.D. dissertation; both were titled "Embodied Energy and Economic Valuation." Costanza presented evidence that market prices of commodities were strongly correlated with the total energy required to produce them. He went on to argue that this correlation showed that, in fact, *economic value derived from energy used, and not solely from utilities*.[20] Here he drew to some extent on Daly, but most crucially on Daly's own mentor, the heterodox Romanian economist Nicholas Georgescu-Roegen (now familiarly known as G-R in ecological economic publications).

Unlike Costanza and Daly, G-R had once been a very highly regarded member of the economic establishment. In the 1930s he had published a series of incisive articles on technical aspects of utility and welfare theory, and he was never subject to the abrupt dismissals that characterize neo-

---

to smash the neoclassical machine. See also Sen's "Rational Fools: A Critique of the Behavioural Foundations of Economic Theory," *Philosophy and Public Affairs* 6 (1977): 317–44.

20. The slide from descriptive to normative here is instructive. Costanza shows that price is largely composed of energy costs. From there he concludes that economic value *really arises out of* energy consumption. Given this result, he then goes back to price with the normative claim: economic systems should be constructed in such a way that prices reflect energy costs more accurately than they already do. The descriptive–normative slide seems to depend crucially on the middle term, which is something like a hypostatization: behind the phenomenal correlation, a deeper truth lurks.

classicist reviews of Daly and Costanza's work. In 1966 his career was revitalized by the publication of his collected works, and in 1971 he authored a long and ramblingly philosophical argument for the relevance of the second law of thermodynamics to economic science. The tendency for entropy to increase, G-R posits, has fundamental consequences for economic theories. All sources of free energy are eventually exhausted; all productive processes lead to the inexorable and irretrievable loss of materials. Economic growth is therefore subject to intrinsic and inflexible limits:

> We need no elaborated argument to see that the maximum of life quantity requires the minimum rate of natural resources depletion. By using these resources too quickly, man throws away that part of solar energy that will still be reaching the earth for a long time after he has departed. . . . There can be no doubt about it: any use of the natural resources for the satisfaction of nonvital needs means a smaller quantity of life in the future.[21]

But this isn't all: the second law also explains what makes things valuable. It is true that economic value only inheres in goods that add to the "enjoyment of life" (p. 282), but the source of this enjoyment is *low entropy:* "the basic nature of the economic process is entropic and the Entropy Law rules supreme over this process and over its evolution" (p. 283). Low entropy (or in some of G-R's formulations, *free energy*) is the base on which the superstructure of utility rests. Costanza's argument follows G-R's precisely, except that Costanza, like the general systems theorist Harold Odum, takes energy consumption directly as his index variable. Calling for "a cost of production theory with all costs carried back to the solar energy necessary to produce them," Costanza provided Daly and his allies with an explicit formulation of a theory of value with which to confront neoclassicism.

Over the course of the 1980s, Costanza and Daly hatched a plan for a "transdiscipline" that would combine the techniques of ecology with the subject matter of economics. Finally, Daly's dream of "economics as a life science"—the title of his first publication, in 1968—would be realized. In 1987 the two edited a special issue of *Ecological Modeling* entitled "Toward an Ecological Economics," in which they declared that "to effect a true synthesis of economics and ecology is the second most important task of our generation, next to avoiding nuclear war." In the same year, the Inter-

---

21. Nicholas Georgescu-Roegen, *The Entropy Law and the Economic Process* (Cambridge: Harvard University Press, 1971), 21.

national Society for Ecological Economics was formed in Barcelona; and in the intervening thirteen years the society has expanded to include some one thousand members worldwide, mostly concentrated in the United States, Western Europe, New Zealand, and Australia. Though miniscule in comparison to the institutions of neoclassical economics (Daly and Costanza were in the year 2000 the only two permanent faculty members at the only American institute for ecological economics, at the University of Maryland), the transdiscipline supports its own journal and biennial international conferences.[22] And while not all contributors to *Ecological Economics* endorse the radical denial of neoclassicism that Daly and Costanza have crafted, a significant number do.

Within this group of diehards, the precise source of the value of nature remains a subject of debate. Fine (and not so fine) distinctions between different measures of energy, measures of entropy, and more complex measures that include distributions of matter are a steady companion to the empirical studies of watershed economics and development trade-offs that grace the pages of *Ecological Economics*. But all of these measures have in common an attempt to pierce the fabric of illusions that markets present us with so that we may see the *true* values that lie beneath monetary prices. These true values then serve as guides to the rational management of the economy away from the unsustainable practices underwritten by neoclassicism. The importance of this shift is beautifully expressed thus by one ecological value theorist:

> Accounting frameworks are more than just a set of bookkeeping rules and conventions. They represent a particular conceptualisation or worldview of how the economy and ecological systems operate. If the accounting framework that is applied to an economic or ecological system is founded on questionable or inappropriate concepts, then it follows that the "prices" or "values" derived from such a framework are also of questionable validity.[23]

The ecological theory of value, then, is the ultimate authorizer of all the truth claims and policy recommendations of the eco-economists. And those theories are, themselves, underwritten by natural law. So nature has its revenge, and "annihilates" the Almighty Dollar (as Daly might say). Almost all the important aspects of utilitarianism are overturned. Value is

22. "About ISEE," 17 July 2000, http://www.ecologicaleconomics.org/about/index .html.

23. Murray Patterson "Commensuration and Theories of Value in Ecological Economics," *Ecological Economics* 25 (1998): 105–25, on 108.

rendered objective, not subjective; empirical, not psychological; public, not private; physical, not hedonic. Despite all this, eco-economic value theory retains a commitment to monism and mathematization. Though there are some interesting exceptions and modifications, ecological economics remains on the whole committed to algorithmic solutions. By reducing all economic and ecological phenomena to a single scale, eco-economics aspires to rankings of all proposed courses of social action, just like its neoclassical rival. Nature—in the sense of natural law—is the basis of a formal machine that spits out unambiguous answers to the question "What should we do?"

It should be noted that in this arrangement, nature still does not have any *moral* authority of its own. The moral imperative of ecological economic policy recommendations results from a chain of reasoning something like the following. (1) Eco-economics models show what courses of action can, under the constraint of natural law, be sustained over the long or at least the medium run; (2) we have an obligation to future generations (or to our future selves) not to deprive them of the means to sustain life; (3) therefore, we will take courses of action that ecological economics tells us are sustainable. In order to have moral force, natural necessity needs to be hooked up to a more-or-less independent system of ethical judgment; but once the two are properly attached, they constitute a sort of machine for the automatic production of ethical judgments. This algorithmicity of ethical judgment is one of the ways in which ecological economics differs from the other main challenger to neoclassical valuations of nature: the theory of intrinsic value provided by environmental ethicists.

Incommensurability and Moral Obligation

Ecological economics, despite its radical disagreements with neoclassicism, shares with it two critical features. The first, as pointed out above, is faith in algorithmic solutions to conflicts over the disposition of natural goods. The second is the aspiration to provide a unified scientific account of a huge range of human activity. Neoclassicists are famous for their attempts to explain all manner of cultural phenomena in economic terms; and eco-economists are perhaps even more extreme, with their aspirations to "transdisciplinarity."[24] In any case, *neither* neoclassicism nor ecologicism shows much interest in limiting the scope of economic reasoning.

24. On neoclassicist expansionism, cf. Nancy Cartwright's parallel with physics (in *The Dappled World* [Cambridge: Cambridge University Press, 1999], 1):

The discipline of environmental ethics has challenged neoclassicism on precisely the latter grounds. In effect, ethicists claim that nature's value is of an entirely different kind than is economic value; where the protection of wild nature is concerned, economic reasoning does not apply. This effective walling off of nature from economy is accomplished by establishing a direct moral claim that nature exerts on us; thus, the "moral authority of nature" is nowhere so immediately applied as it is here.

Like ecological economics, the subdiscipline of environmental ethics has its origins in the general flourishing of environmental thought and action during the early 1970s. While the claim that "[p]rofessional environmental ethics arose directly out of the interest in the environment created by Earth Day in 1970"[25] may be exaggerated, it seems unlikely that the field could have supported itself without the growing institutionalization of environmentalism of which Earth Day was representative. But for the most part, professional philosophers in the 1970s steered clear of the culture of scientific expertise that characterized much of the new environmentalism.[26] Though their writings often uncritically embraced (and continue to embrace) the findings of ecologists, ethicists for the most part disdained to build their theories directly on natural scientific grounds, looking instead for some *a priori* foundation for nature's claims on us.[27]

I shall focus on physics and economics, for these are both disciplines with imperialist tendencies: they repeatedly aspire to account for almost everything, the first in the natural world, the second in the social. Since at least the time of the Mechanical Philosophy, physicists have been busy at work on a theory of everything. For its part, contemporary economics provides models not just for the prices of the rights for off-shore oil drilling, where the market meets very nice conditions, but also for the effects of liberal abortion policies on teenage pregnancies, for whom we marry and when we divorce and for the rationale of political lobbies.

The environmental ethicists I discuss here would agree but go further, denying the legitimacy of economic analysis even of offshore drilling.

25. Eugene C. Hargrove, "Weak Anthropocentric Intrinsic Value," *Monist* 75, no. 2 (1992): 183–207, on 183.

26. In particular, away from the regulatory apparatus and policy think tanks established in this era. On the refashioning of American environmentalism in the 1970s, see Robert Gottlieb, *Forcing the Spring: The Transformation of the American Environmental Movement* (Washington, D.C.: Island Press, 1993), esp. 117–205. It should be noted that, at the same time, environmental ethicists seem rather distant from the direct action tactics of the more radical wing of environmentalism, or from any concern with the issues of environmental justice. In many respects, "intrinsic value" theories seem more than anything to be formulations in the language of professional philosophy of sentiments derived from the pre-1960s conservation groups.

27. This is in contrast to the 1960s notion of ecology as a "subversive science," a notion that survives today in ecological economics, among other places.

This is true even of the "deep ecology" approaches, which claimed to build primarily on insights from ecology, not directly on scientific results.[28] The earliest philosophical accounts attempted to build arguments on a notion of *rights* for nature, often in parallel or even conflated with arguments for animal rights. Echoing Aldo Leopold, whose 1948 essay "The Land Ethic" called for an "extension of ethics" to include "the land," these philosophers first sought to extend the liberal framework of rights to wild creatures and plants. At their most radical, they advocated a kind of limited personhood for natural entities.[29] But as the subdiscipline gradually institutionalized itself in the English-speaking world (with the founding of *Environmental Ethics* in 1979 and *Environmental Values* in 1990), "rights" gave way mostly to various versions of a theory of "intrinsic value." In the remainder of this section, I therefore first indicate briefly the shape of intrinsic value arguments. I then shift gears slightly to look more closely at how one particular ethicist uses moral obligation to build up a barrier between economics and nature.

A single, general characterization of intrinsic value theories of nature is somewhat difficult, as they are not immune to the successive (obsessive?) refinements of both terms and concepts that characterize most of Anglo-American philosophy. At one end of the field are "nonanthropocentric" value theorists (of both strong and weak varieties), who generally want to claim that the value of nature is both intrinsic to it and independent of humans. They differentiate themselves from the "weak anthropocentrists," who, by various tortured arguments, attempt to wriggle out of the *prima facie* incoherence of affirming that nature's value is intrinsic to it, but still dependent on humans. Recently, a small group of "pragmatists" has abandoned the language of intrinsic value, but for the most part these philosophers also hold that nature exerts a moral claim on us.[30] Despite important distinctions between different versions of intrinsic value, they share in

---

28. For "deep ecology," see Arne Naess, "The Shallow and the Deep, Long-Range Ecology Movements," *Inquiry* 16 (1973): 95–100, and Naess, "A Defence of the Deep Ecology Movement," *Environmental Ethics* 6 (1984): 265–70.

29. Aldo Leopold, "The Land Ethic," in *A Sand County Almanac* (New York: Oxford University Press, 1949), 201. On personhood, see Christopher D. Stone, *Should Trees Have Standing? Toward Legal Rights for Natural Objects* (Los Altos, Calif.: W. Kaufmann, 1974), and Roderick Frazier Nash, *The Rights of Nature: A History of Environmental Ethics* (Madison: University of Wisconsin Press, 1989).

30. In America, the most prominent representatives of strong nonanthropocentrism are Holmes Rolston III and Peter Taylor; of weak nonanthropocentrism, J. Baird Callicott; of weak anthropocentrism, Eugene C. Hargrove; and of pragmatism, Bryan Norton.

common certain basic features of their arguments. To begin with, all such theories reject the utilitarian notion that the sole source of value is human pleasure, and look to non-utilitarian ethical traditions to provide starting points for attributions of *noninstrumental* value. One common starting point is G. E. Moore's discussions of intrinsic aesthetic value;[31] for non-anthropocentrists, another frequent move is to invoke analogy to Aristotle's arguments for the value of friendship or Kant's quite stringent criterion of ends-in-themselves.

The first great difficulty is to reformulate a definition of intrinsic value so that it can be expanded to include nature and natural entities, for which these starting points make no provision.[32] As J. Baird Callicott puts it in the particular case of nonanthropocentrism:

> An adequate value theory for nonanthropocentric environmental ethics must provide for the intrinsic value of both individual organisms and a hierarchy of superorganismic entities—populations, species, biocoenoses, biomes, and the biosphere. In should provide differential intrinsic value for wild and domestic organisms and species . . . and it must provide for the intrinsic value of our present ecosystem, its component parts and complements of species.[33]

This is, as might be imagined, a Herculean task, and even within environmental ethics some have argued that it is impossible—that no coherent theory of intrinsic value can be formulated that has all these characteristics.[34] Still, there are all manner of clever tricks to be tried, sometimes involving escape-artist moves like invoking quantum indeterminacy or indexicality. If these sometimes stretch the definition of "intrinsic" far from the OED definition of "belonging to the thing in itself, or by its very nature," well, that's not uncommon for technical terms.

The second difficulty, though, is in linking intrinsic value (of whatever stripe) to moral obligation. Intrinsic value, especially in its most tortured versions, does not self-evidently bestow moral standing on an entity. For

31. G. E. Moore, "The Conception of Intrinsic Value," in *Philosophical Studies* (London: Routledge & Kegan Paul, 1922), 253–75.

32. G. E. Moore argues for the intrinsic value of beauty, and includes natural beauty within that category—most famously, sunsets—but most eco-ethicists find this basis too limited.

33. Callicott, "Non-Anthropocentric Value Theory and Environmental Ethics," *American Philosophical Quarterly* 21 (1982): 299–309.

34. See, e.g., Tom Regan, in "Does Environmental Ethics Rest on a Mistake?" *Monist* 75, no. 2 (1992): 161–82.

instance, suppose we interpret "intrinsic value" to mean "having ends, or 'goods' of its own." As John O'Neill points out in a recent critical essay:

> It is a standard at this juncture of the argument to assume that possession of goods entails moral considerability. . . . This is mistaken. It is possible to talk in an objective sense of what constitutes the goods of entities without making any claims that these ought to be realised . . . one can state the conditions for the flourishing of dictatorship and bureaucracy. The anarchist can claim that "war is the health of the state." . . . One can recognize that something has its own goods, and quite consistently be morally indifferent to these goods or [even] believe one has a moral duty to inhibit their development.[35]

So the trick is to find some stronger meaning for intrinsic value, or some more compelling link to obligation. One option is the Kantian criterion of "end-in-itself"; then a ready-made theory of ethical obligation is available. Hence the call from one Australian philosopher for "a recognition that biological systems are items which possess intrinsic value, in Kant's terminology, that they are 'ends in themselves.'"[36] This, of course, has its own problems, since while there may be reasons (analogy?) to call individuals of other species "ends-in-themselves," it's not easy to see how that status transfers up to other higher-order categories, or "nature" as a whole; and a whole host of other problems arise as well.[37] Others look to Aristotelian solutions, still others to pragmatism. However difficult the path, almost all environmental ethicists continue to make the claim that nature possesses a kind of value that exerts a moral obligation on us.

In the conflict with neoclassical economics, this obligation to nature serves to exempt environmental goods from economic calculation. To cite a recent paper on this subject:

> [T]here [is] no reason to assume that judgements about the good of a thing with moral standing could be represented as a quantity amenable to arithmetical treatment alongside other such quantities. This shows that the pro-

35. O'Neill, "Varieties" (note 3), 131.

36. William Godfrey-Smith, "The Value of Wilderness," *Environmental Ethics* 1 (1979): 309–19, on 318.

37. For instance, the conflict between the "intrinsic value" of wolves and deer; or the imperative to provide differential intrinsic value for wild and domestic organisms and species. See Regan, "Does Environmental Ethics Rest on a Mistake?" (note 34).

cedure of adding the values that represent existence value to reach a total for the group is not applicable to the value of the preservation of a thing with moral standing.[38]

Moral decisions simply cannot be discussed in economic terms; obligation weighs directly on us in a manner that is incommensurable with economic ("prudential") reasoning. The legitimate realm of economics is thus diminished, without any arguments being made against its application *in principle* within that smaller purview.

This walling-off of nature from economy is a strategy even in the most "anthropocentric" of the environmental ethicists. One of the most elaborate and forceful defenses of such a division comes, in fact, from law professor Mark Sagoff, who rarely mentions intrinsic value and endorses such a notion at most implicitly.[39] Sagoff's *The Economy of the Earth* vigorously defends two parallel distinctions: between citizens and consumers, and between values and preferences. Politics is the realm of values, where we act as citizens; economics ought to concern itself quite narrowly with preferences, and our actions as consumers. "In my role as a *citizen,*" Sagoff says in the introduction,

> I am concerned with the public interest, rather than my own interest; with the good of the community, rather than simply the well-being of my fam-

---

38. Jeremy Roxbee Cox, "Preservation Value and Existence Value," in Foster, ed., *Valuing Nature?* (note 11), 113.

39. See, e.g., Mark Sagoff, "Carrying Capacity and Ecological Economics," *BioScience* 45 (1995): 610–20, on 610, where he condemns economists for assigning "shadow prices to intrinsic values." In "Do We Consume Too Much?" *Atlantic Monthly,* June 1997, available online at http://www.puaf.umd.edu/papers/sagoff.htm, he quotes with obvious approval Ronald Dworkin, to the effect that "we should admire and protect [species] because they are important in themselves" (p. 2); while in "On the Value of Endangered and Other Species," *Environmental Management* 20, no. 6 (1996), available online at http://www.puaf.umd.edu/papers/sagoff.htm, he somewhat vaguely endorses a Kantian framework for environmental ethics. This is one of his characteristically problematic elisions, worth quoting at length:
> The non-utilitarian approach to decision making applies paradigmatically not to consumer goods but to objects which, given their symbolic, historic, and spiritual significance, are valued more because of their meaning than because of their use. Kant (1959 [1785]) draws this distinction as follows; "That which is related to general human inclination and needs has a *market price.* . . . But that which . . . can be an end in itself does not have mere relative worth, i.e., a price, but an intrinsic worth, i.e., 'a dignity.'"
> The rather important distinction between "meanings" and "intrinsic worth" is entirely elided here. In *The Economy of the Earth,* Sagoff wisely confines himself to the much more neutral term "intangible values."

ily. . . . In my role as a *consumer* . . . I concern myself with personal or self-regarding wants and interests; I pursue the goals I have as an individual. I put aside the community-regarding values I take seriously as a citizen, and I look out for Number One instead. I act upon those preferences on which my personal welfare depends.[40]

When decisions truly involve no ethical dimension—when no *value* insists that we take it into account—then economic reasoning is called for. When, however, any kind of moral issue is at stake, economics must be put on hold, and political processes engaged. To treat values or beliefs (Sagoff often equates the two) as "preferences" in the economic sense is a category mistake, which compares incommensurables. Contingent valuation misunderstands this fact fundamentally. While Sagoff concedes that "[m]arkets generally succeed . . . in maximizing the welfare or well-being of individuals within resource constraints,"[41] that is not the goal of environmental policy. Environmental policy is not about what we want, but what is right. Trying to put a price on this is absurd, just as it would be to ask how much we would be willing to pay for our favorite team to win a football game, or giving a trial verdict based on how much jurors are willing to pay for the verdict they endorse. "Those willing to pay the most, for all intents and purposes, have the right views; theirs is the better judgment, the deeper insight, and the more informed decision."[42] This notion is a pernicious confusion, and economists who go about computing contingent valuations and shadow prices "may be compared to the Japanese soldiers who were found on islands in the Pacific years after the end of the Second World War, still fighting although the mainland had surrendered and the cause had long since been lost."[43]

This diagnosis, of course, applies just as much to ecological economists as to neoclassical ones; contingent valuation is just a particularly obvious violation of the distinction in principle between economy and nature. Any

40. Mark Sagoff, *Economy of the Earth* (Cambridge: Cambridge University Press, 1988), 8.

41. Sagoff, "On the Value" (note 39), 4.

42. Sagoff, *Economy of the Earth* (note 40), 42; see also 113, 87. The choice of analogies is interesting. Professional sports competitions *are* determined to a large extent by willingness to pay; and the impact of defendants' wealth on trial verdicts is well-known. This is one of many moments when Sagoff's platitudes suggest that the distinction he draws is nowhere so clear as he supposes.

43. Mark Sagoff, "Four Dogmas of Environmental Economics," *Environmental Values* 3 (1994): 285–310, available online at http://www.puaf.umd.edu/papers/sagoff.htm.

attempt to reduce the value of nature to a measure, and then use that measure to derive algorithmic solutions to the disposition of environmental goods, also misses the point. In 1995 Sagoff admonished Daly and Costanza on these very grounds:

> The reasons to protect nature are moral, religious, and cultural far more often than they are economic. . . . To argue for environmental protection on utilitarian grounds—because of carrying capacity or sources of raw materials and sinks for wastes—is therefore to erect only a fragile and temporary defense for the spontaneous wonder and glory of the natural world.[44]

Though the ethicists and the ecologists often support similar policies, they do so under the auspices of radically different kinds of authority. The ethicists pooh-pooh natural law, replacing it with the towering figure of Nature; but at the same time, they are perfectly content to give neoclassical accounting free rein in the inconsequential spheres of human activity that lie outside the cathedral of nature. Small wonder that Daly responds with accusations of "pantheistic sentimentality about the divinity of nature."[45] Sagoff and the intrinsic value theorists deny commensurability of value and algorithmic applications, but they replace economic machinery with an absolutism that admits of little compromise with other human concerns—and only vague recommendations for resolving conflicts among them. Though both ecological economics and environmental ethics starkly oppose neoclassicism, and especially the technique of contingent valuation, they seem to be just as far removed from each other as they are from their common enemy.

Some Closing Thoughts

Contemporary conflicts over nature's value are framed by two sets of oppositions: between formal and nonformal accounts of value, and between subjectivism and objectivism. Both neoclassical and eco-economics have formal theories of value and attempt to develop formalisms that can give algorithmic instructions for the disposition of natural goods. True, the

44. Sagoff, "Carrying Capacity" (note 39), 618.
45. Herman Daly, "Reply to Mark Sagoff's 'Carrying Capacity and Ecological Economics,'" *Bioscience* 45 (1995): 621–24, on 624.

content of those instructions tend to differ radically; but behind both traditions is the old technocratic faith in the possibility of optimal management. And both contend that by optimizing the economy, the disposition of natural goods can also be optimized. In this management ethos both economic schools stand in stark contrast to the ethicists, who decry the extension to nature of any formalized valuations as illegitimate. The ethicists and eco-economists stand together, though, in their rejection of subjectivism. Both look beyond desire to ground value—the ethicists to Nature, the eco-economists to Natural Law.

As a general view of the seething debates over our environment, the picture I've drawn is doubtless skewed and myopic; but though we may have lost sweeping vista, I am hopeful we have gained in clarity (not to imply that those two virtues are commensurable!). Like Kant in the First Critique, we see that our little playing field has an empty square. Subjectivist, nonformal theories of value play no important role in appraisals of nature. Perhaps they should; and in the last few years scholars have tentatively made moves in that direction. I have no well-worked out arguments against such efforts; but as regards all the major positions, I have strong reservations.

My uneasiness starts with all the manifold signs that Sagoff's division between consumer and citizen roles is untenable. Even consumption decisions are always about far more than narrow utility maximization; they're about major corporations versus petty bourgeois merchants, about organic versus conventional, about local versus global, even—still—about artisans versus industry. All of these are concerns with moral elements that intrude deeply into the workings of the market, concerns that may or may not be commensurable with each other. Nor is this critique of utilitarian economics new (though, like others who have leveled it, I find it devastating). What I want to point out here is that it is almost as dangerous for intrinsic value theorists as it is for utilitarians. If the legitimate application of utility calculation shrinks to the null set, then we are left with a field of incommensurate and conflicting moral obligations, and no clear guidance on how to work them out. Unless one holds to a sort of totalitarian theory of the intrinsic value of nature, under which nature's value simply trumps all other values regardless of circumstance, the environmental ethicists have brought us no further than we were. Incommensurability cuts both ways.

Eco-economics escapes that trap insofar as it trumps authority with necessity. When gravity says go down, you don't call a conference on it, you drop. This line of argument is oddly reminiscent of medieval complaints about people who break the laws of nature, especially since eco-economists do sometimes seem to be carrying around great big signposts reading "Paradise, this way; apocalypse, that way."[46] Nonetheless, it would be too hasty to dismiss Daly, Costanza, and company on grounds of comical analogy. But their ham-fisted deployment of thermodynamics does, I think, merit some philosophical objection. Eco-economics relies on a vision of the unity of science that only the most unreconstructed positivist could love, where economic laws not only reduce to physical laws, but are replaced by them altogether. My objection here is not simply that I find "the disorder of things" a more pleasing picture, but that there are very strong philosophical, and yes, *scientific,* reasons to avoid extending laws from one domain into another. Neither the economy nor the natural environment are thermodynamics laboratories, and treating them as such is deeply suspect.[47]

If there is a way forward, I suspect it does not lie precisely in the terrain mapped out here. For in opposing subjective value theories with objective ones, environmentalists have participated in a philosophical fallacy with profound practical consequences. To imagine that the subjective/objective dichotomy is exhaustive is to ignore not only a huge philosophical literature, but the whole huge realm of the social—to deny that values are formed and held in collectivities, in social situations, *in context.* If the inner and natural world both have contributions to make to value, so too does the social world; we need, I think, a theory of value that makes space for collective judgments. Indeed, the need is quite pressing. The fact is that institutions for the regulation of world economic activity exist; that they have explicit policies based on theories of efficiency and justice; and that these policies have concrete effects on the lives of people all over the world. It is hardly irrelevant that the World Bank is staffed by neoclassicist

46. Cf. Robert Costanza, "Visions of Alternative (Unpredictable) Futures and Their Use in Policy Analysis," *Conservation Ecology* 4, no. 1 (2000), available online at http://www.consecol.org/vol4/iss1/art5.

47. On problems with unity, see John Dupré, *The Disorder of Things: Metaphysical Foundations of the Disunity of Science* (Cambridge: Harvard University Press, 1993); Peter Galison and David J. Stump, eds., *The Disunity of Science: Boundaries, Contexts, and Power* (Stanford: Stanford University Press, 1996). To my mind, the most powerful critique of reductionism is Nancy Cartwright's. Her notion of nomological machines—sets of shielding conditions under which scientific laws can legitimately be said to function—goes a long way toward making sense of the power of scientific knowledge while setting quite strict limits on its application. See Cartwright, *The Dappled World* (note 24).

economists, or that Herman Daly's name appeared on flyers for the teach-ins preceding mass demonstrations against the Bank.[48] Whether we heed it or not, a struggle is raging over how to weigh justice, economic efficiency, and the environment; and the value of nature has everything to do with it.[49]

48. For an audio recording of Daly's contribution, see "Panel 1," *A16—IFG Teach-in 4/14/2000*, available online at http://www.radio4all.net/proginfo.php?id=1826. Both the Direct Action Network and 50 Years is Enough!, the two central organs of the Seattle Movement against globalization, list environmental devastation high among the Bank's crimes: Fifty Years is Enough!, "Platform," 15 October 1998, available online at http://www.50years.org/platform.html; Direct Action Network, "Mission Statement," 8 August 2000, available online at http://www.directactionnetwork.org/missionprinciples2.htm.

49. Certainly this struggle continues in the recent debates surrounding Danish statistician Bjørn Lomborg's *The Skeptical Environmentalist* (Cambridge: Cambridge University Press, 2001). Lomborg's main argument, expressed in a rambling, polemical style over some 350 text-book-sized pages, is that the *cost* of preserving natural goods (such as biodiversity and global climate patterns) far outweighs the *value* of those goods themselves. Frustrated critics , while ready enough to respond to Lomborg's calculations, seem not to grasp that their most fundamental disagreement lies not in mere empirical details, but in Lomborg's calculus of value. For one example of many, see Stuart Pimm and Jeff Harvey, "No Need to Worry about the Future," *Nature* 441 (2001): 149–50.

PART TWO

Necessity and Freedom

Debates about the orders of nature are ways of working out the human sense of freedom and perception of necessity. Does nature let us be human in different ways? Or does it dictate to us what being human *is,* beyond which lies only failure—nature's and our own? In the six essays of this section, we explore how physicians and naturalists, philosophers and jurists, rulers and radicals all grapple with the tensions between freedom and constraint.

"Nature" is characteristically expressed as self-evident and given, never invented or produced. This gives it the force of necessity, at different levels. Nature sometimes designates each person's makeup, whether reflected in the balance of bodily humors or, as some feminist activists assert, in a unique cervix (see Murphy, essay 13). Each individual is also a member of a species endowed with its own characteristics—humans, for example, with reason and ants with cooperative instincts (as studied, respectively, by Cadden [essay 8] and Lustig [essay 11]). Nature has also presided over social arrangements marked by diverse hierarchies and divisions (of labor, classes, sexes), or forms of social and legal organization. Finally, individuals, species, and societies must all fit into the general order of the universe.

Every possible omnipotence has been ascribed to nature, but not all at once: we seem to be uneasily placed among these interlocking, potentially conflicting orders. People have sought them in various places. Nature may give order to us directly, as in medieval poetry (Cadden), or indirectly, as in the case of natural historical observation, of ants (Lustig) or onanists (see Vidal, essay 10). Or people may assert nature's power only to deny it as an act of political will, by subverting childbirth (Murphy) or indeed their entire physical selves (see Thomas, essay 12). The essays in this section all demonstrate how much energy has gone into asserting nature's power, justifying it, maintaining it, contesting it, controverting it, and denying it.

Are there ways to escape that power altogether? The realm outside nature might be thought uninhabitable, but it is in fact densely populated: with escapees, like feminists wielding technoscience to liberate the body (Murphy); exiles, like those whose crimes and very persons were pronounced "against nature" by the state (see Puff, essay 9); artificers, like the Japanese political philosopher who insisted on doing without nature in inventing political order (Thomas); and believers, like the Jesuit evolutionist naturalist for whom nature itself is ordained by the divine (Lustig). Establishing what lies within and what without—defining the natural and its various opposites—are not merely philosophical exercises, but acts of political authority with grave consequences.

But do we have to get outside nature to find freedom, or does our freedom lie within nature? Freedom and nature are not necessarily antagonistic: nature may entail opportunities, not just constrain them. In this section, desire emerges as a force that has long been invested with the ability to open up spaces for agency as well as disrupt nature's orders (Cadden). The expression of erotic desire has at times been perceived as a means to generate freedom within society, as some progressive Japanese postwar thinkers hoped (Thomas), and, in other contexts, as a danger to the heterosexual order that must be policed, as illustrated by early modern legal proceedings against sodomites (Puff). Sometimes it is nature that enforces its will on us, and sometimes it is we who dictate to nature. Sometimes, in contrast, it is failure, ours or nature's, that opens unexpected space for freedom. These six essays examine different forms of this human negotiation with nature—efforts to perfect, disrupt, enforce, or escape nature—and explore the extent to which they represent the conflict of necessity and freedom.

8

Trouble in the Earthly Paradise: The Regime of Nature in Late Medieval Christian Culture

Joan Cadden

[It is] a manifest and detestable error [to hold] that the sin against nature, in particular the abuse in the course of coitus, although it is against the nature of the species, is not, however, against the nature of the individual.

ETIENNE TEMPIER, Bishop of Paris, 1277

In medieval Europe, "necessity" had close ties to the dynamics of nature; "freedom" had close ties to the dynamics of the moral sphere. Like the other categories central to the essays in this volume, however, these concepts embodied specific historical meanings peculiar to their particular contexts. In the thirteenth and fourteenth centuries, nature wielded considerable powers, often expressed in terms of necessity. But the rule of nature, whether construed in prescriptive terms (as its order) or in executive terms (as its dominion) was not absolute. Thus infractions against its rule were not subject to the sort of inexorable retribution by nature itself that Fernando Vidal explores in his essay concerning eighteenth-century medical opinions about the fate of masturbators. Nor did medieval freedom involve the option of choosing, much less creating one's own personal or political reality in ways that became imaginable to moderns—whether the American feminists discussed by Michelle Murphy, who were determined to alter nature, or the Japanese political philosopher discussed by Julia Thomas, who was determined to transcend it. Rather medieval free will offered the choice of aligning oneself (or refusing to align oneself) with divine will, which might be expressed by the providential order of Creation.

I am grateful to Margaret L. Zimansky, as well as to the group that produced this volume, for valuable criticisms and advice.

— — —

Nature was, in many respects, the source, judge, and enforcer of right living and proper social relations in the view of both academic and social elites in the late Middle Ages. It directed the production of desire and its regulation under the auspices of reason, both crucial to the psychological dynamics of virtue and the political dynamics of hierarchy. But, if natural desire and pleasure were necessary elements in the dialectic of moral goodness and perhaps also of justice, they were dangerous forces that necessarily subverted the orders they supported. The cultural powers and weaknesses of nature in the traditions of Christian Europe manifested themselves in sources as diverse as vernacular poetry and scholastic commentaries on Aristotle.[1]

Chaucer's "Parliament of Fowls," written in the last quarter of the fourteenth century, illustrates the late medieval view of moral and political good that arises from the benevolent rule of Nature.[2] In this poem a scholar, interested but inexperienced in love, is reading a book that contains, among other things, a description of the celestial spheres as the sources of earthly harmony and intimations of political virtue. That night, he has a dream in which he finds himself at the entrance to a lush garden. The gateposts announce what he already knows from books: that love is a powerful and ambiguous force, capable of curing mortal ills and administering deadly strokes (ll. 127–40). He is quite paralyzed by the promise/warning but his guide in this dream, a model citizen of the Roman Republic, encourages him to pass through, assuring him that the gates speak only to Love's servants, of whom he is assuredly not one. The garden itself seems at first to be largely untainted by such contradictions and is filled with a profusion of plants and animals. Order and harmony reign—each of the many kinds of trees has its own proper leaves (ll. 173–74) and the sound of the breeze is in tune with the songs of the birds (l. 203). Only the area around the temple of Venus betrays the existence of a problem, harking back to the mixed message at the gate: her entourage includes not only Peace, Patience, and Gentility, but also Lust, Jealousy, and coercive Craft;

1. The literary and philosophical sources have been studied well but separately. See, e.g., George Economou, *The Goddess Natura in Medieval Literature* (Cambridge: Harvard University Press, 1972); Albert Zimmermann and Andreas Speer, eds., *Mensch und Natur im Mittelalter*, 2 vols., Miscellanea Medievalia, Veröffentlichungen des Thomas-Instituts der Universität zu Köln, 21/1–2 (Berlin: Walter de Gruyter, 1991).

2. Geoffrey Chaucer, "Parlement of Foules," in *Works*, ed. F. N. Robinson, 22nd ed. (Boston: Houghton Mifflin at Cambridge, Riverside Press, 1957), 309–18.

Desire keeps company with Youth, Flattery, and Pandering. Within the temple, the dreamer sees Venus herself, flanked by the deities of food and drink, Ceres and Bacchus (ll. 218–77), like Venus, the objects of natural desire. He is disturbed to see wall paintings of so many who came to a bad end because of love, such as Hercules, Dido, Tristan, and Cleopatra.

When he emerges from this disturbed and disturbing space, he and the scene recover their equilibrium in the presence of the surpassingly beautiful and noble queen of the land, the goddess Nature. With her favorite bird, a female eagle, on her hand, Nature calls to order the flocks of every kind of bird that have gathered in her presence, and bids each take its proper and customary place. Chaucer enumerates the pecking order in great detail, with the eagles and hawks occupying the highest perches and the small, common, and ignoble birds in the lower ranks. It is Saint Valentine's Day, when, as she declares, her law decrees that they must choose mates: "by my statut and thrugh my governaunce, / Ye come for to chese—and flee youre way— / Youre makes as I prike you with plesaunce" (ll. 387–89). She herself may not break her own law ("my rightful ordinaunce / May I nat breke" [ll. 390–91]), which dictates that the most worthy male bird will be the first to choose his partner. The bird of highest degree is the tercel eagle—royal, wise, and true—formed in every respect as it pleases Nature. He selects the female favored and held by Nature herself. Two other eagles, however, immediately petition for the right to court her, and each declares why he ought to be favored. Among the various kinds of birds, courting behavior differs according to the dignity of the species and male–female pairing in the lower orders involves a certain amount of squawking. Nature orders her subjects to hold their tongues and has the groups select spokesbirds (the "parliament" of the title) to give counsel. Although she has clear authority to do so, she declines to adjudicate the dispute among the noble suitors. Rather, she gives in to the plea of the female and puts the case off until the following spring. By the end of the poem, with the exception of the eagles, to whose predicament we shall return later, the birds all have mates: "To every fowl Nature yaf his make / By evene accord" (ll. 667–68). After a song, they fly off happily; the still unenlightened scholar awakes from his dream and turns to other books, in hope of improving himself.

Chaucer's poem introduces many of the themes of this study, which aims to elucidate the problematic ways in which notions of the natural come into play in late medieval moral and social values. Similar questions arise in the context of Latin scholastic commentaries on Aristotle's *Nicomachean Ethics,* for learned Latin philosophical discourse and courtly vernacular culture were by no means isolated from each other. "The Parliament of Fowls" sets out a conceptual earthly paradise in which good personal conduct and social order both derive from and are presided over by Nature—an apparent ideal about which both poets and philosophers raised doubts. First, it is Nature that unites the reigning harmonies of the celestial spheres with the political hierarchy of earthly creatures, in which each species occupies its rightful place, and where all counsel and rejoice in concert. This integration evokes the neo-Confucian view of universal correspondences described in this volume's essay by Julia Thomas, but here "nature" is a distinct, if mediating, concept. The connection extends to the (social) mating of individual birds, which Nature accomplishes with "evene accord," resembling the "evene nombres of accord" with which she knits together the (cosmic) qualities of hot, cold, moist, and dry (ll. 379–81). Second, she governs and judges; she is the one whose role it is to set the terms for personal and political relations, and to resolve conflicts. The connection between the individual and the social is explicit here, as it is in the eighteenth-century conceptualization of masturbation described in the essay by Fernando Vidal and in some early modern and modern views of bees and ants as described by Danielle Allen and Abigail Lustig. Third, groups (or "kyndes") and individuals tend to follow Nature's rules without coercion. She permits them to choose mates and elect representatives, a freedom they generally exercise by conforming to the larger order of things. For, in spite of limitations imposed by the structure of the allegory, the birds (and the humans they represent) are of, not simply in, Nature. Thus creatures are constrained by an architecture that extends beyond them and by drives that operate within them. More particularly, the natural order depends on and is mediated by "kinds"—types defined by their specific collective natures—which is why the birds, whether eagle, cuckoo, or turtle dove, speak true to type in the council. Finally, individuals' choices are informed and moved by desire or, more precisely, by incitement to pleasure. They have come, Nature explains to them, to choose mates, "as I prike you with plesaunce" (l. 389).

Unlike the space occupied by Venus and her companions, where such incitement can give rise to discord, suffering, and death, Nature's parliament is a place where desires that give rise to bliss and joy (l. 669) form a part of and conform to the natural order. This peaceable integration of pleasure is signaled by the garden's fertility and celebrated in the final harmonious song honoring its Author, the providential source of all this goodness. The dark side of desire—vices such as lust, jealousy, bribery—so visible in the realm occupied by Venus, seems isolated from the well-ordered feelings and behavior over which Nature presides. Noble avian suitors are inspired to offer service, obedience, kindness, and fidelity as the virtues attendant upon well-ordered love (ll. 420–83). Nature thus has a role in instilling individual virtue, as well as social and political hierarchy and concord.

Such a reading of the "Parliament of Fowls" supports a *prima facie* case for the normative authority of late medieval nature to generate imperatives and enforce constraints regarding individual and social behavior, in other words, to act as an agent or manifestation of necessity. At the same time, it contains some of the machinery by which these norms are confounded and the authority undermined, and thus hints at trouble in Paradise. For as the source and arbiter of value in human life and society, this concept of nature was neither complete nor consistent. First, although reason, civic relations, and heterosexual pairing were all construed as natural, the very concept of virtue and vice implied some sort of freedom—something beyond what was determined by individual or human nature. Second, even when such imperatives did direct human action, as in the commonplace consumption of appropriate food, they were not always powerful enough to coerce compliance, as demonstrated by the distasteful practice of eating dirt and the abomination of cannibalism. Finally, disorder as well as order, unnatural as well as natural behavior, could be the consequences of the most basic principles of nature, since the same appetites necessary to achieve proper ends could also lead toward improper ones.

After suggesting the cultural environments for such concerns, this essay explores these three senses in which late medieval notions of nature and the natural both convey and betray normative control. For although the meanings of nature in late medieval Christian culture derived content and force from a vivid allegorical tradition, a substantial body of respected philosophical writing, and a religious reverence for its Creator, doubts and unease accompanied them as well.

Intersecting Contexts

Vernacular literature produced in a court environment and Latin philosophical works emerging from a university tradition provide the sources for this study. Taken together they reveal that the themes and tensions uncovered here were not just the lush literary play of an insular aristocratic society nor merely the dry philosophical games of isolated academic institutions. Rather, these concerns belonged to a recognizable if not fully integrated social and cultural elite. The resources the two groups drew on and the genres they favored were diverse, perhaps even divergent. In the twelfth century, when the cleric Alain de Lille (d. 1202) had composed an influential Latin discussion of nature's status, he had alternated between poetry and prose, between allegory and philosophical argument. A century later, allegory and poetry belonged largely to the realm of a burgeoning vernacular literature exemplified by Chaucer's work. Philosophy, in the meantime, had become more specialized and had evolved its own technical forms of Latin writing. Newly available texts such as Aristotle's *Ethics* and *Politics* provided the occasion for scholars like Thomas Aquinas or the less-well-known Heinrich von Friemar to explore the constraints within which human virtue and justice operated.

The distance that had grown up between the vernacular and Latin traditions is apparent in the lives of the two authors considered here. Geoffrey Chaucer (ca. 1340–1400) was the son of a London merchant. After some early schooling he entered into a life of service in courts of the English nobility, later holding public office as a member of Parliament and Comptroller of the Custom. His major works (in English) consist of lyric, narrative, and allegorical poetry. By contrast, Heinrich von Friemar (ca. 1245–1340) came from rural Thuringia, had a university education, and made his career as a professor of theology and preacher in Erfurt, occasionally acting as a delegate or officeholder within the Augustinian order. His major works (in Latin prose) consist of moral philosophy, theology, and sermons.[3]

Neither the lives of these men, however, nor the literate traditions in which they participated were as distinct as they might appear. Chaucer made extensive use of Aristotelian natural philosophy, wrote a technical treatise on the astrolabe, and translated Boethius from Latin into English. Indeed, he drew his portrait of Nature from the work of Alain de Lille.[4] Heinrich's writings do not display an equivalent engagement with vernacular culture, but he did have a place in court society as confessor to the local count. Nor was he disengaged from the urban scene or the public events of his time. As a master of the faculty at Paris, he was signatory to findings in the highly politicized heresy trials of the Templars and the mys-

3. Clemens Stroick, *Heinrich von Friemar: Leben, Werke, Philosophisch-theologische Stellung in der Scholastik* (Freiburg: Herder, 1954), esp. 12–20.

4. See Chaucer, "Parlement of Foules" (note 2), l. 316. Chaucer's scholar refers the reader to Alain's *Complaint of Nature.* Nikolaus M. Häring, ed., "Alain of Lille, *De planctu naturae,*" *Studi Medievali,* 3d ser., 19 (1978): 797–879.

tic Margaret Porete, who was burned in 1310 along with her French work, *The Mirror of Simple Souls*. Both men traveled widely. Chaucer spent time on the Continent, picking up works by Dante and other literary figures proficient in Latin as well as prolific in vernaculars. Heinrich, in addition to undertaking voyages on Church business, studied at Bologna and spent a decade at the university of Paris, a cosmopolitan institution rich in texts, opinions, and controversies. Geographically and chronologically separate, the two men did not cross paths, much less influence each other, but they did not live in entirely different worlds.

The more general divide between Latin and vernacular literate cultures was likewise porous, as the history of texts relevant to nature and values illustrates. A version of Aristotle's *Ethics* appeared in Italian in the 1260s, another version in French in 1370.[5] In 1267 Brunetto Latini summarized Aristotelian ethics (among other philosophical subjects) in his *Book of the Treasure* that circulated widely throughout Europe in various languages. Nor was the flow of texts and ideas all in one direction. The long French allegorical poem on love, *The Romance of the Rose,* in which the figures of Reason and Nature play significant roles, was widely read by university students. Participants in an exchange of polemics on its moral stance included not only the court-oriented French author Christine de Pizan but also the prominent theologian Jean Gerson, chancellor of the University of Paris. It is because of this intellectual and social commerce that an exploration of a variety of late medieval texts promises to yield a map not of influences but of some shared questions, concerns, and cultural resources relating to the ways in which ideas about nature were implicated in the constitution of moral and social values.

These examples also suggest crosscurrents between the religious or theological and lay or secular environments. Although the texts examined here relate to what were sometimes called "natural" or "rational" virtues and vices, the presence and weight of Christian faith and doctrine color the questions, methods, and responses of the authors considered. In particular, for many, grace was far more relevant to the origins of goodness than nature. And while Aristotle's ideas about human freedom and the individual's cultivation of a just disposition could be harmonized with Christian notions of responsibility, the doctrine of free will had status and

5. Nicole Oresme, *Le Livre de Ethiques d'Aristote: Published from the Text of Ms. 2902 Bibliothèque Royale de Belgique,* ed. Albert Douglas Menut (New York: G. E. Stechert, 1940); on other translations, see the introduction to ibid., 39.

implications of its own.[6] Furthermore, the Creator of nature as a whole and of the various natural kinds always lay behind any power nature might have to produce goodness (or the conditions for it) and to judge evil. It was as "vicaire of the almyghty Lord" that Chaucer's Nature secured the harmony of pairs of opposite qualities (ll. 379–81). To the extent that Nature had any normative authority, it derived from the divine; to the extent that she failed to advance the norms or exercise the authority, she would seem to have departed not simply from her own principles of governance but from her responsibility to higher, divine laws.

For poets like Geoffrey Chaucer or the authors of the *Romance of the Rose,* as for theologians like Albertus Magnus and Thomas Aquinas, divinity and faith were never irrelevant; but medieval Christian authors themselves considered, often as a sort of thought experiment, whether and to what extent a pagan or a Jew could be virtuous. Thus historical examination of the nontheological dimensions of good and bad actions is, up to a point, legitimate. The great honor accorded Virgil by Dante is but the most famous medieval variation on this theme. An anonymous thirteenth-century author, asking about the origins of moral virtue, answers first, "speaking theologically," and continues "one can speak of them in another way, and that solution is according to the philosophers and not according to the theologians."[7] What a pagan like Virgil can achieve is the full realization of his nature as a human being, that is, of his reason. The status of reason, as the definition of human nature, as the foundation of philosophical inquiry, and as the representation of the order of Creation was closely, though not invariably, tied to the status of nature where human affairs were concerned. Thus Thomas Aquinas, reflecting on the foundations of law, declares: "In human affairs, something is said to be just, when it is right according to the rule of reason. And the first rule of reason is the law of nature."[8]

6. Odon Lottin, *Psychologie et morale aux XIIe et XIIIe siècles,* 2nd ed., 6 vols. in 7 (Gembloux, Belgium: J. Duculot, 1957–60). See also Marcia Colish, "*Habitus* Revisited: A Reply to Cary Nederman," *Traditio* 48 (1993): 77–92, here 80–86.

7. Paris, Bibliothèque Nationale, ms. lat. 3804A, fol. 154vb–155ra, cited in Lottin, *Psychologie et morale* (note 6), vol. 1 (1957), 521: "Dicendum est quod loquendo theologice. . . . Aliter potent dici, et ista solutio est secundum philosophos et non secundum theologos."

8. Thomas Aquinas, *Summa Theologiae* (New York: McGraw Hill, and London: Eyre and Spottiswoode), bk. 1a2ae, q. 95, art. 2; vol. 28, p. 104: "In rebus autem humanis dicitur esse aliquid justum ex quo quod est rectum secundum regulam rationis. Rationis autem prima regula est lex naturae."

Virtue and Justice in the Peaceable Kingdom of Nature

For late medieval thinkers, including Aquinas, there was a snake in this garden of Nature, Reason, and Justice. Necessity and freedom are not harmoniously resolved: if nature dictates, then the just deserve no praise; though nature dictates, it is frequently ignored; and when nature dictates, its messages sometimes conflict. Each of these sources of dissonance is treated below. They derive their significance from the widely articulated and deeply ingrained set of beliefs that bound virtue and justice, on the one hand, to nature and reason on the other. A case in point is the late-twelfth-century theologian-poet Alain de Lille. The Nature he portrays, discussed more fully in this volume by Katharine Park and Helmut Puff, is beset by the unnatural, but her distress depends on an understanding of an ideal order.[9] In another allegorical work, Alain ascribes an active role to virtues that are not explicitly Christian, such as temperance, prudence and piety, in the creation of a new and more perfect type of creature, humankind, under the direction of Nature and her chief counselor Reason. Although Alain's argument is that a being with an immortal soul cannot be constituted without the direct intervention of God and the participation of Christian virtues (notably Faith), each of the more secular virtues contributes gifts to Nature's new production. For example, after Reason has endowed the being with her capacities, "Honesty unlocks her treasure and . . . deposits in him everything she has."[10] Honesty, Prudence, and other virtues thus inhere in humans in the same way reason does—as aspects of their essence and definition, that is, as elements of their nature. In other philosophical and theological works, Alain does not wholly and unambiguously endorse either the notion that virtue is built into human nature or that it is the product of nature in a more general sense, but in his treatise *On Virtues and Vices,* in the context of distinguishing true from apparent virtues, he does assert that "every virtue has its starting point in nature."[11]

9. See Mark D. Jordan, *The Invention of Sodomy in Christian Theology* (Chicago: University of Chicago Press, 1997), 67–91; cf. Jan Ziolkowski, *Alan of Lille's Grammar of Sex: The Meaning of Grammar to a Twelfth-Century Intellectual,* Speculum Anniversary Monographs, 10 (Cambridge: Medieval Academy of America, 1985), where the norms are linguistic.

10. Alain de Lille, *Anticlaudianus,* ed. R. Bossuat (Paris: Librairie Philosophique J. Vrin, 1955), bk. 7, ll. 202–5, p. 163.

11. Alain de Lille, *De virtutibus et viciis,* ed. Johan Huizinga in *Über die Verknüpfung des Poetischen mit dem Theologischen bei Alanus de Insulis,* Mededeelingen der Koninklijke Akademie van Wetenschappen, Afdeeling Letterkunde, vol. 74, ser. B, no. 6 (Amsterdam: North-

Manifested mainly in the valuation of individuals, their desires, and their actions, this association of the natural and the good obtains at the social level also, as the evidence from Chaucer has already suggested. The development of theories of natural law and governance, as well as intellectual efforts to weigh or justify the claims of emperors and popes to political authority, produced a large quantity of literature on the natural and, by that token, rational foundations of the social body. The proper ordering envisioned in this collective context is connected with and sustained by the virtuous and just acts of individuals. Dante's *De monarchia* is typical in justifying a political vision on both theological and philosophical grounds. From the latter perspective: "Nature orders things in consideration of their powers, and this consideration is the foundation of justice in things disposed by nature. From this it follows that the natural order of things cannot be preserved without justice."[12] In a similar vein, the language and conduct of early modern sodomy trials analyzed in the essay by Helmut Puff identify individual transgressions as threats to the community. Those documents resonate with the Bible, whereas in the *Parliament of Fowls* reflections on the ordering of "things according to their powers" often invoked the neo-Platonic celebration of the golden chain connecting celestial and human governance. Ulysses' famous insistence on the observance of properly proportioned hierarchy at the beginning of Shakespeare's *Troilus and Cressida* not only depends on that connection but also warns of the natural and political disasters that result from the abandonment of "priority, and place, insisture, course, proportion, season, form, office, and custom, in all line of order": "Take but degree away, untune that string, and hark what discord follows!"[13]

At an individual and a collective level, in a positive and a negative way, various notions of nature and the natural served to define, motivate, and evaluate the good. Always presumed to derive from a divine Creator, of-

---

Holland, 1932), 99–100 (pp. 187–88 of volume): "Seueritatem et uindictam non esse uirtutes. Cum enim omnis uirtus a natura inicium habeat etsi species naturalis juris est uindicta uel seueritas. potius ex jnfirmitate jnitium habuit quam a natura. Indultum est enim homini ex jnfirmitate ut jllatas propulsaret iniurias." On the complex and transitional character of Alain's position on this subject, see Colish, "*Habitus* Revisited" (note 6).

12. Dante Alighieri, *De monarchia,* ed. Pier Giorgio Ricci, vol. 5 of *Le opere* (Verona: Arnaldo Mondadori, 1965), bk. 2, ch. 6, p. 193: "Propter quod patet quod natura ordinat res cum respectu suarum facultatum, qui respectus est fundamentum iuris in rebus a natura positum. Ex quo sequitur quod ordo naturalis in rebus absque iure servari non possit."

13. Shakespeare, *Troilus and Cressida,* act 1, scene 3, ll. 86–88 and 109–10.

ten modeled and powered by the celestial spheres, regularly associated with hierarchical order and harmonious proportion, nature was, at the very least, a reservoir for the formation of values and norms. Given its prominence as an allegorical queen and goddess, that is, as an explicit, visible, and commonplace legislator and arbiter of right actions and proper social relations, Nature would seem to have served medieval culture as a forceful expression of norms and a powerful medium of their enforcement, especially when formulated in relation to reason—centerpiece of human nature and universal principle of order, proportion, and right. But these apparent powers were circumscribed and undermined in a number of ways. Just as, according to Helmut Puff's essay in this section, Nature turned out not to be a very good lawyer in the early modern period, so, in the Middle Ages, she was not an altogether effective ruler.

Trouble with the Nature of Virtue

An important feature of Nature, of the rationally ordered product and deputy of the providential Mind, is that it constitutes and consists of the natures of distinct types of physical entity. The planets and the elements, plants and animals, medicines and poisons exist and function as distinct, specific kinds. However diverse the individuals in any such category, each species is characterized by a defining essence or form—its nature. It is through these more specific forms that the order of Nature in a broader sense is sustained. Chaucer's goddess does call the birds together on Valentine's Day and directs them to take their proper places, but it is the *nature* of seed-eating fowl to flock together on the ground, water fowl to congregate on the water, and hawks to perch on high, as it is for birds in general to mate in the spring. Thus, though these specific natures are conceptually distinct from the universal and all-encompassing Nature, the two senses of the natural are closely bound up together.

The question of the relationship between norms and nature is mostly, if not exclusively, meaningful in the context of human nature. (Although storms and earthquakes, dark forests and savage beasts might, in some medieval contexts, represent rifts in the goodness of Creation, their normative meanings generally implicated human behavior.) From a Christian theological perspective, human nature was defined first and foremost by the immortal soul; from a philosophical perspective, by reason. From both points of view, the idea that nature, either as a general principle of good order or as a specific defining essence, could be the source of virtue was problematic. The good life that, according to a theologian, might be rewarded with the grace of salvation and that, for a philosopher, might merit the ascription of worldly praise could not simply be construed as the necessary consequence of having human nature or being a part of Nature. In either case something, perhaps free will or reason, had to stand up to necessity. Thus the close cultural connections between what was natural and how things *should* be—and between the unnatural and how they *should not*—was the subject of some systematic reflection in both domains. It is the philosophical that is of concern here.

Even before the massive influx of Aristotelian texts that became the center of the European academic enterprise in the thirteenth century, the idea that virtue was acquired had coexisted—and sometimes interacted— with the notion that it was, in some sense, in human beings by nature. Along with several other late-twelfth-century philosophers and theologians, Alain de Lille, who had depicted the virtues, like reason, bestowing their endowments on humankind and who had declared that the virtues started from nature, had also taken the position that virtue was something people acquired with practice, a *habitus,* to use the Aristotelian term that had begun to enter the Latin literature.[14]

Medieval philosophical inquiry into this question centered on the explication and discussion of Aristotle, especially after the mid-thirteenth century, when Robert Grosseteste made a new and complete translation of the *Nicomachean Ethics* from the Greek.[15] While earlier views on the natural sources of the good persisted, Aristotle's declaration that "moral virtues are not in us by nature"[16] brought home to medieval authors the fundamental problem with positing a natural foundation for ethical behavior. True, like rationality (the hallmark of human nature), morality pertains only to human beings; true, morality regulates our relations as naturally social animals. But the sort of praise and blame with which we qualify true virtue and vice cannot be the result of the actors' innate qualities—of what they *are* as individuals or as human beings. With respect to an individual's particular innate characteristics, in the words of Heinrich von

14. Cary J. Nederman, "Nature, Ethics, and the Doctrine of 'Habitus': Aristotelian Moral Philosophy in the Twelfth Century," *Traditio* 45 (1989–90): 87–110; Colish, *"Habitus"* Revisited" (note 6).

15. *Ethica nicomachea,* ed. René Antoine Gauthier, 5 vols., *Aristoteles latinus* 26/1–5, Corpus Philosophorum Medii Aevi Academiarum Consociatarum Auspiciis et Consilio Editum, Union Académique Internationale (Brussels: Desclée de Brouwer; Leiden: E. J. Brill, 1972– 74). Unless otherwise noted, subsequent references to "*Ethics*" are to fasc. 3 of this edition, the standard medieval Latin version.

16. *Ethics* (note 15), II.i, 1103a18, p. 163.

Friemar in his commentary on the *Ethics*, "We are not praised or censured for those things which are in [us] by nature, since no one condemns a man born blind but rather he will be pitied. But a person is praised for virtue."[17] With respect to an individual's nature (that is, defining characteristics) as a member of a species or kind, medieval commentators followed Aristotle's argument that people are capable of becoming either good or evil in an ethical sense, whereas natures do not change: no matter how many times you throw a stone (by nature heavy) into the air, it will never become capable of rising.[18]

In this sense, virtues (and, for that matter, vices) do not come from nature, since they cannot be the product of necessity. The difficulty would obtain even if we were to suppose that virtue was in a person potentially, in the same way that reason is in a child, since in that case (barring the sorts of accidents discussed below), it would simply develop without effort, though perhaps to a greater or lesser degree of perfection. What *is* natural in humans is a susceptibility to acquire virtues, which is realized by the practice of good actions—not in the way that an intellectual discipline is acquired but in the way that an art, such as building or zither playing, is acquired. The accomplishment is the state or *habitus* of being virtuous. The closely related terms *assuetudo* and *consuetudo* and variants upon them, which refer to becoming accustomed to something or forming a habit, describe the process by which a person arrives at the condition *(habitus)* of virtuousness.[19] This account, which follows Aristotle closely, was in its general outlines accepted by those, like Thomas Aquinas, who commented on his text.[20]

Aristotelian ethics emphasized voluntary action, a feature that made it somewhat compatible with a Christian notion of free will. The framework of individual responsibility in the realm of moral good and evil thus bore no relation to the sort of necessity implicit in the unfortunate individual nature of the blind man or the intractable general nature of any falling

17. Heinrich von Freimar (Henricus de Frimaria), *Commentum super libros Ethicorum Aristoteles,* Berlin, Staatsbibliothek, ms. lat. fol. 584, fol. 47ra: "De his que insult a nature non laudamur nec vituperamur, quia nullus imperat [?] ceco nato sed potius miserebitur . . . . Sed homo laudatur propter virtutem . . . ." Lohr, "Medieval Latin Aristotle Commentaries: Authors G–I," *Traditio* 24 (1968), 221–22; *Ethics* (note 15) III.vii, 1114a22–27, p.189.

18. *Ethics* (note 15), II.i, 1103a21–23, p. 163.

19. *Ethics* (note 15). At II.i the "pure" version of Grosseteste's translation (fasc. 3) uses both terms, the version revised by an anonymous reader-scribe (fasc. 4) sticks to *assuetudo.*

20. Thomas Aquinas, *In decem libros Ethicorum Aristotelis ad Nicomachum expositio,* ed. Raymund M. Spiazzi (Turin: Marietti, 1947), bk. II, lect. 1, §§ 247–52, pp. 69–71.

stone. Indeed, to the extent that nature was the cause of human moral goodness or evil, ethics would cease to exist. As Heinrich von Friemar put it, "For if moral virtue were foreseen in us from nature, then . . . the study of virtues would not pertain to moral [philosophy] but to natural philosophy."[21] At one level, this statement reflects nothing more than the disciplinary (and thus textual) distinction between two branches of learning. At another, however, it projects the substantive implication that what can be understood in terms of natural causes (the subject of natural philosophy) cannot properly be evaluated in the moral terms of praise and blame—whether it relates to the general order of the cosmos, like the disposition of the planets, or to the specific definition of human beings, like their temperate complexions, or to the particular condition of an individual, such as an illness. Human nature must accommodate our acquisition of goodness and justice. Did it not, we would be like the stone that cannot learn to rise: virtue would be outside of our nature *(preter naturam),* if not against it. Beyond this weak sense, it would appear, nature has nothing to do with the sources of moral virtue or vice and thus would seem to have no moral authority.

Accidents of Desire and Imperatives of Necessit

If, by the very definition of moral virtue, nature cannot be the *source* of states or actions that merit praise or blame, it might still provide *standards* by which such states and actions may be judged. Though conformity to nature, precisely because it usually happens naturally—without judgment or choice—did not qualify a person to be called morally good, the contravention of what is natural might qualify one to be called morally bad. The use of the phrase "against nature" to signify particularly abominable acts is one of the persistent reasons for acknowledging that nature did indeed have normative force in the Middle Ages.[22]

21. Heinrich von Freimar, *Super libros Ethicorum* (note 17), fol. 45rb: "Si enim virtus moralis prescise nobis inesset ex nature tunc . . . tractare de virtutibus non spectat [ad] moralem sed ad philosophiam naturalem."

22. The phrase "preter naturam" is frequently distinguishable from "contra naturam," the former referring to something that does not normally happen in the course of natural events, the latter (more strongly) to the sort of assault on the natural order discussed above. In the context of *Ethics* II.i, they appear to be equivalent, though it is possible they convey subtle differences in tone. Thus Aquinas holds that virtues are neither *a natura* nor *contra naturam* (*In decem Ethicorum* [note 20], bk. II, lect. i, § 248), whereas a later compendium says neither *a natura* nor *preter naturam* (*Compendium philosophie* attributed to Albertus Magnus, Berlin, Staatsbibliotek, ms. lat. oct. 142, fol. 109va). On the judicial dimensions of the development of *contra*

Infractions offend against Nature in its universal sense. As Shakespeare's Ulysses warns, they cause the golden chain to be broken, the whole fabric of Creation to be rent, as is Nature's garment in Alain de Lille's allegory.[23] But they are also against more specific natures—the proper balance that constitutes human health; the proper discipline of desire that constitutes human sexual order. As one fifteenth-century Aristotelian explained, "pleasures against nature are not human pleasures."[24] These two senses of the natural, the universal and the specifically human, though they are distinguishable, did not operate separately in the rhetoric of the contranatural. Indeed their convergence added force to the anathema.

The notion of *sins* against nature underscores the element of moral responsibility for the acts so named. Yet not all acts against nature contain the crucial element of free choice. Insofar as such acts occur, they reveal the weakness of universal Nature and specific natures. The universal order of natural place and the heavy specific nature of the element earth may ensure that no stone will move upward of its own accord, but neither can prevent it from moving upward if compelled by a violent motion, such as that of a catapult. Like all medieval rulers, Nature's sway was hardly absolute, a circumstance that, among other things, rendered the unnatural possible.

The fragility of the natural order was evident in the immediate experience of illness and expressed in the medical concept of "contranaturals": diseases, their causes, and their consequences. Here were entirely commonplace situations in which proportion and proper relations had lost their hold. The disruption manifests itself in the so-called naturals, such as the humors (fluids) and organs of the body, but its cause often lies not only outside the individual body but within the power and choice of individual actors, such as those capable of moral and immoral acts. The sources of health and illness largely depended on what were called the "non-naturals," that is, patterns of regimen, especially with respect to eating and drinking, sleeping and waking, sexual activity, and emotions. In and of themselves, these activities are neither natural (in the sense of being

---

naturam, see Jacques Chiffoleau, *"Contra naturam:* Pour une approche casuistique et procédurale de la nature médiévale," *Micrologus* 4 (1996): 265–312, discussed by Puff in this volume.

23. Alain de Lille, *"De planctu naturae"* (note 4), prose 1.

24. Jean Tourneur (Johannes Versoris), *Questiones circa decem libros Ethicorum,* Leipzig, Universitätsbibliotek, ms. lat. 1445, fol. 97v: "Continentia et incontinentia simpliciter sunt circa delectabilia humana . . . sed delectabilia contra naturam non sunt humana." Charles Lohr, "Medieval Latin Aristotle Commentaries: Johnannes de Kanthi-Myngodus," *Traditio* 27, no. 10 (1971): 298.

regulated automatically) nor contranatural (in the sense of disrupting such regulation), and neither good nor bad. Their proper relations and proportions promote health; their excess and deficit incur illness. The judicious supervision of non-naturals is thus analogous to the practical moderation required for the exercise of Aristotelian virtues, each of which has not one but two opposites—a vice of excess and a vice of deficiency. Just as too little courage is cowardly and too much is foolhardy, so too little exercise will weaken the body in one way, too much in another. Too much exercise is not, *per se,* a moral defect, for the subject of health is the body and the subject of virtue is social relations, but medieval concepts of health and disease illustrate both the implications of human choices and the precariousness of natural balance. In this respect and in their shared emphasis on the propriety of moderation, they underscore the possibility of voluntary, violent action against natural proportion.[25]

The proper ordering of human relations is precarious in ways that go beyond human intention and choice, however. It is particularly vulnerable to the effects of inhuman desires called "bestial" by Aristotle and his Latin commentators, not so much because they are desires characteristic of animals in general or of particular species as because they are the desires of individual humans who lack human nature. According to Albertus Magnus, "These are called 'bestial' pleasures since a well disposed human nature abhors such things";[26] Heinrich von Friemar refers to "various inhuman desires, and inhuman and bestial pleasures."[27] In a chapter of the *Nicomachean Ethics* that deals with the concepts of continence and incontinence, as distinguished from virtue and vice proper, Aristotle supplied a short list of examples of bestiality upon which medieval authors occasionally elaborated. Some of these refer to unique individuals, described as insane: the man who ate the liver of his fellow slave and "Xerxes, king of the Persians, who sacrificed and ate his mother on account of madness."[28] In one case

25. *Ethics* (note 15), II.viii and III.viii.

26. Albertus Magnus, *Super Ethicam commentum et quaestiones,* ed. Wilhelm Kübel, 2 pts., vol. 14 of *Opera omnia* (Monasterium Westfalorum: Aschendorff, 1968–72), bk. VII, lect. 5; pt. 2, p. 545: "Et huiusmodi dicitur *bestiales* delectationes, quia natura humana bene disposita abhorret talia."

27. Heinrich von Freimaria, *Super libros Ethicorum* (note 17), fol. 211vb: "Dicitur continentia et incontinentia circa diversas concupinas [sic] inhumanas et delectationes inhumanas et bestiales."

28. Albertus Magnus supplies Xerxes' name and rank, *Super Ethicam* (note 26), bk. VII, lect., 5; pt. 2, p. 545: "sicut propter *maniam* rex Persarum Xerxes *sacrificavit matrem et comedit eam.*" Cf. *Ethics* (note 15), VII.7, 1148b24–26, p. 280.

Albertus Magnus supplies some information relating to circumstances of a dreadful inclination: "There is the story about a certain woman called Lamia, who, after she lost her children, tore open pregnant women and devoured their children."[29] In the other cases no attention is given to causes. Indeed, although madness was sometimes susceptible to systematic causal analysis, it could be triggered by any number of accidents that depended neither on the order of universal Nature nor on the principles of human nature nor on an individual's particular constitution nor on human volition.

The effects of bestial impulses and appetites underscore the weakness of nature—in this case the specific human nature characterized by right reason and proper social relations. They also underscore the belief that there is, indeed, something good about conforming to what is natural at a universal and a specific level, and something bad about contravening it. That is, in their terrible randomness, these cases of "inhuman" and therefore specifically unnatural bestiality validate the patterns and standards of nature as a legitimate basis for making value judgments about behavior. In the cases just mentioned, however, the madness, which not only describes but also informs the state of the actors, takes the actions out of the purview of good and evil in the strict moral sense. In the absence of reason and *natural* desires, and in the presence of passions themselves inexplicable (or at least unexplained) in terms of choice and responsibility, these deeds were, according to the understanding of medieval Aristotelians, not properly speaking morally evil. That is, the acts in question and, as we shall see, less exceptional ones, do not fall along the axis of true virtue and vice. Whether afflicted with the unbridled passions just discussed or prevented by weakness from continent conduct, such people do not intend and freely choose to do evil. And, since intention and choice are required for behavior to be virtuous, so too are they required for behavior to be morally bad. Medieval authors reiterated Aristotle's heuristic observation: we may call a physician or a mime "bad" but (in the absence of sinister intent) they are "bad" only in a loose or analogical sense.[30] Although cannibals and

---

29. Albertus Magnus *Super Ethicam* (note 26), bk. VII, lect. 5, p. 545: "Sicut dicitur de quadam muliere Lamia dicta, *quae* postquam amisit filios, *praegnantes* mulieres *scindebat* ut *devoraret filios* earum." Cf. *Ethics* (note 15), VII.7, 1148b20–21, p. 280.

30. *Ethics* (note 15), VII.7, 1148b1–9, p. 280: "Malitia quidem igitur nulla circa hec est . . . puta malum medicum et malum ipocritam, quem simpliciter utique non dicerent malum." Cf. Aquinas, *In decem Ethicorum* (note 20), bk. VII, lect. 4, § 1367, p. 367; Albertus Magnus, *Super Ethicam* (note 26), bk. VII, lect. 5.

infanticides may be "abominable,"[31] it is the very manner in which and the very extent to which their acts contravene nature that remove them from the possibility of being judged *morally* against the measure of human nature or the larger natural order.

Conflict in Desire

Nature's lack of absolutism, its fragility, and its lapses, while they made moral virtue and vice possible by allowing the exercise of rational choice or free will, thus also permitted the introduction of accidents and their attendant disruptions. But the problem of locating moral authority in Nature or natures went beyond the axiom that virtue and vice cannot follow *automatically* from external order or internal essence; and likewise it went beyond the occasional imperfections of order and essence that occur without the mediation of human choice. It also encompassed the active involvement of natural processes themselves in the subversion of natural orders. Not only may universal principles operating in nature fail to compel the compliance of all individuals, but the particular natures of individuals may impede or prevent such compliance.

One dimension of this subversion is the production of sports, marvels, and monsters that elicited fear, curiosity, and wonder in both learned and lay circles.[32] Because they sometimes did not correspond adequately to a natural kind, such productions, if they might be construed as human, could elicit questions about their legal, social, and, presumably, moral status.[33] More inevitably enmeshed in the web of nature and virtue, however, were aberrations of the mind or spirit, especially those that exhibited identifiable patterns (if not forms or natures) of their own.

Medieval authors were particularly interested in the implications of disordered desires, especially those that appeared to be, in some sense, natural. Like the words on the gates to the garden in Chaucer's poem and the ambiguity of the figures surrounding Venus, desire had an indeterminate

31. Albertus Magnus applies this adjective *(nefarius)* to one Philaris whose behavior is the same as that of Procrustes: *Super Ethicam* (note 26), bk. VII, lect, v; pt. 2, p. 545.

32. See, for example, Lorraine Daston and Katharine Park, *Wonders and the Order of Nature* (New York: Zone, 1998); Bert Hansen, *Nicole Oresme and the Marvels of Nature: A Study of His "De causis mirabilium"* (Toronto: Pontifical Institute of Mediaeval Studies, 1985).

33. See, e.g., Pietro d'Abano (Petrus de Abano), *Expositio Problematum Aristotelis,* Paris, Bibliothèque Nationale, ms. lat. 6540, bk. IV, probl. 13, fol. 58vb.

valence.[34] According to Aristotle's *Ethics,* some physical pleasures, such as the ones arising from the enjoyment of food and sex, are necessary,[35] and these are among the pleasures that are natural—either to all creatures or to particular types of animals or humans.[36] As Albertus Magnus explains, "this food is naturally pleasurable to some [kind of animal] and another to the ass and another to the human being; and similarly, according to the types of humans, some [foods] are naturally pleasurable to melancholics and others to cholerics."[37] Even those pleasures that are not necessary for sustaining and perpetuating human life do not in and of themselves give rise to incontinence (the failure to maintain moderation), as Heinrich von Friemar explains.[38] The pursuit of necessary and natural objects of desire can result in actions that, if not morally good in themselves, may nevertheless serve the moral and social order through the natural mechanisms of desire. As Aquinas explains in his commentary on the *Ethics,* "moral virtue relates to the appetite, which operates according to what is moved toward an apprehended good."[39] Here the parallel between health and virtue is once again suggestive. These two goods have more in common than choice and moderation. Both are fueled by, and yet have the task of regulating, natural desires (including hunger and sexual appetite)—important vectors along which the natural remains involved in the production of praise and blame. A lack of anger can lead to moral imbalance in the same way that a lack of appetite can lead to corporeal imbalance.

The greatest dangers to the proper order of things might not, however, reside in unbalanced passions toward appropriate ends but in appetites for inappropriate objects. Aristotle continues by distinguishing those pleasures that are natural from those that are not. The commentators reinforce the medieval insistence on the importance of conformity with nature by

34. Emerson Brown, "Priapus and the *Parlement of Foulys,*" *Studies in Philology* 72 (1975): 258–74.

35. *Ethics* (note 15), VII.vi, 1147b24–28, p. 278.

36. Ibid., VII.vii, 1148b15–17, p. 280.

37. Albertus Magnus, *Super Ethicam* (note 26), bk. VII, lect. 5; pt. 2, p. 545: "Quia naturaliter alicui est delectabilis hic cibus et alius asino et alius homini, *et* similiter secundum genera *hominum,* quia alia naturaliter sunt delectabilia melancholicis et alia cholericis." See also Jean Tourneur (Johannes Versoris), *Circa decem Ethicorum* (note 24), fols. 96v–97r.

38. Heinrich von Freimaria (Henricus de Frimaria), *Super libros Ethicorum* (note 17), fol. 209vb.

39. Aquinas, *In decem Ethicorum* (note 20), II, lect. 1, § 249, p. 70: "Virtus moralis pertinet ad appetitum, qui operatur secundum quod movetur a bono apprehenso."

labeling such impulses "unnatural" and "against nature,"[40] even though the terms do not occur in the Philosopher's text. This gloss serves to strengthen the tone of disapprobation in their explication of desires and behaviors not proper to human beings. Yet it also highlights the paradoxical character of that disapprobation, when the medieval authors encounter the difficulty that nature itself is implicated in the production of illicit desires.

Aristotle explains that there are three reasons for desires that are not natural: illness, habit, and a bad nature.[41] While these would appear to be sufficiently distinct, it is not always clear which cases Aristotle is attributing to each cause, and the medieval commentators regroup and reinterpret them, with the result that at some times most seem reducible to illnesses and at other times to pernicious individual natures. For example, Heinrich von Friemar comments on a passage that follows the examples of cannibalism and infanticide. In it Aristotle mentions pulling out one's hair, biting one's nails, eating coals or dirt, and men sleeping with other males.[42] To Heinrich, some of these are "pleasures [arising] from a filthy habit" and "they happen on account of an internal sickness and therefore are called illness-related." He seems to make a distinction, when he says that others "happen on account of corruption coming from a bad habit," but he goes on to explain that a habit ingrained in someone through experience from an early age can turn into a "bodily illness" and similarly, "a filthy habit is a certain sickness of the spirit *(egritudo animalis)*."[43]

40. Albertus Magnus, "innaturalis," *Super Ethicam* (note 26), bk. VII, lect. 5; pt. 2, p. 545; Aquinas, *In decem Ethicorum* (note 20), "contra naturam" and "innaturalis" VII, lect. v, §§ 1373 and 1374, p. 369 ; Heinrich von Freimaria (Henricus de Frimaria), "contra naturam" and "innaturalis," *Super libros Ethicorum* (note 17), fol. 212va; Jean Tourneur (Johannes Versoris), *Circa decem Ethicorum* (note 24), "contra naturam," fols. 96v–97v.

41. *Ethics* (note 15), VII.6, 1148b17–18.

42. The inclusion of sex between males in a list of seemingly minor aberrations may have encouraged the application of terms like "against nature." On the process by which sodomy came to occupy a special place among sins, see Jordan, *The Invention of Sodomy* (note 9).

43. Heinrich von Freimaria (Henricus de Frimaria), *Super libros Ethicorum*, Leipzig, Universitätsbibliothek, ms. lat. 1439, 240ra: "*Hii autem egritudinales* Exemplificat de his que fuint [sic] delectabilia contra naturam ex prava consuetudine . . . quibusdam accidunt propter interiorem egritudinem et ideo dicuntur egritudinales vel etiam accidunt propter corruptionem provenientem ex mala consuetudine sicut quosdam videmus ex consuetudine pilos evellere et ungues dentibus corrodere. Adhuc quidam delectantur in commestione carbonum et terre et in abusu venereorum cum ipsis masculis. . . . Quibusdam autem ex consuetudine longa puta quia sunt assuesci ad huiusmodi ex pueris, id est ab ipsa pueritia et ad istos reducuntur illi qui in talia incidunt ex egritudine corporali. Eo quod prava consuetudo est quedam egritudo animalis.*"

Aquinas, though he attributes the acts of cannibalism and infanticide to mania or rage and mentions the notion that some desires are caused by "an interior sickness or corruption originating from habit," concludes that the causes of these desires can be reduced to two, namely, "from the nature of a corporeal complexion" and from long-ingrained habit.[44] Speaking about sodomy, Pietro d'Abano (1257–ca. 1315) reduces even habit to the equivalent of innate nature, using an argument similar to that employed by Heinrich to reduce it to illness.[45] Several commentators take up one example of ungoverned desire based in individuals' inherent nature, that of women's passivity, introduced by Aristotle as an analogy. "On account of the softness of [their] nature they do not lead or rule their feelings according to reason but rather are led by them."[46] In the context of natural philosophy, women have a natural purpose, a final cause,[47] which the other groups, such as those who eat dirt, lack. In the context of these *Ethics* commentaries, the emphasis on inherent qualities deflects explanations of all these conditions away from acts of will.

Furthermore, like nail biters and men who sleep with men, women are a common production of nature, whose desires follow a pattern embedded in their natures. Whether innate or acquired, these natures were often expressed in terms of "complexion," the naturally variable mix of humors that characterizes each individual's constitution. Aquinas says that, "because of something that corrupts [such people's] complexion, they become similar to beasts."[48] Their sources of pleasure, clearly regarded as distorted and labeled pernicious, nevertheless recall the enumeration of "natural pleasures," according to which melancholics and cholerics desire different foods, just as asses and humans do. Indeed, many of the examples of unnatural pleasures fall into the categories of food and sex, held to be not only natural but even necessary. Entire peoples apparently have outlandish appetites, such as the Scythians around the Black Sea who, according to

44. Aquinas, *In decem Ethicorum* (note 20), bk. VII, lect. 5, § 1374, p. 369.

45. Pietro d'Abano (Petrus de Abano), *Problemata* (note 33), bk. IV, probl. 26, fol. 64vb.

46. Heinrich von Freimaria (Henricus de Frimaria), *Super libros Ethicorum* (note 43), fol. 240ra–b: "propter mollitiem nature quod suas affectiones non ducunt vel non regulant secundum rationem sed magis ducuntur ab eis." The sexual meaning of the passage on which Heinrich is commenting was rendered opaque in the Latin translation: *Ethics* (note 15), VII.vi, p. 281, ll. 5–8 (1148b3 ff.).

47. Aristotle, *Generation of Animals* II.i.

48. Cf. Aquinas, *In decem Ethicorum* (note 20), bk. VII, lect. 5, § 1372, p. 369: "Sunt quasi bestiales, quia propter corruptelam complexionis assimilantur bestiis."

Albertus Magnus, enjoy raw fish.[49] These nature-based explanations may in some cases soften the evaluation of the resulting behavior. Considering the status of sodomy as an exceptional category of sin, it is remarkable that Brunetto Latini, commenting on this section of the *Ethics* in the mid-thirteenth century, could declare simply that "Pleasure from bad nature is lying with males and other dishonorable things."[50] The notion of an individual's "bad nature" threatens to undermine the force of the concept "against nature," as the epigraph of this essay suggests. Indeed, when the bishop of Paris forbade the teaching of the idea that the nature of a individual person could contravene human nature, he seemed to have just this challenge in mind.[51] Most important, from the perspective of moral claims based on infractions against nature, is the recognition that these infractions themselves originate *in* nature, albeit a nature burdened by particularity and contingency. As Heinrich acknowledges, "nature is the cause of such bestial pleasures, and thus nobody knowing [that] should call those [susceptible to them] 'incontinent' in the literal sense."[52]

Norms Out of Nature

Nature, like other circumstances, might thus create impediments to living properly. But some of Aristotle's Christian disciples could not leave the

49. Albertus Magnus, *Super Ethicam* (note 26), bk. VII, lect. V; pt. 2, p. 545. Aristotle says "raw meat" and seems to say these are the same people who eat each other's children, whereas Albertus distinguishes different groups. *Ethics* (note 15), VII.vi, p. 280, ll. 22–25 (1148b20ff.).

50. Brunetto Latini, *Li livres dou tresor*, ed. Francis J. Carmody (Berkeley: University of California Press, 1948), bk. 2, ch. 40, p. 207, ll. 16–19: "Delit par male nature est gesir avec les malles, et des autres choses deshonorables." His source for this formulation is a summary of Aristotle's *Ethics* translated into Latin from Arabic: "Il compendio Alessandrino-Arabo *Liber Ethicorum*" edited in Concetto Marchesi, *L'Etica Nicomachea nella tradizione latina medievale (documenti ed appunti)* (Messina: Ant. Trimarchi, 1904), xli–lxxxvi, here lxx. It is intriguing but not necessarily relevant that Dante treats Brunetto as a sodomite (*Inferno* XV). See John Freccero, "The Eternal Image of the Father," in *The Poetry of Allusion: Virgil and Ovid in Dante's Commedia,* ed. Rachel Jacoff and Jeffrey T. Schnapp (Stanford: Stanford University Press, 1991), 62–76 and 264–65; and Michael Camille, "The Pose of the Queer: Dante's Gaze amd Brunetto Latini's Body," in Glenn Burger and Steven F. Kruger, eds., *Queering the Middle Ages* (Minneapolis: University of Minnesota Press, 2001), 57–86.

51. Henri Denifle, *Chartularium universitatis parisiensis,* vol. 1, *Ab anno MCC usque ad annum MCCLXXXVI* (Paris, 1889), no. 473, § 166, pp. 543, 553: "Nonnulli Parisius studentes . . . quosdam manifestos et execrabiles errores . . . tractare et disputare presumunt. . . . Quod peccatum contra naturam, utpote abusus in coitu, licet sit contra naturam speciei, non tamen est contra naturam individui."

52. Heinrich von Freimaria (Henricus de Frimaria), *Super libros Ethicorum* (note 43), fol. 340ra.

matter there. There were philosophers contemplating the fundamentals of ethics who did not accept the necessity implied by the imperatives of individual natures or aberrant types except in the most extreme cases. The medieval solution to this problem returns us to the structural impossibility of Nature's authority in the realm of praiseworthy and blameworthy behavior. Virtue and vice would not be possible without rational reflection and the exercise of will. In the case of those whose natures themselves disrupt the natural order, the answer comes from beyond the (fortunately imperfect) dictates of individual nature and the feeble powers of enforcement that reside in universal Nature. Having considered natural philosophical arguments according to which the inclination to sodomy is caused by a corporeal defect, a habit so ingrained it has become second nature, and an astrological coincidence, the fourteenth-century physician Evrart de Conty appealed to the authority of the *Ethics* in his French commentary on another Aristotelian work: "Aristotle says elsewhere that we are lords of all our actions from the beginning of our life until the end."[53] The lesson was frequently and more colorfully reiterated in the form of the story told by Valentin Groebner in this volume about Hippocrates' resistance to his own nature: his constitution was lascivious but his behavior was not. While it might be tempting to insist that this ability to conquer our innate dispositions is itself in us by nature, such a reconstitution of Nature's status as a source of value would be a breech of the fundamental principle that virtue is *not* in us by nature. Evrart might have appealed to the theological doctrine of free will at this point—a fount of both power and value that had its source outside of created nature. He did not. Instead he compounded the ambiguity by quoting Ptolemy's dictum that the stars determine inclinations but not compulsions. In doing so, he invoked the unsettled medieval rhetoric of astrological causation, in which many of the same questions of necessity and freedom came into play.

Not only is nature barred, by definition, from being a source of virtuous action and social order, not only do its rules lack absolute force, but natural processes themselves corrupt order and set up conflicts among natures, so that full conformity with nature is impossible. Chaucer's goddess Nature is not powerless. She has pricked the birds with pleasure and they have assembled to mate. Indeed, for the most part, they do so by main-

53. Evrart de Conty, *Probleumes*, Paris, Bibliothèque Nationale, ms. fr. 24,281, bk. IV, probl. 26, fol. 117r: "Et ausi dit Aristote ailleurs que nous sommes seigneurs de tous nos fais des le commencement de nostre vie durans en la fin." He cites book 7 of the *Ethics* but the reference is perhaps to III.5, 1113b.

taining the discipline of rank and sex, and by producing the harmony one would expect of a well-governed kingdom. But in the end, her purpose has not been completely accomplished, for the noblest birds will have no mates until next year, when the most perfect female will announce her choice. The prerogative of choice and consent in the constitution of the social order is a necessary limitation on the absolute power of the monarch. Three male eagles have come forward to claim Nature's pet female: desire, nicely got up in rhetorical finery, has disrupted the proceedings. Indeed, the intrusion of the artifices of courtly love suggests agents other than sodomy by which sexual inversion may confound nature. Rather than choosing, these males ask to be chosen, and the most noble male of all promises to subject himself to the female.[54] Furthermore, the desire does not reside in the suitors alone. By her own stated rules, which she says she herself must obey, Nature should have presented the suit of the noblest male to the female he chose, so that she might accept or reject the proposal. The female's rational choice would have been to accept, since the top eagle possesses every natural perfection. That Nature did not do so is a manifestation of her own desire—her particular affection for one of her creatures. Furthermore, as she permits the contest of claims coming from worthy albeit lesser males, she throws the female into a confusion that is assuaged only by the suppression of natural desires and the postponement of mating—the expected and proper social alignment. At the same time, Chaucer's dreaming scholar returns from his journey unimproved, because his desire has, on the contrary, not been aroused by his encounter with Nature.

Seen in this light, the peaceful and well-regulated realm of Nature is not so different from—indeed, necessarily contains within itself—the disordered domain of Venus. For these reasons, nature (whether universal, specific, or individual) can be neither sufficient nor necessary for the generation of right actions or a just order. Conversely, the opprobrium associated with what was "against nature" is never fully clarified or realized, in part because it frequently had natural origins. But ethics and politics were practical sciences, not reducible to unequivocal principles or deducible by rational methods. Like the Aristotelian notion of virtue, the determination of value was more similar to a craft than a mathematical discipline, informed more by prudence than by rigor. Nature, for all its defects and for

54. H. M. Leicester, "The Harmony of Chaucer's *Parlement*: A Dissonant Voice," *Chaucer Review* 9 (1974): 15–34, here 25; Bruce Kent Cowgill, "*The Parlement of Foules* and the Body Politic," *Journal of English and Germanic Philology* 74 (1975): 315–35, here 328.

all its inadequacies as a source and measure of the good, still had a place in the texts in which, and thus in the process by which, medieval norms were pieced together. In Latin philosophy and vernacular poetry, heterosexual pairing, social and political hierarchy, principles of harmony and moderation all derived strength from their association with the natural and by dissociation from the threats to that order emanating from all that was taken to be against nature.

— — —

The persistent association of nature with moral conduct and social order in a wide range of medieval cultural settings may help to explain why, as Helmut Puff explains in the following essay, the phrase "against nature" was uttered at the public executions of sodomites in the early modern period. At the same time, the incoherence and inadequacy of those very associations may help to explain why, in the context of legal practices, it carried so little weight. Later, in the eighteenth-century setting described in the essay by Fernando Vidal, when nature's laws were more invariant and its powers of enforcement more certain, "against nature" would gain far greater force. At present, however, the reign of Nature is less stable. As Abigail Lustig's essay will show, evolutionary ethics did not turn out to be a fully reliable or accepted instrument of Nature's claims to regulate social values. Indeed, those claims have met with radical resistance in the political philosophies analyzed by Julia Thomas and Michelle Murphy. The diverse reactions both among advocates of gay rights and among their opponents to scientific investigations into possible genetic or neurophysiological causes for homosexual desire suggest that we have, as a culture, returned to a state of incoherence and conflict concerning the moral authority of nature.

9

Nature on Trial: Acts "Against Nature" in the Law Courts of Early Modern Germany and Switzerland

Helmut Puff

Les réponses que m'a murmurées ou criées la Nature, je demande qu'on les vérifie

ANDRÉ GIDE, *Corydon*

Orlando could only suppose that some new discovery had been made about the race; they were somehow stuck together, couple after couple, but who had made it, and when, she could not guess. It did not seem to be Nature.

VIRGINIA WOOLF, *Orlando: A Biography*

Nature is a mighty figure. At least this is how theologians, philosophers, and poets presented Her in the wake of Greek philosophy and its resonance in Christian theology of the Middle Ages. According to the twelfth-century theologian Alain de Lille, Nature acts as God's representative *(Dei auctoris vicaria)*.[1] After Bernard Silvestris (d. after 1159), Her rank is often assumed to be that of a goddess, God's divine helpmate, though other writers introduced Her personification more modestly as a lady or a queen.[2] In all Her allegorical emanations, however, Nature's task is identical: to act as God's intermediary and defend the order enshrined in His creation. Her

---

1. Nikolaus M. Häring, ed., "Alan of Lille, 'De Planctu naturae,'" *Studi Medievali: Serie terza* 19 (1978): 797–879, here 825.

2. George D. Economou, *The Goddess Natura in Medieval Literature* (Cambridge: Harvard University Press, 1972). Cf. Peter Dronke, "Bernard Silvestris, Natura and Personification," *Journal of the Warburg and Courtauld Institutes* 43 (1980): 16–31. On the history of Nature's personifications, see Katharine Park, "Nature in Person: Medieval and Renaissance Allegories and Emblems" (in this volume).

grip on the world is predicated on prohibitions and rules—rules that subject both animals and men to an immutable standard of behavior. Together with Reason, She presided over the contest between Ganymede and Helen in order to award the prize to the love that accords with Nature's nature and to condemn the love that violates Her laws (needless to say, Helen wins for her advocacy of heterosexual unions while sexual love between men is condemned).[3] Nature obviously commands a voice to pronounce verdicts—verdicts that are often harsh. For those who do not abide Her norms, Nature holds punishments and disasters in store. While in Her perfection Nature cannot be corrected, She is in a powerful position to correct others.

Yet, as Joan Cadden points out in the preceding essay in this volume, not all is well in Nature's realm. Her stature is riddled by a fundamental paradox: while desire necessitates procreation, this same force threatens to undermine the order it is supposed to sustain by generating undesirable sexual behavior. Alain (1125/30–1203), one of Her most devoted portraitists, therefore draws a picture of Nature as a weakened figure. Her dress is in disarray. It is torn. This outward sign signifies the violation of Her laws. For, endowed with free will humans offend Her and, as Alain insinuates, they do so all too frequently. Following Alain, the allegory of wailing Nature makes a regular appearance in Latin literature. She "rejoices when one obeys her laws; She laments bitterly that Her creatures succumb to a law opposite to Hers by turning the rules of procreation upside down."[4] No doubt, a figure as powerful as Nature has enemies, but same-sex acts offend Her more than most other acts.[5] Yet even Alain's image of a battered

---

3. On the twelfth-century Latin poem *Ganymed and Helen,* see Rolf Lenzen, "Altercatio Ganimedis et Helene: Kritische Edition mit Kommentar," *Mittellateinisches Jahrbuch* 7 (1972): 161–86; John Boswell, *Christianity, Social Tolerance, and Homosexuality: Gay People in Western Europe from the Beginning of the Christian Era to the Fourteenth Century* (Chicago: University of Chicago Press, 1980), 255–60, 381–89 (English translation); C. Stephen Jaeger, *Ennobling Love: In Search of a Lost Sensibility* (Philadelphia: University of Pennsylvania Press, 1999), 25.

4. Metrical *Visio Arislei,* Zentralbibliothek Zürich, Ms. Rh. 172, fol. 97v–99v (p. 186): "Gaudet natura, cum seruantur sua iura; / Et plangit pure, quod succumbunt creature / Contrario iure miscendo modos geniture." Quoted after Sven Limbeck, "Die *Visio Arislei:* Überlieferung, Inhalt und Nachleben einer alchemischen Allegorie: Mit Edition einer Versfassung, in Wilhelm Kühlmann, Wolf-Dieter Müller-Jahnke, ed., *Iliaster: Literatur und Naturkunde in der Frühen Neuzeit: Festgabe für Joachim Telle zum 60. Geburtstag* (Heidelberg: Manutius Verlag, 1999), 167–90, here 186.

5. According to the theologian Jacques de Vitry (1160/70–1240), Mohammed, "the enemy of nature," introduced sodomy to his people. Cf. Jacques de Vitry, *Historia orientalis,* vol. 1 (Douay: n.p., 1597), 18: "vitium sodomiticum hostis naturae in populo suo introduxit."

Nature testifies to Her grandeur—the grandeur She once had and, so the implication, ought to have again.

Scholars of all persuasions have been in awe of Lady Nature. Her rise to power has been the subject of many books and articles. According to these accounts, concepts of nature gained plausibility among proponents of the school of Chartres (eleventh and twelfth centuries) and in Scholastic theology. Inspired by ancient philosophers and Aristotle in particular, theologians like Albert the Great (ca. 1200–80) or Thomas Aquinas (1224/ 1225–74) conferred on nature a central position in their philosophical systems. Both as God's created order *(natura creata)* and as creation in flux *(natura creans),* the concept of "nature" occupies a middle ground between the divine and the human. Positioned in a complex web of references to the particular and the universal, nature stimulated debate in a vast array of disciplines. Subsequently, it entered the corpus of Latin as well as vernacular poetry. Nature's triumph became even more evident, when, at the end of the Middle Ages, natural law emerged as a major source for the development of the early modern state.[6]

Nature's other side, behavior *contra naturam,* accompanied Nature's rise to power. In a recent article, Jacques Chiffoleau traced the diffusion of the formula "against nature" from late antiquity to the waning of the Middle Ages.[7] Invented enigmatically by Plato (427–348/347 B.C.) (*"para physin"*)[8] and cited infrequently in ancient philosophy, the formula *contra naturam* gained importance when, in late antiquity, Christian teachings were fused with the Roman philosophical and legal tradition. Chiffoleau elaborates that the formula usually appeared where God's order suffered insults. Yet offenses to sexual decorum occasioned citation most frequently. St. Paul set a precedent in his letter to the Romans (1:26–7), lamenting that "women did change the natural use into that which is against nature; And

6. *La filosofia della natura nel medioevo: Atti del terzo congresso internazionale di filosofia medio-evale* (Milan: Società editrice vita e pensiero, 1966); F. Kaulbach, "Natur," in *Historisches Wörterbuch der Philosophie,* vol. 6 (Basel: Schwabe, 1984), cols. 441–62; *Mensch und Natur im Mittelater,* Albert Zimmermann and Andreas Speer, eds., 2 vols. (Berlin: de Gruyter, 1991); Jacques Krynen, "Naturel: Essai sur l'argument de la nature dans la pensée politique française à la fin du moyen âge," *Journal des savants* (April–June 1982): 162–90.

7. Jacques Chiffoleau, "'Contra naturam': Pour une approche casuistique et procédurale de la nature médiévale," *Micrologus* 4 (1996): 265–312.

8. Plato, *Laws* LCL 187, vol. 1 (Cambridge: Harvard University Press, 1994), 40 (i.636B–C). Cf. ibid. LCL 192, vol. 2 (Cambridge: Harvard University Press, 1984), 148–52 (viii.836A–837A).

likewise, also the men, leaving the natural use of the woman, burned in their lust one toward another."[9] Following St. Paul, same-sex activity among women and among men frequently held the status of the *contra naturam* offense *par excellence* for Christian thinkers. It was Emperor Justinian's (527–65) law code (528/542) that first linked same-sex acts among men to the destruction of the city of Sodom as well as natural disasters such as epidemics or earthquakes. This provision thus indicted individual sexual behavior described as *contra naturam* as cause for the fearful fate of larger populations and whole communities.[10] Yet *contra naturam*'s range was not limited to same-sex transgressions. The term covers constantly shifting conceptual grounds. Chiffoleau even observes a proliferation of the offenses grouped together as *contra naturam*. When Thomas Aquinas worked previous references into a more systematic treatment in his *Summa theologiae,* the "vice against nature" comprised a number of sexual acts all said to resist Nature's call for procreation (for example, masturbation, bestiality). Moreover, he characterized this sin, or rather sins, as the most severe of all sexual transgressions.[11]

Like *natura, contra naturam* does not circumscribe a clearly defined essence. Rather, the "antithesis to nature"[12] is an unstable concept whose precise meaning changes according to the context of its use.[13] What Mark Jordan has shown for *sodomia, contra naturam*'s twin, holds true for medieval discussions of the category "against nature": despite their semblance of system, discussions of *contra naturam* and the term's applications were "vitiated . . . by fundamental confusions and contradictions."[14] Against the backdrop of these semantic fluctuations, *contra naturam* proved a useful argument in polemical contexts. Theologians marshaled Nature's powers against simonists, heretics, sorcerers, tyrants, and others typed as traitors to the divine cause. When the "persecuting society" emerged in the eleventh

9. Bernadette J. Brooten, *Love between Women: Early Christian Responses to Female Homoeroticism* (Chicago: University of Chicago Press, 1996), 239–66 and passim.

10. *Corpus iuris civilis* (part v, *novellae,* n. 77), ed. Christoph Heinrich Freiesleben (Neuköln: Emanuel Thurneysen, 1775), cols. 1041–42 *(Edictum Iustiniani ad Constantinopolitanos de luxuriantibus contra naturam).*

11. Thomas Aquinas, *Summa theologiae,* ed. Petrus Caramellus, *Pars secunda secundae* (Turin: Marietti, 1986), 674–76 (q. 154, arts. 11 and 12).

12. Mark D. Jordan, *The Invention of Sodomy in Christian Theology* (Chicago: University of Chicago Press, 1997), 156.

13. Chiffoleau, "Contra naturam" (note 7), 265.

14. Jordan, *Invention* (note 12), 9.

and twelfth centuries, *contra naturam* locutions acquired a deadly force in medieval theology, heresiology, as well as jurisdiction.[15]

Chiffoleau and other scholars have been so impressed with Nature's all-encompassing stature (to which *contra naturam* descriptions belong prominently) that the limits of Her power have rarely been taken into consideration. This essay seeks to address this deficit with regard to one of the most prominent transgressions described as inimical to Nature, sodomy, understood here as same-sex sexual acts. I survey a vast array of materials, mostly from early modern German-speaking countries, to investigate whether arguments pertaining to nature were integral to the early modern "sexual system" and its legal enforcement.[16] Two kinds of texts provide answers to this question, trial records and legal manuals. My argument, therefore, is an argument from below. Instead of taking for granted that discussions of nature and natural law influenced the mentality of practitioners with regard to the "sin" or "crime against nature," we need to research the manifold ways in which *contra naturam* arguments figured and functioned in early modern jurisdiction. In her probing analysis of medieval philosophy and poetry, Cadden's essay carves out the contradictions inherent in nature's reach. By contrast, in this essay I bring to the fore the tensions that arose when concepts of nature were applied to judicial praxis. There, the dialectic between Nature's neutrality and Her penchant to get involved in moral issues gave way to pure interventionism. Nonetheless, these tensions and contradictions appeared in the courtroom, namely as conflicts in the uneven encounter between defendants and their judges.

Chronologically, my inquiry stretches from the onset of civic sodomy trials in the later Middle Ages into the eighteenth-century, when natural philosophy exercised an unprecedented influence on debates over human society. In the course of this investigation, we will traverse erudite as well as vernacular culture and delve into the unforgiving language of disciplining and its development in the early modern period. We will encounter practitioners of the law as well as an individual who attempted to escape

15. Robert Ian Moore, *The Formation of a Persecuting Society: Power and Deviance in Western Europe, 950–1250* (Oxford: Basil Blackwell, 1987); Bernd-Ulrich Hergemöller, "Sodomiter— Erscheinungsformen und Kausalfaktoren des spätmittelalterlichen Kampfes gegen Homosexuelle," in Bernd-Ulrich Hergemöller, ed., *Randgruppen der spätmittelalterlichen Gesellschaft* (Warendorf: Fahlbusch, 1994), 361–403.

16. I have adopted this term from Isabel V. Hull, *Sexuality, State, and Civil Society in Germany, 1700–1815* (Ithaca: Cornell University Press, 1996). Systematically, I have investigated sodomy trials in Switzerland and Germany until 1600. After 1600, I rely on an expanding body of sources collected by myself but also on an ever growing amount of published materials.

Nature's fierce grip. In other words, this essay approaches "against nature" from the vantage point of the term's social history, its shifting meanings and functions in legal practice. I show that the reach of nature arguments prevalent in theology and philosophy was limited when it came to offenses "against nature." During the early modern period, sodomy rarely was a locus where ideas of nature were discussed. Unlike nature, "against nature" attracted few contestations. Throughout the early modern period, *contra naturam* remained closely linked to the idea of divine punishment. Therefore, neither the concept nor the offense experienced secularization, understood here as dissociation of natural thinking from its religious sources.

"Against Nature" in Judicial Practice

Contra naturam is positioned in a nexus of related terms. When persecuting sodomites, the authorities had a number of expressions at their disposal to denote what in modernity would be called homosexual behavior ("sodomy," "heresy," "to florence," "to commit unchastity with," and so on).[17] The "crime" or "sin against nature" was just one among many expressions used in early modern court proceedings. In Germany and Switzerland, the unequivocal "unchastity against nature" *(widernatürliche Unzucht)* had not yet acquired by that time the wide acceptance it would gain in nineteenth- and twentieth-century legal discourse. Its later prominence is the result of a process of standardization whose roots, as we will see, lie before 1800. Yet, at the same time, "against nature" was set apart from this constellation of terms. Whereas "to florence" was used in conversations as well as in court records, whereas "heretic" *(ketzer)* or sodomite, besides its legal function, frequently served as an insult, *contra naturam* was a formulation whose usage the authorities monopolized exclusively. Once applied, the term expressed a sentence—a sentence to which there was no recourse. As an act of speech, it encapsuled a licence to act and execute. Like few other terms, enunciations of *contra naturam* thus amounted to nothing less than a death sentence.

One of the earliest trial records from the empire to feature the formula "against nature" are from Regensburg. In 1471, the council sentenced four culprits to death for having committed "the sin of unchastity which is

17. Bernd-Ulrich Hergemöller, "Grundfragen zum Verständnis gleichgeschlechtlichen Verhaltens im späten Mittelalter," in Rüdiger Lautmann and Angela Taeger, eds., *Männerliebe im alten Deutschland* (Berlin: Rosa Winkel, 1992), 9–38; Helmut Puff, *Sodomy in Reformation Germany and Switzerland, 1400–1600* (Chicago: University of Chicago Press, 2003).

called the mute sin against human nature." This form.1.1a was repeated almost word by word in all four verdicts. It was supplemented by a reference to God's wrath that had struck five cities (including ᾿ ᾿dom and Gomorrah) as retaliation for "such sins." [18]

Let us take another look at this scenario: wher. humans act against their own nature—understood as the species' not their individual nature, to stretch the formula as it appears in the document—offenders deserve the harshest of treatments for their so-called unnatural acts. Penal authority in capital offenses like sodomy derived directly from God who, according to Genesis 19, had destroyed the city of Sodom in retaliation for its transgressions. Whatever breach of the divine order the Old Testament story referred to originally, medieval tradition saw the offense as sexual and, since Justinian's law code, frequently as male homoeroticism. Genesis 19 thus amounted to a divine mandate—a law binding all Christians. If the representatives and guardians of God's order on earth acted against offenders, a divine response similar to the destruction of Sodom (and the other four cities) might be avoided and, unlike Sodom, their own city possibly saved from its ruin or the death of innocent inhabitants. The biblical reference thus provided legitimation for the city's judges.

Linguistically, the Regensburg entry fuses expressions with distinct genealogies to one compelling cluster: "unchastity" as an umbrella term for illegitimate sexual acts (described in more detail in the proceedings); "the mute sin," an equivalent for sodomy, masturbation, and—rarely—bestiality; sodomy as the sin of the inhabitants of Sodom; and "the sin against nature." In their entanglement, these phrases might have been compiled to convey unequivocally the delict in question, sexual acts involving several men. Since all locutions refer to a number of acts, officials might have felt the need for terminological clarification. In this and other trials, profuse narrations of the events served the same purpose. The complex formula from Regensburg gives evidence to an unusual degree of learning for a fifteenth-century court—a time when trained experts handled judicial affairs in imperial cities only rarely. Yet it also deviates from

18. Christine Reinle, "Zur Rechtspraxis gegenüber Homosexuellen: Eine Fallstudie au. dem Regensburg des 15. Jahrhunderts," *Zeitschrift für Geschichtswissenschaft* 44 (1996): 307–26, 323–26, here 325: "sunde de unkeusch, die man nennet die stumenden sunde wider die menschlichen nature, und sie mit im gesundet haben, darumb den gott umb dergleich sunden vor die funff stett vertilget hat." Other examples for usage of "contre nature" and "onnaturlyke" are reported from Bruges (1490–1515), cf. Marc Boone, "State Power and Illicit Sexuality: The Persecution of Sodomy in Late Medieval Bruges," *Journal of Medieval History* 22 (1996): 135–53, here 137, 143.

conventional terminology in qualifying nature with human ("the mute sin against human nature").[19]

Before the sixteenth century, references to nature surfaced rarely in German or Swiss sodomy trials before urban courts.[20] According to a case brought before the council of Lucerne in 1520, Johannes Nusser confessed to having had sex with women and men, boys and adult men, Italians and Germans, laypeople and monks while serving in the papal guard in Rome. Yet this impostor of a priest who sermonized and heard confession is also said to have acted "inordinately and against nature with women"—a description referring most likely to heterosexual anal intercourse.[21] In Augsburg in 1534, Michael Will, like his male companions *in sexualibus,* faced the death sentence. According to the document's wording, the accused had committed the "abominable evil and vice against nature with himself and with others."[22] In Zurich, the term "unnatural" appears in a trial from 1537. The formula "shameful, unnatural heresy [sodomy]" epitomized the accused's sexual transgression with a boy of twelve.[23]

As a rule, expressions pertaining to nature surfaced where sexual behavior was at stake. Therefore, the term resonated with an everyday understanding of nature as man's sexual drive.[24] It would be difficult, however, to render precisely what magistrates and officials imagined as nature's

19. Cf. the fifteenth-century sermonizer Johannes Herolt, who used the same phrase; see his *Sermones discipuli* (Mainz: Johannes Albinus, 1612), 488.

20. Ecclesiastical courts are not part of my investigation. Evidence suggests that these courts prosecuted sodomy only rarely.

21. Staatsarchiv Luzern RP 11, fol. 141r/v (Johannes Nusser), here 141r: "Aber het Er veriehen / wie Er zu° Rom jn der gwardi ein zit gewesen vnd dem gmeinen man prediget vnd bicht ghört. Daselbs hab Er mit tütschen vnnd wälschen knaben zu° schaffen gehept, des glich ouch vnordenlich vnnd wider die Natur mit den wiben."

22. Stadtarchiv Augsburg, Reichsstadt, Urg. 19 March 1534, Michael Will: "hat das grewlich vbel vnnd laster wieder die natur für sich selbs bei vnnd mit anndern . . . vilmaln bewegt, gevbt vnd begangen." On the concept of "against nature" with regard to masturbation, see Vidal's essay in this volume.

23. Staatsarchiv Zürich (hereafter StAZ) B VI 253 (*Rats- und Richtbuch*), fol. 216r/v (Marx Anthoni 1537), here 216v: "schandtliche vnnatürliche kätzerÿ." For St. Gallen, see Stefanie Krings, "Sodomie am Bodensee: Vom gesellschaftlichen Umgang mit sexueller Abartigkeit in spätem Mittelalter und früher Neuzeit auf St. Galler Quellengrundlage," *Schriften des Vereins für Geschichte des Bodensees und seiner Umgebung* 113 (1995): 1–45, here 22. The 1596 verdict of Franciscus Rouiere stated that he had acted "zu ettlichen underschidlichen maalen durch ingäbung und anraizung deß boßen vyendts uß der christenhaidt wider alle nattur."

24. Cf. Matthias Lexer, *Mittelhochdeutsches Handwörterbuch,* vol. 2 (Leipzig: S. Hirzel, 1876), col. 41; Jakob Grimm and Wilhelm Grimm, *Deutsches Wörterbuch,* vol. 13 (Leipzig: S. Hirzel, 1889), col. 436.

"other" when they employed *contra naturam* in its vernacular variants. As a phrase, the concept "against nature" invoked the supreme authority of God in order to justify the harshest of responses to an act deemed criminal. Yet by content, the concept was redundant. It replicated provisions in the Bible on the plain of God's creation, nature.

The concept "against nature" was rarely cited in isolation. The double expression "against divine and natural law" became a frequently utilized formula of justice in sodomy trials. In Lucerne in 1629, Melchior Brütschlin was sentenced to death for having committed an "abominable misdeed" "against the laws of God the Almighty and of nature."[25] Like in Protestant Zurich, the judges embedded this phrase in a host of legal formulas. It is precisely this web of interlocking, hierarchically arranged authorities that early modern judges were interested in to legitimate their decisions. Nature had no primacy in this context. Though Her proximity to God may be the source of Nature's power, this is also the reason why She could be disposed of so easily. As *natura genitrix,* an autonomous and autonomously acting being, She never made much of a career in legal practice. As God's double, God could always fill in for Nature.

Though officials adopted the expression "against nature" in early modern sodomy trials, they did so infrequently. While its incidence increased after the end of the sixteenth century, many trials were conducted without the judges taking resort to *contra naturam* formulas. The genre of legal advice, written by experts of the law or theology and characterized by argumentative explicitness, was no exception. In the eighteenth-century trial against Catharina Margaretha Linck and Catharina Margaretha Mühlhahn, nature does not loom large. In a lengthy legal opinion from 1721 addressed to the king of Prussia, Friedrich Wilhelm I (1713–44), nature comes up only once: "all the interpreters of Romans I,26 expound on the unnatural sexual intercourse and libido of the female sex."[26] It is nobody but St. Paul whose authority led legal experts to include a reference to "against nature" in this case. This lack is surprising not least because the same *consilium* is a receptacle for many fields of knowledge, such as ancient literature, anthropology, medical discourse, and legal commentaries.

In another *consilium,* references to the unnatural are more frequent.

25. Staatsarchiv Luzern A1 F6 SCH 826 (Melchior Brütschlin 1629): "wider Gottes des Allmechtigen und der natur gesatz."

26. Brigitte Eriksson, "A Lesbian Execution in Germany, 1721: The Trial Records," in Salvatore J. Licata and Robert P. Petersen, eds., *The Gay Past: A Collection of Historical Essays* (New York: Harrington Park Press, 1985), 39.

When fifteen young men were accused of sodomy in the county of Kyburg near Zurich (1688), a group of ministers drafted a counsel with frequent references to the "(un)natural." This term takes precedence even over "sodomy": the unnatural is explained by reference to Sodom rather than vice versa, and God is said to punish according to natural law. Yet this counsel is all but a document of secularization or the steady influx of natural law into the experts' thinking. Only biblical authorities feature in this context.[27]

Significantly, when officials issued sentences, they deployed the term "against nature." *Contra naturam* was cited in verdicts to license one of the state's most violent actions, executing culprits. During interrogations and their protocols, usage of the phrase was rare, however. If the term appears at all before the verdict was issued, "nature" mostly denoted a man's ejaculate, at least until the seventeenth century, when the word *samen,* German for "semen," came to replace "nature" in this regard. This divide between the wording of protocols and that of the verdicts points to a split between lay and elite cultures in the language of sex.[28] "Against nature" was a concept of experts and state authorities. It was cited to support severe disciplinary measures over the authorities' subjects. It was not a common or popular expression.

The decades after the Reformation witnessed the dissemination of a professional legal discourse in the vernacular. This emerging idiom provided linguistic models to capture criminal offenses—a language replete with overlapping formulas and terms. Due to the German reception of Roman law, the extended training of government officials, the professionalization among lawyers, legal advisors, and civil servants as well as the dissemination of written or published standards, a novel idiom in the vernacular was put into use in chanceries and offices. This process of legalization was, to a large degree, stimulated by an increasing reliance on textual models and literary forms of communication.

Viewed in this context, the rise of *contra naturam* formulas reflects, among other developments, the reception of the *Constitutio Criminalis*

27. StAZ A 10 (*Bestialität und Sodomiterei* 1561–1765), 1688.

28. For other areas of Europe similar discrepancies are reported. Apparently, ordinary citizens hardly used the phrases such as "sin against nature" or sodomy. Cf. Michael Rocke, *Forbidden Friendships: Homosexuality and Male Culture in Renaissance Florence* (New York: Oxford University Press, 1996), 90; Jonas Liliequist, "State Policy, Popular Discourse, and the Silence on Homosexual Acts in Early Modern Sweden," *Journal of Homosexuality* 35 (1998): 15–52, here 25.

Carolina of 1532, the German Empire's criminal law code. The code's relevant article 116 described sodomy and bestiality as the "unchastity against nature."[29] While cities in the medieval empire had relied on common law for verdicts against sodomites, now, for the first time in German history, criminal persecution of both male and female sodomites gained a clear terminological as well as legal basis. Yet the *Carolina* was subsidiary law and its acceptance advanced only slowly.

Legalization of affairs of state led to a marked increase in inquiries, epistolary exchanges, and legal counsels. The discourse that evolved from these practices is, however, more than a modernized linguistic receptacle in which legal facts were cast. These linguistic patterns referred to and reflected practices that shaped the course of the proceedings. In the sixteenth and seventeenth centuries, for instance, investigators increasingly subjected defendants to questions about whether they had committed unchaste acts against nature. But they added "with cattle" (or, in other cases, "with men") to make sure suspects understood.[30]

This new language of legal affairs gave expression to a particular understanding of rulership. The political elites emphasized patriarchal hold over a state's subjects and their moral behavior—a notion reinforced in communications with lower-level authorities. Yet this idiom also influenced the ways in which subjects presented their cause in court. In 1695, Johannes Blass, a minister in a rural parish, became the subject of an investigation before the council of Zurich. Unspecified rumors of sexual illrepute forced the authorities to act in his case. As it turned out, Blass had never had sex with his wife of nine years. He developed emotional attachments to adolescent farmboys, with whom he shared his bed when his wife

---

29. *Die Peinliche Gerichtsordnung Kaiser Karls V. von 1532 (Carolina)*, ed. Gustav Radbruch and Arthur Kaufmann (Stuttgart: Reclam 1996), S. 81 § 116 *(Straff der vnkeusch, so wider die Natur beschicht);* Louis Crompton, "The Myth of Lesbian Impunity: Capital Laws from 1270 to 1791," in Licata and Petersen, eds., *The Gay Past* (note 26), 18; Warren Johansson, "Sixteenth-Century Legislation," in *Encyclopedia of Homosexuality*, ed. Wayne R. Dynes, vol. 2 (New York: Garland, 1990), 1198–1200; Heinrich Mitteis, *Deutsche Rechtsgeschichte: Ein Studienbuch,* ed. Heinz Lieberich, 19th ed. (Munich: Beck, 1992), 334. The best contextualization in English remains John H. Langbein, *Prosecuting Crime in the Renaissance: England, Germany, France* (Cambridge: Harvard Univerity Press, 1974).

30. For a sample document featuring the questions asked during an interrogation, cf. Wien, Haus-, Hof- und Staatsarchiv Wien, Reichshofrat, Alte Prager Akten 156 (139, 140) (Wolf Resch contra Sachsen-Coburg 1587–1609) (the date of the trial was January 21, 1588): "Ob er nicht wieder die Natúr mit Viehe vndt Kelbern sich vermischt vnd vnzúcht getrieben?"

was absent. From time to time, he would fondle their genitals. According to all parties interrogated, however, no emission of semen had occurred.

As a strategy of defense and as explanation for his doings with persons much below his social status or education, Blass claimed a weak constitution that left him unable to sire offspring—a condition he described as his "nature."[31] When Blass meditated on nature, he reflected on his own nature and how it came that certain of his acts were at odds with the nature of others or with human nature. He wondered whether one could change one's nature over the course of time. He claimed to have checked this change's progress when he repeatedly scrutinized his or the body of others. By entering temporality into the equation, Blass explored the conundrum called nature in ways different from legal scholars. Whereas the latter invoked nature in the sense of an immutable order, the defendant posited nature closer to an individual disposition that was open to change. Yet his somewhat naive usage also points to connotations of the term that were never fully shed in the legal everyday.[32] In many contexts, "nature" was closely associated with the physical, the sexual, and one's biological sex. Whereas "nature" attracted a host of appropriations (including Blass's), no defendant ever adopted "against nature" to refer to his own doings, so foreign, severe, and deadly were the term's associations.[33]

The idiom that shaped the protocols of early modern court proceedings was not exclusively inspired by legal models. Rather, legal discourse used a medley of registers and styles. When describing sodomy, the authorities drew on the language of religious didacticism, for example. Religious manuals in the vernacular, composed to inform the laity about the teachings of the church, had popularized the phrase "unnatural" since the fifteenth century.[34] As in these publications, drastic language was avoided and euphemistic ways of speaking were instituted also in legal documents.[35] Thus, a rhetoric replete with gestures of revulsion became a regular feature of court records. Such wording attempted to solicit disgust

---

31. Cf. Groebner's essay in this volume.

32. StAZ A 10 (*Bestialität und Sodomiterei* 1561–1765) (Johannes Blass 1695).

33. The realm beyond nature was thus anything but desirable as it was for the radical feminists, as explored in Murphy's essay in this volume.

34. Cf. Johannes Geffcken, *Der Bildercatechismus* (Leipzig: n.p., 1855); P. Egino Weidenhiller, *Untersuchungen zur deutschsprachigen katechetischen Literatur des späten Mittelalters* (Munich: Beck, 1965); Grimm and Grimm, *Deutsches Wörterbuch* (note 24), vol. 24, cols. 1205–7.

35. David Warren Sabean, "Soziale Distanzierungen: Ritualisierte Gestik in deutscher bürokratischer Prosa der Frühen Neuzeit," *Historische Anthropologie* 4 (1996): 216–33.

from the reader or listener. This is surprising since civic archives were not accessible beyond the political elite. Yet, as a rule, verdicts were read aloud before the executions. Since convicts were publicly executed, the idiom of legal discipline also addressed the urban public at large and helped to disseminate phrases *contra naturam* (though in some sodomy cases, early modern authorities sometimes mandated specifically against any public mention of this crime for fear that words about sodomy induced people to emulate the practice encoded in the terms.)[36]

Clusters of stark and condemnatory phrases set the culprit apart from his or her community, at least rhetorically.[37] Under the aegis of this rhetorical regime on sex, the expression "unnatural" became common coinage in many contexts ranging from homiletics to the letter of the law. In Zurich and other Swiss cities, the phrase "unnatural and unchristian acts" reigned supreme as a circumlocution for sodomy and bestiality—a novel wording without an equivalent in fifteenth-century trial records of the same cities.[38]

The term "unnatural" itself is an example of a proliferating group of words in early modern German. These expressions preface a particular concept by the prefix "un-."[39] Obviously, this class of words contains the notion that is being negated. The un-natural conjures up Lady Nature as the arbiter of human actions. Yet in the case of un-natural (as opposed to

36. Staatsarchiv Basek-Stadt, Criminalia 4 12 (Elisabeth Hertner 1647), 6r. Cf. Krings, "Sodomie" (note 23), 21–22; Liliequist, "State Policy" (note 28), 18–20; Jakob Michelsen, "Von Kaufleuten, Waisenknaben und Frauen in Männerkleidern: Sodomie im Hamburg des 18. Jahrhunderts," *Zeitschrift für Sexualforschung* 9 (1996): 205–37, here 216.

37. StAZ B VI 245 (*Rats- und Richtbuch* 1513–1519), fol. 232r–233r (Bläßj Hipold, 1519); StAZ B VI 255 (*Rats- und Richtbuch* 1538–1544) (Uli Rügger, 1540); StAZ A 27.15 (*Kundschaften und Nachgänge* 1544–1545) (Jacob Müller, 1545); StAZ A 27.13 (*Kundschaften und Nachgänge* ca. 1530–1570) (Hans Mötsch, s.a.) (bestiality); StAZ A 27.35 (*Kundschaften und Nachgänge* 1579) (Wilhelm von Mühlhausen, 1579); StAZ B VI 262 (*Rats- und Richtbuch),* fol. 203r/v (Wilhelm von Mühlhausen, 1579).

38. StAZ A 27.15 (*Kundschaften und Nachgänge* 1544–1545) (Jacob Müller, 1545); StAZ B VI 257 (*Rats- und Richtbuch* 1545–1552), fol. 11r/v (Jacob Müller, 1545); StAZ B VI 259 (*Rats- und Richtbuch),* fol. 271r (Rudolf Bachmann and Uli Frei, 1567); StAZ A 27.35 (*Kundschaften und Nachgänge* 1579) (Wilhelm von Mühlhausen, 1579); StAZ B VI 262 (*Rats- und Richtbuch),* fol. 203r (Wilhelm von Mühlhausen, 1579). Similar expressions including the word "unnatural" are reported from other parts of Europe, see Boone, "State Power" (note 18); Liliequist, "State Policy" (note 28), 26. With regard to bestiality, see Jonas Liliequist, "Peasants against Nature: Crossing the Boundaries between Man and Animal in Seventeenth- and Eighteenth-Century Sweden," *Journal of the History of Sexuality* 1 (1991): 393–422, here 402, 406.

39. See, for instance, Susanna Burghartz, *Zeiten der Reinheit— Orte der Unzucht: Ehe und Sexualität in Basel während der Frühen Neuzeit* (Paderborn: Schöningh, 1999).

un-painted or un-interesting, for instance) the prefix "un-" also exceeds the semantic function of denoting the opposite. Un-natural is not simply non-natural, the opposite of natural. By the sheer weight of the rhetorical tradition and frequent usage in moralizing contexts, "un-" words take on additional connotations, the other side of the norm. From the point of view of the speaker, "un-natural" articulates a polemical stance. "Un-" enunciations condemn that which is expressed, declare it as dangerous, treacherous ground (like "un-American" in House Committee on Un-American Activities). It is a word that polices the dangerous boundary between the normative and the nonnormative, the pure and the impure.

"Unnatural" is therefore not a descriptive term. As a word in context, this expression of "symbolic extremities"[40] expresses an ought, a call for action. Unnatural connotes a wretched state that ought to bring about the most vocal condemnation. It is meant to activate, though the precise nature of the implied action remains undefined. Yet the emotional response solicited from the listener/reader by means of this wording is clear: horror.

— — —

In medieval and early modern Europe, shoring up the state's authority in religious matters resulted in the necessity to intervene when subjects were thought to have acted against God's will. This necessity left little room for subjects to respond other than by excusing or defending themselves. Yet once in the dock, few defendants were able to extricate their lives from the interlocking authorities represented by the state. Blass's recourse to his physical nature reminds us vividly of the conceptual and power hierarchies involved that left many accused without an effective defense. Most likely, it was not the argument that he had acted according to his own nature that saved Blass from execution, but the muddled evidence in the case. In theory, Nature was omnipotent; nobody could escape Her grip. Yet Nature's agent and earthly helper, the state, was not as potent as its rhetoric or its references to nature suggested. Blass's case therefore is also a reminder that many people must have managed to live quite comfortably in a realm the state declared "against nature." In the texts of medieval philosophy, just as in the practices of persecution, nature thus emerged as a complex and often contradictory terrain.

Our survey of the sources has brought up some evidence to support the

---

40. Peter Stallybrass and Allon White, *Politics and Poetics of Transgression* (Ithaca: Cornell University Press, 1986), 3.

view that *contra naturam* arguments affected everyday legal activities during the early modern period. Increasingly, court documents featured brief references to nature as a guiding principle for trying sexual behavior. Typically, however, the concept appeared in clusters, surrounded by other concepts that, like nature, bestowed authority on actions of the early modern state. In these contexts, the formulaic expression "against nature" usually mediated between two other elements, divine and secular law. It provided little persuasive force of its own. Rather, it drew its powers from a nexus of interlocking authorities. Yet references to the condemnation of sodomy in divine and secular law sufficed entirely to legitimate a death sentence. Many a time, nature arguments therefore proved an addendum. They provided more of the same according to the motto: the more arguments, the better. Only toward the end of the early modern period, when court records became more extensive in general, did discussions of the "counternatural" appear more frequently.[41] The book of nature is obviously not a law code in which legal provisions can be found. In an age when experts of the law professionalized their practice, scribes, counselors, and commentators were in search for citations to underpin their counsels or decisions. While the laws of nature were suitable material for speculative philosophy, they lacked the force of clear-cut provisions that were the hallmark of professional legal discourse.

"Against Nature" in Legal Manuals

When looking for help in dealing with the quandaries of legal problems, early modern practitioners of the law turned to manuals for information. Often entitled *practica* or *practica rerum criminalium,* compendiums of criminal law were published on a rather grand scale after the mid-sixteenth century. The professionalization of jurisdiction and lawmaking fueled demand for these publications, while in turn these same tools greatly advanced professional standards of wording. Experts used printed manuals for the purpose of reference and for the training of future lawyers.[42] Because of their

41. Theo van der Meer, "Sodom's Seed in The Netherlands: The Emergence of Homosexuality in the Early Modern Period," *Journal of Homosexuality* 34 (1997): 1–16, here 8–9.

42. Joachim Knape, *Dichtung, Recht und Freiheit: Studien zu Leben und Werk Sebastian Brants 1457–1521* (Baden-Baden: Valentin Koerner, 1992). Cf. Heinrich Rüping, "Die Carolina in der strafrechtlichen Kommentarliteratur: Zum Verhältnis von Gesetz und Wissenschaft im gemeinen deutschen Strafrecht," in Peter Landau and Friedrich-Christian Schroeder, eds., *Strafrecht, Strafprozess und Rezeption: Grundlagen, Entwicklung und Wirkung der Constitutio Criminalis Carolina* (Frankfurt am Main: Vittorio Klostermann, 1984), 161–76.

systematic structure, we can expect these handbooks to be more comprehensive and reflexive than verdicts or legal opinions in actual cases. At the same time, as manuals, they never stood wholly apart from the pragmatics of legal issues.

Early modern experts of the law did not dissociate a secular legal sphere from its religious superstructure well into the eighteenth century. They embedded their opinions in divine law and ultimately in the Bible. We should not be surprised, therefore, to find that commentators switched between the terms "crime" and "sin" or "sodomy" and "sin against nature" indiscriminately. In fact, legal manuals contained information from manifold disciplines such as history, medicine, and, importantly, theology. These oscillations, however, also increased the demand for clarification.

With regard to the tribulations of terminology, the Dutch legal scholar Jost Damhouder (1507–81) suggested to divide the crimes (or sins) *contra naturam* along the traditional lines of sexual acts with oneself, among humans, and between humans and animals.[43] By content and by rhetorics, *sodomia*—defined as masturbation, sodomy (referring to both men and women as well as illegitimate intercourse between men and women), and bestiality—occupied a prominent status in Damhouder's edifice. The author shed a flow of distancing formulas and expressions of disgust on his reader, as if this reader would have to be protected from being contaminated. Whereas all other sexual activities were described as "natural," that is "they emerge from nature's instinct, rule, and dictate (but not from reason), only this sin [note the singular for what to us is a number of sins]—

---

43. I use the 1570 edition of the *Praxis rerum criminalium,* printed by Ioannes Bellerus in Antwerp, 1570, at 308: "Tripliciter enim coitus seu peccatum contra naturam (quod Sodomiticum vocant) committitur: nempe quum quis venere abutitur aut secum, aut cum hominibus, aut cum animantibus brutis. ex his speciebus primum peccatum graue est: alterum grauius: tertium grauissimum." I have chosen to foreground this manual for two reasons. First, Damhouder treats questions of nature more extensively than other manuals such as Egidio Bossi's *Practica* (Basel: Henricpetri, 1578), Christoph Crusius's *Tractatus de indiciis delictorum specialibus* (Frankfurt am Main: Wolfgang Hoffmann, 1635), Jakob Döpler's *Theatripoenarum suppliciorum et executionum criminalium* (Leipzig: Friedrich Lanckisch, 1697), or Justinus Goebler's *Der Rechten Spiegel / Auß Natürlichen / den Beschribnen / Gaistlichen / Weltlichen / vnd andern gebreuchlichen Rechten* (Frankfurt am Main: Christian Egenolff Erben, 1558). Second, Damhouder's text provided an important model for the most acclaimed *Practica rerum criminalium,* Benedict Carpzov's 1635 manual with a title similar to Damhouder's, *Practica nova imperialis Saxonica rerum criminalium* (Wittenberg: Zacharias Schürer Erben, 1635). Carpzov's *Practica* went through numerous editions and was still reedited in the eighteenth century. All these manuals show the advanced degree of professionalization as compared to one of the earliest manuals for a German audience, Ulrich Tenngler's *Laÿen Spiegel: Von rechtmässigen ordnungen in Burgerlichen vnd peinlichen regimenten* ([Augsburg], 1509).

having rejected the laws of nature—breaks forth violently beyond the limits of nature." Moreover, "in open war, it fights, violates, confounds this same nature."[44] By introducing reason, Damhouder moved beyond marshalling nature solely for the purpose of moral discipline and hinted at philosophical treatments of the subject matter. Yet the chapter on the "sin against nature" never wholly disentagles itself from moral politics. Casting sin as an agent in this context obfuscates the legal issues at hand while it perpetuates an allegorical understanding of nature as a vicar of God. Far and above that, reason is also identified with the Christian faith. Not accidentally, therefore, Damhouder proceeds to stereotype (Protestant) Christianity's outsiders as proselityzers for the cause of the "unnatural": the inhabitants of Rome, the Turks, and other heathens. This is only one of many examples for the polemical thrust of Damhouder's *Practica*. Ultimately, God was thought to legitimate these stark emotions: "nature's architect and guard declared this sin of nature . . . [to be] detestable."[45]

Geared toward broad circulation, Michael Beuter's (1522–87) German translation of Damhouder's manual translated revulsion into abridgment.[46] All the specific information with regard to masturbation, sodomy, anal intercourse, and bestiality was deleted, not least for reasons of brevity. Yet by casting nature as "common nature" *(gmeyne Natur),* the translation also added an explanatory element.[47] For an audience untrained in the implications of natural philosophy, this wording indicated that individual nature, or rather an individual's sexual drive, was to be differentiated from the nature of the human species.

As we have seen earlier, "crimes against nature" exemplified the religious foundation of early modern jurisdiction ideally. When an act of

44. Damhouder, *Praxis* (note 43), 308–9: "Aliae enim omnes libidinum species naturales sunt, et a naturae instinctu, et regula, et quodammodo a naturae dictamine (licet non a ratione) proficiscuntur, hoc vero solum peccatum spretis naturae legibus, imo extra naturae cancellos violente erumpens, ipsam naturam aperto Marte oppugnat, violat, confundit, et naturae permissis regulis abutitur." Carpzov, *Practica* (note 43), 271, would add that sodomy "hinders" the procreation of mankind and that this sin turns Christians against their faith.

45. Damhouder, *Praxis* (note 43), 308: "hoc ipsum peccatum naturae conditor & custos Deus tam aperte sibi detestabile passim proclamat [. . .]."

46. Jost Damhouder, *Praxis rerum criminalium: Gründlicher Bericht und Anweisung,* trans. Michael Beuther (Frankfurt am Main: Johannes Wolf, 1565); Jost Damhouder, *Practica, Gerichtlicher Handlunge in Bürgerlichen Sachen,* trans. Johannes Vetter and Michael Beuter (Frankfurt am Main: Nicolaus Basse, 1575).

47. Damhouder, *Praxis* (trans. Beuther) (note 46), 182v. Cf. L. Honnefelder, "Natura communis," in *Historisches Wörterbuch der Philosophie,* vol. 6 (Basel: Schwabe, 1984), cols. 494–504.

sodomy occurred, the authorities only had to follow in God's footsteps. By burning the culprits, divine wrath had set the example for early modern rulers. Commentators presented the stake on which sodomites were to be burnt as reminiscent of the destruction of Sodom. A woodcut of the burning Sodom adorns the relevant chapter in Damhouder and other authors. Surprisingly, however, commentators did not elaborate on Genesis 19 as an emblem of early modern statehood with regard to criminal matters. Instead, the story of Sodom was quoted to demonstrate this crime's exceptional severity: so much had this sexual behavior provoked God that in response He struck mankind with disaster.

The result of our brief investigation into early modern manuals of criminal law resonates well with Isabel Hull's finding with regard to the writings of early modern cameralists, trained experts in state bureaucracy and "the first theorists of German civil society."[48] She observed that "the catalog of bad sexual acts was a largely inherited one which the cameralists outfitted with new justifications when the older religious or sentimental justifications needed shoring up."[49] The eminent Samuel Pufendorf (1632–94), professor of natural law and the law of nations, is a good case in point. According to Pufendorf, sexual acts posed a great threat to civil society. Particularly potent stimuli for human actions, erotic desires endangered the social fabric. In *De jure naturae* (1672), Pufendorf asserted that "all lusts which have no other end but obscene titillation" do not accord with natural law.[50] Having posited natural law as guarantor of the public order, he concluded that sexual activity was only legitimate in matrimony. By contrast, all sexual activities outside of marriage were folded into one category of illegitimate sex acts. No further thought was given to acts *contra naturam*.

Early modern legal commentators only occasionally exploited the potential for challenging received wisdom with regard to *natura/contra naturam*. Let us therefore conclude our exploration with the meditations of a speculative thinker who, to an unusual degree, delved into *natura's* equivocations in the context of his discussion of *contra naturam* behavior. In his "Moral Treatise on Natural Shame and the Dignity of Man" (1676),

---

48. Hull, *Sexuality, State, and Civil Society* (note 16), 155.

49. Ibid., 176.

50. Samuel von Pufendorf, *De jure naturae et gentium libri octo* (Frankfurt am Main: Friedrich Koch, 1726), 797–98, here 797: "illud . . . constare arbitramur, juri naturali non congruere illas libidines quae nullum alium finem praeter obscoenam titillationem habent."

Lambert Velthuysen (1622–85) set out to demonstrate that shame is foundational to human nature and society.[51] A student of theology, law, and medicine, Velthuysen practiced as a doctor before he took the post of church supervisor in his native Utrecht. Working at the intersection of several disciplines, Velthuysen conferred on the laws of nature a supreme authority. He decried that in his own time all ethics was drawn from human actions, not from their foundation in human nature. Carving out concepts of nature and natural law in their specificity, Velthuysen differentiated between vices, sins, breaches of criminal law, and breaches of natural law. In his view, such differentiations were anything but academic. He deplored that judges, for instance, were enmeshed in practicing their profession while neglecting the scientific basis of their activities.[52] For reformers of the body politic like Velthuysen, nature was a preferred point of reference. The reflections of the divine in nature were open to human understanding. While the Godhead in its perfection remained inaccessible to the human mind, nature exuded the promise that it could be investigated and understood, and provisions regulating human society issued accordingly.

In his treatise on shame, Velthuysen worked toward a society predicated on reason and nature—a well-ordered universe whose center was matrimony. Yet in order to provide society with a rational foundation in nature, Velthuysen explored offenses to sexual decorum, to which the "sodomitical sins" belonged prominently (though no definition of sodomy is given). For him as for so many other writers before and after, offenses *contra naturam* constitute the most severe transgression against the nature of humankind and, in the context of his treatise, against the laws of shame. But Velthuysen wanted to prove this claim beyond any doubt where other writers conjectured or relied on scholarly authorities. By positing sodomy as an offense whose degree of offensiveness is said to be greater than that of other crimes, the author does not break with traditional condemnations of sodomy. Yet he also challenges this stance by contextualizing sodomy with offenses of a lesser degree, usually not described as counternatural. Is it a crime, he asks, if humans use bodily members against nature, that is, against their primary purpose (like the mouth, the body's organ for taking in food, to do the reverse, vomit)? Even in the case of behavior as seem-

---

51. Lambertus Velthuysius, "Tractatus moralis de naturali pudore et dignitat hominis," in *Opera omnia* (Rotterdam: Reiner Leers, 1680), 161–240.

52. Ibid., 162. The same dynamic of sorting out disciplines drove Enlightenment discussions of masturbation; see the essay by Vidal in this volume.

ingly counternatural as infanticide, he suggests, we might find motivations such as "fear of disgrace or fear of poverty" in hindsight.[53] Velthuysen inundates his reader with such questions, only to break off this line of reasoning when it comes to sodomy. Regarding sodomy, he finds an "alteration of nature against [its] laws" and a "complete aberration from the laws of nature."[54] In the case of sodomy, the superiority of natural law metamorphosed into proof for sodomy's status as the ultimate offense. This conclusion notwithstanding, a theorist à la mode had to provide a rationale along the way. Velthuysen thereby created an exceptional venue for discussion of contra naturam—a discussion that ended abruptly, though.

Despite the fact that the category "against nature" challenges the very notion of Nature's all-encompassing rule, mentions of contra naturam, as far as I can see, did not become a stimulus for discussions of nature's ambivalent intentions, the general applicability of ideas of natural law, or the ruptures between individual nature and human nature. To be sure, the descriptive and the prescriptive were never wholly congruent. Since nature functioned as an argumentative weapon in the arsenal of early modern judiciary discipline, tensions were often set aside and left unexplored. Taking up Velthuysen, the author of the article on sodomy in Johann Heinrich Zedler's (1706–63) encyclopedia (1732–54), one of the leading dictionaries of the early Enlightenment, focused on the argument rather than on the conclusion. He pointed to the different ways in which the authorities used the nature argument. What is more, the entry called attention to the fact that many assumptions about human nature and shame remained to be proven scientifically.[55]

In sodomy proceedings from early modern Germany, nature and natural law played only a minor role in lending ideological weight to the persecution of sodomites. Nature and its terminological allies filtered into early modern trial records only slowly. Even the many legal commentaries that accompanied the growth of the early modern state had little to offer beyond using natura for standard descriptions of sexual acts. In other

53. Velthuysius, "Tractatus moralis" (note 51), 198: "Mater quae infanticidium committit etsi nihil magis contra naturam esse putetur, quam crudelis esse in suum sanguinem, tamen aut metu infamiae, aut paupertate, aut alia aliqua spe impulsa, illud flagitium aggreditur."

54. Ibid., 198: "illud hujus peccati foeditatem supra omnia declarat, et in aperto ponit, quod in illis est non solum totalis aberratio a naturae legibus, sed inflexio naturae contra illas leges." This problem as a whole is reminiscent of Thomas Aquinas, Summa theologiae, ed. Petrus Caramellus, Pars prima secundae (Torin: Marietti, 1952), 425–30 (q. 94, arts. 1–6).

55. "Sodomie," in Grosses Vollständiges Universal-Lexikon, vol. 38 (Leipzig, 1743; reprint, Graz: Akademische Druck- und Verlagsanstalt, 1997), col. 328.

words, in the discussion of *contra naturam* offenses, legal commentators did not probe the limits of Nature's reach. Legal experts rarely embarked on thought experiments with regard to Nature's nature by reference to Her other side, acts *contra naturam*. *Contra naturam* arguments therefore resonated only dimly with discussions of *natura* in disciplines such as theology, jurisprudence, or philosophy during the early modern era. If same-sex acts were mentioned, be it in trial records or in legal commentaries, they provided little occasion for meditating on the workings of *natura* in individuals. Apparently, the condemnation of same-sex acts as "against nature" needed little exploration.

Instead, consensus ruled about the abject status of behavior described as offending nature. If, as Jonathan Goldberg, a leading Renaissance scholar and queer theorist, once postulated, "[s]odomy . . . fully negates the world, law, nature,"[56] early modern lawmakers chose to ignore this wholesale attack. To them, the condemnation of sodomy was well based in the interlocking authorities of an order that rested in the Creator, whose great design for the world was thought to resound in nature, secular law, and the common good. Was it fear of contagion or a concern about learning the sexual practice by hearing about it that kept scholars from exploring the practices conventionally circumscribed by their opposition to nature's laws? Whatever the answer, it follows from my inquiry that arguments pertaining to the unnatural probably led an existence somewhat apart from the efflorescence of natural philosophy during the early modern period, especially the Enlightenment, when nature's uses peaked in intellectual debates in western Europe. As Lorraine Daston and Fernando Vidal show in their contributions to this volume, Nature's nature shifted significantly in the course of the eighteenth century: physicotheologians offered new ways of seeing the natural world to the ever growing army of nature lovers who often recorded their personal experiences in writing. Samuel Tissot, the famous theorist of masturbation, wedded medical empiricism to moral philosophy when he revealed the masturbator's individual *phusis* as nature's battleground. These shifts and reformulations, though never constituting a complete break with previous understandings of nature as God's playground, transformed natural philosophy profoundly.

In a kind of contiguity of that which is not contiguous, concepts of *con-*

56. Jonathan Goldberg, *Sodometries: Renaissance Texts, Modern Sexualities* (Stanford: Stanford University Press, 1992), 19; quoting Alan Bray (*Homosexuality in Renaissance England* [London: Gay Men's Press, 1982], and "Homosexuality and the Signs of Male Friendship in Elizabethan England," *History Workshop Journal* 19 [1990]: 1–19) as a reference.

tra naturam did not follow the same reformist trajectory. Originating in ancient philosophy and taken up in Scholastic theology, the pedigree of the concept "against nature" was too eminent to call for revisionist thinking. Gestures of disgust were too firmly entrenched in the rhetoric of sodomy to call for rigid questioning. More than anything else, it is this striking and reflex-like consensus among scholars and practitioners of the law where *contra naturam*'s deadly force lay.

— — —

How could one dispose of such a powerful tradition? Arguing in favor of the decriminalization of sodomy (ca. 1785), Jeremy Bentham (1748–1832) wisely refrained from opening the Pandora's box that had "nature" inscribed on its lid. Unlike his adversary, Voltaire (who had shed harsh words on *l'amour socratique* in his *Dictionnaire philosophique*),[57] Bentham left nature out of the picture when advocating for a reform of sexual politics. For the social reformer, sexual activity was a matter of taste. On this level of intervention, the Gordian knot tying individual behavior to the common good and the species' interests dissolved in the mutual pleasure of sexual agents, no matter what their sex (though Bentham was far from condoning such pleasures in the case of sex between men or between women).[58] Bentham never published his essay, however. It saw the light of day only *post mortem*.

57. Voltaire, "Amour socratique," in *Dictionnaire philosophique*, vol. 1 (Paris: Frères, 1829), 220–25.

58. Louis Crompton, "Jeremy Bentham's Essay on 'Paederasty,'" *Journal of Homosexuality* 3 (summer 1978): 383–87; 4 (fall 1978): 91–107. For intellectual context, see Miriam Williford, "Bentham on the Rights of Women," *Journal of the History of Ideas* 36 (1975): 167–76; G. S. Rousseau and Roy Porter, eds., *Sexual Underworlds of the Enlightenment* (Manchester: Manchester University Press, 1987).

10

Onanism, Enlightenment Medicine, and the Immanent Justice of Nature

Fernando Vidal

Ecoutez donc la nature, elle ne se contredit jamais.

PAUL-HENRI THIRY D'HOLBACH, *Système de la nature* (1770)

Wenn es Todsünden gibt, so sind es zuverlässig die Sünden gegen die Natur.

CHRISTOPH WILHELM HUFELAND, *Makrobiotik* (1797)

"You are right," declares a character at the end of a Voltairian dialogue, "there is a natural law; but it is even more natural for many people to forget it." And the other replies, "It is natural too to be one-eyed, hump-backed, lame, deformed, unhealthy; but we prefer well-formed and healthy people."[1] Their conversation is typical of the *philosophic* eighteenth century in that "facts" drawn from nature cast doubts on established convictions that are also said to be founded on nature. Like a Freudian dream, the idea of nature was made up of displaced and condensed elements, its meaning overdetermined and variable, its interpretation subject to endless and undecidable debate. Its huge success in the French Enlightenment came from its capacity to unite opposites and apparently answer every possible question about the world, humanity, and the place of the latter within the former.[2]

The Swiss physician Samuel Tissot, whose work is examined below,

Preparation of this chapter was supported in part by a grant from the Athena program of the Swiss National Science Foundation.

1. *Dictionnaire philosophique*, article "Loi naturelle," in *Oeuvres complètes de Voltaire*, vol. 34 (Paris: Armand-Ambrée, 1829), 157–60, here 160. Following the Kehl edition (1784–89), these *Oeuvres* collect under one title all of Voltaire's dictionary-like works.

2. Jean Ehrard, *L'idée de nature en France dans la première moitié du XVIIIe siècle* (Paris: Albin Michel, 1994 [1963]).

made imagination responsible for counternatural sexuality. Imagination obviously belonged in human nature, but its exercise frequently counteracted the purposes of human life, also defined by nature itself. As in the medieval context studied in this volume by Joan Cadden, although nature remained associated with moral conduct and social order, natural processes themselves could corrupt natural order. For most *philosophes,* "returning" to nature was the means to solving such conflicts, and to emancipate humanity from unenlightened and antinatural traditions. This return, however, also yoked humanity to physical and moral necessity. In this essay I illustrate these tensions of eighteenth-century naturalistic ethics through Tissot's ideas about public health, focusing on his treatise on onanism and the notion of an immanent justice of nature.

Public Health and Obedience to Nature

Samuel Auguste André David Tissot (1728–97) was one of the most famous physicians of the Enlightenment. He was born near Lausanne, in the Pays de Vaud (then ruled by Bern, and today a Swiss canton), studied medicine in Montpellier, and except for three years as professor at the University of Pavia, spent the rest of his life in Lausanne as private doctor to an affluent local and European clientele. He wrote on various medical subjects, including an early defense of innoculation. His best known works, discussed below, turned him into a celebrated advocate of public health and a paradigm of the enlightened friend of humanity, especially concerned with the conditions of the poor.

As most physicians of the day, Tissot adopted a Galenic view of the causes of health and disease, based on the exercise of the "nonnaturals." These were environmental, acquired causes of health and disease—food and drink; air; movement and rest; sleep and waking; retention and evacuation (including sexual activity); and the passions—and were distinguished from the "naturals" (constitutional and innate causes) and the "contranaturals" (pathological causes). Tissot combined Galenism with eighteenth-century notions of irritability and the nerves. Health required a "strong fiber," capable of producing sufficient activity in vessels and viscera; transpiration abundant enough to eliminate acrid humors; and firm nerves, whose sensibility was not such that they disturb the rest of the body during the transmission of even light sensory impressions.[3]

3. Samuel Auguste Tissot, *Essai sur les maladies des gens du monde,* 3rd ed. (Paris: P. Fr. Didot le Jeune, 1771), §§ 6–8.

In Tissot's epistemology, as long as imagination predominated over empirical science, nature remained unknown. Imagination was as much a nuisance to health as to scientific progress.[4] As science and art, adhering to nature and following imagination were exact opposites. In the arts, an imperfect sketch is useful; in science, a "failed system, especially if based on fantastic ideas presented as accurate observations," constitutes an obstacle.[5] "Some naturalists treat the book of nature as theologians treated the Bible. They do not consult it to know what it contains, but to find materials that legitimize their ideas. They do not interrogate nature: they imagine oracles, and boldly deliver them as her decisions . . . ."[6] In Tissot's typically Enlightenment description of intellectual history, post-Baconian science triumphantly led to the discovery of the properties of air, the circulation of the blood, and electricity. The time had come, he thought, for irritability.[7]

The discovery of this hitherto unknown property of living tissues was due to the Bernese physiologist Albrecht von Haller (1708–77), longtime professor in Göttingen. According to Tissot, Haller found irritability by getting rid of the "rubbish of a mass of imaginary systems."[8] On the basis of over five hundred experiments (generally vivisections), many cited in the *Dissertation on the Sensible and Irritable Parts of Animals* translated by Tissot, Haller established that muscles contract when a stimulus is applied directly to them. He also showed that a stimulus applied to a nerve does not affect the nerve itself, but produces the contraction of the muscle connected to it. Irritability was therefore attributed to muscles, sensibility to nerves.

Throughout the eighteenth century, nerves were considered the intermediaries between soul and body. In believing this, Tissot held standard views. For him, nerves may suffer from excessively strong bodily or psychical impressions; but since they are "more immediately exposed to the action of the soul," they are affected more by it than by the external senses.[9] Moreover, in Tissot's largely fluidistic view of the body, "vessels" and the fluids circulating in them are of primary importance. We are in a

---

4. On the latter, see Lorraine Daston, "Fear and Loathing of the Imagination in Science," *Daedalus* 127 (1998): 73–95.

5. Samuel Auguste Tissot, "Discours préliminaire du traducteur," in Albert de Haller, *Mémoires sur la nature sensible et irritable des parties du corps animal,* vol. 1 (Lausanne: Marc-Michel Bousquet, 1756), xiv.

6. Ibid., xxxvii.

7. Ibid., iii.

8. Ibid., xiv.

9. Samuel Auguste Tissot, *Traité des nerfs et de leurs maladies,* vol. 3 (Paris: P.-F. Didot le jeune, 1778–80), 280.

state of health when vessels are neither too strong nor too weak, and when fluids are neither too thick nor too thin, and move neither too much nor too little.[10] The key to health was an appropriate regimen, based on conformity to nature and the balanced use of the nonnaturals. Literary persons's habit of working at night, for example, runs against "the laws of Nature, which mark the beginning of night as that of rest."[11]

The convergence of morals and medicine on the appeal to nature's authority characterizes Tissot's books on nervous maladies, the diseases of fashionable and literary persons, and the pathological consequences of masturbation. The influential *Advice to People in General, with Respect to Their Health* is less moralizing and hardly invokes nature. The reason is that, in a treatise aimed at improving the lot of a social class whose ill health he did not attribute to antinatural practices, but largely to bad medicine, Tissot did not need to renaturalize, and thereby remoralize his audience. Peasant life was morally and medically preferable. Rural youths' entering commerce or domestic service was particularly threatening because it involved living in the city and acquiring urban habits. The rural poor tried to imitate the urban rich; their expenditures prevented them from marrying, or diminished their resources for bringing up children. As a consequence, fewer families were formed. Luxury, debauchery, and idleness bred new habits, and the individuals who returned to the countryside were no longer adapted to it. Women became more delicate and less capable of having children. At the same time, the number of artisans required to produce luxury goods increased, and with them, the number of sedentary individuals prone (because sedentary) to become busied with their fantasy—"a new very real loss for agriculture and population."[12]

Tissot's praise of rural life reaches its climax in his *Essay on the Disorders of People of Fashion,* where the peasant's practice of the nonnaturals is presented as a model of hygiene. In contrast to the idle, pleasure-seeking classes, laborers's lives are ruled by need and occupation. These breed "natural pleasures" and "natural usages."[13] Since they exert their bodies very much, and their minds very little, peasants are content with basic nourishing food and do not stray into the realms of thought and imagination. They can do all their work "as a true automaton, without any reflection. Such paucity of ideas," Tissot commented, "is one of the surest preserva-

10. Samuel Auguste Tissot, *De la santé des gens de lettres* (1768; Geneva: Slatkine, 1981), 58.

11. Ibid., 85–86.

12. Samuel Auguste Tissot, *Avis au peuple sur sa santé,* 3rd ed., vol. 1 (Paris: P.-F. Didot le jeune, 1782 [1761]), 6–10, quotation on 9.

13. Tissot, *Maladies* (note 3), § 9.

tives of health, which is almost always inverse to the faculties of mind and their exercise."[14] Limited to satisfying their needs, and thus avoiding excess, peasants feel no passions such as fear, defiance, jealousy, and sadness, which accompany the urban rich's pursuit of luxury and prestige, and weaken the fibers, irritate the nerves, increase sensibility, excite the imagination, prevent the circulation of the humors, disturb sleep, and end up producing the maladies characteristic of the *gens du monde*.

Although deprived of the islander's extreme freedom, Tissot's male peasant sexually resembles the utopian Tahitian of Diderot's *Supplément au Voyage de Bougainville*. In both cases, as the doctor observed of the peasant, *l'homme de la nature* is superior.[15] In both, moreover, sexual desire and activity were said to begin late, when the body produces an overabundance of semen, and thus a purely natural, rather than imaginary, need for sex. The elders of Diderot's island released the young from chastity only after they observed the persistence of "virile symptoms," and were assured of the "frequent emission and quality of seminal liquor."[16] Similarly, Tissot's peasant, preserved from inactivity and boredom, "dangerous readings," and "objects of seduction," knows desire only in full maturity, once he feels the reproductive urge. In healthy individuals, "nature inspires [sexual] desires only when the seminal vesicles are filled with a certain amount of liquid too thick to be reabsorbed" in the body.[17] Prompted by purely mechanical and quantitative factors, sexual desire in Tissot's natural men lacks the impetuousness due more to imagination than to need. Since desire "almost never goes beyond necessity," it contributes to health; city youth, in contrast, are weakened as a consequence of their early initiation to debauchery.[18]

---

14. Ibid., § 20.

15. Ibid., § 28.

16. Denis Diderot, *Supplément au Voyage de Bougainville, ou Dialogue entre A. et B. sur l'inconvénient d'attacher des idées morales à certaines actions physiques qui n'en comportent pas* (written 1772), in *Oeuvres*, ed. Laurent Versini, vol. 2 (Paris: Robert Laffont, 1994), 559.

17. Samuel Auguste Tissot, *L'onanisme: Dissertation sur les maladies produites par la masturbation*, preface Théodore Tarczylo (1768; Paris: Le Sycomore, 1980), 72–73 (II.3). To help in locating passages, after the page reference I add the article (four in total) and section numbers (which start anew in each article). In the only reprinted English translation, sections are numbered continuously: *Onanism: or, a Treatise upon the Disorders produced by Masturbation: or, the Dangerous Effects of Secret and Excessive Venery*, trans. A. Hume (1766; New York: Garland, 1985; facsimile series Marriage, Sex, and the Family in England 1600–1800, vol. 37). Translations here are mine, and more literal than Hume's.

18. Tissot, *Maladies* (note 3), § 28.

As an instance of peasant sexual mores, the polygraph Rétif de la Bretonne's graphic depiction of rural boys masturbating together in the open air—a scene he claimed to have witnessed at age seven, and which he later took as proof that there is almost as much corruption in the country as in the city—is more representative than Tissot's chaste countryside.[19] Sexuality, however, was only one factor among others that, for the physician, made "degeneration" inevitable: the love of sciences and the culture of letters; the excessive use of warm baths; the propagation of luxury and city life, and the ensuing increase in the intensity of the passions; the spiciness of foods; and "secret maladies." Tissot complained that by the fourth generation, strength and health were known to exist only in octogenarians or by hearsay. The only ways of recovering them were unlikely (more reasonable behavior), or undesirable (several centuries of barbarism).[20] Degeneracy and unhealthiness were directly proportional to one's rank in the social scale and distance from rural customs.[21]

Although Tissot's Rousseauian sensibility made him oppose the natural simplicity of the countryside to the corruption of cities, his perspective on society and civilized life was "redemptive."[22] He did not want Europeans to live like "savages."[23] Indeed, even if he thought peasants's life "happier," he was far from idealizing it entirely. In the *Advice,* he exposed the high mortality and morbidity rates of the rural poor, and attributed them to excessive work, insufficient food, poor hygienic conditions, and bad medicine. Still, *gens du monde* pay their toll for not contributing to human so-

19. Nicolas-Edme Rétif de la Bretonne, *Monsieur Nicolas, ou le coeur humain dévoilé,* vol. 1, ed. Pierre Testud (1796–97; Paris: Gallimard, 1989), 39–40. In a Latin note Rétif describes the onanists's gesture, says he does not know whether they ejaculated, and adds, "but I did not see any of them blush" (p. 40, n. A). It should be mentioned that Rétif authored two novels on the corruption of virtuous peasants in the city. Although in the *Causes célèbres* (1734–89), a 253-volume collection of legal cases, the scenario of sexual crimes often conforms to the city/country opposition, numerous cases contradict the image of a sexually idyllic countryside. Hans-Jürgen Lüsebrink, "Les crimes sexuels dans les *Causes célèbres,*" *Dix-huitième siècle,* 12 (1980): 153–62, here 157–58. More details are found in Lüsebrink, *Kriminalität und Literatur im Frankreich des 18. Jahrhunderts: literarische Formen, soziale Funktionen und Wissenskonstituenten von Kriminalitätsdarstellung im Zeitalter der Aufklärung* (Munich: R. Oldenburg, 1983). Cf. Jean-Louis Flandrin, *Les amours paysannes (XVIe–XIXe siècle)* (Paris: Gallimard/Juillard, 1975), 160–65 on onanism.

20. Tissot, *Santé* (note 10), 182, n. l.

21. Tissot, *Maladies* (note 3), § 9.

22. Anne C. Vila, *Enlightenment and Pathology: Sensibility in the Literature and Medicine of Eighteenth-century France* (Baltimore: Johns Hopkins University Press, 1998), 195–96.

23. Tissot, *Maladies* (note 3), § 66.

ciety. Tissot's approach to onanism and the diseases of *gens de lettres* further highlights the morally legislative role he assigned to nature, and the subversive character he attributed to the imagination.

Literary illnesses were predominantly "nervous," that is, they affected the link between soul and body. Tissot understood nerves to be hollow tubes or elastic solid fibers (both common eighteenth-century theories). As tubes, they carry the "animal spirits" prepared in the brain, and take them to the other organs, whose functioning depends on such innervation. Disease therefore results from the loss, insufficient production, or inadequate distribution of those subtle and pure fluids. As solid fibers, healthy nerves are characterized by firmness and tension; illness results from degrees of stiffness, softness, or shaking *(ébranlement)* that preclude adequate innervation of the organs. Tissot compared the effects of immoderately intense thinking both to a ligature that binds the nerves and thereby suspends their operation, and to excessive evacuations that weaken the organism.[24] These analogies had physiological bases. *Gens de lettres,* Tissot noted, often suffer from digestive disorders. If the nerves that innervate the stomach are ligated, food will rot without being digested. Similarly, if the soul monopolizes the nerves, animal spirits are not distributed throughout the body, the stomach is insufficiently innervated, and digestion does not take place.[25] In general, too much intellectual work is an obstacle to the circulation of the nervous fluid.

The remedy is simple: return to nature. Like the long-lived Fontenelle, the healthy man of letters exercises his body and restores his mental forces by gardening.[26] Those who fail to establish such a balance should stop thinking, "forget the existence of the sciences and books, . . . and become what Nature made men, a peasant or a gardener."[27] But nature is not wilderness. On the contrary, Tissot emphasized that persons of fashion and letters are able to enjoy higher pleasures, derived from the "culture of the mind" and "the exercise of sentiment," to which country folk have no access.[28]

Following nature implies perfecting it in conformity with its own laws. The conviction that underlies Tissot's prescriptions is constitutive of Enlightenment discourses about the normative import of nature. Gardens

24. Tissot, *Santé* (note 10), 36.
25. Ibid., 91–92.
26. Ibid., 56.
27. Ibid., 221.
28. Tissot, *Maladies* (note 3), §§ 66, 68.

have long been a resource for thinking about nature and society. When, around 1760, the Physiocrats proposed an order founded on physical necessity and natural-scientific evidence, they designated agriculture as the economic means of enforcing it. At the end of the century, the *médecin-philosophe* Cabanis (1757–1808) claimed that to "deduce" the rules of conduct from the laws of nature is to base morality on "its unique and true source."[29] Steering humanity toward happiness entails revising and correcting nature's work—a "[b]old enterprise," wrote Cabanis, "which nature itself seems to have particularly recommended."[30] Physiology, he claimed, shows that we are all born with different temperaments and inclinations, yet equally endowed with the power of sympathy. The universality of this faculty points toward a new aim: to "produce a sort of equality of means that is not in our original makeup," and that would be the physiological equivalent of the French Revolution.[31] In Cabanis's view, the natural character of this goal followed from our capacity for sympathy, and from the imperative that we must correct nature because it made us perfectible. The line is therefore thin between using scientific knowledge to reach moral and political ends, and formulating these ends as if they were compulsory norms, dictated by nature itself. The empirically ascertained existence of sympathy may indicate that we are able to work for equality or that we must pursue it because nature intends us to be equal.

Onanism and the Punishment of Sin and Crime

Tissot's *Onanism* is the culmination of a system characterized by the coalescence of the descriptive and the normative. In the preface, Tissot claimed not to deal with the moral aspects of masturbation, and, indeed, he focused on physiology, regimen, and the presentation of clinical cases. Regarding the causes of onanism's dangerousness, he explained:

> Convinced that, since their creation, bodies are subjected to laws that govern all their operations necessarily, and whose economy the Divinity modifies only in a small number of special cases, I would resort to miraculous

29. Pierre-Jean-George Cabanis, "Lettre à M. F. sur les causes premières" (written ca. 1806–7), in *Oeuvres philosophiques,* vol .2, ed. Claude Lehec and Jean Cazeneuve (Paris: Presses Universitaires de France, 1956), 294.

30. Cabanis, *Rapports du physique et du moral de l'homme* (1802), in *Oeuvres* (note 29), vol. 1, 357.

31. Ibid.

causes only when one finds an obvious opposition to physical causes. This is by no means the case here: everything can be very well explained by the laws of the mechanics of the body, or by those of its union with the soul.[32]

Moreover, Tissot distanced himself from the anonymous *Onania* of 1716 (an English publication from which he borrowed clinical cases), by scorning it as a "rhapsody" and describing its "reflections" as "theological and moral trivialities."[33] This has led commentators to write both that *Onanism* was "a thoroughly medical book" and that it is filled with exaggeration and sophistry.[34] Both judgments take Tissot's writing and ideas out of their textual and historical context. As Hufeland (1762–1836) explained in *The Art of Prolonging Human Life,* his popular systematization of the ideology of natural living, physical and moral perfection are as tightly connected as the body and the soul, and only together can they bring about human perfection.[35] This was a basic principle of Enlightenment anthropology, which should not be confused with the professional moral-theological discourse Tissot wished to demarcate from his.

The traditional Christian view is that, as the Roman Catholic Church still affirms, to the extent that masturbation aims at procuring sexual pleasure outside marriage, independently of love and procreation, it is "an intrinsically and gravely disordered action."[36] Tissot said that much. For theologians, however, when onanism is committed deliberately and in full cognizance of its gravity, it instantiates lust, one of the seven deadly sins. This doctrine is supposed to express God-given moral and natural orders. In Genesis 38, God kills Onan for spilling his seed to avoid making his brother's childless widow pregnant. It is unclear whether Onan was punished primarily for committing *coitus interruptus* or for refusing levirate, that is, to father a child who would be regarded as legal descendant of his dead brother. Jewish and Christian exegetes, however, consistently condemned the destruction of the male seed.

The order thus affirmed is "natural" primarily in the sense of a belief

---

32. Tissot, *L'onanisme* (note 17), 72 (II.3).

33. Ibid., 21, 39 (preface, I.3).

34. Karl Braun, *Die Krankheit Onania: Körperangst und die Anfänge moderner Sexualität im 18. Jahrhundert* (Frankfurt: Campus, 1995), 79; Karl Heinz Bloch, *Masturbation und Sexualerziehung in Vergangenheit und Gegenwart: Ein kritischer Literaturbericht* (Frankfurt am Main: Peter Lang, 1989), 177.

35. Christoph Wilhelm Hufeland, *Makrobiotik oder die Kunst das menschliche Leben zu verlängern,* ed. K. E. Rothschuh (1797; Stuttgart: Hippokrates Verlag, 1975), 19 (foreword).

36. *Catechism of the Catholic Church* (1992), § 2352.

in the existence of norms that transcend individual will, cultural variation, and historical circumstances. These norms, as explained for example in Jean-Jacques Burlamaqui's *Principes du droit naturel* (1747), one of the most widely used law textbooks of the eighteenth century, are knowable with irrecusable certainty through reason, by means of the empirical study of human nature. Nevertheless, their ultimate moral and legal authority comes from their having been ordained by God. It is in such a framework that, since the Middle Ages, masturbation qualifies as a crime "against nature," susceptible of being legally repressed. Jurists recommended punishing it with exile or fines.[37] Yet their advice was purely theoretical since, as they realized, the secrecy of the crime made its discovery impossible.[38] Pleasure was in the onanist's hand; punishment in God's.

What did change fundamentally in the history of the theological approach to masturbation is the role of subjective factors.[39] By its new emphasis on the moral actor's secret intentions, motives, fantasies, and feelings, the moral theology developed in the framework of the Council of Trent and the Catholic Reformation in the mid-sixteenth century constitutes a link between the Scholastic natural law approach to nonprocreative sexual acts and the nineteenth-century medical models of sexual pathology. Desire is henceforth crucial. Thus, wrote the Franciscan Jean Benedicti in his *Somme des péchés et le remède d'iceux comprenant tous les cas de conscience,* first published in 1599, the man who masturbates while desiring a married woman commits adultery in addition to onanism; if he desires a virgin, stupre; if a relative, incest; if a nun, sacrilege; if another male, sodomy. The source of sin was desire, sexual fantasy, and the loss of self-control—not the wasting of sperm. The dangers of onanism were spiritual, not medical. It was only in the eighteenth century that masturbation was seen as fundamentally pathogenic, and that the imperative of the internal regulation of desire foregrounded by post-Tridentine theology was legitimized by medical reasons.

Medical considerations about masturbation had been made before the eighteenth century, and in two different directions. On the one hand, insofar as Galenic physiology noted the beneficial effects of seminal dis-

---

37. Jean Stengers and Anne van Neck, *Histoire d'une grande peur: la masturbation* (1984; Paris: Pocket, 2000), 35–36.

38. Neither masturbation nor sodomy figure in the *Causes célèbres;* see Lüsebrink, "Crimes" (note 19), 154 (table).

39. This paragraph follows Pierre Hurteau, "Catholic Moral Discourse on Male Sodomy and Masturbation in the Seventeenth and Eighteenth Centuries," *Journal of the History of Sexuality* 4 (1993): 1–26, esp. 16–24.

charge, moral theologians insisted that onanism was unacceptable even if carried out *propter sanitatem*. In the mid-seventeenth century, a Spanish Cistercian went as far as claiming that, if God had not explicitly forbidden masturbation, it would be obligatory as a means to prevent certain illnesses. When Pope Innocent XI condemned the proposition, he recalled that natural law forbade onanism.[40] On the other hand, sin *(crimen onaniticum)* and crime (offences *contra naturam*) sometimes coalesced with pathogenic behavior.[41] Since the sixteenth century, physicians affirmed that onanism could cause severe illness. It remains the case, however, that it was only in the Enlightenment that masturbation was widely perceived as an urgent medical threat, and as a dangerous epidemic of momentous individual and social consequences.

As for eighteenth-century moral theology, it remained essentially independent of medicine. The most directly relevant example from Tissot's own cultural context, and just the kind of book from which he wanted to distance himself, is a 1707 *Treatise against Impurity* by the Neuchâtel theologian Jean-Frédéric Ostervald (1663–1747), author of an influential annotated translation of the Bible and a reformer of Protestantism who emphasized good deeds and moral conduct over dogma. While Ostervald was explicit about adultery and fornication, he refused to name the other forms of impurity, formulated criteria (such as lascivious gestures or lack of modesty in dressing) for recognizing them, and asked his readers to examine themselves and, as he put it, tell themselves what he himself did not dare speak about. It was perhaps Ostervald's discretion that prompted his English translator to enumerate various other "Unnatural and Dangerous Species of Uncleanness," including self-pollution. Ostervald elaborated "considerations from nature"—but in them he emphasized the agreement of Revelation with right reason, argued that man should not submit himself to instincts and passions, and mainly defended matrimony as the means to realizing God's will for individuals and society. To these and other specifically theological arguments, he added a lengthy moral-theological defense of chastity.[42]

40. Instances from Stengers and van Neck, *Histoire* (note 37), chap. 2.

41. Michael Stolberg, "Self-pollution, Moral Reform, and the Venereal Trade: Notes on the Sources and Historical Context of *Onania* (1716)," *Journal of the History of Sexuality* 9 (2000): 37–61.

42. Jean-Frédéric Ostervald, *Traité contre l'impureté* (Amsterdam: Thomas Lombrail, 1707), 1.1.1 (Considérations tirées de la Nature), 1.1.7 (Des autres espèces d'Impureté), IIe partie (De la chasteté). On the species of impurity, see "To the Reader," in Ostervald, *The Nature of Un-*

A crucial role in launching the medicalizing trend was played by *Onania, or the Heinous Sin of Self-Pollution*.[43] Published anonymously in London in 1716, it treated masturbation as sin, but also popularized the concept of postmasturbatory disease, and has been described as "the first work exclusively devoted to the discussion of masturbation that gave a detailed account of the manifold physical consequences, illustrated them with individual cases, and offered medical help to those who could identify with the description."[44] The book immediately inspired critiques and imitations, went through several editions (fifteen by 1730, and there were later printings), and was translated into German. The clinical material grew, especially in the form of readers's testimonial letters. Later judgments about *Onania* (and to a lesser extent Tissot's *Onanism*) as irrational, pseudo-medical, or sadistic ignore that both fit into contemporary physiology and medicine, and gave the impression of being based on reliable observations. Physicians found in the notion of postmasturbatory disease "a convenient frame of reference for the interpretation of a wide range of disease cases"—not the least because each case validated the concept, and made it immune to empirical falsification.[45]

In contrast to Tissot's *Onanism*, a "dissertation on the illnesses produced by masturbation," *Onania* was a book on "the heinous sin of self-pollution." While the former emphasized physiology and the criterion of bodily need, the latter highlighted moral transgression and its connection to the voluntary or involuntary character of self-pollution (what Tissot decried as "theological and moral trivialities"). *Onania*, however, also described self-pollution as producing disease, and could thus be usefully incorporated into a predominantly moral and juridical frame of reference. The article on "Self-pollution" of a major encyclopedia of the *Aufklärung* is exemplary of this process. In it, masturbation is defined as the "unnatural habit" whereby persons of both sexes "apply themselves to counterfeit nature, and obtain the same sensation God prescribed so as to make more pleasant the carnal commerce of man and woman for the uninterrupted propagation of the human species." All the classical themes are present. Onanism

cleanness consider'd: Wherein is discoursed of the Causes and Consequences of this Sin, and the Duties of such as under the Guilt of it (London: printed for R. Bonwicke, 1708), xiv–xv on onanism.

43. *Onania, or, The Heinous Sin of Self-Pollution* (1716; New York: Garland, 1986; facsimile series Marriage, Sex, and the Family in England 1600–1800, vol. 12).

44. Stolberg, "Self-pollution" (note 41), 50–51.

45. Ibid., 56. See also Michael Stolberg, "An Unmanly Vice: Self-Pollution, Anxiety, and the Body in the Eighteenth Century," *Social History of Medicine* 13 (2000): 1–21.

is against nature, and an imitation thereof. It contravenes divine will, natural law, and a legal order that upholds reproduction as essential to society. Natural religion and reason demonstrate that refusing to reproduce is contrary to God's wisdom. Psychologically, masturbation is a habit—not an innate vice, but a voluntary behavior. As a sin, it is not merely *contra naturam:* it "inverts and, so to say, exterminates nature. The individual who renders himself guilty of it works for the ruin of his species, and . . . tries to damage Creation itself." In so doing, he "contradicts himself," and calls forth, as shown by examples from *Onania,* "terrifying consequences and evils." In conclusion, masturbation "grossly injuriates divine and natural laws, as well as civil society." [46]

In addition to lending support to earlier moral and juridical views, the medicalization of masturbation solved the problem of impunity. The Boston Calvinist Cotton Mather (1663–1728) apocalyptically announced onanists's punishment. After being raised for Judgment, their bodies, "which have been kept like an *Oven,* by the *Burning* of *Unchastity,*" will be themselves made "as a fiery Oven," and subjected to "the Torments of a *Devouring Fire, and Everlasting Burnings.*" In case such distant prospects were not enough, Mather warned that "there is a yet nearer *Destruction from God,*" and that onanists "bring such a *Destruction* upon their *Health* . . . as to render themselves a sort of *Self-Murderers.*" [47] Mather, whose interest in medicine was as intense as his belief in witches and demons, recommended combining repentance and the "*Methods* of the *Christian Warfare*" against the devil with work, fasting, abstinence (or marriage), temperance, and plenty of water. [48] Thanks to this regimen, young men could avoid the fate of the "Wretch Putrifying, and Emaciated under the Arrest of that *Foul Disease*" whose image the preacher remembered from the publication of an unnamed Italian doctor. [49] The sin included its own punishment—and a rapid one at that.

Tissot took this penal immanence for granted. A classmate of his who died of consumption after two years of intense masturbation "did not wait

46. "Selbst-Befleckung," in Johann Heinrich Zedler, *Großes vollständiges Universallexikon aller Wissenschaften und Künste* (1732–50; Graz: Akademische Druck- und Verlagsanstalt, 1993), vol. 36 (1743), cols. 1586–90, quotations cols. 1586–89.

47. [Cotton Mather], *The pure Nazarite. Advice to a Young Man, concerning An Impiety and Impurity (not easily to be spoken of) which many Young Men are to their perpetual Sorrow, too easily drawn into. A letter forced into the Press, by the Discoveries which are made, that Sad Occasions multiply, for the Communication of it* (Boston: T. Fleet, [1723]), 6.

48. Ibid., 11, 16.

49. Ibid., 6–7.

long for his punishment."[50] In general, symptoms are the "punishments" onanists "find in their own crimes."[51] One might try to fight one's criminal inclinations, but (as a patient put it) the "difficulties of victory" are considerable; the threat is such that "the opportune time for amendment is short."[52] Prevention is of course the best approach, since, warned the Enlightenment's most famous confessed onanist, Jean-Jacques Rousseau, "if once [the young man] knows this dangerous supplement, he is lost."[53] Full cure, states the last paragraph of *Onanism,* is exceptional; grave sequels the rule.

Tissot's language of crime and punishment has been considered a residue of moral theology.[54] This interpretation reinforces a certain simplistic image of the Swiss physician as agent of secularization. For example, one of his biographers writes, "As a sin onanism had to be endured; as a disease, Tissot turned it into a curable condition."[55] Such a description is inaccurate, and disregards the crucial element of style. For theologians, sin was an act for which one was responsible, not something to be endured; for doctors, masturbation was a pathogenic cause, not a disease. Moreover, Tissot's intensely judgmental vocabulary about onanism is embedded in vivid descriptions of its circumstances and consequences. His depiction of the "frightening scene" or the "awful sights" of patients's sufferings were meant to produce the same emotions he allegedly felt when witnessing them or reading about them.[56] Of the first encounter with the subject of his most spectacular case, the watchmaker L. D★★★★, Tissot reports, "I was myself terrified."[57] One "does not read [physicians's accounts] without horror;" the doctor "shudders" as he reads in *Onania* a patient's depiction of his suicidal impulses.[58] These reactions are moral. The sight of symp-

---

50. Tissot, *L'onanisme* (note 17), 75 (II.3).

51. Ibid., 39 (I.3).

52. Ibid., 76, 75 (II.3).

53. Jean-Jacques Rousseau, *Emile ou de l'Education* (1762), in *Oeuvres complètes,* vol. 4, ed. Bernard Gagnebin and Marcel Raymond (Paris: Gallimard, 1969), 663. For Rousseau's avowal of the *dangereux supplément qui trompe la nature,* see *Confessions,* bk. 3 (1782), in *Oeuvres complètes,* vol. 1 (Paris: Gallimard, 1959), 109. On puberty, sex, and masturbation, compare Tissot's ideas with *Emile,* bk. 4.

54. Théodore Tarczylo, *Sexe et liberté au siècle des Lumières* (Paris: Presses de la Renaissance, 1983), 126.

55. Antoinette Emch-Dériaz, *Tissot, Physician of the Enlightenment* (New York: Lang, 1992), 53.

56. Tissot, *L'onanisme* (note 17), 35 (I.1), 51 (I.4).

57. Ibid., 40 (I.4).

58. Ibid., 31 (I.1), 43 (I.4).

FIGURE 10.1. Plates from *Le livre sans titre* (Paris: Audin, Libraire-éditeur, 1830). From the copy at the Bibliothèque publique et universitaire, Geneva. Photo: Jean Marc Meylan. The small *Book without a Title* is "dedicated to young people, and to fathers and mothers," and epigraphed with Tissot's pronouncement, "This fatal habit kills more young people than all the diseases of the world." The first plate is the frontispiece, and the others illustrate one chapter each. The short chapters (three to four pages each), poorly written in sentimentalistic tones, stereotypically follow the symptoms that take the masturbator from health to tomb, and deal successively with dorsal consumption, stomach pains, redness of eyes, extreme weakness, disturbed sleep and nightmares, spitting of blood, loss of hair, vomiting of food, vomiting of blood, pustules on the entire body, fever and pallor, rigidity of the body, and agony and death. In the Genevan copy, the booklet is bound with a ten-page "Advice on the Means of Correcting Young Convicts from the Habit of Onanism, by an Administrator of Prisons." The author's method consists of presenting the *Book without a Title* to the prisoners whose masturbatory habit has been ascertained. The text, he comments, "is mediocre; its entire utility is in the plates. . . . After having duly prepared the young convict you wish to reform, show him the plates one after the other, while making sure to simultaneously influence him by means of a skilfully graduated morality, so that he will be convinced that those ravages that frighten him will be—if he does not correct himself immediately—the inevitable outcome of the excesses he commits" (pp. 4–5). The method, a crude but faithful application of late Enlightenment medico-moral pedagogy, was said to have already proven successful in the prisons of Geneva.

toms could have produced compassion; but since at least the Renaissance, the sin inscribed in them occasioned horror.[59] As Helmut Puff highlights in his essay for this volume, a study of early modern German and Swiss judicial practices and discourses, the sin appeared all the more hideous that it was *contra naturam*. Tissot's medico-moral strategy of persuasion would work for decades—to the point of tragic caricature in the 1830 *Book without a Title* that depicts the young onanist's ghastly and fatal decline (fig. 10.1).

By unveiling his feelings, Tissot confirmed the truth of natural law, told his audience how to respond to the text, and gave his book maximal efficacy. It is therefore primarily through a style, aimed at producing the emotional coincidence of writer and reader, that he blended medical argument and public health objectives with legal and moral-theological principles.[60] Tissot's declaration of intention reveals the coherence of con-

59. Arnold I. Davidson, "The Horror of Monsters," in James J. Sheehan and Morton Sosna, eds., *The Boundaries of Humanity: Humans, Animals, Machines* (Berkeley: University of California Press, 1991), 36–67, 57–58 on Tissot; also in Davidson, *The Emergence of Sexuality: Historical Epistemology and the Formation of Concepts* (Cambridge: Harvard University Press, 2001.)

60. For other considerations about Tissot's writing, see François Rosset, "Samuel-Auguste Tissot: le docteur écrivain," in Vincent Barras and Micheline Louis-Courvoisier, eds., *La médecine des Lumières: tout autour de Tissot* (Geneva: Georg, 2001), 258 on *Onanism*.

Il était jeune, bon, il faisait l'espoir de sa mère...

Il est cacochyme... bientôt il porte la peine de ses... tombe vieux avant l'âge... son dos se courbe...

Un feu dévorant embrâse ses entrailles, il souffre d'horribles douleurs d'estomac...

Voyez ces yeux naguères si purs, si brillants, de combien d'une bande de feu les entoure...

Il ne peut plus marcher, ses jambes fléchissent...

Des songes affreux agitent son sommeil... il ne peut dormir...

Ses dents se gâtent et tombent...

Sa poitrine s'enflamme... il crache le sang...

Ses cheveux, si beaux tombent comme dans la vieillesse, sa tête se dépouille avant l'âge...

Il a faim, il veut apaiser sa faim, les aliments ne peuvent séjourner dans son estomac...

Sa poitrine s'agrandit... il vomit le sang...

Tout son corps se couvre de pustules... il est horrible à voir...

Une fièvre lente le consume, il languit, tout son corps brûle...

Tout son corps se roidit... ses membres usent d'eau...

Il délire, il se roidit contre la mort, la mort en plus forte...

Il expire, et dans des tourments horribles...

tent and style. "My goal," he explained, "was to write about the disorders produced by masturbation, not about masturbation as a crime. But is it not enough, to prove its criminal nature, to demonstrate that it is an act of suicide?"[61] Onanists were no less responsible of their death than those who shot or drowned themselves.[62] Theologically a sin, legally a crime, suicide was against the law of nature, a God-given instinct of self-preservation, the duty of pursuing perfection and happiness, and one's obligations to society. Masturbation thus qualified as an instance of what the *Encyclopédie* called "indirect suicide," one carried out without self-homicidal intention, but resulting from a disordered lifestyle or imprudent exposure to obvious dangers.[63] Such ideas were not functionless vestiges left over in Tissot's language. On the contrary, as postulates of his interpretation of onanism, they rooted medical, naturalistic discourse in the moral domain, and reinforced the epistemological and normative authority of both.

Tissot believed in an essential link between God's will and natural laws.[64] He wrote: "As subjects to God's laws, we must study in His works the principles of our conduct"; to attain happiness ("the end of man"), we must know the laws of human nature, which show that "the more reasonable man is thereby the more virtuous."[65] Morality is not arbitrary, "since its precepts are based on the general laws of nature." It follows that there is an inherent link between virtue and happiness, vice and misery. Perfection, Tissot explained, "is the obedience and acquiescence to order and nature"; the desire to reach perfection is therefore the "great principle of man," and humanity's "purest homage . . . to the Author of nature." In the realm of love and pleasure, following the laws of "our reasonable and social nature" rather than instinct leads to monogamous unions, and thus

61. Tissot, *L'onanisme* (note 17), 20 (preface).

62. Ibid., 37 (I.2).

63. "Suicide *(Morale),*" in *Encyclopédie,* vol. 5, 639–41, 640. "*Encyclopédie*" refers to Denis Diderot and Jean d'Alembert, eds., *Encyclopédie ou dictionnaire raisonné des sciences, des arts et des métiers* (Paris: Briasson . . . , text vols. 1–7, 1751–65; 8–17, 1765, first ed. (in-4°) accessed through the site of the ARTFL project (University of Chicago/CNRS): http://encyclopedie .inalf.fr/.

64. In examining John Wesley's 1767 *Thoughts on the Sin of Onan* (largely made up of excerpts from Tissot, but emphasizing the power of God over the physician's authority and the will of the patient), James G. Donat aptly speaks of Tissot's "medical theology." Donat, "The Rev. John Wesley's Extractions from Dr Tissot: a Methodist *Imprimatur,*" *History of Science* 39 (2001): 285–98, here 291.

65. Samuel Auguste Tissot, *Principes de Philosophie morale. Seconde partie. Des Loix naturelles de l'Homme,* manuscript department, Bibliothèque cantonale et universitaire, Lausanne, fols. 4, 43. Many thanks to Patrick Singy, who lent me his notes before I could consult the manuscript.

to fulfilling "the goal of nature" and satisfying "the need of the heart without depressing reason." In sum, pleasure is legitimate when guided by the principle that it "has been made for man, not man for pleasure."[66]

Tissot's moral philosophy, natural religion, medicine, and principles about public health are integral and complementary parts of each other. In his public health treatises, he insisted that his approach was purely medical; in *Onanism,* he adhered to that principle to the point of not hiding the fact that, sometimes (in a Galenic humoral perspective), masturbation was beneficial to health.[67] In his manuscript *Principles of Moral Philosophy,* which he described as a treatise of natural law,[68] he approached health "more as moralist than as physician," and defined it as "[a] disposition of the body such that, by making man happy, allows him to fulfill all the duties imposed on him by his nature, and by the circumstances in which he happens to be have been placed" (fol. 39). Tissot intended to define moral and social obligations by going back to indubitable natural facts (fol. 1). But when he wrote that "[t]rue science gives us knowledge of ourselves and of the foundations of our duties" (fol. 58), he implied that one of the criteria for asserting the truth of science is its conformity to otherwise proven moral principles. Nature is known through the sciences, but it "speaks" to each of us through reason and sentiment (fol. 86). The mutual dependency of the medical and moral viewpoints is best understood in the context of theistic iusnaturalism and natural religion. The laws of religion, Tissot wrote, "belong in the laws of nature," and "must therefore be subjected to the supreme law of nature, the law of the greatest good."[69]

The Enlightenment entomologists Lorraine Daston explores in her essay "beatified" lowly creatures, and placed them through sustained attention within a complex of theological, esthetic, moral, and economic values. Inversely, but through equivalent empirical practices that entailed observation, description, diagnosis, and prescription, physicians damned contranatural sexuality and placed it, but with negative valence, at the heart of an analogous set of interconnected values. As Ludmilla Jordanova

66. Ibid., fols. 4, 6, 70, 51, 55.

67. Tissot, *L'onanisme* (note 17), 130–33 (IV.1).

68. Tissot, *Principes* (note 65), fol. 22.

69. Ibid., fol. 108. Tissot connects natural religion to each person's obligation to work for the good of others (fol. 104), and theism to the relations between the existence of God and the problem of evil, the laws of sociability, and conflicts among duties (fol. 6, 83, 107). "Theism" (belief in the existence of God) was sometimes distinguished from "deism" (forms of natural religion that exclude revelation and external cult); but since word usage was unstable, I leave Tissot's deism the name of *théisme* that he gave it.

points out, Tissot's approach to onanism is characterized by its inclusiveness; "[n]otions of sin, evil, crime and punishment are all incorporated into a larger vision, which sets improper sexual activity in the context of class relations, family dynamics, responsibility and dependency."[70] Especially significant is the continuity of medicine and moral theology.

In Tissot's own city of Lausanne, a Protestant minister of mystical inclinations wrote a "moral and philosophical" companion to *Onanism*. In it, he recalled that masturbation produced "terrifying infirmities," but emphasized (as Mather before him) that it was a violation of the body as temple of the Holy Spirit, and tried to guide young people more through exhortatory advice and hopeful glimpses of eternal life than through the rhetoric of fear favored by physicians and educators.[71] In Spain, a translation of *Onanism* appeared only in 1807; four earlier official refusals were motivated by the fear that it might promote the vice, not deny or supersede religion. On the contrary. In 1791 the ecumenical vicar of Madrid had recommended its publication as a book that, "far from opposing faith and morality, may contribute to the salutary instruction of physicians, and teach patients a Christian lesson."[72] And in Rio de Janeiro, a full century after the publication of *Onanism,* a medical thesis on school hygiene advised that, in addition to exercise and strict measures of surveillance, pupils found to masturbate be read by a priest "Tissot's work on the dangers of onanism."[73]

Onanism was a bestseller. Twelve authorized editions appeared between

70. Ludmilla Jordanova, "The Popularization of Medicine: Tissot on Onanism," *Textual Practice* 1 (1987): 68–79, 77.

71. [Philippe Dutoit-Membrini,] *De l'onanisme; ou Discours philosophique et moral sur la luxure artificielle, & sur tous les Crimes relatifs* (Lausanne: Antoine Chapuis, 1760).

72. Quoted in Enrique Perdiguero Gil and Angel González de Pablo, "Los valores morales de la higiene. El concepto de onanismo como enfermedad según Tissot y su tardía penetración en España," *Dynamis* 10 (1990): 131–62, 154.

73. "Os alumnos que em causa emmagrecerem, ficarem palidos e indolentes, devem ser vigiados; e quando convictos do onanismo, devem os directores mandar fazer pelo sacerdote a leitura da obra de Tissot sobre os perigos do onanismo, e obrigados a exercicios corporaes . . . ." Candido Teixeira de Azeredo Coutinho, *Esboço de uma hygiene dos collegios applicavel aos nossos. Regras principaes tendentes á conservação da saude e ao desenvolvimento das forças physicas e intellectuaes, segundo as quaes se devem regular os nossos collegios* (Rio de Janeiro: Typographia Universal de Laemmert, 1854), unpaginated. The recommendation is found in the last paragraph of the thesis, in a section on precepts for literary, moral, and religious education. This material came to my attention through José G. Gondra, *Artes de Civilizar: Medicina, Higiene e Educação Escolar na Corte Imperial,* Ph.D. diss., Universidade de São Paulo, 2000. I thank Dr. Gondra for a copy of Coutinho's thesis.

1760 and 1799, six revised by Tissot.[74] Unauthorized reprints bring the total to thirty-one eighteenth-century printings in French. In addition, there were eight German translations, six English, and four Italian, and, in the nineteenth century, thirty-two French printings, four German, one English, six Spanish, four Italian, and one Russian.[75] *Onanism* became the basic textbook of a highly repetitive, alarmist, and self-perpetuating anti-masturbation campaign. Armed with Tissot and Rousseau, the partisans of pedagogical naturalism (such as those Eckhardt Fuchs presents in his essay in this volume) aimed not only at preserving individual health, but also the health of the state and society at large.

According to Isabel Hull's brilliant study of the sexual system in the development of postabsolutist civil society in Germany, debates from around 1760 to 1790 on such themes as masturbation and incest played the role of "thought experiments that organized reflections on difficult issues encountered in the transition from state tutelage to civil freedom," and helped compare the social principles of Enlightenment and absolutism.[76] The ensuing mechanisms of sexual regulation concerned less the repression of sexuality than its satisfaction within marriage, for the orderly expansion of society, and under the guidance of willed, reasonable desire. Pedagogues, however, were caught in a "self-critical tautology" that made onanism appear as "the negative exemplar of Enlightenment," as the consequence of new lifestyles characterized by conviviality, mobility, luxury, education, reading, and the formation of a private sphere.[77]

Among Tissot's public health books, *Onanism* is the one most patently aimed at the internalization of self-control. As would be the case among enlightened German educators, medicine appeared for that purpose more

---

74. List by Silvio Corsini in *A l'ombre des Lumières: Un médecin lausannois et ses patients. Auguste Tissot 1728–1797* (Lausanne: Bibliothèque cantonale et universitaire, 1997), 14–16. *Onanism* first appeared in 1758 as an eighty-eight-page *Tentamen de morbis ex manustupratione* appended to Tissot's *Dissertatio de febribus biliosis*.

75. List by Tarczylo in *Sexe et liberté*, app. 4 (note 54).

76. Isabel V. Hull, *Sexuality, State, and Civil Society in Germany, 1700–1815* (Ithaca: Cornell University Press, 1996), 5. Cf. Uwe Rohlje, *Autoerotik und Gesundheit: Untersuchungen zur gesellschaftlichen Entstehung und Funktion der Masturbationsbekämpfung im 18. Jahrhundert* (Münster: Waxmann, 1991).

77. Hull, *Sexuality* (note 79), 270, 269. For overviews of later developments, see Stengers and van Neck, *Histoire* (note 37); Bloch, *Masturbation* (note 34); E. H. Hare, "Masturbatory Insanity: The History of an Idea," *Journal of Mental Science* 108 (1962): 1–25; Freddy Mortier, Willem Colen, and Frank Simon, "Inner-scientific Reconstructions in the Discourse on Masturbation (1760–1950)," *Paedagogica Historica* 39 (1994): 817–47.

effective than moral theology. Tissot himself remarked that fear of a present evil had more impact than the appeal to religious principles whose truth could be doubted.[78] One de Bienville, author of a treatise on nymphomania, stated it even more openly than his admired Tissot:

> All those threats that we constantly address to libertines of both sexes would be totally incapable of returning them, however so slightly, to moral or Christian virtue, if numerous reasonings founded on nature, as well as countless known experiences that strongly confirm those reasonings, did not impress them with their evidentiary character, and convinced them only because they terrify them.
>
> It would be in vain that a mass of Christian philosophers would constantly cry at them that incontinence, especially of the kind dealt with here, is an absolutely abominable crime, if a good naturalist did not show how that crime will kill them with as much cruelty as speed.[79]

Medical evidence about life-threatening effects did not convince by itself, as an instance of the light of reason, but also because it generated horror and fear. As we saw, these emotions respond to the moral violation expressed in the body, not to physical symptoms in and by themselves. Scientific rationality does not replace religious or moral injunctions, but joins them in a novel fashion. The efficacy of the argument comes from its penal weight, made inseparable from its supposed evidentiary value. Thus, the relevant article of Diderot's *Encyclopédie,* based on *Onanism,* lets theologians "decide and make known the enormity of the crime," and asserts that infrequent masturbation "has no consequences, and is not an evil (in medicine)."[80] But it then reproduces Tissot's vocabulary of crime, punishment, guilt, and their tormenting outcomes.

Onanism assumes the mutual implication of morality and nature as

78. Tissot, *L'onanisme* (note 17), 20, 22 (preface).

79. M.-D.-T. de Bienville, *La nymphomanie ou Traité de la fureur utérine: Dans lequel on explique avec autant de clarté que de méthode, les commencements et les progrès de cette cruelle maladie, dont on développe les différentes causes* (1771; Paris: Office de librairie, 1886 [1778]), 36–37. For Tissot on the consequences of masturbation in women, see *L'onanisme* (note 17), 52–56 (I.5). My translation here is more literal than the one in *Nymphomania, or, A Dissertation Concerning the Furor Uterinus . . .* , trans. Edward Sloane Wilmot (London, 1775), facsimile published with Tissot, *Onanism* (note 17).

80. [Jean-Joseph Ménuret de Chambaud,] "Manstupration ou Manustupration (*Médec. Pathol.*)," in *Encyclopédie* (note 63), vol. 10, 51–54, here 51.

known through science. Patients's expressions of remorse, Tissot wrote, cannot weaken the "impression of horror" conveyed by their condition "because this impression depends on facts."[81] Conversely, the empirical study of nature necessarily confirms moral truth. Self-diagnosis was not merely a medical act: it entailed recognizing one's crimes. It thus corroborated the etiological significance of masturbation. Tissot himself described this self-enlarging (but not self-correcting) circle when he explained that, by establishing resemblances between their ills and those described in *Onanism,* "guilty" readers discover the cause of their ailments, and become the first to disclose it to physicians.[82] The patient who declares, "nature opened my eyes," or the dying man who confesses that masturbation is the true cause of his downfall are brought to self-diagnosis more by books and doctors than by nature.[83] Their path to enlightenment parallels that of the Englishman beseeched by Tissot to tell him "if he had never polluted himself with Onan's abominable crime."[84] The medical and especially the pedagogical strategies set up in the context of the antimasturbation campaign from the last third of the eighteenth century onward generated, rather than repressed, abundant discourse about sex, thereby instituting the very object they would seek to discipline.[85] At the same time, in order to prevent the body from speaking the unspeakable, it was imperative to obtain the transparency of hearts and consciences. Especially in Germany, late-eighteenth-century educators strived toward that goal, and developed for that purpose an arsenal of procedures, including dramatic fictions such as the one illustrated here in figure 10.2. The young narrator and co-protagonist (with the elderly pedagogue) of the depicted scene says that the episode (a lesson against masturbation) ended with a prayer "that included

81. Tissot, *L'onanisme* (note 17), 39 (I.3).

82. Ibid., 22 (preface).

83. Ibid., 44, 49 (I.4). Patrick Singy has argued that, although Tissot's medical rationalism concealed a religious morality, the efficacy of the book derived from its naturalistic argumentation. The prohibition of masturbation no longer emanated from an external authority (God or His terrestrial representatives), but from the patients's themselves, who compared the descriptions they read with observations they made of themselves. Contrary to what I attempt to do here, Singy isolates the naturalistic contents of Tissot's discourse both from the style in which they are couched, and from the complex of social, religious, and epistemological values in which (as he himself acknowledges) they are embedded. Singy, "Le pouvoir de la science dans *L'Onanisme* de Tissot," *Gesnerus* 57 (2000): 27–41.

84. Tissot, *L'onanisme* (note 17), 88 (III.1).

85. Michel Foucault, *Histoire de la sexualité,* vol. 1, *La volonté de savoir* (Paris: Gallimard, 1976).

FIGURE 10.2. Frontispiece from *Ueber Kinderunzucht und Selbstbefleckung: Ein Buch bloß für Aeltern, Erzieher und Jugendfreunde, von einem Schulmanne* (Züllichau/Freystadt: Nathanael Sigismund Frommans Erben, 1787). From the copy at the Staats- und Universitätsbibliothek, Dresden. This engraving is the only illustration of the anonymous *On Children's Unchastity and Self-Pollution: A Book Only for Parents, Educators, and Friends of Youth, by a Schoolteacher.* It depicts the crucial scene of a lengthy "pedagogical dream" (pp. 241–51), told in the first person, from the standpoint of the young man, and in the style of eighteenth-century "sensibility." The old man, described as a Socrates figure with youthful eyes, is the youth's *Schulvater,* his pedagogical mentor. One day, they are both looking at the plates of Buffon's *Natural History.* When they come across an image of apes (notorious for their alleged lasciviousness, and even for masturbating, as Buffon reports of the baboon), the teacher starts bemoaning the death of his only son at age eighteen. He recalls that, before dying after a debilitating illness, the son declared that a few words of warning would have sufficed to prevent him from becoming a victim of a sin he did not know, and an object of divine vengeance. The teacher then decided that none of his *Schulsöhne* would have to blame him in similar terms. Having told this story, he asks God to touch his pupil's heart, and leads him by the hand into a chapel. He kneels in front of a black curtain that seems

to hide an altarpiece, asks God's forgiveness, and thanks Him for allowing him to save other youngsters. He then opens the curtain, thus revealing a skeleton covered with mourning crepe—the skeleton of his own son. While the pedagogue weeps, the pupil reads a marble plate placed under the skeleton and engraved with black letters: "Wenn schnöde Wollust dich erfüllt, / So werde durch dies Schreckenbild / Verdorrter Todenknochen / Der Kitzel unterbrochen" (When you are filled with filthy lust, the itch shall be stopped short by this frightful image of a dead man's withered bones). He also notices the "excellent verses" inscribed in gold on a blue medallion above the skeleton: "Flieh, Jüngling, vor der Wollust Pfade / Und wach und rufe Gott um Gnade . . ." (Flee, young man, the path of lust, and watch, and call to God for grace . . .). With a slight modification on the first line, the verses turn out to be the last stanza of "Warnung vor der Wollust" (Warning against lust) of the *Geistliche Oden und Lieder* by Christian Fürchtegott Gellert, one of the most renowned poets of the German Enlightenment. As teacher and pupil come again together, the former looks directly into the latter's eyes, and tells him that he does not expect pledges about moral behavior (God and his dead son already demand a solemn oath), but only a promise to respond with "openness" to his questions. He then takes the skeleton's hand, the young man joins his, and the promise is made.

everything which went on in my soul, as accurately as if the good old man had seen in my heart."[86]

Even the detail of Tissot's scientific argumentation betrays the moral and legal authority he attributed to nature under the species of medicine. In treating of *gens du monde,* and especially of *gens de lettres,* Tissot focused on neuroanatomy. In examining masturbation, he gave more etiological significance to fluids. The ominous opening words of *Onanism* announce

86. *Ueber Kinderunzucht und Selbstbefleckung: Ein Buch bloß für Aeltern, Erzieher und Jugendfreunde, von einem Schulmanne* (Züllichau/Freystadt: Nathanael Sigismund Frommans Erben, 1787), 251.

Wenn schnöde Wollust dich erfüllt,
So werde durch dies Schreckenbild.
Verdorrter Todtenknochen
Der Kitzel unterbrochen.

the fatal theme of liquid loss, waste, leakage: "Our bodies lose continuously . . . ."[87] If we did not recover the substance we lose, we would rapidly fall victims to mortal weakness. Why, however, is loss of a certain quantity of semen by masturbation worse than equivalent losses by intercourse?

As we saw, Tissot explained natural sexual desire in terms of the quantity and density of semen. Individuals whose desire is governed by imagination and habit will seek to satisfy it beyond necessity, and therefore squander valuable fluids. The more this mode of gratification is practiced,

87. Tissot, *L'onanisme* (note 17), 25 (introduction).

the more it produces irritations that need to be assuaged. The onanist gradually self-destructs by subjecting himself to the artificial needs he creates.[88] Like the man of letters obsessed with one topic, he remains fixed on his "foul meditations"; incapable of attending to anything else, he falls prey to melancholy, catalepsy, epilepsy, imbecility. The waste of useful spirits and the accompanying diminution of mental faculties render him incapable of "becoming anything in society," lower him below brutes, and justly (because of his responsibility) turn him into an object of scorn rather than pity.[89]

Tissot suggested that the same processes—dissipation of fluids, subjection to imaginary needs and fixed ideas, nervous disorders, exclusion from society, miserable death—can result from excessive sexual intercourse. And yet, masturbation remains worse. In an effort to bring distant theological punishments into line with immediate medical consequences, Mather calculated: "It is by a strange Vengeance of GOD, upon this *Impiety,* that a Tenth Part of the *Expense* made in this *Lustful way,* does ten times the Damage to the *Health* of the Trespassers, that is felt, by what is made in the *Lawful way.*"[90] In this way—strange indeed!—God is let to discipline through natural laws that do not require His direct intervention. As Tissot put it, onanists are "guilty of a crime whose sanction divine justice did not want to suspend, and which it punishes instantly with death."[91]

After asking for God's blessings in the preface to *Onanism,* Tissot placed Him in parentheses, and went into the details. Of several physiological causes, only one specifically justified the belief that onanism is unhealthier than sexual promiscuity. Confirmed observations, Tissot explained, show that our skins not only transpire, but also inspire. During coitus, what one individual gives up, the other gains. This reciprocal compensation of fluids does not take place in masturbation: "the onanist loses, and recovers nothing." Moral and psychological factors also play a role. On the one hand, the regrets and shame felt by the onanist who has realized the extent of his crime actually increase its danger. On the other, the masturbator ignores the joys of legitimate love that help restore the strength lost during intercourse. "And can we doubt that nature has attached more joy to the pleasures gained according to her ways than to those contrary to them?"[92]

88. Ibid., 73 (II.3).
89. Ibid., 74 (II.3).
90. Mather, *Nazarite* (note 47), 6.
91. Tissot, *L'onanisme* (note 17), 80 (II.3).
92. Ibid., 79 (II.3) for both quotations.

Tissot, a modern man of *coeur* and *sensibilité,* emphatically did not. The essence of the onanist's fault was not a failure to follow a particular regimen, but a refusal to accept nature's offers and comply with its injunctions. Like masturbation, the excesses of literary and fashionable people could also cause illness and death; but only masturbation was, in Tissot's wording, an "abominable crime," that is, both a mortal sin and an action punishable by law. *Gens de lettres* and *gens du monde* embodied, respectively, a denatured way of life and a denatured social class. Neither, however, were defined by activities to which moral condemnation was attached. They certainly behaved in unnatural ways, but their positive qualities made them capable of enjoying the highest delights of culture and sentiment. They never abandoned the human universe.

The onanist, in contrast, did not just deviate from nature's prescriptions, but lived directly against its fundamental goals. Deserting nature meant forsaking humanity and society. Masturbating was the most radical form of betraying nature because it led, by natural mechanisms, to an individual's dehumanization. Yet in the process of placing himself, like an animal, outside the reach of moral judgment and human laws, the onanist submitted himself to the immanent justice of nature. As Hufeland noted in his (very standard) account of masturbation, "nature avenges nothing in a more dreadful manner than that through which one sins against her. If there are mortal sins, then they are certainly the sins against nature."[93]

Nature and the Freedom of Enlightenment

Ein Jahrhundert, das an Brefeiung denkt und Gefängnisse phantasiert.
HANS MAGNUS ENZENSBERGER, *Mausoleum* (1975)

Coupled with the appeal to nature as cognitive and moral authority, the Enlightenment return to nature was supposed to help liberate humanity from passions and false traditions. The prize to be paid for the newly acquired freedom could be high—in the case of onanism, years of terror, obsession, and anxiety following Tissot's paradigmatic treatise.[94] Onanism

93. Hufeland, *Makrobiotik* (note 35), 231 (Praktischer Teil, I.2). As a reviewer of *Ueber Kinderunzucht* commented, "it appears as if nature wanted to avenge itself in this awful fashion for the transgressions against its laws." Quoted in Hull, *Sexuality* (note 76), 265.

94. For German examples, see sources in the anthology by Katharina Rutschky, ed., *Schwarze Pädagogik: Quellen zur Naturgeschichte der bürgerlichen Erziehung* (Frankfurt am Main: Ullstein, 1997 [1977]), part 7.

became Enlightenment's own self-pollution, both a consequence of its achievements and a sign of its demise.

Eighteenth-century arguments for freedom in the name of nature created opportunities, but also implied inescapable constraints and inevitable destinies. Unlike the berated Christian God, or even human justice, the *philosophe*'s Nature knew neither charity nor forgiveness. More fatally than other crimes, masturbation realized the utopia of self-enforcing laws that punish without accusation or defense. Yet the principle of an immanent justice is emblematic of the paradoxical outcomes of Enlightenment naturalistic ethics. Nature, like the unfathomable Freudian dream, ignores contradiction. It is a first principle, against which it is impossible to behave.[95]

In d'Holbach's materialist system, for example, man "never acts otherwise than according to the laws of his organization and the matter from which nature made him."[96] The "imaginary" and "invisible" powers priests invented to oppress humanity, or the "fictions" and "empty words" that hindered the progress of knowledge result from the natural constitution of the mind: nothing can deviate a single instant from the laws "according to which nature herself acts."[97] The predicament of individuals acting against (their) nature illustrates a common pattern of tension between necessity and freedom, between improving the lot of humanity by taking it back to nature and postulating that nothing escapes natural laws. Tissot was aware of this "apparent contradiction," and tried to solve it by distinguishing two concepts of human nature: one that considers it as a set of abstract and general laws willed by the Creator, but constantly broken; another that represents the individual natures persons invariably follow in the particulars of their behavior.[98]

The naturalistic ethics of the eighteenth century had a radical critic in Immanuel Kant. In his lectures on pedagogy and ethics of the 1770s and 1780s, Kant reproduced current juridical, medical, and pedagogical arguments against masturbation, considered onanism contranatural for the same reasons as his contemporaries, and, like the author of *Ueber Kinderun-*

95. Günther Mensching, "La nature et le premier principe de la métaphysique chez d'Holbach et Diderot," *Dix-huitième siècle* 24 (1992): 117–36.

96. Paul-Henri Thiry d'Holbach, *Système de la nature ou des lois du monde physique & du monde moral* (1770), in *Oeuvres philosophiques,* vol. 2, ed. Jean-Pierre Jackson (Paris: Alive, 1999), 169.

97. Ibid., 621–22, 197.

98. Tissot, *Principes* (note 65), fols. 85–86.

zucht (fig. 10.2), advocated being open about it and its consequences.[99] Insofar as sexual love is destined "by nature" to preserve the human species, he explained, lust is "unnatural" when aroused not by a real object, but by one's imagining it. Yet the specifically moral reason for limiting sex to its "natural end" is not that it would be otherwise unnatural, but that nonprocreative sex violates one's duty to oneself. Onanists use themselves "merely as a means to satisfy an animal impulse," and thereby surrender their personality and violate their humanity. That is why masturbation exceeds suicide: to the extent that it requires courage, suicide leaves room "for respect for the humanity in one's own person."[100]

Naturalistic arguments thus play no substantive role in the Kantian condemnation of onanism. Human free choice can be affected, but not determined by sensible impulses. Actions based on desires, feelings, or interests are only hypothetical imperatives, to be carried out as means to something else, and *if* certain subjective conditions are met. In contrast, the metaphysical principle of morals is categorical, and states that we should act as if the maxim of our action were to become by our will a universal law.[101] Contrary to the laws of nature, moral laws are laws of freedom, dependent for their accomplishment on the ability of pure reason to subject the principle of every action to the categorical imperative.[102] In short, moral norms must be autonomous from empirical rules, motivations, or consequences, and ethics must be independent from the empirical sciences. Kant thus reversed the *philosophes*'s morals. From atheists to deists, there were many who equated the realization of perfectibility with returns to nature and obedience to natural norms. Yet in doing so, they subjected humanity to the inexorable authority of nature; and while inventing freedom, they turned the Enlightenment into "[a] century that thinks about liberation and phantasizes prisons."[103]

99. See *Über Pädagogik* (based on Kant's lecture notes) and *Moralphilosophie Collins* (notes taken in 1784–85), in Immanuel Kant, *Gesammelte Schriften* [GS] (Berlin: W. de Gruyter, 1966–), 9:496–98 and 27:390–92 for the passages on onanism.

100. Immanuel Kant, *The Metaphysics of Morals* (1797), in Mary J. Gregor, trans. and ed., Kant, *Practical Philosophy* (Cambridge: Cambridge University Press, 1996), 549 (*GS* 6:425).

101. Immanuel Kant, *Groundwork of the Metaphysics of Morals* (1785), in *Practical Philosophy* (note 100), 73 (*GS* 4:421).

102. Kant, *Metaphysics of Morals* (note 100), 375 (*GS* 6:214).

103. Jean Starobinski, *L'invention de la liberté, 1700–1789* (Geneva: Skira, 1964); Hans Magnus Enzensberger, *Mausoleum. Thirty-seven Ballads from the History of Progress,* trans. Joachim Neugroschel (New York: Urizen Books, 1976 [1975]), 41.

11

Ants and the Nature of Nature in Auguste Forel, Erich Wasmann, and William Morton Wheeler

A. J. Lustig

Myrmecology has been more fortunate than many other branches of entomology in the men who have contributed to its development. These have
been actuated, almost without exception . . . by a temperate and philosophical interest in the increase of our knowledge.

WILLIAM MORTON WHEELER

Scientists, if not sluggards, go to the ant to study her ways and be wise.
The social insects—ants, bees, wasps, and termites—perpetually fascinate:
surely their small polities must, by design or chance or history, have something to tell us about our own? From the last third of the nineteenth century through the first third of the twentieth, the success of evolutionary
theories changed the relationship of the social insects to human beings.
In a static cosmos, whether created by chance or design, they could serve
as exemplars, paragons, utopians or dystopians; but their value was metaphoric. In this volume Danielle Allen discusses the roles that honeybees,
for example, played in Western political discourse. In a Darwinian world,
the social insects' significance could be more direct. If not quite cousins of
ours, the ants might nevertheless be able to tell us something general:
about the meaning of social behavior, its origins, its plasticity and limits,
the connection of altruism with morality (if any), even, perhaps, general
laws that transform species from the war of all against all to the solidarity
of community.

Social behavior thus naturalized in the ants conveys complex messages

Many thanks to Sarah Jansen, Charlotte Sleigh, and the Moral Authorities for their suggestions and advice.

about the naturalness of social behavior in humans, with ramifying implications. If ants divide their labor, communicate, cultivate crops, gather their harvests, raise cattle, war on one another, prey on one another, parasitize one another, even support arrays of unrelated, sometimes detrimental species, then how far can similar behaviors be unique products of human rather than natural nature in humans who have likewise evolved? How *natural,* in short, a category is *society?* And is human freedom within societies a product of nature, or a human attempt to deny it?

The biologists Auguste Forel (1848–1931), Erich Wasmann (1859–1931), and William Morton Wheeler (1865–1937) each grappled with these problems, finding different tensions and resolutions. The three premier "pure" myrmecologists of their day (as opposed to applied entomologists whose interest in ants lay primarily in devising ways of stamping them out rather than celebrating their marvels), Forel, Wasmann, and Wheeler laid down a foundation of observations, terminology, and theory on the social insects that continue to shape modern biology. Their careers overlapped for several decades after 1900, as they corresponded and wrote in response to one another's ideas in an ongoing conversation that, while usually—superficially—confined to ants, in fact also addressed deeper questions of the structure and naturalness of societies, of the proper domain of scientific authority, and of the levels on which ideas of individuality, autonomy, sociality, and freedom operated.[1]

The period in which these myrmecologists worked was one of great theoretical diversity within natural history and evolutionary theory: all three framed their theories in terms of evolution, but evolution's mechanisms in this period were very much a matter of debate, and various hypotheses were advanced to explain the origins of social behavior, from teleological or vitalist speculations about directional drives within lineages, to various Lamarckian effects, to neo-Darwinian natural selection. At the same time, evolutionary entomology in this period was still firmly a discipline of natural history depending on long-term observation and personal knowledge of individuals, colonies, and species. Forel, Wasmann, and Wheeler all practiced the same intensive disciplines of attention that Lorraine Daston discusses in her essay, following individual insects and colonies hour after hour, year after year.

---

1. For a thorough analysis of myrmecology in the period, including discussions of Forel and Wheeler complementary to those here, see Charlotte Sleigh, "Six Legs Better: A Cultural History of Entomology in the Late Nineteenth and Early Twentieth Century." (Ph.D. diss., University of Cambridge, 2000).

They likewise succumbed to—indeed, gloried in—the "creature love," as Daston puts it, that marked the emergence of modern observational natural history from the eighteenth century; while they would destroy and dissect ants in the service of experiment and investigation, they nevertheless also wrote movingly of individual insects, and both Forel and Wheeler wrote of long-cherished domesticated colonies at last released to freedom. This creature love was fiercely loyal and territorial: virtually no monograph concerning the ants begins or ends without an encomium from the enthusiast as to the superiority of his chosen subject above all other social insects (myrmecologists' attitude to honeybees, for example, closely resembles that of Jane Austen's landed gentry to men who have made their money in trade), if not all other social organisms, or, indeed, the entirety of Creation.

Recognition of this intensely personal knowledge of and identification with their subjects is vital to appreciating the complex integration of the formic and the human in Forel's, Wasmann's, and Wheeler's thought: this period in evolutionary biology was the last to be fully framed by the disciplines and aesthetics of natural history, not yet transformed by the abstract apparatuses of population genetics, kin selection, and game theory that came to frame the language of evolution from the 1930s to the 1960s, as natural selection came to eclipse all other evolutionary mechanisms.

– – –

Cataloguing of ant species proceeded at a breakneck pace throughout the nineteenth century, as amateur and professional entomologists collected assiduously in Europe, North America, and in outposts of trade and empire throughout the world. This myrmecoid industry made the diversity of ant lifestyles ever more manifest, with, for example, the validation of long-dismissed biblical and classical observations of seed-husbanding ants in the Near East (no northern European ants exhibit these behaviors); the discovery of the cultivation of fungus gardens among the American Attinidae; and description of the voracious nests-on-the-move of the tropical American and African army ants. Each of these discoveries broadened the plasticity ants were known to exhibit, both morphologically and behaviorally, opening up still further questions about the evolutionary origins and maintenance of sociality.

Even stranger things came to light. Pierre Huber's discovery of "slavery" in ants in 1810 caused a sensation, and it was already a well-known phenomenon by the time of Darwin's *Origin of Species*. Workers of the two European slavemakers, *Formica sanguinea* and *Polyergus rufescens,* and of a few related American and Asian species, raid the nests of various other species of ants (in Europe usually the common *Formica fusca*), seize their brood, and carry them back to their own nests, where the larvae and pupae hatch out normally and the emergent workers proceed to act exactly as they would in a nest of their own species: tending the brood both of the slavemaking and of the slave species, feeding each other, constructing and cleaning the nest, defending it against predators and other ant nests. In a century that saw a wide European and colonial debate about the "naturalness" of the institution of slavery among humans, the occurrence of apparently parallel behaviors in ants was seized upon by all sides; the happenstance that the European "slave" species are smallish black ants, while the slavemakers are (more or less symbolically, depending on the rhetorical use to which they were put) larger, fiercer, and in the case of *Polyergus rufescens* blood-red in color, only hastened the probably inevitable use of ants in that political and cultural war.[2]

Huber's observations were only the first to demonstrate that ants of different species could be found living together in apparent harmony, and the ways in which these compound nests were formed provided one focus for studies of ant society. Another was the collection and classification of a myriad of inquiline species—non-ant arthropods, mainly other insects, that live in and around ant colonies as scavengers, predators, and individual and social parasites. These "myrmecophiles" provided perhaps the strongest challenge to theories about evolution and behavior in the social insects: why would ants take assiduous care—indeed, sometimes even more assiduous care than they took of their own sisters—of alien creatures that often were killing them and their brood individually, and crippling or killing their nests as a whole as well? Close attention to the natural history of ant behavior revealed numerous variations on the theme of communal living and the structure of societies, many of which came to light in Forel's work on the ants of Switzerland.

Auguste Forel

Auguste Forel, born in 1848 to a conservative Swiss-French family in the canton of Vaud, led an isolated childhood, prevented by his mother for social reasons from associating with local children. With little other re-

2. See J. F. M. Clark, "'The Complete Biography of Every Animal': Ants, Bees, and Humanity in Nineteenth-Century England," *Studies in the History and Philosophy of the Biological and Biomedical Sciences* 29C (1998): 249–67.

course, natural history, and particularly the study of ants, fascinated him from an early age; he related in his autobiography the story of watching ants *(Formica fusca)* living in the front steps of his parents' house while his mother played Beethoven sonatas on the piano. From his early adolescence, Forel came to abandon the strictly Calvinist religious tenets of his parents (while nevertheless retaining most of their psychological and moral influences): at sixteen he refused confirmation, and at his marriage in 1883 some jiggery-pokery was required with the (fortunately accommodating) presiding minister so that Forel would not be required to make or assent to any statement of faith.[3]

In a common nineteenth-century trajectory, Forel studied medicine in order to indulge his taste for scientific pursuits; while still a medical student he was invited to publish a compilation of his observations and experiments on ants, *Les fourmis de la Suisse* (1874), a work that gave him a European currency in entomology and brought him into correspondence with Charles Darwin. Forel did eventually acquire an interest in medicine, which led him to brain anatomy and eventually the study of psychiatry; he spent the greatest portion of his career as the director of Switzerland's largest insane asylum, the Burghölzli Institute in Zurich. Here he concentrated his efforts on the treatment of alcoholism, eventually both promulgating and practicing total abstinence, and on the prevention and treatment of sexual disorders (Forel's *Sexuelle Frage,* a bestseller translated into a dozen or more languages, was by far his best known work outside the myrmecological world).[4]

Forel was interested in all aspects of ant natural history, but particularly in the social relations subsisting between ants of different species. Forel's ants were far removed from, and increasingly in his work superior to, humans in their essential nature. He saw ants as relatively plastic and unmechanical in their behavior, in contrast to the theories of the German physiologist Albrecht Bethe, who maintained that ants were essentially reflex machines.[5] In 1874 Forel discussed the differences and similarities between ant and human societies in the closing chapter of the *Fourmis de la Suisse.*

3. Auguste Forel, *Out of my Life and Work,* trans. Bernard Miall (London: George Allen and Unwin, 1937).

4. Auguste Forel, *Die Sexuelle Frage, eine naturwissenschaftliche, psychologische, hygienische und soziologische Studie für Gebildete* (Munich: Ernst Reinhardt, 1905).

5. Albrecht Bethe, "Dürfen wir den Ameisen und Bienen psychische Qualitäten zuschreiben?" *Archiv für die gesamte Physiologie* 70 (1898): 15–100; Bethe, "Noch einmal über die psychischen Qualitäten der Ameisen," *Archiv für die gesamte Physiologie des Menschen* 79 (1900): 39–52.

Ants, he asserted, like humans, were motivated by social instincts, but unlike humans their instincts were egalitarian and impersonal. Consequently, drawing from the political movements of his day, Forel claimed ants were socialists by nature, on much the same economic terms as humans: society depended on the accumulation of resources and the division of labor. But at the same time, they demonstrated the *impossibility* of such societies among humans, because their society depended on a drastically different set of instincts:

> One might say that these insects represent the perfect type of socialism, taken to its final limits. They simultaneously show us both what man lacks and what he has too much of (individuality) to be able to govern himself in that way. . . . The *family* of the ants is their ant colony [*fourmilière*]; this is thus identified with the whole *collective society*. Every individual lavishes care equally on all others (on all those which it recognizes as belonging to its society, to its *fourmilière*) in direct relation to their size and their utility to the community. . . . Ants do this work freely; there are no chiefs; therefore they work out of instinct . . . and pleasure, without which their society could not exist.[6]

Even the existence of slavery among certain ants did not diminish the essentially egalitarian nature of the ant colony as a unit; the *fourmilière* was defined not by species, nor even by kinship, but by behavior, and since the "slave" individuals behaved as though they belonged to the nest in which they had their adult emergence, their individual condition was essentially the same as those of their "so-called masters." The slaves had been kidnapped as pupae in violent raids on their maternal nests, but it was only in the adult phase that social instincts came into play:

> In effect, the slaves of *P. rufescens, F. sanguinea*, etc. never doubt their origins. They work voluntarily, by inclination and by instinct, in the society where they were born. . . . If they wished it, nothing prevents them from removing themselves from their so-called masters and letting these die of hunger. . . . It is exactly by profiting, at first more or less consciously (*F. sanguinea*), and then unconsciously (*P. rufescens*), by the work instinct of weaker species, that the idle ants have gradually come, by selection and mutation, to lose their natural inclination for work. . . . An amazon *four-*

6. Auguste Forel, *Les fourmis de la Suisse*, 2nd ed. (1874; La Chaux-de-Fonds: Imprimerie coopérative, 1920), 313.

milière is as republican as any other; it only contains two sorts of individuals, of which the function of one is defense of the nest and looting, and the other work.[7]

In an 1898 paper, Forel introduced the concept of "social parasitism" alongside the category of slavery. Social parasitism differed from slavery in that workers were not removed from their home colony; rather, an invading queen took up residence in a host colony and either did away with or suppressed the full reproduction of its queen, so that the invader's own offspring were reared in the place of the host queen's, as in the case of *Strongylognathus testaceus* or *Anergates atratulus,* parasitic on *Tetramorium caespitum.* Social parasitism was nevertheless parallel to the case of slavery in that the host species "cares for this parasite for its own pleasure, by a perverted instinct [par instinct dévoyé]."[8] Moreover, in the more extreme cases of either, the "master" species was utterly dependent on the labor of alien workers; workers of the slavemaking species *Polyergus rufescens* had both their mandibles and their instincts so modified that they were incapable of nest construction or even feeding themselves, and the queens of the social parasite *Anergates* produced only fertile females and males, with no workers at all.

The violence and depravity of World War I impressed Forel deeply, and in his entomological as well as his psychiatric work he came increasingly to concentrate on the differences between humans and ants, concentrating on the human frailties that in his view resulted from inherited paradoxes of conflicting instincts, cooperative and aggressive. While humans appeared worse, Forel's views of ants ameliorated significantly, as though to make the contrast stronger. In the foreword to the second edition of the *Fourmis de la Suisse,* written in 1919, he drew a sharp contrast between the state of the human world and the *fourmilière* of ants, whether of single or mixed species.

For ants, questions of morality no longer came into play; while Forel still insisted on the plasticity of their intelligence and their capacity for individual agency, their behavior was nevertheless entirely drained of moral consequence, whether in their own context or in ours. The actions of humans, however, whether impelled by instinct or chosen by intelligence (he rated ant behavior as 97 percent instinctive and 3 percent intelligent,

7. Ibid.
8. Auguste Forel, "La parabiose chez les fourmis," *Bulletin de la Société Vaudoise des Sciences* 34 (1898): 380–84, 383.

whereas humans were more nearly equal), Forel painted in the direst terms. The wars of ants, between species and species, or even between colony and colony of the same species, were excusable in that they always took place between self and other, among creatures too small to be capable of global alliance. Moreover, in 1874 Forel had already insisted that slavery among ants was less pernicious than that of humans, given that "enslaved" ants worked for the good of the only hive they knew. He now strengthened that assertion. Enslaved ants were in fact *better off* than their helpless masters:

> If some species make slaves, these slaves are also free, even more free, because less dependent than their "masters" (Work is freedom!) [le travail rend libre!].[9]

Even the parasitism of one ant species on another was morally innocent, since the parasites "instinctively subvert the instinct of other species that never doubt them."[10]

For humans, on the other hand, the case was nearly hopeless—the aggression that in ants was an "instinct" was rather in humans a "thirst":

> . . . *the growing mental opposition that results in ourselves on the one hand from our inherited cruel instincts for domination, thus natural, (instincts far more cruel than those of apes), and on the other hand from the increasingly urgent necessity for good social organization between men.* . . . The human individual has an instinctive thirst for freedom for himself, but for domination and exploitation of others. He cries out when he is oppressed, but himself dreams of nothing but exploiting and dominating his own kind, individually and collectively.[11]

The only hope for the future of humanity lay in making ourselves as much like ants as possible—uphill work:

> It is impossible for us to imitate the anarchistic organization of ants, because our inherited instinct is something completely different from theirs. . . . We can only change the warlike and individualistic education of our children and give them a social and egalitarian education in work, an education which alone could show man the true freedom towards which his ideal can

9. Forel, *Fourmis de la Suisse* (note 6), xiv.
10. Ibid.
11. Ibid., xiv–xv; emphasis in original.

aspire without hurting that of his own kind. Social work is freedom; the ants show it to us.[12]

The ant colony is a utopia achieved *because of* formic nature—indeed a utopia even in its most apparently dubious cases, slavery and social parasitism—whereas humans can only achieve utopia *in spite of* human nature, in the unlikely event that they subordinate, indeed negate their instinctive nature in favor of artificial social constructs. Forel campaigned through the last decades of his life for the social measures that he thought would allow humans to transcend their biological failings and bring them closer to the utopian version of the ant world, "to grow nearer to the ants and yet remain men,"[13]: the League of Nations, disarmament, Esperanto, female suffrage, temperance, and eugenics. We need civilization to protect us from ourselves, because at present it is precisely when human beings are at their most highly organized, in wars between states, that they display "egotism in its highest manifestation."[14]

Forel died in 1931, before seeing over which gates the motto of his happily oblivious myrmecic slaves—*Arbeit macht frei*—would be inscribed. He would doubtless have been appalled, and yet his utopia and the Nazis have something in common. On the one side, Forel speaks for the power of human reason and social perfectibility, in the best tradition of the Enlightenment. On the other, his perfect society must be achieved—in fact, *could* be achieved—only through total social control.[15] Necessity, not freedom (beyond the freedom that is work), is the governing law of Forel's utopia. Individual freedom is no concern of his. His society free from war and poverty and idiocy and disease is not to be derived from citizens who are themselves free in any sense of innate natural rights or according to natural law; and here Forel, who presided over involuntary eugenic sterilizations, practiced what he preached.[16] Only by work to overcome our own

12. Ibid., xv.

13. Auguste Forel, *The Social World of the Ants, Compared with that of Man,* 2 vols., trans. C. K. Ogden (London: Putnam, 1928 [1921–23]), 2:339.

14. Ibid., 1:xxxviii.

15. See Sarah Jansen, "Ameisenhügel, Irrenhaus und Bordell: Insektenkunde und Degenerationsdiskurs bei August Forel (1848–1931), Entomologe, Psychiater und Sexualreformer," in Norbert Haas, Rainer Nägele, and Hans-Jörg Rheinberger, eds., *Kontamination* (Eggingen: Edition Isele, 2001): 141–84.

16. See, e.g., Forel, *Sexuelle Frage* (note 4), 382: "I also had a hysterical fourteen-year-old girl sterilized, whose mother and grandmother were procuresses and prostitutes, and who already enjoyed giving herself to any boy off the streets, because I wanted thereby to prevent the origin of wretched descendants. At the time it was the fashion to treat hysteria by sterili-

FIGURE 11.1. Frontispiece to Auguste Forel, *Le monde social des fourmis* (Geneva: Kundig, 1921).

nature could we achieve what is second nature to the ants, perfect community (fig. 11.1).

Erich Wasmann

Erich Wasmann was born in 1859 to Catholic converts living in the South Tyrol; he received his education in Catholic schools, then attended a Jesuit seminary with the intention of being ordained and pursuing a career in a Jesuit university. But from 1879 to 1883, frail health (a lung disorder that plagued him throughout life) forced him to take a break from his theological course, and Wasmann spent a period working as a contributor to a Jesuit cultural periodical.

Although he had evinced a youthful enthusiasm for natural history, the young Wasmann was, like the young Darwin, a beetle fan; in 1884 he published a work on the instinct of the birch leaf roller beetle.[17] In the same year, the editor of the journal asked Wasmann to write a review essay on the subject of ant intelligence, then under discussion due to British entomologist John Lubbock's *Ants, Bees, and Wasps,* a work recently translated

zation, and I took this fashion as a pretext for my action, which in reality had only a social object." See also Forel, *Life and Work* (note 3), 193; Willi Wottreng, *Hinriss: Wie die Irrenärzte August Forel und Eugen Bleuler das Menschengeschlecht retten wollten* (Basel: Weltwocke-ABC-Verlag, 1999).

17. Erich Wasmann, *Der Trichterwickler, eine naturwissenschaftliche Studie über den Tierinstinkt* (Münster: Aschendorff, 1884).

into German and making great claims for the intelligence of the social insects (Lubbock claimed among other things to have tamed a wasp). Wasmann set out to acquire a basis of personal knowledge in the subject, observing the ant species to be found in Holland around the Jesuit theologate of which he was a member, and he was bitten by the ant bug at last. He devoted the rest of his life to work on ants (and as a corollary, termites) and the problems posed by their behavior, the characteristics of their colonies, and the other arthropods that live in association with them. His first major work on the subject was *Die zusammengesetzten Nester und gemischten Kolonien der Ameisen* (The Compound Nests and Mixed Colonies of Ants) in 1891, a work that took up the questions of mixed ant colonies, including the origins of ant slavery.

Wasmann was ordained in 1888, and throughout his life he integrated his priestly with his biological calling, a mixture with which many of his colleagues (both scientific and theological) were uncomfortable.[18] Wasmann's 1884 book on the leaf-roller beetle had concluded with an orthodox anti-Darwinian design argument for the divine origins of this instinct. But between then and his ant work, Wasmann became a convinced evolutionist, due partly perhaps to two years spent studying zoology at the German University in Prague; and partly, apparently, on the basis of his own studies of four species of beetles of the genus *Dinarda* that live in ant nests and demonstrate morphological and behavioral variance according to geographical distribution, which Wasmann took as inferential demonstration of an adaptive radiation from a single ancestor. For the rest of his career, Father Wasmann was a very visible advocate of evolutionary theory, a public advocate of the reconcilability of evolutionary theory with Catholic doctrine, and a source of frustration and suspicion to the ultra-Darwinist establishment of the period, who felt that a Jesuit who promulgated evolutionary theory must be the spokesman for the kind of devious cabal for which that order was so well known, with its object to lead the public astray from true evolutionary doctrine.

Wasmann's chief love was less the behavior of ants themselves than the behavior of a hitherto neglected group of organisms—the numerous species of other arthropods, particularly beetles, that live in close association with ant colonies. These "Ameisengäste" or "myrmecophiles," as he called them, exhibited numerous peculiar adaptations, both morphologi-

---

18. See, e.g., Simon FitzSimons, "Father Wasmann on Evolution," *American Catholic Quarterly Review* 35 (1910): 12–48.

cal and behavioral, to their specialized roles, and in an 1894 work, *Kritisches Verzeichnis der myrmekophilen und termitophilen Arthropoden* (Critical Classification of Myrmecophile and Termitophile Arthropods), Wasmann classified them according to their various relationships with their hosts.[19]

Most interesting to Wasmann were the *symphiles,* organisms that are apparently accepted by their hosts as part of their colony, to the extent, on occasion, of rearing their larvae, feeding them as they would nestmates, grooming them, or carrying them from place to place within the nest or even in moves between nest sites. What could induce creatures to take such care of organisms that at the very least were no relation to them, and that in some cases actively preyed on them or their brood, or had further debilitating effects on the ant colony? Wasmann devoted much attention to these questions, making them simultaneously the crux of his understanding of evolutionary theory, the test case for the rejection of the theory of natural selection, and the vehicle for the development of an alternative mechanism bearing on this and possibly other cases of social evolution. The cases to which he repeatedly returned were those of the staphylinid beetles of the Lomechusini group, members of the genera *Lomechusa* and *Atemeles,* symphiles living in the nests of common European ants of the genera *Formica,* particularly *Formica sanguinea* and *Myrmica.*

Adult lomechusines exist in an apparently amicable and mutually beneficial relationship with their hosts; the former solicit food from the ants and are fed by them just as they feed their own nestmates (fig. 11.2). The beetles are strikingly modified from their nonmyrmecophile cousins, having become in color and partially in shape—particularly in the mouthparts and antennae, necessary for stimulating the feeding response—visual and tactile mimics of their host species. The ants in return are apparently rewarded by exudate droplets found on small tufts of bristle between the segments of the beetle's abdomen, which they lick and gnaw.

The lomechusine larvae are another story. They are exclusively carnivorous, and their prey is the ant brood in the host nest. In spite of the fact, however, that the beetle larvae are busily devouring their own larval nestmates, the workers tend to the beetle larvae as assiduously as they do their own. Finally, *Lomechusa*-infested nests suffer in another fashion. For somewhat obscure reasons, the production of normal workers and fertile females is interfered with; instead of normal workers, pupae increasingly develop

19. Erich Wasmann, *Kritisches Verzeichnis der myrmekophilen und termitophilen Arthropoden* (Berlin: Felix L. Dames, 1894).

FIGURE 11.2. Erich Wasmann, *Modern Biology and the Theory of Evolution,* 3d ed. (London: Kegan Paul, 1910), 336.

into *pseudogynes,* an abnormal form both behaviorally and morphologically intermediate between the sterile workers and the fertile females, unable either to do the normal day-to-day work of the colony or to reproduce, thereby crippling the colony as a whole.

Why should ants thus care for an alien species of insect that both directly and indirectly destroys their nest? Wasmann proposed in 1901 an evolutionary mechanism, which he called "amical selection," to explain these phenomena.

The process of evolving from occasional ant predator like most of their relatives, to colony denizen, to full symphile inquiline, began in the lomechusine beetles, Wasmann hypothesized, with the attraction of the ants to the volatile exudates that they fortuitously found so tasty. The ants treated best and were least hostile to those beetles that produced the most exudate, so that these were more successful; their success then fostered the development of ever-more elaborated adoption and caretaking instincts in the ants, finally extending to the care of the beetle brood. In effect, the ants were engaging in an active selection process of the beetles rather like the artificial selection practiced by human beings.

Wasmann explicitly opposed amical selection to Darwinian natural selection as an explanation for the origin of symphilic instincts.[20] The latter, he said, could not account for the appearance of positive instincts in the ants to care for their symphiles, exactly because caring for them harmed the

20. Erich Wasmann, "Giebt es thatsächlich Arten, die heute noch in der Stammesentwicklung begriffen sind?" *Biologisches Centralblatt* 21 (1901): 689–711, 737–52; 738 ff.

nest and so should be eliminated. In any case, Wasmann agreed with his Prague professor, Hans Driesch, and many biologists of the time that natural selection as a general rule could play only a subordinate, negative role, weeding out the unfit, but could not account for the appearance of novelty. Positive lines of development arose from innate developmental tendencies within lineages.[21] A different process must therefore have taken place:

> This antagonism between amical selection and natural selection and the victory of the former over the latter explains the seeming contradiction that among some of their actual guests (particularly of the genera *Lomechusa* and *Atemeles*) *the ants have bred and are still actually breeding their greatest enemies.*[22]

Or, as Wasmann summed up in *Modern Biology and the Theory of Evolution:* "Speaking from the point of view of supporters of the evolution theory, we may justly say: Amical selection has triumphed over natural selection, which, in this case, far from being all-powerful, is powerless."[23]

Ant society was for Wasmann a construct of instincts, rather than the materially bounded nest. These instincts for selflessness and mutual care existed, in a sense, independent of their objects, so that workers could lavish care on apparently "foreign" guests and derive the same gratification from doing so that they would from caring for their own. Amical selection is opposed to natural selection, rather than being merely, as Wheeler would claim, a facet of it, precisely as a result of this autonomous instinct: within their psychical limits, ants are capable of discrete choice of their social partners, of free will, and it is this that makes the colony.

When it came to discussing the psychic capabilities of ants, Wasmann, like Forel, disavowed Albrecht Bethe's purely mechanistic conception of ant behavior; for him also they were very far from tiny reflex machines. But in spite of the autonomy he granted them in the case of amical selection, they were even farther from humans than if they had been machines. Wasmann's sharpest clashes with his fellow biologists came not over the issues of ant taxonomy or behavior or even of evolutionary interpreta-

21. See, e.g., Erich Wasmann, *Modern Biology and the Theory of Evolution,* 3d ed., trans. A. M. Buchanan (London: Kegan Paul, 1910), 324; Wasmann, *The Berlin Discussion of the Problem of Evolution; Full Report of the Lectures Given in February, 1907, and of the Evening Discussion* (London: Kegan Paul, 1912), 42.

22. Wasmann, "Noch in Stammesentwicklung begriffene Arten?" (note 20), 739–40.

23. Wasmann, *Modern Biology* (note 21), 339.

tion, but in disputes about what science could or could not explain, and how. Wasmann tirelessly opposed monistic philosophy, preached by Ernst Haeckel and espoused by Forel, which insisted that all phenomena were reducible to the material and ultimately explicable through science.

Haeckel laid down the tenets of monism in numerous publications, particularly *Der Monismus als Band zwischen Religion und Wissenschaft, Glaubensbekenntniss eines Naturforschers* in 1892, translated as *Monism as Connecting Religion and Science: The Confession of Faith of a Man of Science* in 1895, and the enormously influential *Die Welträthsel* (1899), translated as *The Riddle of the Universe* in 1900.[24] In these works he attested that the seven everlasting cosmic enigmas enumerated by Emil du Bois-Reymond in 1880—ranging from the nature of matter through the origins of life, sensation, and rational thought to the existence of free will—had all but the last been solved by the combined application of monistic, that is, materialist, philosophy and of evolutionary biology; while the last, free will, is merely "a pure dogma, based on an illusion, and has no real existence," and thus is not a fit subject for science.[25]

Haeckelian monism held that all apparent dualisms—matter/energy, organic/inorganic, body/soul—were properly to be seen as manifestations of one immutable set of laws governing matter and energy alike. Three particularly inflammatory consequences followed. The first was the insistence that if matter and spirit were only one thing, then God had to be either everywhere, fully identified with the world, a form of pantheism; or nowhere, atheism. The options were to Haeckel, at least, logically equivalent.[26] The second was the unity of psychic and physiological phenomena, which meant that "the soul is . . . a natural phenomenon" and psychology no more than a branch of physiology, an opinion with which Forel concurred: "As long as we see in every detail the psychic and the physiological appear and disappear together, and be identically destroyed, as psychophysiology and all the brain sciences demonstrate, we must maintain their identity, until someone can show us a separate apparition of the

24. Ernst Haeckel, *Der Monismus als Band zwischen Religion und Wissenschaft, Glaubensbekenntniss eines Naturforschers* (Bonn: Strauss, 1892); Haeckel, *Monism as Connecting Religion and Science: The Confession of Faith of a Man of Science,* trans. J. Gilchrist (London: A. & C. Black, 1895); Haeckel, *Die Welträthsel, Gemeinverständlich Studien über monistische Philosophie* (Bonn: Strauss, 1899); Haeckel, *The Riddle of the Universe at the Close of the Nineteenth Century,* trans. Joseph McCabe (London: Watts, 1900).

25. Haeckel, *Riddle of the Universe* (note 24), 16.

26. See, e.g., ibid., 298.

soul."[27] Third was the fact of human evolution as an article of faith, such that "it is a matter of comparative indifference how the succession of our animal predecessors may be confirmed in detail. Sufficient for us, as an incontestable historical fact, is the important thesis that man descends immediately from the ape."[28]

Haeckel, the E. O. Wilson of his day, summed up thus:

> Monistic cosmology proved . . . that there is no personal God; comparative and genetic psychology showed that there cannot be an immortal soul; and monistic physiology proved the futility of the assumption of "free will." Finally, the science of evolution made it clear that the same eternal iron laws that rule in the inorganic world are valid, too, in the organic and moral world.[29]

All of these propositions were proven facts: "It is only the ignorant or narrow-minded who can now doubt their truth." Any scientific efforts to maintain a philosophical dualism must therefore "rest on confusion or sophistry—when they are honest."[30]

Enter a Jesuit, specially trained, as everyone knew, in sophistry and confusion. Wasmann insisted at every point on the weaknesses of monistic theory, the sound philosophical basis for psychological dualism, and the reconcilability of scientific materialism with theology. Forel and Wasmann spatted for years in the pages of the *Biologisches Centralblatt* ("Science or Superstition?" versus "Scientific Evidence or Intolerance?"), and Wasmann and Haeckel clashed even more spectacularly in the course of two public lecture series given in Berlin in 1905 and 1907.[31] Wasmann admitted the logical possibility of human descent from animals, which was for him an empirical question, although so far inconclusively demonstrated, unlike

27. Ibid., 91; see Auguste Forel, "Naturwissenschaft oder Köhlerglaube?" *Biologisches Centralblatt* 25 (1905): 485–93, 519–27; 520.

28. Haeckel, *Riddle of the Universe* (note 24), 86. For further discussion of debates on human evolution, see the essay by Proctor in this volume.

29. Haeckel, *Riddle of the Universe* (note 24), 357.

30. Haeckel, *Confession of Faith* (note 24), 39; Haeckel, *Riddle of the Universe* (note 24), 296.

31. Forel, "Naturwissenschaft oder Köhlerglaube?" (note 27); Erich Wasmann, "Wissenschaftliche Beweisführung oder Intoleranz?" *Biologisches Centralblatt* 25 (1905): 621–24; Ernst Haeckel, *Der Kampf um den Entwickelungs-Gedanken, Drei Vorträge, gehalten am 14., 16. und 19. April 1905 im Saale der Sing-Akademie zu Berlin* (Berlin: G. Reimer, 1905); Erich Wasmann, *Der Kampf um das Entwicklungsproblem in Berlin* (Freiburg: Herder, 1907); Wasmann, *Berlin Discussion* (note 21).

the article of faith it was to Haeckel. But Wasmann insisted that human beings were *qualitatively* different from all animals, in possessing a spiritual nature or soul, not reducible to material law although certainly correlated with it, that was indubitably the work of God, and might have been breathed into human beings when their corporeal ancestors had become sufficiently complex to house it. Over the question of the *content* of spiritual existence, rather than its description, science had no jurisdiction:

> Biology has considered the question from its point of view *in only one, and that a material, aspect. The other, i.e. spiritual, aspect* of the same problem falls outside its scope, and the results of biological investigation do not touch the existence in man of a soul created by God. . . .
>
> Similar remarks will apply to the *hypothetical history of the human race*. It may on its material side originate in the dust of the earth . . . [But] man would have become man completely only when the organised matter had so far developed through natural causes, as to be capable of being animated with a human soul. The creation of the first human soul marks the *real creation of the human race*, although we might assume that a natural development lasting millions of years had preceded it.[32]

To return to his special study, myrmecology, the likeness between ant and human societies for Wasmann could not possibly be more than metaphoric, however tempting metaphor might be. Ants had psychical lives, their community was held together by "instinctive sympathy," and was even capable of accommodating alien species in perfect amity, but "even the intelligence of an ant would be sufficient to understand, that animal and human societies are as far apart as heaven and earth."[33]

William Morton Wheeler

William Morton Wheeler was born in 1865 in Milwaukee, Wisconsin, a city, he said in an autobiographical reflection, whose "cerevisiacal fame . . . unfortunately quite eclipsed the fame of its temperate and highly intellectual German population and excellent school system."[34] Wheeler attended

---

32. Wasmann, *Berlin Discussion* (note 21), 50–51.

33. Erich Wasmann, *Comparative Studies in the Psychology of Ants and of Higher Animals*, 2nd ed. (St. Louis: Herder, 1905), 89.

34. Quoted in George Howard Parker, "William Morton Wheeler, 1865–1937," *Biographical Memoirs of the National Academy of Sciences* 19 (1938): 201–41, here 203. See also Mary

a German academy in Milwaukee through high school, which gave him a comprehensive humanist education. Unlike Forel and Wasmann, who both confessed that the arts and *belles lettres* were closed books to them, Wheeler was exceptionally well read. He wrote a correspondingly pungent, often wry or droll prose, laced with quotations in Greek, Latin, French, German, or English.

Wheeler was interested in science from a young age, and found part-time work in various natural history museums. He then was the child of good fortune: Milwaukee in the 1880s became something of a center for cutting-edge zoology, as Charles Otis Whitman and two of his students, fresh from a tour of European biology, established themselves at the Allis Lake Laboratory. During the next few years, Wheeler became a "hard-boiled morphologist," working in marine biology and insect embryology, receiving a Ph.D. from Clark University in Massachusetts.[35] He subsequently spent five years at the University of Chicago, first as instructor, then as assistant professor of embryology. Wheeler was most disingenuous, therefore, when he once apologized in a lecture for being "a zoologist, reared among what are now rapidly coming to be regarded as antiquated ideals."[36]

Wheeler caught the ant bug in 1900, after leaving the laboratories of Chicago for a professorship of zoology at the University of Texas; the biology of the social insects and its evolutionary interpretation were to occupy the rest of his life. In dialogue with Forel and Wasmann, Wheeler moved from straightforward natural history and taxonomy to a fundamental reassessment of social insect biology.

In 1905, Wheeler and Wasmann sparred over the origins of ant slavery in the pages of the *Biologisches Centralblatt,* just as Wasmann was defending himself on another front in the same volume in his ongoing spat over monism with Forel. Both linked slavery to temporary social parasitism.[37] Wheeler saw this process, however, in quite different terms from Wasmann: where Wasmann saw the action of an internal tendency to beneficial variation inherent in all living things, Wheeler's explanation followed

Alice Evans and Howard Ensign Evans, *William Morton Wheeler, Biologist* (Cambridge: Harvard University Press, 1970).

35. Parker, "William Morton Wheeler" (note 34), 210.

36. William Morton Wheeler, "The Ant-Colony as an Organism," *Journal of Morphology* 22 (1911): 307–25, here 307.

37. William Morton Wheeler, "An Interpretation of the Slave-Making Instincts in Ants," *Bulletin of the American Museum of Natural History* 21 (1905): 1–16.

Darwin's—a preexisting instinct shaped by fortuitous variation, then "perpetuated and intensified by natural selection"; he brushed off Wasmann's appeal to internal laws of variation as "merely scholastic formulae."[38] In its origins, the instinct was the collection of a rich, portable, storable food supply, which could then be converted from energy into workers.[39] This view was borne out, Wheeler claimed, by the observation that large, old, *sanguinea* colonies often dispensed with the capture and raising of slaves altogether, just as the temporary parasites like *F. consocians* eventually grew out of their compound nest phase.[40]

Wheeler here for the first time adumbrated a holistic view of the ant colony that he developed over the next decade and a half. The first step, in 1906, was his suggestion that biologists should shift their theoretical emphasis from workers to the fertile female, albeit in a different framework from the seventeenth or eighteenth centuries' "Feminine Monarchy." The fertile female was now, rather, "the epitome of the species . . . to which we must more and more resort in tracing the worker instincts back to their origins and meanings."[41] In a 1911 article, "The Ant-Colony as an Organism," Wheeler brought the ontogeny and phylogeny of the ant colony into harmony with that of the individual founding queen, and both of these into accord with Haeckel's biogenetic law, bringing himself also full circle from the cellular ontogeny and morphology of his early training through organismal natural history, and back again to ontogeny.[42]

If, Wheeler concluded, the ant colony behaves in every respect like an individual, with an individual ontogeny and phylogeny, then in every respect it can be *treated* as an individual, a shift in the meaning of the metaphor of the social organism. Three consequences followed. The division between queen and workers Wheeler now identified with German evo-

38. Ibid., 8, 16.

39. Ibid., 9–10.

40. Wheeler later entirely changed his opinions of the evolutionary origins of dulosis, attributing its origins to the instincts of the queen to take over a slave species nest, defeating the queen and adults, and appropriating the unhatched pupae, which on emergence will rear her first brood. Her own offspring then inherit the brood-appropriation instincts. See William Morton Wheeler, "On the Founding of Colonies by Queen Ants, with Special Reference to the Parasitic and Slave-Making Species," *Bulletin of the American Museum of Natural History* 22, no. 4 (1906): 33–105; Wheeler, "Origins of Slavery among Ants," *Popular Science Monthly* 70 (1907): 550–59, 554.

41. Wheeler, "Founding of Colonies by Queen Ants" (note 40), 102.

42. Wheeler, "Ant-Colony as Organism" (note 36), 313–14. On the history of the biogenetic law, see Stephen Jay Gould, *Ontogeny and Phylogeny* (Cambridge: Harvard University Press, Belknap Press, 1977).

lutionary theorist August Weismann's differentiation between germ and soma, immortal undifferentiated lineage and differentiated mortal body.[43] Gone were the metaphors of the ant colony as perfect socialism or as republic. The social parasites, whether other ants as in the case of temporary or permanent social parasitism or of dulosis, or other arthropods entirely such as *Lomechusa,* could now also be seen as parasites in a literal and not merely a figurative sense, in which the colony–organism was castrated or eventually killed (as in the case of social parasitism) or damaged and diseased beyond the colony's ability to regenerate (as with *Lomechusa*).[44] Finally, with the ant colony redefined as an organism in its own right, the question of altruism among its constituent parts took on an entirely new cast. Individuals within the colony had no particular importance except as they performed essential tasks. Does one ask why cells in the hand aid those in the foot? What mattered was the relationships of "organisms" with each other:

> In all of these phenomena our attention is arrested not so much by the struggle for existence, which used to be painted in such lurid colors, as by the ability of the organism to temporize and compromise with other organisms, to inhibit certain activities of the aequipotential unit in the interests of the unit itself and of other organisms; in a word, to secure survival through a kind of egoistic altruism.[45]

The apparent "altruism" of workers was thus, for perhaps the first time, restated in terms of egoism, of survival of the "self."

In 1918 Wheeler introduced the idea of *trophallaxis* (from the Greek for "exchange of nutriment") for the persistent sharing of food between members of an ant colony, whether workers with workers, with the queen or queens, or with the brood. This instinctive behavior, originally derived from the ancestral hymenopteran feeding of larvae by mothers, had gradually come, Wheeler asserted, to encompass the whole hive and to serve as its chief means of internal regulation, expanding, "like an ever-widening vortex," to include all members of the colony. Trophallaxis had several dimensions. Bodies and energy might be quite interconvertible: in times of dearth the adult workers might devour their own eggs, larvae, and pupae, and even, among some species, the soldiers or some of the workers,

43. Wheeler, "Ant-Colony as Organism" (note 36), 312.
44. Ibid., 317–19.
45. Ibid., 325.

as "so much potential or stored nutriment available for the adult ants" then to be distributed equally among the remaining workers by means of trophallaxis.[46]

Lastly, Wheeler was to become increasingly interested in the question of the ontogenetic and phylogenetic origins of sociality as a general phenomenon. Linking individual and colony ontogeny and phylogeny together once again, he hypothesized that sociality began with trophallaxis between female and offspring, "as a mutual trophic relation between the mother insect and her larval brood." From this familial beginning,

> the ants have drawn their living environment, so far as this was possible, into a trophic relationship, which, though imperfect or one-sided in [some cases] has nevertheless some of the peculiarities of trophallaxis[47] [fig. 11.3].

The root of this relationship, however, the center of the vortex, remained always the relationship between mother and offspring; all the other possible permutations of relationships were somehow extensions or subversions of this primal instinct. The "problem" of symphily in *Lomechusa* and *Atemeles* thus disappeared: not an *explanandum* requiring a particular set of

---

46. William Morton Wheeler, "A Study of Some Ant Larvae with a Consideration of the Origin and Meaning of Social Habits among Insects," *Proceedings of the American Philosophical Society* 57 (1918): 293–343, here 325.

In a 1919 address to the American Society of Naturalists, "The Termitodoxa, or Biology and Society," Wheeler had King Wee-Wee, 43d Neotenic King of the 8429th Dynasty of the Bellicose Termites, extol the benefits of trophallaxis in the following manner:

> So perfectly socialized have we now become that not infrequently a termite who has a slight indisposition, such as a sore throat or a headache, or has developed some antisocial habit of thought, or is merely growing old, will voluntarily resort to the committee of biochemists and beg them to stamp him. He then walks forth with a radiant countenance, stridulating a refrain which is strangely like George Eliot's "O, may I join the choir invisible!" and forthwith becomes the fat and proteid "Bausteine" of the crowd that assembles on hearing the first notes of his petition. If you regard this as an even more horrible exhibition of our mores, because it adds suicide to murder and cannibalism, I can only insist that you are viewing the matter from a purely human standpoint. To the perfectly socialized termite nothing can be more blissful or exalted than feeling the precious fats and proteids which he has amassed with so much labor, melting, without the slightest loss of their vital values, into the constitutions of his more vigorous and socially more efficient fellow beings.

William Morton Wheeler, "The Termitodoxa, or, Biology and Society" (1920), in Wheeler, *Essays in Philosophical Biology* (Cambridge: Harvard University Press, 1939), 71–88, here 79–80.

47. Wheeler, "Ant Larvae" (note 46), 326.

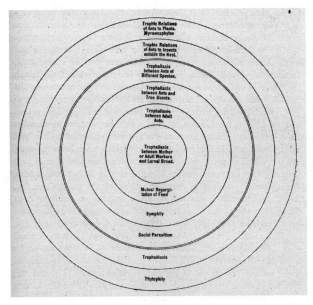

FIGURE 11.3.
William Morton
Wheeler, "A Study of
Some Ant Larvae, with
a Consideration of the
Origin and Meaning of
the Social Habit among
Insects," *Proceedings of
the American Philosophical
Society* 57 (1918): 293–
343, here 327.

instincts on the part of the ant hosts, as for Wasmann, but rather a perversion of the hosts' normal trophallactic instincts to one another.[48] No genuine altruism therefore existed on the part of *F. sanguinea* workers who fed adult and larval *Lomechusa*s and who let their larvae be devoured by them. They were, rather, the victims of a con game, insufficiently able to distinguish *self* from *other;* or, otherwise seen, the ant-colony-as-organism was subject to a type of internal parasitism little different from those suffered by more unitary bodies.[49] No "amical" interest whatsoever led ants to care for these would-be offspring.

— — —

What did this radical reconception of sociality imply for the mammalian parallel of the social insects, that "paragon of social animals," knowledge of which "is after all one of the chief aims of existence," man?[50] Wheeler's concept of the evolution of the social colony as an organism, or *superorganism,* as he came to call it, itself evolved from 1911 to the end of Wheeler's career in the mid-1930s, through two different, incompatible

48. Ibid., 333–34.
49. Ibid., 336–37.
50. William Morton Wheeler, *Ants; Their Structure, Development, and Behavior* (New York: Columbia University Press, 1910), 503.

futures, each with different implications for the potentials of freedom within biological constraints.

The superorganism's first aspect was hopeful. The tension between the individual and the society of which it is an inextricable part, and without which its existence is effectively meaningless, can be creative. The colony-as-organism concept as Wheeler proposed it in 1911, moreover, was an implicit argument against a narrow, deterministic reading of Darwinism in which heredity and competition determined all fate.[51] Cooperation and plasticity were the necessary characteristics of the social organism's, and the superorganism's, success, in a way that was consistent with the tenets of the Deweyan communitarian democracy underpinning the biological milieu of Wheeler's Chicago and Woods Hole colleagues in this period, in which individual liberty was fully realizable only by creating the greatest overall common social good. As Wheeler put it in 1910, "to live in societies, like those of man and the social insects, implies a shifting of proclivities from the egocentric to the sociocentric plane through a remarkable increase in the amplitude and precision of the individual's responses to all the normal environmental stimuli."[52]

Over the next decade, however, the decade of World War I, Wheeler's view darkened. The superorganism now led rather to a bleak *Brave New World* future of human evolutionary degeneration and decay. The hypertrophy of complexity might bear within it the seeds of its own downfall. The social insects, pinnacle of the Arthropoda though they might be, nevertheless represent an evolutionary dead end, in which individuality is entirely suppressed in favor of community. Faith in evolutionary progress is unwarranted, perhaps "illusory." Sociality may well represent an evolutionary cul-de-sac. We have seen the future, and the ants are us:

> Will this prospective, more intensive socialization be analogous to that of the highest social insects, a condition in which specialization and constraint of the single organism are so extreme that its independent viability is sacrificed to a system of communal bonds, just as happens with the individual

51. See Sandra D. Mitchell., "The Superorganism Metaphor: Then And Now," in Sabine Maasen et al., eds., *Biology as Society, Society as Biology: Metaphors* (Dordrecht: Kluwer, 1995): 231–47.

52. Wheeler, *Ants* (note 50), 2. For the Deweyan context of American biology in this period, see Gregg Mitman, "Defining the Organism in the Welfare State: The Politics of Individuality in American Culture, 1890–1950," in Maasen et al., eds., *Biology as Society* (note 51), 249–78.

cell in the whole organism? . . . [E]volution by atrophy certainly accompanies an advance in social integration in the insects. Turning to man we notice a similar regressive development of the individual as civilization proceeds. There is a decline in the sense-organs . . . the absence of any demonstrable improvement in the brain cortex and intelligence during historic time (possibly even some deterioration!), the greater activity of the visceral nervous system and endocrine glands as shown by the higher emotivity, increasing insanity, criminality, and mob-psychology in our larger cities, etc. . . . [W]e can hardly fail to suspect that the eventual state of human society may be somewhat like that of the social insects—a society of very low intelligence combined with an intense and pugnacious solidarity of the whole. . . . A society of the type towards which we may be drifting might be quite as viable and quite as stable through long periods of time as the societies of ants and termites, provided it maintained a sufficient control of the food supply.[53]

The only hope for humanity was to be found in the very fault that prevented humans from achieving ant-like harmony, which Wheeler called "the problem of the male." In an address to the American Society of Naturalists in 1933, Wheeler drew a comparison between "the high degree of integration and stability of the insect society and the extraordinarily harmonious and self-sacrificing cooperation of its individual members, as contrasted with the mobility, instability, and mutual aggressiveness so conspicuous among the members of our own society."[54] This assessment of human society sounded remarkably like Forel's equally pessimistic postwar view. But unlike Forel, Wheeler attributed humanity's failures not to general human instincts for dominance and aggression, but to specifically *male* ones:

The "problem of the male" . . . has been successfully solved by the social insects but not by mammal or human societies. . . .

[T]he male was originally and remains throughout the evolution of the Arthropod and Chordate phyla . . . the unsocial sex. In many animals, in fact, he might more properly be called the antisocial sex.[55]

53. William Morton Wheeler, *Emergent Evolution and the Social* (London: Kegan Paul, 1927), 37

54. William Morton Wheeler, "Animal Societies" (1934), in Wheeler, *Essays in Philosophical Biology* (Cambridge: Harvard University Press, 1939), 231–61, here 239.

55. Ibid., 241.

This problem was, according to Wheeler, males' comparative superfluity. As most of the female Hymenoptera were able to store a lifetime supply of sperm from a single mating or mating period, there was no reason for the trophallactic society to maintain males once they had fulfilled this brief function, hence the well-known expulsion of drone bees and like phenomena. The ants solved their problem by neutralizing their males and neutering most of their females.

Mammals, however, which require males at discrete but regular intervals for the increase of their societies, have a different problem: "Owing... to the decidedly unsocial character of his behavior, which manifests itself almost exclusively in voracity, pairing, and fighting with other males, [the male] is always, so to speak, socially more or less indigestible."[56]

While civilized humans may "at first sight" appear to have solved the problem of the male, this solution is illusory. At best, a majority of men are sufficiently socialized in collaboration with women to maintain a peaceful status quo; this state, however, is inevitably wrecked by antisocial "criminals, or individuals of low mental age and with unbalanced endocrines," whom Wheeler saw at work as he spoke in 1933.[57] But on the other hand, it is also to precisely this male instability that we owe modern civilization's achievements: "all progress in our civilized societies is initiated by a relatively small portion of the male population, whose restlessly questing intellects are really driven by the unsocial dominance impulses of their male mammalian constitution and not by any intense desire to improve society," and who, while "less socialized," manifest their dominance "mainly in the intellectual and emotional fields . . . . [of] sciences, arts, technologies, [and] philosophies, theologies, social utopias."[58] The resemblance of this invaluable third class to professors may not have been entirely coincidental.

Conclusion

So having refracted our view of human society through the lens of ant society, where does freedom rest, and how can we attain it—if, indeed, we can? For Forel, our only freedom lay in contradicting our biology to mold ourselves after a formic possibility: our own biology pushes us one way; our intellect, sharpened by myrmecological observation, should pull us

56. Ibid., 246.
57. Ibid., 250.
58. Ibid., 250–51.

another. But Forel's freedom lies only in the means, and not in the end; his myrmecoid utopia has little space for human liberty. For Wasmann, we have already been freed, not by biology, but by the divine gift that has permitted us to transcend it: having been given a soul, we are not wholly tied to the material sublunary world. Nature both constrains and frees, as ant societies demonstrate, but we have a freedom beyond it, because our psychic lives cannot be reduced to our physical selves. We have heaven, while the ants have only earth. Wheeler's view of society, in contrast, plunged Lucifer-like from heaven to earth. In his earlier ant career he built up a picture of successive emergences of novelty, complexity, and potential degrees of action from the evolution of sociality (though he balked at postulating God as the final emergent phenomenon, as some other philosophers of the time did).[59] From this early position of optimism, Wheeler came to see humans and ants on disastrous parallel courses of convergent evolution, leading only to stagnation, degeneration, and the inertia of the mob.

The irony is that while all three myrmecologists, Forel, Wasmann, and Wheeler, agreed that societies, including (even for Wasmann, to a degree) human societies, were natural phenomena, naturally evolved, they nevertheless all finally held, each in a different way, that civil behavior could be the product only of the most agonizing struggle *against* nature. The example of World War I was fresh in their minds, just as the World War II, as Julia Thomas discusses in the following essay, convinced Maruyama Masao that the only hope for Japanese modernity lay in leaving nature entirely behind. Having gone to the ant and considered her ways, they found her the fount of truth perhaps, but not of wisdom. Wisdom, such as it is, is evidently an attainment *contra naturam*.

59. See Wheeler, *Emergent Evolution and the Social* (note 53), esp. 16–17.

12

"To Become As One Dead": Nature and the Political Subject in Modern Japan

Julia Adeney Thomas

How splendid it is:
Even as one lives, to become as one dead,
Completely dead,
And to work as the heart desires!

ZEN POEM

In the summer of 1944, Maruyama Masao (1914–96) hurriedly finished the third and final essay of the collection *Nihon seiji shisō shi kenkyū*[1] and left Tokyo to join the Imperial army. It was not ardent adherence to the militaristic Japanese state that prompted Maruyama's obedience to his draft notice. Indeed, through his essays, Maruyama, who would ultimately become twentieth-century Japan's leading political theorist,[2] had launched a

My profound thanks go to Lorraine Daston, Fernando Vidal, and the rest of our colleagues for their help in shaping this essay. I also thank Martin Collcutt, who prompted me to ground my abstract questions about nature in Maruyama's texts; Michael Geyer, who saw potential in this project from the first; and Alan G. Thomas, whose nature it is to combine laughter with intellectual intensity. A version of this essay with the title "The Cage of Nature: Modernity's History in Japan" appeared in *History and Theory* 40, no. 1 (2001).

1. These pieces were originally published separately in the magazine *Kokka gakkai zasshi* (1940–44) before being collected in *Nihon seiji shi shisō kenkyū* (Research in Japanese political thought) (Tokyo Daigaku Shuppankai, 1952). This volume was translated by Mikiso Hane as *Studies in the Intellectual History of Tokugawa Japan* (Tokyo: University of Tokyo Press; Princeton: Princeton University Press, 1974). All quotations unless otherwise indicated are from Hane's translation.

2. Testaments to Maruyama's importance and his independent stance during the heated debates of the postwar period can be found in, for instance, Sasakura Hideo, *Maruyama Masao ron nōto* (Misuzu Shobō, 1988); *Gendai shisō: Tokushū: Maruyama Masao* (22, no. 1, 1994) a special issue of *Contemporary Thought* devoted entirely to commentary on Maruyama's ideas; Andrew

partial critique, despite wartime censorship, of the system he would serve. Central to his criticism of Japan's totalitarian system was the idea that Japan had not yet escaped nature's hegemony. Indeed, prewar and wartime ideology made the Japanese nation—its politics, culture, values, and people—the embodiment of nature, equating the existing national community with nature itself.[3] In such a system, Maruyama argued, autonomous individuals could never hope to flourish because of the extraordinary difficulty of imagining their world other than how they found it. If nature was defined as Japanese culture and Japanese culture as nature, there was no authority for challenging the status quo unless one turned, subversively as Maruyama did, to resources outside Japanese tradition, resources suspect as unpatriotic as well as lacking the justificatory force of either nature or culture. In short, according to Maruyama, nature still dominated Japanese ideology, deforming the modernity and the freedom for which he somewhat ambiguously yearned.

The very same year, in the more secure surroundings of Los Angeles, California, Max Horkheimer (1895–1973) and Theodor Adorno (1903–69) were engaged in a similar project.[4] They too sought to understand the foundation of the immense, destructive state power that had emerged in the twentieth century, particularly in their German homeland. They too made nature an important category in their analysis of the failed hope for freedom. However, contrary to Maruyama's analysis, the triumph of totalitarianism rested in their view on human mastery of nature, not on nature's mastery of the human. Horkheimer and Adorno bemoaned nature's utter subordination to the apparatuses of human reason, arguing in *Dialectic of Enlightenment* that systems of knowledge originating in ancient Greek ideas of reason had conquered "terrifying nature, which was finally wholly mastered."[5] Even the natural pleasures of the body, they asserted, have

Barshay, "Imagining Democracy in Postwar Japan: Reflections on Maruyama Masao and Modernism," *Journal of Japanese Studies* 18, no. 2 (Summer 1992): 365–406; and Rikki Kersten, *Democracy in Postwar Japan: Maruyama Masao and the Search for Autonomy* (London: Routledge, 1996).

3. As Fa-Ti Fan's essay in this volume makes clear, Japan was hardly *sui generis* in its impulse to link nature and nation, since this form of national ideology emerged in many places.

4. Jürgen Habermas rightly describes *Dialectic of Enlightenment* as "an odd book," cobbled together in large part from notes taken by Gretel Adorno during discussions between Horkheimer and Adorno. It was eventually published in 1947 three years after its completion by Querido Press in Amsterdam. See Jürgen Habermas, *The Philosophical Discourse on Modernity,* trans. Frederick G. Lawrence (Cambridge: MIT Press, 1990), 106–7, for a brief history of Horkheimer and Adorno's text and its extensive influence, despite few sales.

5. Max Horkheimer and Theodor W. Adorno, *Dialectic of Enlightenment,* trans. John Cumming (New York: Continuum, 1987), 105.

been commandeered by the state that administers and corrupts them. No longer an active subject, mysterious and beyond human control, nature is made pure object, a dissected corpse that only too late we discover to be our own.

In coming to terms with modernity in the mid-twentieth century, both the Japanese and the German writers depict the eradication of nature from political consciousness as the sign of the modern. Maruyama longs for this not-yet-realized modernity in Japan while Horkheimer and Adorno shrink from modernity in California. For Horkheimer and Adorno, nature's absence, at least its absence as an independent realm separable from the apparatuses of reason, technology, and state power, produces nothing but oppression. They link the subordination of nature to the totalitarian power that reigns over all aspects of life in the modern state. Reason has triumphed over nature; humankind is liberated from the natural realm to the realm of pure reason, a reason that, paradoxically, becomes as fearsome as nature ever was. Horkheimer and Adorno warn that "it is as if the final result of civilization were a return to the terrors of nature."[6]

Maruyama, on the other hand, judges modernity's triumph over nature in the opposite way. For him, true modernity provides individuals with autonomy by liberating them completely from nature. Totalitarian power is aligned not with nature's overcoming but with its continued presence. It is the lack of overcoming, the lack of complete liberation from nature, that he holds responsible for Japan's plight. If Japan had achieved modernity, if nature had been completely subordinated, a different, independent idea of the subject would have emerged, and thus Japan might never have found itself engaged in the ghastly fifteen-year war for dominance on the Asian continent.

Having identified much the same nexus between nature, power, and freedom,[7] the Tokyo professor *cum* military man sees nature as oppressive while the Frankfurt school refugees see its subordination as oppressive. For

6. Ibid., 113. Slavoj Žižek observes the same Hegelian paradox in postmodern Europe where "elevating a contingent Other . . . into an absolute Other" breeds excessive violence. Žižek writes, "[T]he final arrival of the truly rational concrete universality—the abolition of antagonisms, the mature universe of negotiated coexistence of different groups—coincides with its radical opposite, with thoroughly contingent outbursts of excessive violence." Slavoj Žižek, "A Leftist Plea for Eurocentrism," *Critical Inquiry* 24 (Summer 1998), 1000.

7. For the purposes of this essay, following Isaiah Berlin, I do not attempt to distinguish between liberty and freedom. See Berlin, *Four Essays on Liberty* (New York: Oxford University Press, 1960), 121.

Maruyama, Japan is totalitarian because it has *not* eradicated nature from politics; for Horkheimer and Adorno, Germany is totalitarian because it has.

How are we to explain this difference? It is certainly not that wartime Japan was totally dissimilar from wartime Germany. Indeed, much of Horkheimer and Adorno's account of totalitarian politics would describe the Japanese state just as well as those European states suffering from what they define as "the extremes of Enlightenment." For instance, Horkheimer and Adorno portray "the West's" situation as one in which "the individual is wholly devalued in relation to the economic powers, which at the same time press the control of society over nature to hitherto unsuspected heights. Even though the individual disappears before the apparatus which he serves, that apparatus provides for him as never before."[8] Japan also had developed systems of control that subordinated individuals and raw materials to collective goals. In Japan, too, the standard of living had risen to unprecedented heights. As this comparison indicates, modern industrial uses of nature's material resources do not automatically determine nature's ideological use. Harnessing nature to the purposes of the state can be done as a natural activity in the name of nature, as it was in Japan, or as a rational activity objectifying nature, as described by Horkheimer and Adorno.

These divergent evaluations of the ideological roots of mid-twentieth-century totalitarianism stem, I argue, not from differences in the physical, political, and economic environments of Japan and Germany, nor from the technological capacities of these societies, but from the different place each nation was assumed to hold in a history of modernity. These wartime writings share the largely unanalyzed assumption that modernity consists in overcoming nature. Although twentieth-century theories of modernity rarely recognize their indebtedness to a conception of nature as the starting point of history and as the antithesis of freedom, it is obvious in the writings under discussion here, at least, that concepts of nature structure concepts of modernity and its promise of liberation. For Maruyama, Japan still hesitates at the threshold of this great historical adventure from nature to freedom, while, for Horkheimer and Adorno, Germany—or, rather, modern Europe—has already traversed this promising trajectory only to discover horror rather than liberty at its end.

In coming to terms with the largely unrecognized potency of concepts of nature in discussions of modernity and freedom, I will focus on the

8. Horkheimer and Adorno, *Dialectic* (note 5), xiv.

wartime and immediate postwar writings of Maruyama Masao. For Maruyama and many of his generation, discovering the intellectual underpin-
nings of autonomous subjectivity became the principal desideratum. In
this quest, nature was the enemy in at least three ways. First, nature figured
as the past that Japan needed to transcend to become modern. In Maruyama's intellectual history of the Tokugawa period (1603–1868), discussed
in the first section below, he describes what he sees as the failure in Japan
of "invention" *(sakui)* to emerge as the primary political value, a failure
that results in the continued premodern assurance that existing institutions
are manifestations of nature. Second, nature was a mode of speaking about
the difference between Europe and Japan. The well-worn dichotomies of
Orientalist discourse work here to distinguish "the West" and "the East"
on the basis that "the East" remains in the thrall of oppressive nature, unable to free itself from its physical environment and its traditions. Third,
nature in the form of the body and its sensuality obstructs the formation
of autonomous political subjectivity, because carnal indulgence ensures the
political conformity of the individual too immersed in physical pleasures
to develop the necessary awareness of his or her political position. According to Maruyama's critique then, nature is Japan's deformed past, the
mark of Japan's tragic difference from "the West," and Japan's accursed sensuality shackling it to uncritical bodily pleasures.

With this enormous burden of guilt, it is no wonder that nature is
anathema. While I fully recognize the intellectual power of this analysis, I
will argue that Maruyama and other mid-twentieth-century thinkers, in
conceiving of modernity and modern subjectivity as absolutely antinatural,
ill-served the liberty they sought. In the essay's conclusion, I will consider
another, more productive way to think of modernity's relationship with
nature.

Terms and Context

Before plunging into an analysis of Maruyama's complex argument, it is
perhaps best to begin with some preliminary notes about Japanese terms
and context. When Maruyama targets "nature," the term he employs is
shizen, the modern meaning of which ranges, as does its English equivalent, from the concrete ("the natural environment" [*shizen kankyō*]) to the
abstract ("natural law" [*shizenhō*]) to spontaneous human instincts, emotions, or characteristics (*shizen na*). Although crucial to Maruyama's analysis of intellectual history from the seventeenth century, the term *shizen* did

not become standard in Japanese until the 1890s,[9] acquiring its modern meaning at the same time as its Chinese counterpart, *ziran* (discussed in Fa-ti Fan's essay). Before the 1890s, *shizen* appears to have been rather uncommon, and certainly it was not a preoccupation in the Chu Hsi Confucianism[10] that dominated Tokugawa scholarship. As historian Hino Tatsuo comments, "In the nine classics [of Confucianism], you can not find one example of the use of the word 'shizen.'"[11] Instead, the writers whom Maruyama studies relied on an array of terms and phrases to express ideas of nature. Among the most important of these were *manbutsu* referring to "the myriad objects" of the physical universe and *kaibutsu,* which emphasized "the opening up of things,"[12] as well as *tenchi* (heaven and earth), *tenten* (the truth of heaven), *tenka* (all under heaven), and *tenri* (heaven or nature's law), and *honzen no sei* (human nature). "The way of heaven and earth" was expressed with terms such as *tendō* and *tenchi shizen no michi,* and "the principle of nature" with *jōri, tenchi seibutsu no ri,* and *tenchi no jōri.* Given the capacity of Japanese orthography to represent the sound "ten" with *kanji* other than that for "heaven," opposition to the orthodox Chu Hsi Confucian political and social hierarchy could be expressed through the creative choice of other *kanji,* sounding like "ten" but meaning something else such as "to revolve" or "to change."[13]

Besides these Confucian terms, Buddhist, Taoist, and Shinto traditions each had their own eclectic vocabularies. Furthermore, in the sixteenth

9. For a discussion of when *shizen* becomes the standard term for "nature" in Japanese, see Yoshida Tadashi, "Shizen to kagaku" (p. 342) and Sagara Tōru "Preface" (p. iii), both in Sagara Tōru, Bitō Masahide, and Akiyama Ken, eds., *Shizen,* vol. 1 of *Kōza: Nihon shisō* (Tokyo Daigaku Shuppankai, 1983); and also Minamoto Ryōen, "Komento," *Shizen no shisō* (Tokyo: Kenkyūsha, 1974), 42–55.

10. "Chu Hsi" is the older romanization of what is also referred to as "Zhu Xi" or Neo-Confucianism. I use "Chu Hsi" here in order to be consistent with the standard translation of Maruyama's work.

11. Hino Tatsuo, "Soraigaku ni okeru shizen to sakui," in Sagara Tōru, Bitō Masahide, and Akiyama Ken, eds., *Shizen* (note 9), 193.

12. Tessa Morris-Suzuki, "Concepts of Nature and Technology in Pre-Industrial Japan," *East Asian History* no. 1 (June 1991), 83–84.

13. The physician and political philosopher Andō Shōeki (1703/6–1762?), who envisioned a world of small egalitarian communities where everyone worked the land, emphasized nature's cycles and the constant ebb and flow of life by writing *tenchi* and *tenten* with altered orthography. Maruyama, *Studies in the Intellectual History of Tokugawa Japan* (note 1), 256 n. 29. Neo-Confucian naturalist Miura Baien (1723–89) also used such *kanji* in writing *tenchi* to impress upon his readers the fluctuations of nature. Minamoto Ryōen, "Komento," *Shizen no shisō* (Tokyo: Kenkyūsha, 1974), 46.

and early seventeenth centuries, Japanese terminology was enriched by European missionaries, particularly those of the Jesuit order, who sought to translate the Catholic doctrine that nature is the creation of an omnipotent single deity. The Latin *natura,* for instance, was expressed as *tenchi banbutsu* and *tenchi jitsugetsu.* Although contact with European nations diminished to almost nothing for two centuries during the Tokugawa period, Japan again embraced the full range of European ideas in the mid-nineteenth century, and more conceptions of nature were introduced as democrats urged individual "natural rights" or *tenpu jinken,* while social Darwinists thought "evolution" *(shinka)* and "natural selection" *(shizen tōta)* provided a natural basis for understanding national development.[14] As this diverse nomenclature indicates, nature was multivalent during the period studied by Maruyama.

Why then, despite this plethora of "natures" in Tokugawa and early Meiji Japan, did Maruyama rely on the term *shizen* to make his argument? There are, I think, two answers to this question. First, *shizen* was the term of choice for many of his antagonists, the ultranationalist intellectuals such as Watsuji Tetsurō (1889–1960), Tanabe Hajime (1885–1962), and others from Kyoto Imperial University whose work blended strands of Zen Buddhism and German phenomenology in their support of Greater Japan and its military empire. *Shizen's* propagandistic uses are evident in *Kokutai no hongi* (The essential principles of the nation), written, in part, by Watsuji and published in March of 1937 by the Ministry of Education for use in all schools. Its references range from the physical to the metaphysical. Japan's superlative physical environment "surpasses that of all other nations"[15] and the country's "temperate climate," its "beautiful mountains and rivers," its "spring flowers, autumn tints, and the scenic changes accompanying the seasons" all garner praise. From time to time, the authors admit, natural calamities *(shizen no saika)* such as earthquakes and typhoons do occur in Japan, but the Japanese people respond to such disasters with fortitude, never with fear, despair, or a desire for conquest over nature as

---

14. In the early 1880s, political activists Baba Tatsui and Katō Hiroyuki conducted a fierce debate over the relative value of "natural rights" and "social Darwinism." See Baba Tatsui, *Tenpu jinkenron,* reprinted in Yoshino Sakuzō, ed., *Meiji bunka zenshū,* vol. 5 (Tokyo: Nihon Hyōronsha, 1927); and Katō Hiroyuki, *Jinken shinsetsu,* republished in Uete Michiari, ed., *Nishi Amane, Katō Hiroyuki* (Tokyo: Chūō Kōronsha, 1984). I discuss this debate in "Nationalizing Nature: Ideology and Practice in Early Twentieth-Century Japan," in Sharon Minichiello, ed., *Japan's Competing Modernities: Issues in Culture and Politics, 1900–1930* (Honolulu: University of Hawaii Press, 1998), 114–32, and in *Reconfiguring Modernity: Concepts of Nature in Japanese Political Ideology* (Berkeley: University of California Press, 2001).

15. *Kokutai no Hongi* (Tokyo: Monbusho, 1937), 91.

in "the West." "Love" is used to describe the intensity of the relationship between all Japanese people and *shizen (shizen o aisuru),* but even this seems an inadequate characterization.[16] Indeed, *Kokutai no hongi* pushes the ascribed attachment beyond love to an even more intense state of faithful intimacy, manifest in the annual festivals, family crests, architecture, and gardens, but ultimately transcending these cultural expressions. Indeed, beyond the surface of daily life, the coalescent devotion between the Japanese people and nature unites consciousness with physical experience to such an extent that one cannot be separated from the other.[17] At some mystical level, then, the nature of these islands and the nature of the awareness of those who live on them are the same thing. Moving from the physical environment to aesthetic practices to consciousness itself, nature *(shizen)* unifies all aspects of Japanese existence. Maruyama never refers explicitly to this wartime view of nature as national consciousness. Instead, for obvious reasons, he focuses on the ideological uses of nature more than two centuries earlier, but in choosing the term *shizen* he implies the contemporary object of his critique.

The second reason that Maruyama masks the plethora of "natures" in Japanese history is that he wants to attack a monolithic entity, antithetical to freedom at all times. In other words, nature is not only bad in the ideology expressed in *Kokutai no hongi,* but it has always been bad in exactly the same way. By making "nature" the same throughout his history, Maruyama is not called upon to explore the possibility that different conceptions of nature might lead to different political possibilities, or that some forms of nature might even create a space for liberty. In short, Maruyama is committed to the conception of modernity as the universal historical arc from nature to freedom, and he does not wish to complicate this story by implying that there were multiple forms of nature since such a view might imply multiple forms of modernity. This being the case, Maruyama relies on the single, standard term *shizen,* which unifies all things of a Japanese nature, past and present, from the physical environment to popular consciousness.

Japan's Deformed Past: Nature versus Invention

Let me now return to Maruyama's wartime essay "Kindai Nihon seiji shisō ni okeru 'shizen' to 'sakui.'" This investigation of the internecine quarrels

16. Ibid., 54.
17. Ibid., 55.

of Tokugawa Chu Hsi Confucianists proceeds "in terms of two concepts, nature *(shizen)* and invention *(sakui)*."[18] These two concepts are pitted against one another, "nature" being embraced by orthodox Chu Hsi Confucianism and "invention" by Ogyū Sorai and his followers. As Maruyama says, the aim of both schools of thought was to support the Tokugawa shogunate and to quell disorder and disobedience, but the two schools' theoretical justifications were antithetical. While Chu Hsi Confucianism equated the existing feudal hierarchy of the bakufu with "the *natural order itself*,"[19] Ogyū Sorai insisted that political institutions were not natural at all but rather the products of creative political leadership.

In Maruyama's reading, Chu Hsi Confucian philosophy encourages entrenchment by holding up the mirror of nature to the feudal hierarchy. Nature's mirror produces a double reflection of shogunal authority: one image on a cosmological level in "the order of the universe (the Principle of Heaven)," or *tenri,* and a duplicate image in "man's original nature," or *honzen no sei.*[20] Thus, despite all their superficial differences, the bakufu, the cosmos, and inner human spirit become homologous manifestations of identical *li* (or *ri*), often translated as "principle." The hierarchical order found in nature, such as the relationship between heaven and earth, is the same as the hierarchical order found in human society. Just as heaven is above and earth below, so too the ruler is above and the people below. It is crucial to understand that nature and culture are not opposing realms, not even analogous realms, but the same realm because the same metaphysical essence inheres in both the physical world and human society, giving order to each despite different superficial manifestations (*chi* or *ki*). Within human society, the five human relationships—ruler/subject, father/son, elder brother/younger brother, husband/wife, and older friend/younger friend—all partake of the same natural, hierarchical order. The virtue of benevolence displayed by those of greater status and the virtues of loyalty and obedience expected from those below help maintain this natural hierarchy, but these virtues are themselves natural, not disciplines imposed consciously on the self in order to accord with nature. The result is an outlook that makes all difference, all strife, and all critique unnatural without being antinatural or perverse.

18. Maruyama Masao, "Kindai Nihon seiji shisō ni okeru 'shizen' to 'sakui,'" translated by Mikiso Hane as "'Nature' and 'Invention' in Modern Japanese Political Thought," in *Studies in the Intellectual History of Tokugawa Japan* (note 1), 191. This essay was one of the three that Maruyama finished just before joining the army in 1944.

19. Ibid., 201.

20. Ibid., 198.

For those accustomed to various European systems of thought where the dichotomies of culture and nature, mind and body, public and private, sacred and profane prevail, the issues of will, belief, and desire loom large. By contrast, within the seamless world of nature-culture in much of Japanese thought, no particular emphasis falls on will, belief, or desire, since no impetus is necessary to mediate between two halves of a fractured universe. Without the profound difference created by opposing the primitive, the sexual, or God, on the one hand, to culture, reason, or "man's" fallen nature on the other, the structure of longing is transformed. Acceptance— sublime unquestioning acceptance—of natural order is the highest form of wisdom; absolute harmony alone is natural. But, harmony is not attained through effort. Instead, the melting away of desire, will, and the self itself is what is required. As Maruyama sees it, the hierarchical, ethical, and political order is thus locked in a moribund rigidity by reliance on nature for justification. Chu Hsi Confucianism offers nothing but increasingly impotent resistance to change as Japan moves toward a protocapitalist economy, while it encourages a supine and uncritical form of political subjectivity.

Against this failure of Chu Hsi Confucianism, Maruyama posits as inevitable a turn away from nature to invention as the ideological support the Tokugawa regime. He writes that in an increasingly unstable political situation, "when social relations lose their natural balance . . . a body of thought is bound to emerge that stresses the idea of the autonomous personality *(shutaiteki jinkaku)* whose task it is to strengthen the foundations that uphold the social norms and to bring political disorder under control."[21] Ogyū Sorai (1666–1728) and his followers in the *kogaku-sha* (school of ancient learning) fulfill this role by valorizing the acts of the Ancient Sages who, in the time before time, somewhere in China, invented the social forms of rites and music and the economic activities of agriculture and weaving. The Sages' inventiveness becomes the standard by which subsequent rulers are properly judged.

In Maruyama's consideration of Japan's failure to achieve modernity, the struggle between "invention" as championed by Ogyū Sorai and "nature" as championed by the Chu Hsi Confucianists takes on an almost allegorical significance. Maruyama hopes to discover in Ogyū's position a dichotomy between nature and invention that is as complete and absolute as his own. However, by Maruyama's lights, Ogyū fails to achieve a sufficiently definitive break with nature. His reliance on "an agricultural liveli-

21. Ibid., 206–7.

hood, a natural economy, a family-based master–servant relationship and so on"[22] restricts the powers of the autonomous rulers to create new institutions. Maruyama writes, disappointedly, that "the Sorai school's position is, really, in the last analysis, an attempt to produce nature by the logic of invention."[23] Ogyū's failure becomes that of a tragic hero repulsed by his evil adversary. The sense of lost hope and the betrayal of narrative necessity is made even stronger because Maruyama describes exactly what that victory should have consisted of: "If the theory of natural order was to be completely overcome, no normative standards of any kind would be present in the background as the premise; instead, the starting point had to be human beings who, for the first time, invented norms and endowed them with their validity."[24] Maruyama is emphatic about the need to ground autonomy in the complete rejection of nature and the complete acceptance of invention. Although Maruyama himself subsequently questioned the early Tokugawa hegemony of Chu Hsi Confucian thinking asserted in this early work, and consequently undermined the drama of Ogyū's struggle,[25] the polemical difference between nature and invention is crucial to his critique of wartime and postwar Japan.

Maruyama's disappointment with Ogyū is, in my view, ironic because Maruyama himself steps back from invention at the critical moment, unable, it would seem, to embrace it as the culmination of the historical process. For all Maruyama's scathing critique of nature and his seemingly whole-hearted embrace of invention, he too distrusts the contingency that results from introducing pure invention into political life. Maruyama's ambivalence on this crucial point emerges in the key passage: "I have argued above that when any really existing order is justified by the idea of a natural order, that existing order is in its stage of ascendancy or stability, whereas when, on the contrary, *if justified in terms of autonomous personalities, it is in its period of decline or crisis.*"[26] Under this schema, the very modernity for which Maruyama strives must be a moment of "decline or crisis," and autonomous individuals—the agents of invention—do little more than mark the transition. Like the Ancient Sages, they arise and act

22. Ibid., 222.

23. Ibid.

24. Ibid., 210.

25. See Maruyama's introduction in Mikiso Hane's translation, *Studies in the Intellectual History of Tokugawa Japan* (note 1), xxxiv. See also Herman Ooms, *Tokugawa Ideology: Early Constructs, 1570–1680* (Princeton: Princeton University Press, 1985).

26. Maruyama, "'Nature' and 'Invention'" (note 18), 228–29 (emphasis mine).

at the start of an era, only to sink back into the history-less miasma of a naturally justified order. The inventing subject that Maruyama describes is so unbounded that his or her continued autonomy promises only chaos. Indeed, Maruyama paradoxically refuses to countenance continually inventive subjects while at the same time blaming Ogyū Sorai for doing likewise. His semiconscious ambivalence toward *sakui* implies an incipient critique of modernity not unlike that of Horkheimer and Adorno. The purely autonomous subject appears to threaten chaos no less surely than "the fully enlightened earth radiates disaster triumphant." [27]

Whatever Maruyama's doubts about *sakui* and, perhaps, modernity, he never returns to nature for solace. His is a resolutely antiromantic stance: nature is not the site of self-discovery as Goethe would have it (in Robert Richard's essay), nor the basis of proper pedagogy (as analyzed in Eckhardt Fuchs's essay). Instead, because of the failure of invention to overcome nature in the Tokugawa period, Japan's subsequent history is marred. Even in the twentieth century, Maruyama argues, Japanese political institutions are apprehended as both completely natural and wholly cultural. This seamless totality absorbs all possibility of critique since there is no basis outside Japanese nature-and-culture (except foreign places) from which to criticize the state.

Japan's Tragic Difference from "the West"

For Maruyama, nature serves not only to mark the contrast between premodern Japan and modernity, but also to mark the contrast between "the East" and "the West." By his reading, Japanese history is not only developmentally deformed in not having espoused "invention" during the Tokugawa era, but also deformed in comparison with Europe where reason and creativity triumphed. The horrors of Japanese militarism are therefore the result both of Japan's past history and its current identity as an Oriental nation. In this analysis, Maruyama joins a widespread Orientalist discourse that made "the East" closer to nature. This understanding of "the East" served powerfully as the basis for distinguishing Japanese "fascism" [28] (Maruyama's term) from that of Germany. As historian John Dower points out,

27. Horkheimer and Adorno, *Dialectic of Enlightenment* (note 5), 1.

28. I use the term "fascism" advisedly. Although some have argued against applying this term to Japan (see Stanley Payne, *A History of Fascism, 1914–1945* [Madison: University of Wisconsin Press, 1995]), the term "fashizumu" was widely used in postwar Japanese discussions,

while Nazism was perceived as a cancer in a fundamentally mature "Western" society, Japanese militarism and ultranationalism were construed as reflecting the essence of a feudalistic, Oriental culture that was cancerous in and of itself. To American reformers [of the Allied Occupation after the war], much of the almost sensual excitement involved in promoting their democratic revolution from above derived from the feeling that this involved *denaturing* an Oriental adversary and turning it into at least an approximation of an acceptable, healthy, westernized nation.[29]

Maruyama, like the American occupiers, sought to denature Japan to make it more like "the West."

How does Maruyama measure Japan's tragic distance from Occidental freedom? Some scholars have insisted that Maruyama's vision is fundamentally Hegelian. From this perspective, the distance between "the West" and Japan is a matter, principally, of time. Europe exemplifies the successful completion of the historical dialectic that ends in the consciousness and reality of freedom. As Hegel puts it, "the History of the World begins with its general aim—the realization of the Idea of Spirit—only in an *implicit* form *(an sich)* that is, as Nature; a hidden, most profoundly hidden, unconscious instinct; and the whole process of History (as already observed), is directed to rendering this unconscious impulse a conscious one."[30] By this reasoning, Japan is not excluded from this dialectic, although it is delayed. Interpreted as Hegelian, Maruyama's investigation

---

including Maruyama's essay "The Ideology and Dynamics of Japanese Fascism," in Ivan Morris, ed., *Thought and Behavior in Modern Japanese Politics* (New York: Oxford University Press, 1969), 25–83, originally delivered as a lecture in June 1947. Japan's particular form of fascism is sometimes termed *tennōsei fashizumu* (emperor system fascism), prompting a long-running debate. See, for instance, Komatsu Shigeo, "Nihon gata fashizumu" (Japanese style fascism), *Shidō sha to taishō,* vol. 5 of *Kindai Nihon shisō shi kōza* (Tokyo: Chikuma Shobō, 1960), 277–326; Nakamura Kikuo, "Tennōsei fashizumu wa atta ka" (Was there emperor system fascism?), *Jiyū* 7, no. 12 (December 1965): 50–59; Ishida Takeshi, "'Fashizumu ki' Nihon no okeru dentō to 'kakushin'" (Tradition and 'renovation' in Japan's 'fascist era'), *Shisō* no. 619 (January 1976): 1–20; Yamaguchi Yasushi, *Fashizumu* (Fascism) (Tokyo: Yuhikaku Sensho, 1979); Amano Keiichi, "'*Tennōsei fashizumu ron' no genzai*" (The current "emperor system fascism debate"), *Ryūdō* (January 1980): 56–65; Yoshimi Yoshiaki, *Kusa no ne no fashizumu: Nihon minzoku no sensō taiken* (The roots of fascism: the wartime experience of the Japanese folk) (Tokyo: Tokyo Daigaku Shuppankai, 1987).

29. John W. Dower, *Embracing Defeat: Japan in the Wake of World War II* (New York: Norton, 1999), 79–80.

30. Georg Wilhelm Friedrich Hegel, *The Philosophy of History,* trans. J. Sibree (New York: Dover, 1956), 25.

of Tokugawa thought becomes an attempt to excavate the "profoundly hidden" instinct for freedom in Japanese history through the work of Ogyū Sorai, only to discover that the necessary dialectic was never initiated. Instead, Japan remains in the first stage of History, namely, Nature.

While there is certainly Hegelian coloring to the form of freedom Maruyama desires, I believe his mode of analysis rests on less lofty sources. As his own references attest, Maruyama relies most heavily on the arguments of German political scientist Hans Kelsen, whose positivist influence is felt throughout the essay. This distinction is crucial. While Hegel's History proceeds as a dialectic that ultimately absorbs and expresses Nature within the triumphant Spirit, Kelsen is not interested in dialectical synthesis. Instead, Kelsen asserts the claims of positive law against natural law. For him, politics and history present "either–or" choices, not possible synthesis. In this regard, I think Maruyama is a true Kelsenian: a choice must be made between nature and invention; compromise or synthesis is illegitimate. By this measure, the distance between Japan and "the West" is categorical.

Maruyama pursues this "either-or" reasoning in considering nature law and the social order:

> Generally speaking, as soon as natural law is related to the actual social order, it encounters an "either–or" *(Entweder–oder)* characteristic. Either by rigid adherence to pure doctrine it becomes a revolutionary principle directed against the concrete social order, or by its complete identification with the actual social relations it becomes an ideology guaranteeing the permanence of the existing order.[31]

Later in his argument, the revolutionary possibility of natural law is muted and ultimately dismissed when nature is exclusively tied to the conservative task of justifying an "existing order in its stage of ascendancy or stability."[32] Indeed, although both Maruyama and Kelsen hint that nature occasionally represents a point from which to critique existing structures of power, they quickly retreat from this recognition of nature's dual political valences to insist that it always justifies the existing state.[33]

31. Maruyama, "'Nature' and 'Invention'" (note 18), 199 (German in the original).

32. Ibid., 228.

33. On this point, Maruyama overstates his case, insisting that "Kelsen denies the revolutionary character of natural law" (ibid., 199n). In fact, Kelsen does at times concede that natural justification can be sought for a range of political purposes. In *General Theory of Law and*

This diametrical opposition between nature and invention locks political action into a rigid pattern where its only options are revolution or acquiescence without room for liberal compromise. No latitude for continual negotiation between the two is provided. Ultimately, natural law is completely absorbed by positive law and the state of nature by the State. Even the difference between nature and invention disappears during the State's ascendancy when it invents a form of nature congenial to itself, as exemplified in the family–state form of community (*kyōdōtai,* the Japanese translation of *gemeinschaft*) propagated in Imperial Japan.

Maruyama even relies on this "either–or" methodology in his reading of the political philosophers of the European Enlightenment.[34] Maruyama wants to ally himself with writers in this tradition, especially in their concern for individual freedom, but they pose a significant problem for him because of the prominence of nature and its cognates in their theories. Unlike Kelsen who dismisses all Enlightenment references to nature as irrational, sublimated allusions to the divine,[35] Maruyama performs the curious operation of transferring "natural rights," "natural law," and "the state of nature" to his own category of "invention." He declares that "[i]nsofar as the logical core of the natural law of the Enlightenment was the 'theory of social contract,' it belongs clearly to the category of invention in my classification."[36] He later elaborates this point:

> Let us stop here long enough to note that the theoretical basis for this doctrine of liberty and popular rights is the natural law of the Enlightenment.

State, Kelsen writes that "the doctrine of natural law is at times conservative, at times reformatory or revolutionary in character." Hans Kelsen, *General Theory of Law and State,* trans. Anders Wedburg (New York: Russell & Russell, 1961), 11. In other writings, however, Kelsen restricts natural law's general reformatory or revolutionary powers: "[T]he character of natural law doctrine in general, and of its main current, was strictly conservative. Natural law as posited by the theory was essentially an ideology which served to support, justify, and make an absolute of positive law, or, what is the same thing, of the authority of the State." Hans Kelsen, *Natural Law Doctrine and Legal Positivism,* trans. Wolfgang Herbert Kraus (New York: Russell & Russell, 1961), 416–17.

34. Maruyama considers European Enlightenment figures in "'Nature' and 'Invention'" (note 18) and in later works such as "Nihon no okeru jiyū ishiki no keisei to tokushitsu" (The formation and characteristics of liberal consciousness in Japan), originally published in *Teikoku daigaku shinbun* (August 1947), reproduced in *Senchū to sengo no aida* (Between war and postwar) (Tokyo: Mizusu Shobō, 1976), 297–306.

35. Kelsen argues that to appeal to nature suggests an underdeveloped "personality-type" given to a "fundamentally pessimistic mood of self-consciousness, not weak in itself but directed, so to speak, against itself." Kelsen, *Natural Law Doctrine* (note 33), 424.

36. Maruyama, "'Nature' and 'Invention'" (note 18), 249 n. 15.

Since the latter taught that the rights of men are natural rights, it would seem superficially that we should classify it as a theory of natural order. But a more careful examination shows directly that the opposite is true. The "rights of man" in question are not rights embedded in any actually existing social order. On the contrary, they are concrete embodiments of the autonomy of man, who can establish a positive social order. Thus the theory's insistence on the a priori character of natural law necessarily implies the view that any positive law derives its validity from its original establishment by man.[37]

The main thrust of this statement underscores the artificiality of any social order, but, in the course of this argument, Maruyama interprets Enlightenment thought such that natural law and natural rights (the passage slips between them without distinction) become the same thing as individual autonomy. He claims that to speak of nature in the European Enlightenment tradition is simply to use a code word for "invention," albeit a misleading one. Unlike Kelsen, Maruyama does not dismiss these elements of European political thought as remnants of religious prejudice, but embraces them as equivalents of modern subjectivity.

In the end, despite differences on Enlightenment thought, Maruyama and Kelsen share, indeed advocate, the same antinatural version of modernity. They both seek to unbind the Prometheus of human reason from the chains of tradition, religion, social context, and nature. In so doing, they exemplify a powerful strand of social scientific thought, and stand at the culmination of an intellectual tradition that gradually shed all sources of authority except the self. They also envision European history as the standard model against which to measure the progress of other regions. With their "either–or" criteria, Japan will not achieve freedom through dialectical development from within itself over time, but must instead choose against the anathema of its Oriental identity.

On a more immediate political level, this analysis also came to provide Maruyama with an excuse for not having opposed the war more vigorously. In reflecting later on why he and others did not challenge Japanese "fascism," he suggests that in an "either–or" world, the middle ground provides no leverage. Japanese intellectuals never actively struggled against fascism, Maruyama insists, because they were never sufficiently Westernized, never sufficiently denatured. Instead, Japan's intelligentsia was caught

37. Ibid., 313.

between two worlds, sad, weak hybrids capable at best of "passive resistance." As Maruyama explains in 1947:

> [I]n Japan the intelligentsia is essentially European in culture, and unlike its counterpart in Germany, could not find enough in traditional Japanese culture to appeal to its level of sophistication. In the case of Germany to exalt nationalism meant also to take pride in the tradition of Bach, Beethoven, Goethe, and Schiller, who at the same time provide the culture of the intelligentsia. These conditions did not exist in Japan; inasmuch as the European culture of the Japanese intelligentsia remained a culture of the brain, filling only an ornamental function, it was not deeply rooted in thinking or feeling. Hence it lacked the moral courage to make a resolute defense of its inner individuality against fascism. On the other hand, its European culture would never permit it to respond to the low tone of the fascist movement and to its shallow intelligence. Such a lack of thoroughness, coupled with the intellectual detachment and isolation of the intelligentsia in general drove it to a hesitant and impotent existence.[38]

In such a Manichean world, organized through the opposition of East and West, Nature and Culture, Fascism and Freedom—rather than along developmental or dialectical models—hybrids are not inherently more flexible and resistant than pure strains. Quite the reverse. For Maruyama, occupying the middle ground does not provide the strength of synthesis; it provides an excuse for paralysis.

Japan's Accursed Sensuality

In the aftermath of the war, Maruyama did not abandon his attack on nature. He not only republished his wartime essay "'Nature' and 'Invention,'" but broadened his critique of nature to include eroticism and sensuality. In the October 1949 issue of the magazine *Tenbō*, Maruyama defined the faults of postwar literature and postwar politics in terms of *nikutai* or carnality. While some novelists, notably Sakaguchi Ango and Tamura Taijirō, argued that sensuality provided a foundation for individuality and that the "gate of flesh" *(nikutai no mon)* was the "gate to modernity,"[39]

---

38. Maruyama, "The Ideology and Dynamics of Japanese Fascism" (note 28), 59–60.

39. Tamura made this claim, titling one of his novels *Nikutai no mon*. See Jay Rubin, "From Wholesomeness to Decadence: The Censorship of Literature under the Allied Occupation," *Journal of Japanese Studies* 11.1 (1985): 71–103, here 77.

Maruyama strenuously denounced attempts to find liberty through erotic expression. This 1949 essay, "Nikutai bungaku kara nikutai seiji e" (From carnal literature to carnal politics), bemoans the obsession with sex and sensuality in Japanese literature, arguing that the overwhelming concentration on the body denies the subject's free agency or spirit. "In Japan," Maruyama writes, "the spirit is neither differentiated nor independent from perceptible nature—of course I include the human body as a part of nature—and so the mediating force of the spirit is weak."[40]

Maruyama does not view bodily pleasure as a possible counterpoint to authority as it so emphatically is in early modern European sodomy cases (see Helmut Puff's essay) or in eighteenth-century efforts to suppress masturbation (see Fernando Vidal's essay). Indeed, given the lack of tension between mind and body in Japanese thought, the pleasures of the body were never considered in themselves a particular source of sinfulness, and therefore never opposed quite so directly to goodness or spirit or society as in cultures dominated by Christian thought. Even today, there is comparatively little consternation in Japan over the wide range of human sexual practices and appetites, including pornography. The realms of erotic desire and fleshly longing do not axiomatically serve as a liminal space where the boundaries of social authority over the individual can be observed and tested. Instead, for Maruyama and others, embracing "the body" and entering the "gate of flesh" is simply another evasion of political and artistic responsibility for the indolence of unself-reflective nature.

It might be useful to pause here and compare Maruyama's position with that of the other twentieth-century figures analyzed in this volume. Like the radical American feminists described in Michelle Murphy's essay, Maruyama seeks absolute freedom from the body, yet he is uninterested in reordering our physical nature through speculative scientific techniques such as ectogenesis or in achieving local control over our bodies through practices like menstrual extraction. Maruyama's is a purely intellectual transcendence, devoid of technological apparatus. In his ethereal world, he does not even concede the modified naturalism of the myrmecologists Forel, Wasmann, and Wheeler (portrayed in Abigail Lustig's essay), who allowed that all human societies are to some extent a natural phenomenon.

40. Maruyama, "From Carnal Literature to Carnal Politics," in Morris, ed., *Thought and Behavior in Modern Japanese Politics* (note 28), 251. Maruyama's negative evaluation of the body is directly contrary to Horkheimer and Adorno who argue that the separation of the intellectual from the sensuous "means the impoverishment of thought and of experience: the separation of both areas leaves both impaired." *Dialectic of Enlightenment* (note 5), 36.

These ant specialists retained a painful respect for the way societies were constrained by nature, even as they urged humanity to contend against it to achieve civilization in the wake of World War I. In the case of the feminists, the individual (female) body and, in the case of the myrmecologists, the social body continually limited the possibilities for liberty, and those who wished to achieve freedom were compelled to adopt techniques of scientific or social discipline to fight against nature. Maruyama was uninterested in such techniques.

Again, Maruyama's approach is "either–or." Using much the same logic that he employed in analyzing the failure of Tokugawa political thought, Maruyama suggests that where "the spirit is not functionally independent of nature," the possibilities for creative politics are dead.[41] Postwar literature produced by truly independent spirits would forgo disjointed images of flesh in order to project a world beyond the givens of the current social environment. This fictional world would serve to remind readers that "the public order, institutions, *mores,* in short the whole social environment" are created by human beings, a fiction, *not* a natural reality.[42] It is important to recognize here that Maruyama does not fault fiction for its fictionality. Fiction or "making things up" is allied to "invention" in its capacity to project other worlds and to recognize the contingency of established institutions. Indeed, Maruyama famously calls democracy itself a "fiction," not, of course, to dismiss it but to insist that it requires willful imagination to bring it into existence.

Given this sophisticated conception of fiction, it is puzzling that Maruyama did not consider the possibility that nature itself might serve as a fiction or indeed that "the state of nature" was just such a fiction in Enlightenment thought. If Maruyama had deemed "nature" a fictional construct, he might have viewed the "state of nature" as a desirable fiction that illuminates the created quality of society. As a fiction in this sense, nature might serve not just as that which we needed to be liberated *from* but also something that itself might require liberation, the conceptual liberation *of* nature so that it can refer to a realm outside the current bonds of society. If it referred to something at the edges of civilization,[43] it could illuminate

41. Maruyama, "From Carnal Literature" (note 40), 252.

42. Ibid., 255.

43. Robert Pogue Harrison suggests that the most likely origin of the term "forest" "is the Latin *foris,* meaning 'outside'" and argues that forests in Europe provide an external perspective on society. Robert Pogue Harrison, *Forests: The Shadow of Civilization* (Chicago: University of Chicago Press, 1992), 69.

the boundaries of the "status quo." Maruyama might even have spoken of the liberation *to* nature where an autonomous individual could remove him or herself in part from the confines of society in order to rethink its perimeters.

However, as Maruyama's analysis in "Nikutai bungaku kara nikutai seiji e" suggests, he continued to celebrate a form of invention totally divorced from nature. For him, nature has no redeeming political functions: Nature is not the presocial condition of individuals in the state of nature before they form cohesive social groups; nature is not a set of laws or rights that impinge on the positive statutes, constitutions, and the institutional configuration of the state; nature as the body is not an erotic playground outside of and, possibly, contesting the boundaries of society and the state; nature is not a "fiction" against which to measure and critique current circumstances. Nature is a cage, and the freedom Maruyama seeks from this cramped enclosure is absolute. For its sake, the autonomous political subject must strip itself of its past, its physical and social environment, and its very body, leaving us disembodied in a denuded world. In his distaste for nature in all forms, Maruyama comes perilously close to investing life itself with the iron chains of unfreedom. Negating nature in his terms results in the absolute freedom found only in death. Indeed, only when "completely dead," as the Zen poem paradoxically suggests, can one "work as the heart desires!"[44] Maruyama is undoubtedly a "utopian pessimist" as historian Andrew Barshay has so aptly called him.[45]

The Enlightenment from the Far Side of Modernity

From our current perspective, we now bear witness to the dangers of this immodest mid-twentieth-century vision of freedom, not, I think, because of our superior wisdom, but simply because we stand at a different place. The critique of the rational subject launched by Horkheimer and Adorno has hit home in our experience; the sorrows of pure subjectivity are manifest in our power to destroy our societies, ourselves, and our environment. While Horkheimer and Adorno forecast universal disaster and barbarity, Louis Dupré describes the loss in individual terms: "In becoming pure

44. Ironically, this poem is quoted in *Philosophy as Metanoetics,* written in 1944 by state ideologue Tanabe Hajime, one of Maruyama's antagonists. Tanabe, *Philosophy as Metanoetics,* trans. Takeuchi Yoshinori (Berkeley: University of California Press, 1986), p.172.

45. Andrew Barshay, "Imagining Democracy in Postwar Japan: Reflections on Maruyama Masao and Modernism," *Journal of Japanese Studies* 18, no. 2 (Summer 1992): 406.

project, the modern self has become severed from those sources that once provided its content. The metaphysics of the ego isolates the self. It narrows selfhood to individual solitude and reduces the other to the status of object."[46] Even Maruyama, for all his defense of modernity, hints at this problem when he suggests that if an existing order is "justified in terms of autonomous personalities, it is in its period of decline or crisis."[47]

In the face of modernity's societal and personal predicament, what resources are available to us to make good our losses? Was modernity doomed to spawn disaster? Was its only possible history a deterministic trajectory away from nature? Horkheimer and Adorno answer in the affirmative, excoriating the entire thrust of European history from the ancient Greeks to the Final Solution. In *Dialectic of Enlightenment,* the only recourse they suggest borders on the reembrace of mysticism. Maruyama, as we have seen, argues that insufficient modernity also spawns disaster. More than half a century later, the dilemma is still before us, but we are also free to find other avenues of approach. We can seek neither to condemn nor praise modernity, but to reinterpret it. Dupré suggests as much when he writes:

> While [earlier writers] exalted rational objectivity, moral tolerance, and individual choice as cultural absolutes, we now regard these principles with some suspicion. Undoubtedly there are good reasons to distrust the equation of the real with the objectifiable, progress with technological advances, and liberty of thought and action with detachment from tradition and social bonds. But should we attribute all such excesses to the original principles of modern culture?[48]

Dupré returns to a pre-Cartesian moment to find an alternative origin for modernity in the late fourteenth and early fifteenth centuries before the Enlightenment. Likewise, Stephen Toulmin lays claim to the late Renaissance as the initial literary and humanistic phase of modernity, preceding the scientific and philosophical quest for certainty that began in the second half of the seventeenth century.[49] Both authors broaden the possibilities of modernity by emphasizing its partnership with certain forms of nature.

46. Louis Dupré, *Passage to Modernity: An Essay in the Hermeneutics of Nature and Culture* (New Haven: Yale University Press, 1993), 119.

47. Maruyama, "'Nature' and 'Invention'" (note 18), 228–29.

48. Dupré, *Passage to Modernity* (note 46), 1.

49. Stephen Toulmin, *Cosmopolis: The Hidden Agenda of Modernity* (Chicago: University of Chicago Press, 1990), 23.

Taking this lead from Dupré and Toulmin, we can develop an alternative history of modernity where nature is no longer the starting point or antithesis of the history of freedom, but instead continually balances some of the excesses of reason and invention. Freedom is thus redefined: it need not be absolute to be valuable. The existentialist philosopher Paul Ricoeur has suggested that "[f]reedom is not a pure act, it is in each of its moments, activity and receptivity. It constitutes itself in receiving what it does not produce: values, capacities, and sheer nature."[50] If we accept this view, it may be possible to carve out a limited space for liberty while accepting nature, defined as the past, the joys and limitations of bodily existence, and the physical environment. Such a "nature" would circumscribe, but not wholly determine, the course of action, creating a limited, but nonetheless worthwhile freedom.

There are indeed losses in adopting this position. As critics of modernity (and particularly of democracy and liberalism) have vociferously pointed out, the result is a form of split subjectivity, a product of both nature and culture. In *An Intellectual History of Liberalism,* Pierre Manent laments this split subjectivity, which he terms "duplicity." He argues that, since Montesquieu, "we remain radically *divided,* the dividing line between natural man and the citizen is now within us."[51] Manent sets out to describe this divided modern self within the European tradition and to denounce the misfortune and corruption that he believes "the democratic project" has spawned in the face of "man's natural desire is to bring this duplicity into unity." As he says, "This division or duplicity guarantees that no end, no good, can require anything of man. What nature gives, cannot be ordered by it; what sovereignty orders, it cannot give."[52] In short, the ultimate responsibility for responding to an end or a good remains with the individual, a circumstance that Manent abhors. Ironically, Maruyama's wartime Japanese state resembles the world Manent desires (without its Christian overtones). As we have seen, Japanese ideology brought nature and sovereignty together, and Japan's people, in achieving fascist unity, did not suffer the strains of a divided consciousness, though they suffered nonetheless.

The situation of the "duplicitous" modern self has its rigors, it is true.

---

50. Paul Ricoeur, *Freedom and Nature* (Evanston: Northwestern University Press, 1966), 484.

51. Pierre Manent, *An Intellectual History of Liberalism,* trans. Rebecca Balinski (Princeton: Princeton University Press, 1994), 64.

52. Ibid., 115.

It may even be admitted, as Ricoeur does, that "there is no logical procedure by which nature could be derived from freedom (the involuntary from the voluntary), or freedom from nature. There is no *system* of nature and freedom."[53] We may be caught perpetually straddling this ungainly divide. This may appear insecure, exacting, and exhausting, but it may still be that continual renegotiation between nature and invention is our best bet for liberty. Freedom has never been synonymous with comfort.

Not only does renaturalizing modernity create a different form of subjectivity and a different form of freedom, it also suggests a different form of modern history, no longer a trajectory from nature to freedom and from East to West, nor certainly an "either–or" choice. Focusing on the continuing presence of nature in modern thought would allow us to describe more precisely Japan's particular confrontation with modernity. Perhaps Japan's modernity was of a different rather than a later or lesser type than "the West's"; perhaps Japan's modernity reveals that the problems of modernity within "the West" cannot be solved without reference to "the East." If, as Fredric Jameson comments, "in different historical circumstances the idea of nature was once a subversive concept with a genuinely revolutionary function,"[54] a return to modern history with an expansive view of nature's multiple ideological possibilities seems in order.

53. Ricoeur, *Freedom and Nature* (note 50), 19.
54. Fredric Jameson, "Reflections in Conclusion," in Ernst Bloch et al., *Aesthetics and Politics* (London: Verso, 1986), 207.

13

Liberation through Control in the Body Politics of U.S. Radical Feminism

Michelle Murphy

After the feminist revolution, women no longer gave birth from their bodies. The severing of reproduction from corporeality was one of the revolution's last steps:

> Finally there was that one thing we had to give up too, the only power we ever had, in return for no more power for anyone. The original production: the power to give birth. Cause as long as we were biologically enchained, we'd never be equal.[1]

Within the postrevolutionary societies forged on the former East Coast of the United States, human babies were grown in "brooders," mushroom-shaped buildings in which fetuses floated upside down in sacs coddled by floral imagery and wafting music. Once born, delivered from the brooder through a contracting canal amid a ritual of drumming and flowers, babies were raised by three biologically unrelated "mothers," who could be either women or men as both sexes were hormonally modified to lactate. Not only did postrevolutionary society unchain reproduction from the female body, it also disconnected genetics from kinship. Though each community had a distinct cultural identity—Wampanoag Indian or Harlem Black, for example—the individuals within a community held diverse genetic heritages that had been collected and mixed in the brooder, cutting the bonds between genes and culture, birth and motherhood, so there could be no more racism or sexism again.

---

1. Marge Piercy, *Woman on the Edge of Time* (New York: Fawcett Crest, 1976), 105.

This utopian feminist society that disentangled reproduction from bodies and human kinds through the technical development of ectogenesis—generation outside the body—was a political thought experiment from the widely read science fiction novel *Woman on the Edge of Time* (1976) written by feminist activist and novelist Marge Piercy.[2] With the hope, and for some the expectation, that revolution was around the corner, Piercy's description of ecotogenesis was not simply an exercise in imagination, but a possible blueprint for the radical feminist goal of women "taking control" of their own bodies.

The radical feminism Piercy's novel participated in was a specific brand of feminism that emerged in late 1960s America primarily among white college-educated women, many of whom had become disenchanted with the New Left.[3] Unlike the professional and liberal National Organization of Women (NOW), radical feminism was practiced in small, local, independently formed cells organized around consciousness-raising groups.[4] Cell 16 in Boston, Redstockings in New York, The Furies in Washington, Sudsoflopson in San Francisco, Women's Liberation Group in Gainsville, and Westside Group in Chicago all formed in a crescendo of radical feminisms nourished by a lively national traffic of mimeographs and newsletters that seemed to proliferate exponentially. Cells split, fused, and propagated, forming a shifting multiplicity of feminist positions that gave birth to diverse strands of feminist thought, from ecofeminism to feminist businesses,

2. Like many other white, college-educated women of the early 1960s who would later in the decade become radical feminists, Piercy had been an activist in the New Left, but then rejected it in a devastating critique. Marge Piercy, "The Grand Coolie Damn," in Robin Morgan, ed., *Sisterhood Is Powerful: An Anthology of Writings from the Women's Liberation Movement* (New York: Random House, 1970), 421–37. Piercy participated in the civil rights and antiwar movements of the early 1960s, helped found the North American Congress on Latin America in 1966, worked in the Students for a Democratic Society in Brooklyn, and then turned to organizing consciousness-raising groups within the burgeoning New York City radical feminist scene in the late 1960s.

3. On radical feminism, see Alice Echols, *Daring to Be Bad: Radical Feminism in America, 1967–75* (Minneapolis: University of Minnesota Press, 1989); Diane Bell and Renate Klein, eds., *Radically Speaking: Feminism Reclaimed* (London: Zed Books, 1996); Alison Jaggar, *Feminist Politics and Human Nature* (Totowa, N.J.: Rowman & Allanheld, 1983); Anne Koedt, Ellen Levine, and Anita Rapone, eds., *Radical Feminism* (New York: Quadrangle Books, 1973); Sara Evans, *Personal Politics: The Roots of Women's Liberation in the Civil Rights Movement & the New Left* (New York: Vintage, 1979).

4. On the small group format, see Jo Freeman, *The Politics of Women's Liberation: A Case Study of an Emerging Social Movement and Its Relation to the Policy Process* (New York: David McKay, 1975).

from battered women's shelters to feminist theology. Within this diversity, radical feminists tended to share the organizational and epistemological method of consciousness raising, an ideology that rooted women's oppression (commonly held to be universal) in patriarchy (and not capitalism) and, lastly, the political strategy of tearing down patriarchal structures, not reforming them. According to much of radical feminist thought, patriarchy manifested itself most perniciously in the institution of the family and thus in women's personal lives. Radical feminists were particularly concerned with the body as the locus for personalized patriarchy from which women had to liberate themselves.

Expressing the centrality of the body in radical feminism was the call for women *to take control of their own bodies,* a credo widely applied to issues ranging from freedom of sexuality, violence against women, pornography, and rape. Most significantly the credo was used to articulate feminist stances toward reproduction and reproductive technology. Why was the language of "control" used by so many twentieth-century feminists to express their political stance toward reproduction? How precisely was taking control of one's body conceived as an act of liberation? What was it about the body that women needed liberation from? What assumptions about the relations between freedom and constraint gave form to this feminist credo?

The twentieth-century call for women to control their bodies has been historically structured by a paradox that runs through Western feminist thought: a feminist must speak as a woman, thereby invoking a supposed "natural" difference, while at the same time seeking to undermine that difference.[5] In the most contradictory articulation of this paradox, feminists spoke as women in order to make the category "woman" disappear. At the other extreme, another tradition of feminism affirmed a positive value for womanness, thereby reinstantiating the difference between male and female. Both positions, as well as a panoply in between, can be found in radical feminism.

Chained to the expression of this paradox was an association between the categories "woman" and "nature," a robust and long-lived association, as Katharine Park's essay in this volume shows. Feminists who spoke as "women" were citing, whether they intended to or not, the category's his-

5. For discussions of this paradox, see Nancy Cott, *The Grounding of Modern Feminism* (New Haven: Yale University Press, 1987); Joan Scott, *Only Paradoxes to Offer* (Cambridge: Harvard University Press, 1996); Ann Snitow, "A Gender Diary," in Marianne Hirsch and Evelyn Fox Keller, eds., *Conflicts in Feminism* (New York: Routledge, 1990), 9–43.

torical association with a corporeal capacity to bear children, a capacity typically understood to be a "natural" biological function. Since the eighteenth century, an important strand of Western feminist thought has been directly concerned with constraining "nature's" authority to determine what makes a woman. Thus, the route to freedom was theorized on a conceptual terrain that saw difference as something necessary to both deny and affirm. To this end, feminists have argued in sundry particulars and through diverse methods that the body, as the site where "nature" adheres to "woman," has to be transcended, ignored, proven irrelevant, revalued, or, as Piercy suggested, physically altered. Within Western feminism's irreducible paradox, the twentieth-century call for women to control their own bodies expressed a desire to intervene in the link between oppression and reproduction by technically redirecting, conceptually disavowing, or even physically abolishing reproductive capacities, yet in doing so radical feminists were required to speak as subjects embodying a natural difference in order to change that difference.

If the concept of *nature* has been a pivot on which this paradox teeters, what does this term "nature" connote? What have feminists understood themselves to grapple with when they constrain or invoke nature? Even when feminists argued that there is no female nature, we can still ask what was understood to inhere in the domain of nature that had to be disavowed? Or, put another way, how do different feminist positions constitute the fundamental terms of the paradox? While this paradox may have structured Western feminist thought, it was not simply the prior condition of possibility for modern feminism; the paradox itself was always in a state of becoming because its terms were always under historical revision: what "nature" meant in different historical conditions changed the paradox itself. Thus, the feminist calls for women to control their bodies played a part in revising the very terms it was built on. When using the rhetoric of "control," radical feminists were directly challenging the authority of nature by invoking their own authority over nature, and thereby authoring "nature."

To understand the discourse of control in radical feminist politics, this essay focuses on two examples drawn from the rich heterogeneity of radical feminism: Shulamith Firestone's call for ectogenesis and the Feminist Self Help Movement's practice of menstrual extraction. I have chosen these two examples, not because they are the most measured or representative radical feminist positions, but because they take the credo I am interested in, and the paradox that structures it, in extremely different directions. Further, they are both utopian and revolutionary feminist efforts that employed technology in the service of women's liberation. In tracing this slogan—that women should take control of their bodies—through divergent feminist settings, one thread of my narrative leads to an understanding of the role granted to nature in the causes of women's oppression and the path to liberation. Another narrative thread describes changes in the way the domain "nature" was presupposed by different feminist positions. What various qualities were invested in "nature"—from a force, to a biological function, to variegated individuality—and how did these diverse "natures" affect political positions?

Structuring the place "nature" holds in the feminist call for women to take control of their bodies is a series of assumptions that have governed the use of this phrase, operating as a common and unexamined internal logic, though enunciated in different historical specificities. Calls to control our bodies have assumed, first, that biology, and more specifically reproduction, is not fixed but malleable, and second, that reproductive capacities must be acted on, and not just passively accepted, because they play a role in women's oppression. That is, if nature constrains us, we must take action to constrain it. This response to nature's constraint is only one of many that could have been assumed. As Helmut Puff's essay in this volume shows, another response could have been a moral injunction to accept nature's bounds rather than go against it. Finally, there is a third self-evident grammar lurking within the history of this important credo: taking "control" is an action that constrains, curbs, or circumscribes—all actions that have their force through negation. Using technologies, from birth control to ectogenesis to menstrual extraction, were acts intended to limit nature through the management of biology.

The call for women to take control of their bodies at the same time invoked and challenged historically specific conceptions of biological bodies held within biomedicine. Thus, to better understand how ectogenesis and menstrual extraction were fashioned as feminist acts, this essay also sketches two important twentieth-century developments in the conceptualization of nature and human kinds: first, the engineering approach to biology that began in the late nineteenth century, when the modern scientific discipline of biology was itself just being institutionalized, and second, the mid-twentieth-century efforts to separate social categories of human kinds, both race and gender, from questions of biological difference and judgments about relative worth. The partitioning of the biological and the social into distinctive, and at times opposing, realms framed not only the feminist articulation of a "sex/gender system" in the 1970s, but also set the conditions of possibility for thinking about the relationship between biology, humanity, and liberation more generally. Undergirding feminist calls for women to control their bodies was the larger question of how constraining nature, either physically or conceptually, could effect human difference and social equality.

Birth Control, Population Control, and the Engineering of Life

The earliest uses by feminists of the term "control" to articulate a stance toward the body tied women's ability to control their bodies to a modernist scientific project of engineering life and a socialist hope of engineering society. The knot was made tight by Margaret Sanger, cunning leader of the American birth control movement in the first half of the twentieth century, a time when birth control was as controversial in the United States as abortion is now. As a young radical working for the Wobblies and experienced as a visiting nurse among the urban poor, Sanger became increasingly critical of the Industrial Workers of the World's narrow focus on the work place, which did little to relieve the plight of women who suffered as mothers. Beginning in 1914, with the publication of her rabble-rousing and law-breaking feminist newsletter *The Woman Rebel,* Sanger took the rhetoric of "control" from socialist calls for workers to collectively control production and applied it to contraceptive politics, introducing and coining the very term "birth control."[6] In the pages of *The Woman Rebel,* Sanger provoked prosecution by asserting that contraception was a potent and even necessary weapon against capitalism because with it women could refuse to give birth to its future drones, while at the same time liberating themselves from the slavery and danger of undesired motherhood. Women, Sanger proclaimed, "cannot be on equal footing with men until they have full and complete control of their reproduction function."[7] Yet before the 1960s, the technologies available for women to enact this control were severely limited. On a practical level, Sanger could only advise women to use diaphragms and condoms, and provide them with unreliable recipes for douches (the most common was a teaspoonful of the household cleaner Lysol) and suppositories (made of cocoa butter or gelatin with

6. Ellen Chesler, *Woman of Valor: Margaret Sanger and the Birth Control Movement* (New York: Simon & Schuster, 1992), 97. On Control in American labor movements, see David Montgomery, *Worker's Control In America* (Cambridge: Cambridge University Press, 1979).

7. Margaret Sanger, *The Woman Rebel,* July 1914, 7.

a touch of quinine, salicylic acid, and boric acid added).[8] During the hey-day of the Comstock Laws, merely publishing such practical instructions made Sanger an outlaw.

Five years later, after her political exile in England, Sanger was deeply enmeshed, both personally and intellectually, with elite Neo-Malthusian free-love circles that drastically altered the political valence of her use of the term "control."[9] Sanger fashioned her own unique form of eugenics in which she coupled the creation of a better "race" with the liberation of women's sexuality through birth control. For Sanger, birth control tamed the wild reproductive element of "sex," thereby "elevat[ing] sex into an-other sphere."[10] Sex acts, assumed heterosexual, could then become a form of willful human expression or, in Sanger's words, "self-realization." In Sanger's pre-"gender" feminism of the early twentieth century, "sex" was not strictly a biological category, but rather held within it the inter-twining of biological, social, and spiritual qualities, all of which might be affected by birth control.[11]

In Margaret Sanger's feminist eugenics, individual self-realization was joined to the engineering of humanity; through population control scien-tists were to "exercise a directing, guiding, or restraining influence" over the "blind play of instinct." Sanger's approach to reproductive politics be-came an expression of the modernist engineering approach to biology that saw life, not as something for scientists to passively observe, but rather to control and improve upon. According to the American biologist Jacques Loeb, spokesperson for this "engineering standpoint," what was important was not whether a phenomenon was normal or natural, but instead the ef-fect a given manipulation of organisms could produce. To much fanfare, Loeb applied this engineering standpoint to reproduction through the in-vention of "artificial parthenogenesis" in a sea urchin in 1899, foreshad-owing the possibility of artificial human reproduction and raising specula-tions over the power of science to rationally reconstruct the natural order

8. Joan Jensen, "The Evolution of Margaret Sanger's *Family Limitation* Pamphlet, 1914–21," *Signs* 6, no. 3 (1981): 548–67, here 562.

9. Linda Gordon, *Women's Body, Woman's Right: A Social History of Birth Control* (New York: Penguin Books, 1976); Carole McCann, *Birth Control Politics in the United States, 1916–1945* (Ithaca: Cornell University Press, 1994); Donald Pickens, *Eugenics and the Progressives* (Nashville: Vanderbilt University Press, 1968). For more on the history of control and eugen-ics as they apply to the moral authority of nature, see the essay by Mitman in this volume.

10. Margaret Sanger, *The Pivot of Civilization* (New York: Brentano, 1922), 212.

11. Bernice Hausman, "Sex before Gender: Charlotte Perkins Gilman and the Evolution-ary Paradigm of Utopia," *Feminist Studies* 24, no. 3 (1998): 489–510.

and thereby the social order.[12] The engineering standpoint became the dominant frame for the development of modern reproductive science.[13]

Like Loeb, Sanger wanted a garden, not a jungle. Birth control was no longer a revolutionary weapon in the hands of the exploited, but an "instrument of liberation" used by scientist who applied "intelligent guidance over the reproductive powers" and thereby engineered social change.[14] "Birth control" encapsulated in two small words Sanger's vision of the scientific management of population and races, a means to restrain the "choking human undergrowth" that "threatens to overrun the whole garden of humanity."[15] By constraining the "primordial force" of sex, redirecting it from blind chaos toward loving sexual expression, the practice of "control" would liberate women's humanity from biological slavery and "remodel the race."[16] Sanger's vision of science-as-civilizer was not simply rhetoric. As founder of the American Birth Control League (later renamed the Planned Parenthood Federation), which set up clinics as models for the medical administration and scientific investigation of birth control, as an influential architect of the international population control movement, and as fundraiser for research into hormonal contraceptives, Sanger institutionalized her project to change what it meant to live in a sexed body. Through her tangled feminist, racist, and engineering approach, Sanger's work set boundaries for reproductive politics that later radical feminists would both reinstantiate and tear down.

Two Distinct Domains and the Location of Malleability

Fundamental to Margaret Sanger's project was the belief that different humans embodied different levels of worth. By midcentury, however, following the crimes committed under the banner of eugenics in World War II and during African American civil rights struggles against racial segregation, the assumption of a deterministic relationship between hu-

---

12. Philip Pauly, *Controlling Life: Jacques Loeb and the Engineering Ideal in Biology* (New York: Oxford University Press, 1987). On the history of science fiction considerations of ectogenesis in the early twentieth century, see Susan Merrill Squier, *Babies in Bottles: Twentieth-Century Visions of Reproductive Technologies* (New York: Rutgers University Press, 1994).

13. Adele Clarke, *Disciplining Reproduction: Modernity, American Life Sciences, and the Problems of Sex* (Berkeley: University of California Press, 1998).

14. Sanger, *The Pivot of Civilization* (note 10), 239, 13.

15. Ibid., 265.

16. Ibid., 229, 261.

man nature and social status was increasingly questioned. A typical anti-racist intellectual position was to limit the significance of "nature" by stressing the importance of "culture," as if the two terms balanced in a zero-sum game. This strategic move of defining nature against an opposite that encompasses all it is not is as old as the concept of "nature" itself, as this volume shows. Moreover, in this dichotomy, what was shared among humans was more likely to be understood as natural, and what was different as cultural. Such a dichotomy further functioned to provide something to work against, with the goal of bringing about its collapse; it presented a choice to be made between two alternatives, it called on a person to take sides. At stake in these efforts to parcel out phenomena between the natural and the social domains was the question of malleability. What was fixed about humans, and what could be changed? What role did human biology play in making race or sex, and thus what role could human biology play in impeding or instigating social equality?

Two documents written in Paris in the years after World War II—both of which were to have profound impacts on U.S. politics—were representative of a discourse that sought to diminish the relevance of naturalized human differences, both sexed and raced. The first, the UNESCO "Statement on Race" (1950), written by a committee of anthropologists and sociologists assisted by geneticists and other life scientists, concluded that though humans certainly varied, their "genetic differences are not important in determining the social and cultural differences."[17] The other text was Simone de Beauvoir's *The Second Sex* (1949), a book explicitly influenced by the discussions about race preceding the UNESCO statement.[18] She took as a model for her project Gunnar Myrdal's *An American Dilemma: The Negro Problem and Modern Democracy* (1944) and drew on contemporary debates about race, which, in her words, "no longer admit the existence of unchangeable fixed entities that determine given characteristics, such as those ascribed to woman, the Jew, or the Negro."[19] Her

17. Ashley Montagu, ed., *Statement on Race: An Annotated Elaboration and Exposition of the Four Statements on Race Issued by the United Nations Educational, Scientific, and Cultural Organization* (Westport, Conn.: Greenwood Press, 1972), 10.

18. On the importance of Beauvoir's argument for later gender theory, see Judith Butler, "Gendering the Body: Beauvoir's Philosophical Contribution," in Ann Garry and Marilyn Pearsall, eds., *Women, Knowledge, and Reality: Explorations in Feminist Philosophy* (Boston: Unwin Hyman, 1989), 253–62.

19. Simone de Beauvoir, *The Second Sex*, trans. H. M. Parshley (New York: Vintage, 1972), xvi; Gunnar Myrdal, *An American Dilemma: The Negro Problem and Modern Democracy* (New York: Harper & Brothers, 1944).

most basic argument was captured in her pithy statement that "one is not born, but rather becomes a woman."[20] Yet well before her book expressed this famous sentence, Beauvoir first engaged in an extensive, though less noticed, account of female biology.[21]

Published in France just five years after women attained the right to vote and when birth control and abortion were illegal yet illicitly practiced, *The Second Sex* began with an elaborate account of female reproductive anatomy that, far from minimizing the role of biology, portrayed it as enslaving women: female biology placed the interests of the species ahead of those of the individual woman. In her chapter titled "Destiny: The Data of Biology," written in the distanced and authoritative tone that characterized the book as a whole (and that was reinforced by the publisher's choice of English translator, the zoologist H. M. Parshley), Beauvoir worked her way successively up the animal kingdom, from the reproductive biology of amoeba to insects, fish, birds, and finally mammals. For Beauvoir, human reproduction was just one form of animality that happened to have two sexes. In humans, this mammalian two-sex reproductive system took a heavy toll on women, for "in truth menstruation is a burden," "gestation is a fatiguing task," and "childbirth itself is painful and dangerous," creating for women a "servitude imposed by her female nature."[22] Though Beauvoir claimed that understanding the facts of female biology was essential to examining the condition of women, she went on to assert, "I deny that they establish for her a fixed and inevitable destiny. They are insufficient for setting up a hierarchy of the sexes; they fail to explain why woman is the Other; they do not condemn her to remain in this subordinate role forever."[23] Thus, she tempered biology's significance with historical materialism and existentialism:

> [H]umanity is not an animal species, it is a historical reality. Human society is antiphysis—in a sense it is against nature; it does not passively submit to the presence of nature but rather takes over the control of nature on its own behalf.[24]

20. Beauvoir, *The Second Sex* (note 19), 301.

21. On the circumstances surrounding the production of the book, see Deirdre Bair, *Simone de Beauvoir* (New York: Simon and Schuster, 1990).

22. Beauvoir, *The Second Sex* (note 19), 31, 33, 34, 35.

23. Ibid., 36.

24. Ibid., 58.

In short, what made one human was history, not biology. Biology alone could not determine the fate of humans, for humans were historical creatures who demonstrated their humanity through their ability to control and not blindly submit to nature. Thus, the female body was not a fixed thing, but a historical and existential "situation" that could be changed through social change.[25]

Birth control was not given a prominent place in *The Second Sex*. Nonetheless, in its brief mention of the subject, Beauvoir maintained that the evolution of technological advancements in childbirth, artificial insemination, and birth control has provided woman with the capacity to "emancipat[e] herself from nature," and "gain mastery of her own body."[26] When left alone "nature," figured as animal procreation, was both fixed and oppressive. However, nature could be overcome mechanically through birth control, itself a product of human historical development.

Feminists and antiracists were far from the only scholars trying to detach biology from human difference. In 1955, John Money, writing with Joan and John Hampson, lifted the term "gender" from philology and placed it into psychiatry in the first published distinction between a domain of biological "sex" and socially conditioned "gender."[27] Based on his work creating medical intervention protocols for ambiguously genitaled newborns, Money argued, first, that biology does not always adhere to the strict difference between male and female that American culture desired, and, second, that whether one thought of oneself as a boy or girl not a matter of biological sex, a point of particular significance to the feminists who took up his term. The acquisition of gender, according to Money, was not open-ended but a psychological process set within the narrow window of the first eighteen months of life. Once set, gender could not be altered without severe trauma. Money's treatment protocols, therefore, insisted on quick medical and surgical interventions so that ambiguously genitaled infants could be physically recognized as male or female, and thus reared to bring their sex and gender into harmony with each other.

Following Money's work, "gender" quickly became a subject of sci-

25. Here she drew on the phenomenology of her friend Maurice Merleau-Ponty.

26. Beauvoir, *The Second Sex* (note 19), 136.

27. John Money, Joan Hampson, and John Hampson, "Hermaphroditism: Recommendations concerning Assignment of Sex, Change of Sex, and Psychologic Management," *Bulletin Johns Hopkins Hospital* 97 (1955): 284–300; John Money, "Hermaphroditism, Gender and Precocity in Hyperadrenocorticism: Psychological Findings," *Bulletin of Johns Hopkins Hospital* 96 (1955): 253–64.

entific study. Gender identity clinics proliferated, most prominently at Johns Hopkins and UCLA, in which the sex of infants and transsexuals were surgically and hormonally altered, behavior modification treatments developed, and tests to measure gender engineered.[28] What was happening to "sex" as "gender" began to attain an ontological purchase? The term "sex" was no longer understood to imply both physical and social qualities, but was now strictly used to denote the ambit of "biology," a domain that was outside of, even prior to, the social processes of gender. "Sex" became a biological phenomenon of hormones, chromosomes, and genitals to be addressed by the injections of endocrinologists, the karyotyping of geneticists, and the scalpels of surgeons. "Gender" became a phenomenon of socialization, development, and identity for the study of sociologists, psychologists, and psychiatrists. Thus, one could now think of two distinct domains that did not necessarily correspond with each other and worked by different processes. The flesh of sex was profoundly alterable, even in adult life, through medical intervention, while the psychological state of gender was fixed early in life.

Academic feminists soon adopted the term "gender," though at first with an uncritical assessment of the research it came out of and its relationship to informed consent and human experimentation that later intersex activists and feminists would critique.[29] In 1969, the New York radical feminist Kate Millett referred to this burgeoning "gender identity" literature in her best-selling book *Sexual Politics*.[30] Yet, it was not until feminist anthropologist Gayle Rubin described a "sex/gender system" in 1975 that "gender" was enthusiastically embraced by feminists who reconceived it in terms of cultural relativism.[31] "Gender" provided a locus for conceptual-

28. P. Burke, *Gender Shock: Exploding the Myths of Male and Female* (New York: Doubleday, 1996); John Money and Anke Ehrhardt, *Man & Woman, Boy & Girl* (Baltimore: Johns Hopkins University Press, 1972); Robert Stoller, "A Contribution to the Study of Gender Identity," *International Journal of Psycho-Analysis* 45 (1964): 220–26; Robert Stoller, *Sex and Gender: On the Development of Masculinity and Femininity* (London: Hogarth Press, 1968).

29. Suzanne Kessler, *Lessons from The Intersexed* (New Brunswick: Rutgers University Press, 1998); Alice Domurat Dreger, *Hermaphrodites and the Medical Invention of Sex* (Cambridge: Harvard University Press, 1998); Anne Fausto-Sterling, *Sexing the Body: Gender Politics and the Construction of Sexuality* (New York: Basic Books, 2000); Bernice Hausman, *Changing Sex: Transsexualism, Technology, and the Idea of Gender* (Durham: Duke University Press, 1995).

30. Kate Millett, *Sexual Politics* (Garden City, N.Y.: Doubleday, 1970).

31. Gayle Rubin, "The Traffic in Women: Notes toward a Political Economy of Sex," in Rayna Reiter, ed., *Toward an Anthropology of Women* (New York: Monthly Review Press, 1975), 157–210. For an earlier use, see Ann Oakley, *Sex, Gender & Society* (London: Maurice Temple Smith, 1972). On some attempts to write the history of the term "gender," see Joan Scott, "Gender: A Useful Category of Historical Analysis," in her *Gender and the Politics of History*

izing the malleability of "woman" or "man" by analyzing these roles as socially produced phenomena. In the tradition of Margaret Mead and the Boasian school of anthropology, which argued that so-called racial differences were cultural rather than inherent, feminist cultural anthropologists of the 1970s began marking as "gender" those attributes of women that differed among cultures.[32] While gender was a site of human plasticity, sex as biology tended to be invoked as an explanation for human universals, which often included the supposed universal existence of patriarchy.

The fixed, inherent, and universal qualities that the domain of sex affiliated with biology in the sex/gender system rubbed uneasily with the material conditions that were simultaneously and dramatically altering the possibilities for shaping reproduction. As Money's work demonstrated, it was now possible to profoundly manipulate the organs, hormones, and genes attributed to biological sex. Childbirth was, by 1952 (the year the translated *Second Sex* arrived on North American shores), a medically managed event, especially among the white middle class. In 1961, the first oral contraceptive, the "Pill," hit the U.S. market, joined in 1964 by the Lippes Loop, the first FDA-approved intrauterine device, which together provided women with more effective birth control. What did the call for women to control their own bodies mean in light of the reproductive technologies that seemed to be flooding from laboratories and into women's lives, from birth control and abortion here and now, to the potential future technologies of in vitro fertilization and ectogenesis? What role could nature play in women's liberation when it was not only separable from the social category "woman," but also seen as pliable rather than fixed?

(New York: Columbia University Press, 1988), 28–50; Linda Nicholson, "Interpreting *Gender*," *Signs* 20, no. 1 (1994): 79–105; Ann Oakley, "A Brief History of Gender," in Ann Oakley and Juliet Mitchell, eds., *Who's Afraid of Feminism? Seeing through the Backlash* (New York: New Press, 1997); Donna Haraway, "'Gender' for a Marxist Dictionary: The Sexual Politics of a Word," in her *Simians, Cyborgs, and Women: The Reinvention of Nature* (New York: Routledge, 1991), 127–48.

32. For example, Sherry Ortner, "Is Female to Male as Nature Is to Culture?" *Feminist Studies* 1 (1972): 5–31; Michelle Zimbalist Rosaldo and Louise Lamphere, eds., *Woman, Culture, and Society* (Stanford: Standford University Press, 1974); Carol MacCormack and Marilyn Strathern, eds., *Nature, Culture and Gender* (Cambridge: Cambridge University Press, 1980); Sherry Ortner and Harriet Whitehead, ed., *Sexual Meanings: The Cultural Construction of Gender and Sexuality* (Cambridge: Cambridge University Press, 1981). For a commentary on these efforts, see Michelle Zimbalist Rosaldo, "The Use and Abuse of Anthropology: Reflections on Feminism and Cross-Cultural Understanding," *Signs* 5, no. 3 (1980): 389–417; Louise Michelle Newman, *White Women's Rights* (New York: Oxford University Press, 1999).

Changing the Organization of Nature

If patriarchy was universal, rooted in the biological two-sex division of reproduction, yet socially changeable, could feminists use science to liberate
women by removing reproduction from bodies? The possibility of ectogenesis seemed to offer reality to the fantasy of separating out sex from
gender. Ectogenesis, thus, brought together the engineering approach to
reproduction and the midcentury desire to separate biology from hu-
mankinds. In the late 1960s a handful of East Coast radical feminists idealized ecotogenesis as a solution that blocked the possibility of sexism by removing its biological referent from women's bodies. As the most vocal
proponent of this line of thought, Shulamith Firestone wrote her widely
read book *The Dialectic of Sex: The Case for Feminist Revolution* (1970) as a
manifesto for ectogenesis. Its first demand for the postrevolutionary society was the *"freeing of women from the tyranny of their reproductive biology by
every means available, and the diffusion of the childbearing and childrearing role to
the society as a whole, men as well as women."* [33] Firestone was not satisfied
with taming sex as Sanger wanted or modifying the physical presentation
of sex as Money had, but instead she sought to bring an end to sexed bodies entirely.

Firestone was a high-profile, militant leader within the radical feminist scene. The number of radical feminist groups she was instrumental in
founding are both a testimonial to her importance in the movement and
her own difficulties within it. She helped to start the first radical feminist
cell, the Westside group in Chicago, which formed in reaction to the sexism at the 1967 National Conference of New Politics, a meeting that
brought the New Left together with the Black Panthers in a who's who
of radicalism. Within a month, Firestone moved to New York City and
founded New York Radical Women, which first promoted the method of
consciousness raising. From there she founded New York Radical Feminists in 1969 and the Redstockings, whose name revealed their traffic with
socialism, three months later. [34]

Written in a voice of ecstatic prophecy modeled on Marx and Engels's
"Manifesto for the Communist Party," Firestone's book called for a feminist revolution explicitly based on a modified historical materialism that
found its ultimate cause to be not economic relations, but the "first divi-

---

33. Shulamith Firestone, *The Dialectic of Sex: The Case for Feminist Revolution* (New York:
Bantam Books, 1970), 206.

34. On Firestone's role in radical feminism, see Echols, *Daring to be Bad* (note 3).

sion of labor," the "dialectic of sex: the division of society into two distinct biological classes for procreative reproduction, and the struggles of these classes with one another."[35] Even racism, she pompously argued, was ultimately a manifestation of the dialectic of sex. Directly building on Beauvoir, yet rejecting her existentialism in favor of historical materialism, Firestone proclaimed that women's oppression was indeed biologically derived—without birth control women were at the "continual mercy of their biology"—yet humans were historical creatures who could escape from nature by controlling it.[36] Unlike the modernist Japanese political theorist Maruyama Masao in Julia Thomas's essay, who saw liberation in an autonomy from the physical self gained through intellectual transcendence, Firestone's route to liberation comes through the technological alteration of the flesh. Reiterating Engels's prophecy that revolution would put nature under the "dominion and control" of man, Firestone believed that a feminist revolution required the technological mastery of reproduction, which she believed was not only possible but close at hand.[37]

Firestone argued, explicitly paraphrasing Marx and Engels, that just as the elimination of the economic classes required a temporary dictatorship of the proletariat and their seizure of the means of production,

> so to assure the elimination of sexual classes requires the revolt of the underclass (women) and the seizure of the control of reproduction: not only the full restoration to women of ownership of their own bodies, but also their (temporary) seizure of control of human fertility—the new population biology as well as all the social institutions of childbearing and childrearing.[38]

Firestone called for nothing less than the end of "woman" as a natural category through technology. The feminist revolution would bring about not just the end of patriarchal culture, but also a shift in "the very organization of nature."[39]

As with Sanger, Firestone's socialism was a form of progressivism that

35. Firestone, *Dialectic of Sex* (note 33), 12.

36. Ibid., 8.

37. Friedrich Engels, *The Origin of the Family, Private Property, and the State,* trans. Eleanor Burke Leacock (New York: International Publishes, 1975).

38. Firestone, *Dialectic of Sex* (note 33), 11.

39. Ibid., 1.

saw an ever-improving science as setting preconditions for an evolution from animality to full humanity. The text presented her as a prophetic witness to the unfolding of these sweeping historical changes, from the beginning of humanity to the eve of the revolution, which she summarized in elaborate diagrams. According to Firestone, in the true revolution science will "free humanity from the tyranny of its biology" by replacing childbirth with artificial reproduction, working with "cybernation," and overcoming ecological imbalance with population control.[40] Thus, Firestone, like Sanger, sought to ground an idea of progress scientifically and to engineer an improved humanity based on racist white middle-class assumptions. For Firestone, the reproductive technologies entering the market place or percolating in the lab during the late 1960s were harbingers that "a certain level of evolution had been reached."[41] The revolution was not brought about, but occurred by virtue of the underlying force of the dialectic of sex. " [I]ndeed," Firestone proclaimed, "the situation is beginning to *demand* such a revolution."[42] Not only did Piercy join her in considering the prospects for ectogenesis, but so too did the radical feminist group "The Feminists: A Political Organization to Annihilate Sex Roles," founded in 1968 by former New York NOW president Ti-Grace Atkinson. Their manifesto proclaimed that not only love and the family should be eliminated, but also that "extra-uterine means of reproduction should be developed."[43] Firestone and her allies did not flinch at prescribing a new nonsexed body for all. Science would engineer artificial reproduction, giving women ownership of their now nonreproductive bodies, and thereby creating a new society without sex.

Taking Back Control and Menstrual Extraction

Firestone's manifesto represented an extreme position within radical feminism that sought to eliminate reproduction from bodies, a position not shared by most feminists. In contrast, the radical feminist thread of the women's health movement, the so-called feminist self-help movement, explicitly rejected the coercive prescriptions associated with the engineering

40. Ibid, 192.
41. Ibid, 1.
42. Ibid.
43. The Feminists, "The Feminists: a Political Organization to Annihilate Sex Roles, 1969," in Anne Koedt, Ellen Levine, and Anita Rapone, eds., *Radical Feminism* (New York: Quadrangle Books, 1973), 367–78, here 376.

approach of both reproductive science and population control. For feminist self-help activists, it was not the result produced by control of their own bodies (for example, less childbearing or nonreproduction) that would liberate women, rather it was the act of exercising control itself. While radical feminists in the Northeast wrote revolutionary manifestos and even bestselling books, the radical feminist self-help movement that began in Los Angeles in 1971 advocated a hands-on approach to taking control over women's bodies, preferring the plastic speculum to the pen.

Feminists of many stripes have called for women to assume control over their own bodies, yet it was the feminist self-help movement that most extensively explored how control was to be practically exerted. The two central artifacts of the movement were the plastic speculum, used to teach women how to perform vaginal self-examinations, and the menstrual extraction kit, a manual suction device that could be used to empty the contents of the uterus. These two technologies represented the movement's political goal of "taking back turf" from patriarchal medicine and giving it to lay women who would themselves practice health care and research on their own bodies.[44] After the legalization of abortion in *Roe v. Wade* in 1973, feminist self-help activists began establishing Feminist Women's Health Centers that provided both abortion and general reproductive health care. In clinics and consciousness-raising groups, women not only learned how to use a speculum, but also how to take a pap smear, palpate for uterus position, administer a urine pregnancy test, observe their menstrual cycle, or artificially inseminate themselves. Moreover, while academic feminists increasingly focused on "gender," as opposed to biology, as their object of study, the grassroots women's health movement provided the most influential feminist reconceptualization of biology.

Far from considering biomedicine an ally in women's liberation as Sanger had, or reproductive technologies as harbingers of the revolution as Firestone had, the feminist women's health movement saw biomedicine as a means by which patriarchy controlled women through their bodies. At consciousness-raising meetings, privileged white women shared stories of being denied birth control and sterilization. Poor women and women of color who both worked at and used feminist clinics, in contrast, protested their limited access to basic medical care and incidents of coerced

---

44. "Taking back turf" was a phrase that circulated among activists, rather than in materials for public consumption. Dido Hasper, Shauna Heckert, and Eilleen Schnitger, Chico Feminist Women's Health Center, group interview, November 1999.

sterilization. Lesbian women who were activists in the movement drew attention to the heterosexual bias within gynecology and complained that doctors routinely thwarted their efforts to become artificially inseminated. Moreover, some feminists began documenting how the products of reproductive science's engineering approach had failed to become the promised panaceas. "The Pill," which Sanger had pinned her hopes on at the end of her life, was shown to have many dangerous side effects; diethylstilbestrol (DES), a synthetic form of estrogen taken orally and prescribed to prevent miscarriages during the 1940s and 1950s, caused vaginal cancer in the daughters of its users; and in 1974, the same year as the first feminist Menstrual Extraction Conference, the Federal Drug Administration acknowledged that a commonly used intrauterine device, the Dalkon Shield, was linked to septic death.[45] Instead of having faith in medical science's ability to liberate humanity, feminists in the women's health movement understood reproductive medicine to increase the exploitation of women through science and questioned the value of rationality estranged from women's lived experiences. Thus, Shelly Farber, director of the Los Angeles Feminist Women's Health Center, declared:

> Over the last century, males have managed to take over the health care of women and have kept that knowledge from us. They enter our systems with harmful chemicals. They shove devices up our uteruses. They examine us with laproscopes and colioscopes all to gain more knowledge of us than we have of ourselves and to control, in fact, to rape our bodies for their personal curiosity and enjoyment and, ultimately, for the control of our lives and the lives of the children we may or may not bear.[46]

In short, patriarchal science controlled women through their bodies and "mystified" knowledge, keeping women ignorant of themselves. Thus, the feminist self-help movement's central goal was to take back control of women's bodies from biomedicine.

The purpose of control was not to change sexed bodies in some way (as Sanger, Money, and Firestone all held), but instead to possess one's own individual body. Access to abortion, considered the minimum necessary procedure for preventing unwanted pregnancies, became the issue that

45. Barbara Seaman and Gideon Seaman, *Women and the Crisis in Sex Hormones* (New York: Bantam Books, 1977).
46. *Proceedings of the Menstrual Extraction Conference,* ed. Shelly Farber (San Francisco: Oakland Feminist Women's Health Center, 1974), 14.

calls for control emphasized. Like other radical feminists, the feminist self-help movement fought for the repeal of abortion laws, not their reform, declaring that there was no reason abortion should not be controlled by lay women.[47] Abortion was not only a necessary procedure for enacting control, it could also, so feminist self-help activists asserted, be a gentle, pleasant, and even consciousness-raising experience. Abortion clinics were nonetheless subjected to state laws and licensing requirements that insisted physicians perform the actual procedure. Menstrual extraction, in contrast, was a technique practiced within self-help groups, not clinics, and thus was entirely out of the hands of professional medicine. Medical abortions were "done *to* women, at the hands of a technician," while menstrual extraction was performed *by* a group of women on each other.[48] The availability of abortion was a necessary requirement for enacting control, yet it was by learning and even developing procedures within self-help groups that women could come closest to attaining control over their bodies.

No detail of a menstrual extraction—from what the participants wore, to how they were addressed, to the size of the cannula inserted into the cervical opening—was too minor for feminist self-help activists to turn into an anti-authoritarian act. The self-help groups that practiced menstrual extraction attempted to prefigure in microcosm the kind of health care the movement wished to bring about. Based on notions of participatory democracy from the civil rights and New Left movements, and inscribed in the practice of consciousness raising, the self-help group sought to practice health care by making the women affected by a study or procedure participants in it. Taking control, then, was enacted by all the tiny details of providing one's own reproductive health care.

The feminist self-help movement insisted, however, that menstrual extraction was not a do-it-yourself abortion. First, it was a group practice in which the woman having the extraction was in charge of the procedure and directed the actions of others.[49] She inserted the speculum herself and pumped the syringe to create a vacuum in the mason jar. Another group member would insert the cannula into the opening of the cervix, all the while narrating her actions and discussing them with the others. The contents of the uterus, either a menses or an early pregnancy, could then be

47. On the history of abortion in radical feminism, see Ninia Baehr, *Abortion without Apologies: A Radical History for the 1990s* (Boston: South End Press, 1990).

48. "Woman Controlled Abortions," *Woman Wise* 2, no. 3 (1979): 6.

49. See the instructions for menstrual extraction in Federation of Feminist Women's Health Centers, *A New View of a Woman's Body* (Los Angeles: Feminist Health Press, 1991).

suctioned out and afterward examined. Most women felt cramping during the extraction and would narrate their sensations to the group as part of the procedure. Because of menstrual extraction's questionable legality, it was performed in secret. Yet, at the same time, similar instruments were being widely used without the feminist procedure under the term "menstrual regulation" by population control programs, especially in Bangladesh.[50] Feminist self-help activists abhorred coercive population control practices because they manipulated women's bodies in the name of nationalism and racism. It was the procedure and its social conventions, not the equipment, that made menstrual extraction an act of liberation.

The feminist self-help movement also insisted that menstrual extraction was not an abortion procedure at all. Instead, it was a means for individual women to control their own menstrual cycles, whether or not a "zygote" was present. Bringing down your menses was likened to blowing your nose. Menstrual extraction was, according to its inventor and the movement's co-founder Lorraine Rothman, a tool for women to "control their reproduction," to " alleviate nuisance or unpleasant aspects of menstruation," and to " gain a greater overall sense of control over one's own body."[51] The ability to empty one's uterus, proclaimed Rothman,

> places each woman in active control of her period. We no longer wait passively for our monthly visitation. We no longer wait for the first days' cramping to pass. We no longer wait the five to seven days for the whole process to stop. We no longer accept the denigrating myths that a woman's monthly period incapacitates her for several days . . . our normal biological functions are not to be used against us. We choose to have or to not have when, where, and how.[52]

50. For overviews of menstrual regulation in Bangladesh, though from a population-control slant, see Ruth Dixon-Mueller, "Innovations in Reproductive Health Care: Menstrual Regulation Policies and Programs in Bangladesh," *Studies in Family Planning* 19, no. 3 (1988): 129–40; Nancy Piet-Pelon, "Menstrual Regulation Impact on Reproductive Health in Bangladesh: A Literature Review," Population Council, 1997. Population control in Bangladesh has been the focus of much feminist interest, including FINRAGE (Feminist International Network of Resistance to Reproductive and Genetic Engineering). There are a variety of feminist positions within Bangladesh on the subject. Santi Rozario, "The Feminist Debate on Reproductive Rights and Contraception in Bangladesh," *Re/productions* 1 (1998); Farida Akhter, "Reproductive Rights: A Critique from the Realities of Bangladeshi Women," *Re/productions* 1 (1998).
   51. *Proceedings of the Menstrual Extraction Conference* (note 46), 6.
   52. Ibid., 12.

What precisely, then, did having "control" over our menses mean? If acts to reclaim bodies were not the domesticating of a wild biology nor a generalized prescription for a new kind of body, neither were they simple promotions of a return to nature, for feminist self-helpers did not romanticize the experience of unwanted pregnancy created by the illegality of abortion and contraception. While childbirth activists and midwives often called for more "natural" childbirth in their efforts to liberate it from medicine, the feminist self-help movement rarely saw their actions as more "natural," but instead called for medicine to be practiced another way—by women themselves.[53] And while many self-helpers advocated "natural" remedies—usually herbal or made from everyday ingredients—for treating common conditions, even these treatments were intended to place women as active participants in the condition of their bodies. Control of one's body was idealized as an active possession of oneself. With menstrual extraction, the right to "choose" to be or not be pregnant, commonly advocated by abortion activists, was extended to the "when, where, and how" of biological functions more generally. In the early 1970s, some feminist self-health activists practiced menstrual extraction so often that they went years without a full period.[54] Thus, "control" within the feminist self-help movement meant taking an active role in managing one's body, thereby asserting possession of it.

Unlike Sanger, Beauvoir, or Firestone, the feminist self-help movement did not develop a general program for circumscribing the role of female biology in women's oppression. Instead, it was concerned with challenging the generalized and abstract accounts of female biology by recharting bodies as instances of *variability*. Through their regimes of health care involving sometimes daily self-examination, women were trained to observe their bodies in lush detail: the pinkish color of the cervix with or without reddish hues, the moisture or dryness of the vaginal canal, the sweet or musky smell of secretions, the look of the curly or toothy flesh of a hymen. A woman might even taste the sticky residue left on the speculum once its was removed. The fine-grain detail of observation marked each woman's anatomy as unique, often likened by self-helpers to the individuality of human faces. In self-help groups, committed participants were

53. Paula Treichler, "Feminism, Medicine, and the Meaning of Childbirth," in Mary Jacobus, Evelyn Fox Keller, and Sally Shuttleworth, eds., *Body Politics: Women and the Discourses of Science* (New York: Routledge, 1990), 113–38.

54. Carol Downer, interview by author, Los Angeles, California, 24 October 1999.

encouraged to take daily observations on their own, perhaps including a quick sketch of what they saw in a journal or calendar that they could later puzzle over with the group. These chronological traces could be assembled into a portrait of minute change over time, further expanding the topography of variation.[55] Rather than comparing themselves to an abstract, universalized norm (as one might find in a medical textbook), the feminist self-help movement relied on comparisons within small groups of women and with each woman's own chronicity. This schooled attention to slight variations in anatomical detail produced a topography through which the feminist self-help movement remapped what was "natural" about women's bodies.[56]

Apprehending bodies as instances of anatomical individualism reinforced the feminist self-help movement's understanding of where "control" should lie—with the individual. Marked as individual acts, rather than as the precondition for liberating women as a class, menstrual extraction was rarely, if ever, considered a tool that would overturn capitalism or patriarchy. Rather the politics of individual control fashioned women as autonomous political actors within a liberal American tradition that held private ownership as the precondition of individual autonomy.[57] Taking back control for self-helpers was a matter of asserting possession of one's own body through actions that placed one in an active relationship to reproduction. Unlike Sanger or Firestone, there was no single feminist prescription—limiting births or removing birth from the body—that all women should follow. Nor was there a universalized female body that explained women's oppression. Instead, the body was where patriarchy attempted to control women through diverse means. Thus the ways that bodies constrained women were also diverse. Only the individual woman could decide how she should use technologies to intervene in her own situation and create the conditions of her liberation.

This practical use of science did not seek a return to the "natural" as opposed to the "artificial," but instead sought to delineate different ways of exercising control that empowered women, or alternately disempow-

---

55. For more on how regimes of attention constitute the authority of nature, see the essay by Daston in this volume.

56. Federation of Feminist Women's Health Centers, *A New View of a Woman's Body* (note 49).

57. On the liberal discourse of property in feminist abortion rhetoric, see Rosalind Petchesky, "Reproductive Freedom: Beyond 'A Woman's Right to Choose,'" *Signs* 5, no. 4 (1980): 663–85.

ered women, to arrange childbearing around their own needs. The feminist self-help movement directed the credo of control in a new direction, and at the same time attached to it another set of historical baggage and was fashioned out of a privileged social location. Control was both confined by liberal notions of individualism and viewed in terms of scientific acts, albeit dramatically modified.

After the Nature of Nature

With the conservative political climate of the Reagan and Bush administrations, U.S. feminist reproductive politics dramatically shifted away from its utopian tendencies of the 1970s. Within radical feminism, Firestone's manifesto for ectogenesis appeared naively optimistic. In 1979, a group of international feminists formed FINRAGE, whose members became the most vocal critics of reproductive technology, envisioning dark futures of gynocide, breeding brothels, and global eugenics.[58] This other strand of radical feminism argued that since reproductive technologies remained in the hands of patriarchy, the best feminist stance was to try to prevent their distribution and foster alternative, often ecofeminist, values. Meanwhile, with the rise of a violent anti-abortion movement among the religious Right, the energy of activists in Feminist Women's Health Centers was by necessity focused on keeping their abortion clinics running rather than the procedural experimentation that characterized their politics in the 1970s. The feminist call for women to control their own bodies became even more tightly linked to defending abortion access and liberal feminist discourse of choice. The credo of control has become ubiquitous, imprinted globally on the feminist face of the international population control industry's promotional literature. Population control in its most recent form connects the control of women's production not only to their liberation, but also to the well-being of national economic development and the state of the global environment.

58. Reproductive technology dystopias were envisioned by several members of the radical feminist organization FINRET (Feminist International Network on the New Reproductive Technologies), later called FINRAGE. See Robyn Rowland, "Reproductive Technologies: The Final Solution to the Woman Question?" Genoveffa Corea, "Egg-Snatchers," and Julie Murphy, "Egg Farming and Women's Future," in Rita Arditti, Renate Duelli Klein, and Shelley Minden, eds., *Test-Tube Women: What Future Motherhood?* (London: Pandora Press, 1984), 356–69, 37–51, and 68–75. For a useful, though stereotyping, account of FINRAGE, see Nancy Lublin, *Pandora's Box: Feminism Confronts Reproductive Technology* (Lanham: Rowan & Littlefield, 1998).

This essay's genealogy of "control" within feminist politics raises questions about how the relationship between biology and freedom is conceived now. In academic feminism, the sex/gender system has clearly been the most influential tool for thinking about the relationship between "nature" and "woman." The term "gender" has had so much success that it has become a synonym for "sex" in everyday speech and even bureaucratic forms. Yet, as gender has achieved this purchase in our cosmologies, what has happened to "sex"? Gender, as the site of human plasticity, increasingly contained more phenomena. "Sex," which connoted natural, inherent, and universal qualities, became increasingly empty in the face of the profound biomedical malleability of bodies and increased valuing of human diversity in the late twentieth century. By 1990, the feminist philosopher Judith Butler's collapsing of sex/gender in her book *Gender Trouble* struck a nerve within academia by arguing that nature and bodies were not prior to social processes, but rather were constituted by them. Sex no longer sustained its ontological purchase and was swallowed by gender.[59] How then does liberation relate to bodies when sex has been evacuated? Where does escaping from "nature" leave one after nature has been disavowed? And how might feminists actualize reproductive bodies without reference to a discredited culture/nature divide? Solving the riddle of the nature of women's nature, for so long a central question in feminism, is no longer an adequate question around which to organize feminist body politics and its relationship to science, nation, and power.

This essay drew attention to the call for control as an alternative to the "sex/gender" conceptual tradition within feminism for grappling with the question of the relationship between "nature" and liberation. It is perhaps not surprising that the word "gender" was rarely used in the feminist self-help literature of the 1970s and 1980s. Emphasizing their practical hands-on ideology, the feminist self-help movement remained agnostic about the metaphysical question of how "natural" women were, even though this question had permeated academic feminist thought. This critical history of the call for control suggests that the "nature," on which the paradox of Western feminism has balanced, is not at all a self-evident and secure domain in the post-sex/gender, postcolonial, global capitalist world of reproductive technologies—neither conceptually nor materially. The contemporary population control industry, for example, has interpolated a

59. Judith Butler, *Gender Trouble: Feminism and the Subversion of Identity* (New York: Routledge, 1990).

new global female subject, who, for the sake of the planet, is enjoined to facilitate the control of her reproduction. The contemporary development argument that controlling birth rates leads to economic progress in developing nations has striking parallels with a feminist project of controlling the body in order to bring about liberation. By thinking about freedom in terms of restraining the natural impulse to reproduction, the credo of control has tended to narrow the relationship between biology and freedom to acts of prevention, removal, limitation, and discipline. Control became a matter of negation, of saying no.

Cracking the conceptual casing marked by our skins puts into question the habit of delineating sexed bodies as an extractable and isolatable site of liberation. If the question of how to exercise control in the name of greater freedom remains unexamined, it is easy to miss that the question is itself predicated on first marking out a domain imbued with the need of being changed by virtue of that control. What is gained and lost in letting the body—or nature or reproduction or sex—be that domain?

PART THREE

Boundaries

Paleolithic hominids, plantation slaves, native weeds—all nomads in the taxonomy of premodern, modern, and postmodern science. But nature's authority has not always been located in transience. According to Aristotle, an object's natural state was rest. Within a Darwinian worldview, the morality play depends on place to begin or end nature's evolutionary narrative. In short, nature is not a place but always in need of placement. It is from this boundedness in place and ease of categorization that much of nature's authority derives.

The rhetorical appeal to nature's authority relies among other things on its ability to provide general categories for the particular. But invoking nature involves more than a move to establish abstract categories. Such categories must be fixed or pinned down concretely to particular places. The body has served as one such concrete locus of nature. For the West, sexual and racial differences have been especially strong organizers of the natural and social order since at least the eighteenth century. Sex and race became grounded in biology as modern science privileged the physical over what was taken to be the intellectual and cultural. The body—seemingly stable, ahistorical, and sexed—became the epistemic foundation for prescriptive claims about social divisions of labor, power, and privilege. The emphasis on exposing nature as a social construction easily loses sight of the fact that the imposition of a natural order is no mere word play: the consequences of invoking a reified Nature are only all too real. This section explores the politics of such categorizations.

To invoke Nature in this way is to establish an allegedly disinterested view from nowhere, to pronounce a claim to scientific objectivity. The effacement of the one speaking on Nature's behalf is among the most powerful effects of the appeal to Nature. Nature becomes an abstract entity and the main agent of the speaker's self-fulfilling prophecies (as the essays in part 2 illustrate). Invoking Nature's authority can therefore be described as a technique of self-formation, of centering: the politics of place are a politics of creating an "us"—of setting up camp, to continue the nomadic imagery. This first-person plural materializes out of the social formations in which it is intended to exert authority. The formations may be as different as a literate sixteenth-century lay vernacular readership (see Groebner, essay 14), revolutionary nineteenth-century Chinese intellectuals (see Fan, essay 16) and British eighteenth- and American twentieth-century medical professionals (see Schiebinger, essay 15, and Mitman, essay 17). It is from such positions that boundaries between inside and outside are drawn; benevolent interior and inhospitable exterior. They allow, for example, the association of a weed with immigrants of unreliable behavior and its relabelling as a "native" benign plant protecting precious national ground against soil erosion. Once an inside is established, the threatening yet strangely familiar invaders from outer, wild spaces are never far away. Simultaneously, the successfully centered encampments provide justifications for the speaker's own expansion into other natural outer regions—unpopulated, unstructured territories awaiting, as it were, to be modified and improved.

First implemented as an abstract and inevitably governing principle beyond the individual's control, Nature can then, in a second step, be located or placed as other people's limited choice. If some people have more nature in (or, as we will see, literally on) their bodies than others, these qualities are up for valorization or disdain. For early modern Europeans, increasing human traffic fueled by mercantile capitalism and the slave trade destabilized place as a fixed referent where administrative officials could look to establish individual identity. The body became a natural landscape upon which the signs of nature's authority might be read (Groebner). While the body became a spokesperson for nature, it never became completely decoupled from the geographic coordinates that marked it at its birth. Enlightenment *philosophes* and politicians, attempting to wipe away the injustices of the *ancien régime,* where European social standing was determined largely by birth, sought to ground a just society on what they considered a higher authority—the laws of nature. Natural law (as distinct from positive law of nations) was held to be immutable, given either by God or inherent in the material universe. The interchangeability of bodies in the Enlightenment may have been consistent with a naturalistic discourse of universal human rights, but the distant outposts of colonial governments encountered new lands, new bodies, and new administrative challenges (Schiebinger).

So, however universal and abstract Nature may be presented as an overall governing principle, the enforcement of her lessons is accompanied by the rise of powerful local interpreters. Their work is centered on one question: How "us" is it? We will therefore encounter molecular geneticists who sought to homogenize and purify the genealogical landscape to make our paleontologic origins meet our boldest aspirations for the future (see Proctor, essay 18). They will be joined by body experts for the qualification and valorization of differences (Groebner). Is the physiological information gained from African slaves, prostitutes, and laborers valid when applied to their superiors' blood and uteruses (Schiebinger)? What kind of *qi* concentrated on the banks of the Yellow River to create Chineseness (Fan)? How can toxic emissions from chemical plants be distinguished from botanical ones (Mitman)?

Encampment is a political act. But when the boundaries of the camp become too rigid, when the systems of taxonomies become stuck in a deadlock of mutually exclusive categories, Nature moves on—for an elsewhere always utopian in character. Every promised land has to be described as a place to which its inhabitants once "naturally" belonged (Fan). Nature, however, cannot always be relied on as an authoritative voice in this return to the "native." Nature too had its own share of wanderers who did not easily map onto the moralized spaces of the country and the city that characterized the early twentieth-century American landscape (Mitman). And what of *Homo sapiens* own ancestral lineage in the larger stage of world history and politics? Africa, central Asia, or multiple sites of origin have figured centrally in the evolutionary migration narratives of *Homo sapiens* out of an Edenic past into the future promised land, where language, culture, and settlement spawned civilization (Proctor). What is at stake in the controversies over the very places where the first human progenitors alleged to set up camp?

Nomads, it seems, are always on their way home.

14

Complexio/Complexion: Categorizing
Individual Natures, 1250–1600

Valentin Groebner

In an anecdote widely used and retold between the thirteenth and six-teenth centuries, pupils of Hippocrates brought the portrait of their fa-mous teacher to Physionomyas (or Philemon, in other accounts), the sup-posed founder of the discipline of physiognomy. Asked for his opinion about the man in the picture based on the outward signs of his appearance, Physionomyas/Philemon replies: "He is a wrangler, lecherous and rude." The young men angrily rebut his claims and return to Hippocrates, who explains that the diagnosis was indeed very correct; but that (I quote here from an English version of 1528) "reason in me ouercometh and ruleth the vyces of my complexion."[1] A German version of the tale published in 1536 nicely differentiates between *physionomey* (the portrait) and *physi-onomi* (the art of deciphering the signs of the body), the latter enabling Philemon to read, in Hippocrates' words, "the true tendencies of my na-ture [*die neiglichkeit meiner natur*]" in the first.[2]

Whatever our notions of the individual's triumphant or destructive

The following is part of a larger project on identity papers and the practices of identification in the Renaissance. Research and writing of this chapter was supported by a grant of the ATHENA-program of the Swiss National Foundation.

1. Quoting the version of Robert Copland, *The Secrete of Secretes of Arystotle* (1528), reprinted in M. A. Manzaloui, ed., *Secretum Secretorum: Nine English Versions* (Oxford: Early English Text Society Publications 276, 1977), 378–79. For a fourteenth-century version in which the story serves as an introduction to the text, see ibid., 11. Other versions of the story, including Cicero's in his *Tusculanes*, attribute it to Socrates and the Egyptian physiognomist Zopyrus. See Richard Förster, *Scriptores physiognomici*, vol. 1 (Leipzig: Hahn, 1894), vii.

2. (Bartolomäus Cocles), *Physionomi vnd chiromanci: Eyn newes complexionbüchlein* (Strass-burg: Cammerlander, 1536; reprint, Hannover: Th. Schaefer, 1980), f. 2r.

abilities for transforming, self-fashioning, and self-inventing, common language use neatly differentiates such metamorphoses from what is supposed to be a stable set of inherent, inborn qualities: a person's "nature." Whether opposed to other slightly intimidating abstract concepts such as "culture" (or "nurture") or highlighted alone as somebody's "true" nature as contrary to disguise, dissimulation, or good manners, nature is generally understood as the fixed, unchangeable basis of a person's physical self, described as a set of aptitudes and inclinations determining his or her abilities. "Natural" categories used in the politics of "place" form the basis for defining and enforcing its boundaries.

In this essay, I explore the background and the uses of some Renaissance notions of the nature of individuals and the boundaries they were placed within, with particular attention to the frameworks of visualization that surrounded and shaped these notions. By what outward signs could the "nature" of an individual be recognized, and by what categories were these signs linked to a person's body and its physical qualities, embodied or incorporated in the literal sense? For this, I focus on the notion Hippocrates uses in our anecdote and describe its journey from the medieval centers of Scholastic learning in Italy and Paris to the sixteenth-century printing presses north of the Alps. Derived from learned Renaissance discourse, it has, to our days, a certain prominence in administrative protocols of a person's outward appearance, a piece of history of science in the identity documents in our pockets and in the passports we need for crossing boundaries. What can "complexion" mean?

Dropping the word into the AltaVista search engine is generously rewarded by 46,715 web pages. "A Beautiful Complexion with Advanced Skincare products," promises the first one, "Advanced Skincare with Oriental Blossom Cleanser and other Skin Care products that really care for your complexion." Whereas "Color of skin or dark complexion cannot alter nature's frame," states the second, set up by the African Methodist Episcopal Church, "skin may differ but ability dwells in black and white the same." Both modern notions highlighted here—race and cosmetics—point to an underlying principle of reading human nature on the skin, on bodily surfaces. They also point to the protocols of physical appearances.[3] It is in this framework—skin and signs, interaction with the body and classifications of (exotic) bodies—that I trace the changing meanings of

---

3. On the *longue durée* of related metaphors, see Jean Michel Massing, "From Greek Proverb to Soap Advert: Washing the Ethiopian", in *Journal of the Warburg and Courtauld Institute* 58 (1995): 180–201.

the term "complexion." Boundaries come in here on several levels. According to what categories were boundaries drawn between invisible and visible qualities of and on human bodies? In what rhetorical contexts were constrasts between individual and collective visual markers defined and enforced? And how did the notion of "complexion," originally designating in medieval texts the balanced proportion of Galenic humors, come to be used as a term for a person's bodily appearance and skin color, crucial for describing the differing bodily natures of inhabitants of Africa and the Americas from the second half of the sixteenth century onward?

Identity Papers

Cosmetics and race aside, individual nature in Renaissance discourse was usually and commonly defined as nonlikeness. "Resemblance," Montaigne writes in his essay "On Experience" in 1588, "does not make things so much alike as difference makes them unlike." For him, it is dissimilarity that forms the basic principle of creation: there are not even two eggs that are completely identical, he ponders, nor the carefully smoothed and whitened backs of cards used in games, let alone humans. "Nature," he adds in the 1592 version of the passage, "has committed herself to make nothing separate that was not different."[4] This "ingenious mixture on the part of Nature" is what constitutes human existence. "If our faces were not similar, we could not distinguish man from beast: if they were not dissimilar, we could not distinguish man from man."[5]

When medieval and Renaissance treatises celebrated a personified (female) Nature for her abundant fertility and remarkable diversity, they pointed to the fact that Nature showed infinite variations in the bodily appearance of mankind. Nature was of course capable of whimsically producing monozygote twins[6]—perfect doubles, albeit only in very rare cases—and more generally physical likenesses of differing degrees between parents and children and among siblings. But however members of the same family or clan might resemble each other, they wore distinc-

---

4. Michel de Montaigne, *Essays,* ed. Donald Frame (Princeton: Princeton University Press, 1965), 815.

5. Ibid., 819.

6. Luigi Gedda, *Twins in History and Science,* trans. M. Milani-Comparetti (Springfield, Ill.: Thomas, 1961); on the different uses of the notion of "nature's perfect doubles," see Hillel Schwartz, *Culture of the Copy* (New York: Zone Books 1996), 19–48. My Basel colleague Claudius Sieber is currently preparing a major study on the interpretations of twins in medieval legal theory and philosophy.

tive individual features characterizing them as unmistakably individual in physique, speech, habits. This individuality was described and registered in very different frameworks—theological, medical, and legal. With what categories were persons described to be identified? How was a person's individual appearance described in the Middle Ages and the Renaissance precisely to "distinguish man from man"?

Between the end of the fourteenth and the end of the seventeenth centuries, papers of identification in different forms and kinds became compulsory throughout Europe—as *passaporti* for soldiers on leave, as *billets de santé* or *bollete di sanità* for possibly plague-infected travelers, as *lettres de conduit* for merchants and diplomats.[7] Identity documents, warrants of apprehension, and similar official descriptions of a person's individual appearance with their detailed and increasingly formalized accounts of clothes, scars, marks, skin color, and bodily signs formed nodal points in a network of disciplines during this period. With the increasing influence of Roman law, legal procedures diminished the role of oath helpers (who attested to the credibility of the accused) and emphasized the importance of eyewitnesses, focusing on outward appearances and visual evidence. As criminal court procedures came to favor visibly verifiable facts, problems of identification played an increasingly prominent role in legal practice and juridical literature. Renaissance learned legal discourse is abundant with parallels to the *cause célèbre* of Martin Guerre. The emerging role of forensic medicine shaped the position of legal and medical experts in the gaze of the law that could penetrate the layers of deception and dissimulation, thereby recognizing a person's "true nature." The texts embodying these tendencies blended fragments of learned tradition with actual administrative information, as did the famous *relazioni* of Italian ambassadors that carefully combined physiognomic allusions to a person's "nature" with detailed descriptions of individual bodies and faces.[8] Nature's moral authority could seamlessly merge into an administrative one.

But if the category Nature was (and is) a resource for producing generalizations, it did (and does) so using a formula implying constraint. Every appeal to the authority of Nature was based on the idea of Nature as given,

7. Daniel Nordman, "Sauf-conduits et passeports, en France, à la Renaissance", in *Voyager à la Renaissance,* Jean Ceard and Jean-Claude Margolin, eds. (Paris: Seuil, 1987), 145–58.

8. For a more detailed account of the project, see my "Describing the Person, Reading the Signs in Medieval and Renaissance Europe: Identity Papers, Vested Figures, and the Limits of Identification 1400–1600," in *Documenting Individual Identity,* ed. Jane Caplan and John Torpey (Princeton: Princeton University Press, 2001), 15–27.

or, to put it differently, as a principle of restricted choice from the individual's side. Nature was always presented as something from which neither the speaker nor the audience she or he addressed could distance themselves. At the same time, Nature was understood as an inherent quality that might be disguised, dissimulated, masked (all techniques of the surface, "cosmetical" ones, as it were) but could not, by definition, be altered in a profound and irreversible way. Is this what complexion stands for?

Physical Qualities

The Galenic texts constituting the doxa of medical learning from the thirteenth century onward taught that all living creatures were composed of hot, cold, wet, and dry qualities. Galen's theoretical treatises—especially his *De Complexionibus* and the *Tegni*—and their medieval commentators used the term *complexio* to describe different kinds of qualitative mixtures and the causes leading to changes of the balance among these qualities. They emphasized above all that the balanced *complexio* was a relative conception. Each creature had its proper mixture of primary qualities: The man, the dog, the lion, the bee were not of the same *complexio,* they stated, even if they were in perfect health, but each had the proper *complexio* suitable to its nature. Not the humors as such but their qualities (active or passive) and their combinations were the vital factors.[9] Nature was both the nature of the individual and the nature of the species, and both had their own complexional range. As a person's behavior, appearance, aptitude, and moral stature were perceived to be intricately connected to his or her physical, humoral, "natural" condition, the term *complexio* was thus employed for a relatively wide range of meanings. It could signify a combination of qualities (cold and dry, hot and moist, cold and moist, and so on) as well a predominant humor (black bile, yellow bile, blood, or phlegm). It also designated an individual's permanent disposition or temperament determined by a governing humor (melancholic, sanguinic, and so on) as

9. R. J. Durling, ed., *Burgundio de Pisa's Translation of Galen's "De complexionibus"* (Berlin: Duncker & Humblot, 1976), 30f. and 37f. For the circulation of Galenic texts, their commentaries, and their uses in university curricula of medicine from the late thirteenth century century on, see Nancy Siraisi, *Taddeo Alderotti and His Pupils* (Princeton: Princeton University Press, 1981), 47, 66, 65, 99–105, and 107; Katharine Park, *Doctors and Medicine in Early Renaissance Florence* (Princeton: Princeton University Press, 1985), 193, 198–220, and 245f.; and Per-Gunnar Ottoson, *Scholastic Medicine and Philosophy: A study of Commentaries on Galen's "Tegni"* (Naples: Bibliopolis, 1984), 132ff.

well as sudden changes in skin color that reflected shifts in temperament—blushing being the classic example. *Complexio* was also held to be age- and sex-linked, men being warmer and drier than women, aged persons colder than adolescents.[10] It involved physical characteristics like the temperature of certain organs or body parts and tactile qualities like the firmness or softness of flesh. Yet it applied as well to a patient's sexual activity and to his or her emotional states. All of these were concerns of the physician because they indicated "change in the *complexio* of the body," as a fourteenth-century Florentine practician put it.[11] An individual's bad complexion *(malicia complexionis)* could thus be targeted as the cause for his or her passions, excessive appetites, and the like. As any disease was understood as a disturbance to the body's *complexio,* and as any particular drug or medicine had its own *complexio* or combination of qualities, the cures prescribed aimed to restore the patient's individual *complexio* back to normal, brought about by another shift in the balance of qualities.[12] Rather than understanding health or illness as bodily qualities, the patient's body was thus treated as literally "being" or embodying health or illness as the material result of subtle interactions.

Scholastic medicine in the thirteenth and fourteenth centuries included complicated philosophical debates in Aristotelian and Averroist terms about perfection or imperfections of elements, of substance and form, about the persistence of elements, and about mixtures as separate entities. The use of the term *complexio* in these writings, however, signified much more a type of analysis than sets of firmly defined hypotheses on causes and effects. Medical writers reminded their readers that the term should be used in practice only for purposes of comparison, not as an absolute.[13] But however elusive the practical applications of the all-embracing theory of *complexio* were, the term provided medieval learned medicine with flexible and satisfying general explanations of many kinds of physical, physiological, and even psychological changes. It figured prominently in the flowering genre of medical *consilia* describing the individual constitution of the patient, usually opening with a detailed description of the patient's per-

10. Joan Cadden, *Meanings of Sex Difference in the Middle Ages, Medicine, Science and Culture* (Cambridge: Cambridge University Press 1993), 170–74.

11. Turisanus, *Plusquam commentum in parvam Galeni artem Turisani Florentini . . .* Venetiis, 1557, f. 12r, quoted in Siraisi, *Taddeo Alderotti* (note 9), 227.

12. Siraisi, *Taddeo Alderotti* (note 9), 130, 139, 204, 257–59.

13. Ottosson, *Scholastic Medicine* (note 9), 142 ff., 146, 153; Siraisi, *Taddeo Alderotti* (note 9), 76–77, 158–60. On definitions of *complexio,* see ibid., 257 f.; and Luke Demaître, "Scholasticism in Compendia of Practical Medicine 1250–1450," *Manuscripta* 20 (1976): 81–95.

sonal *complexio* as the basis for any successful diagnosis and treatment.[14] Not only did each individual human being have his or her own *complexio innata,* the parts of the body themselves—brain, liver, heart, and testicles being the most important—each had *complexiones* of their own that, crucial for any successful medical diagnosis, had to be determined by the doctor's touch and gaze. Jacopo da Forli, teaching in Florence in the 1350s and 1360s, referred at length to Galen's and Avicenna's opinions about human hair being of a warm and humid *complexio,* the brain of a cold and wet one. But the hair cannot be treated as completely determined by the brain, Jacopo then objected. The women in Florence usually had fair hair, whereas those of Padua were dark. Should one therefore state, he polemically asked, that their brains were of a different *complexio* as well? However venerable the teachings of the ancients, he concluded, they have to be supplemented by more and more detailed descriptions of signs indicating individual complexions—bodily signs, of course.[15]

Catalogues of these signs were from the beginning an important of part of the genre, not only in influential and widespread medical handbooks like Niccolò de Falcucci's *Sermones medicinales* or Michele Savanarola's *Practica maior,* but also in more specialized treatises. A thirteenth-century *Liber complexionum,* of which there are numerous fourteenth- and early-fifteenth-century copies, set up long and detailed catalogues where to look for what. After dividing complexions into temperate and intemperate, the author went on to show how the complexion of the brain can be known by the size of the head, its shape, the hair, its actions, the superfluities expelled from it, its temperature, and by signs in the eyes. He provided similar detailed catalogues for the liver, heart, lungs, and, at last, for the body as a whole, whose complexion was to be found from touch—softness or firmness of the flesh—as well as from close scrutiny of the patient's skin, hair, figure, and *actiones.*[16]

As Joan Cadden's article in this volume beautifully makes clear, no firm boundaries separated learned Latin Scholastic discourse from courtly vernacular culture in the Middle Ages. Notions and materials from the one

14. Chiara Crisciani, "L'individuale nella medicina tra Medioevo e Umanesimo: I 'Consilia,'" in Roberto Cardini and Mariangela Regolosi, eds., *Umanesimo e Medicina: Il problema dell' individuale* (Rome: Bulzoni, 1996), 1–32; Karl Sudhoff, "Eine Diätregel für einen Bischof," *Archiv für Geschichte der Medizin* 14 (1923): 184–86; Siraisi, *Taddeo Alderotti* (note 9), 27f.

15. Ottosson, *Scholastic Medicine* (note 9), 214.

16. Niccòlo Falcucci, *Sermones medicinales* (Papiae 1481/1484); Michele Savonarola, *Practica maior* (Venetiis 1559); Lynn Thorndike, "De complexionibus," *Isis* 49 (1958): 398–408, here 401.

realm were transferred and introduced into the other and thereby changed headings, context, and meaning. Learned as well as popular writings from the fourteenth century on described a person's *complexio* not only as a theoretical, abstract concept but, simultaneously, as a term designating the existing composition of actual substances—a mixture of existing, physically present liquids in the body. Astrology was an important element in the medieval discourse about a person's individual constitution. Medical prognostications had to be integrated in a system of causal relationships between the upper and lower worlds, where the qualities of different planets—hot and dry Mars, cold and wet Venus—corresponded with the complexion of a patient's disease, body, and body parts. A *complexio sanguinea,* for example, would therefore be dominated by blood and air, wet and hot elements that determined the bearer's physiological and psychological disposition, making him or her particularly sensitive to the influence of certain planets and stars.[17] The use of these categories for describing personal qualities, affinities, and effects was not limited to medical and astrological writings in the strict sense. "Of his complexion he was sanguyn," Chaucer described in 1386 one of the protagonists in his tales. To summarize Saturn's influence on those born under him, Gower's *Confessio amantis* from 1393 called his complexion *colde*.[18]

However different in their interpretations, all these texts agreed that the exact description of a person's *complexio* and his or her body parts must form the basis of any statement of his or her qualities. Simultaneously, they referred to the infinite variety of possible complexions. The incipit of the *Liber complexionum* read: "The variety of complexions follows the variety of its causes."[19] Nature was thus defined as the generous fertile principle causing the variety and, simultaneously, as its result, a person's individual

17. Roger K. French, "Astrology in Medical Practice," in L. Garcia-Ballester et al., eds. *Practical Medicine from Salerno to the Black Death* (Cambridge: Cambridge University, 1994), 60 – 87, esp. 54; Wolf-Dieter Müller-Jahnke, *Astrologisch-magische Theorie und Praxis in der Heilkunde der frühen Neuzeit* (Stuttgart: Sudhoffs Archiv, Supplement, 1985); Siraisi, *Taddeo Alderotti* (note 9), 139–45, 173 f., 183; W. Seyffert, "Ein Komplexionentext einer Leipziger Inkunabel," *Archiv fuer Geschichte der Medizin* 20 (1928): 272–99, 372–89; Raymond Klibansky, Erwin Panofsky, and Fritz Saxl, *Saturn und Melancholie* (Frankfurt am Main: Suhrkamp, 1990), 115–16. On the debates on astrology between the thirteenth and the end of the fifteenth century, see Laura Smoller, *History, Prophecy, and the Stars: The Christian Astrology of Pierre d'Ailly* (Princeton: Princeton University Press, 1994), 25–42.

18. *Oxford English Dictionary,* vol. 3, 2nd ed. (Oxford: Clarendon Press, 1989), 613; Hans Kurath and Sherman Kuhn, eds., *Middle English Dictionary,* vol. 2 (Ann Arbor: University of Michigan Press, 1959), 468 f.

19. Quoted in Thorndike, "De complexionibus" (note 16), 141.

character, discernible through his or her appearance, features, and signs. Being so strongly tied to nature's variety, the term *complexio* signified paradoxically both an immutable and a mutable characteristic. On the one hand, it was considered as almost identical or identical with the essential determining and active quality that made each complexioned thing (species, compound, organ, or individual human being) what it was. On the other, the complexion of human beings was seen as highly mutable, changing with age, emotions, with the impact of disease and medication, and even with changes of the weather. A young physician from Prato described in 1408 in a letter to a friend his dilemma taking over a patient from a (famous) elder doctor who refused to give him informations about the case. "I did not know," he writes, "what remedies he prescribed or what the patient's condition was, or how his nature and accidents reacted to the treatment. For our nature can change not just over two months but in a day, as you see all the time." [20]

Nature as infinite and ever-changing diversity was a problem to the physician as well as to the judge and the administrator whose authorities rested on their power and abilities to taxonomize. It is in this context that Montaigne, in his essay quoted above, scoffs at the ever-differing shapes and names of our illnesses given to them by medical experts. "They make a description of our diseases like that of a town crier proclaiming a lost horse or dog," he ponders. "Such-and-such a coat, such-a-such a height. But present it to him (the lost animal to the owner), and he does not know it for all that." This is not just irony. In the description of his own physical inclinations Montaigne stresses their variability. "The best of my own bodily complexions are that I am flexible and not stubborn," he writes two pages further on, "some of my inclinations are more proper, more common and more agreeable than others, but with very little effort I can turn away from them and adopt an opposite style." [21] Given nature's infinite variety, even the most intricate taxonomies must produce misrecognitions, Montaigne implies (himself a student of jurisprudence and an experienced administrator), and even more so when confronted with people's subjective patterns of perception. The general descriptions enforced by bodies of systematized knowledge and learning do not necessarily match the cate-

20. Park, *Doctors* (note 9), 116.

21. Montaigne, *Essays* (note 4), 827, 829. I am grateful to Stephen Greenblatt, who drew my attention to this passage. I have slightly altered the wording: cf. Montaigne, *Essais,* ed. Albert Thibaudet (Paris Gallimard, 1950), 1216: "La meilleure de mes complexions corporelles c'est d'estre flexible et peu opinastre."

gories we use to describe our diseases, our animals, and our neighbors, he argues—let alone ourselves.

Inclinations and Appearances

But how, then, could the outward features of a person be read, described, and interpreted? Medieval treatises on complexions show numerous connections and references to physiognomic literature. A wide range of medical writers from Pietro d'Abano (in his *Compilatio physionomiae* from 1295) to Michele Savonarola (in his *Speculum physiognomiae* from 1455/1460) had written on the subject.[22] Together with other Arab sources like Rhazes's *Liber ad Almansorem,* the medieval learned tradition drew most notably from the pseudo-Aristotelian *Secretum Secretorum,* a Latin translation of the ninth-century *Kitab Sirr al-asrar,* of Syrian origin. The *Sirr* purports to be an epistle from Aristotle to Alexander about the arts of government and statecraft, dealing with the ruler's education, the care for his body, court rituals, and physiognomics. Translated in the twelfth century into Latin and reworked into a dialogue between the master philosopher and the king, it quickly became one of the most important political texts in the medieval West. Under the new title *Secretum* it became an influential source for the whole genre of medieval "Mirrors of princes" up to Erasmus's *Institutio principis.* It was translated into all major European vernaculars; more than six hundred manuscripts survive today.[23]

The anecdote about Hippocrates and Philemon with which I began this essay figured prominently in the different versions of the *Secretum.* It was usually accompanied by a reminder to the reader to bear in mind the power of human will over one's inborn nature, recommending that any outward signs be read as mere hints to a person's "natural" (that is, hidden) dispositions and inclinations.[24] But in fact it served as a paradoxical affirmation of the extensive catalogues of alleged visual signals for lechery, sod-

22. Graziella Vescovini, "L'individuale nella medicina tra medioevo e umanesimo: La fisognomia di Michele Savonarola," in Cardini and Regolosi, eds., *Umanesimo* (note 14), 63–87; and Jole Agrimi, "Fisiognomia e 'scolastica,'" in Agostino Paravicini Bagliani, ed., *I discorsi del corpo* (Turnhout: Brepols (=Micrologus 1), 1993), 235–72.

23. Michele Grignaschi, "La diffusion du Secretum Secretorum (Sirr-al-Asrâr) dans l'Europe occidentale," *Archives d'histoire doctrinale et littéraire du Moyen Age* 48 (1981): 1–69; and Walter F. Ryan and Charles B. Schmitt, eds., *Pseudo-Aristotle, The Secret of Secrets: Sources and Influence* (London: Publications of the Warburg Institute, 1982).

24. The famous condemnation of Averroist doctrines and a number of astrological precepts by Etienne Tempier, bishop of Paris, in 1277, explicitly left room open for astrological medicine. See Smoller, *History* (note 17), 33.

omy, or treason given on the pages following the story in the *Secretum Secretorum* or in the numerous treatises on physiognomy compiled on its basis.[25] A black mark on the left hand signified the traitor; other bodily features were no less revealing. "Whan the here of the hede is playne and softe, the man is curteys and jentill, and his brain is colde," a fourteenth-century English version of the *Secretum* states. "Bygge eyes betkeneth to be envyous / unshamefast / slowe and ivnobedyent. A brode nose in the myddes is a grete speker / and a lyer." Red eyes indicate a manly, strong, and bold person, according to an anonymous compendium on complexions printed in Augsburg in 1514. Those who have eyes the color of the sky are of wicked intentions, the above-quoted German compendium from 1536 adds; those whose foreheads show wrinkles and a furrow above the nose are simple-minded, haughty, and have bad luck.[26]

I am less interested here in the mutual borrowings between late medieval and Renaissance treatises on complexions, physiognomics, and astrology than in the ways the dialectics of allegedly visualized "natural inclinations" were presented and used in different contexts. A fourteenth-century manuscript now in St. John's College, Cambridge, engages in a long recapitulation of the marks of a body of hot and humid complexion. It describes the person in question as "having little fat, straight black hair, color between white and ruddy"; he is "faithful, has a round beard, fine black eyes, two large upper teeth, looks at the earth as he walks, talks lightly but is an habitual reader, of medium stature, large face, beautiful or trim eyebrows, fond of clothes, not telling anyone his plans, and not to be fooled." To this extraordinary mix of physical features and attributed behavior it is added: "Has a black spot on one of his teeth." An early fifteenth-century manuscript in Bern offers a very similarly structured catalogue: "So the choleric are generally wrathy, in mind unsettled, fickle, unstable, in body thin and lean, swarthy with dark curly hair, rough and hirsute, hot to the touch, with a strong rapid pulse. In substance their nature is delicate and subtle; in color fiery, glowing and clear."[27]

---

25. The same rhetorical trick to prove the reliability of one's own prognostics through negative analyses of one's "true" individual nature was repeatedly performed by Girolamo Cardano. Publishing his own horoscopes containing devastating statements about his character, he assured his readers that he had overcome these negative predispositions through self-discipline and will power: an all-too-credible expert for himself and others. Anthony Grafton, *Cardano's Cosmos: The Worlds and Works of a Renaissance Astrologer* (Princeton: Princeton University Press, 1999).

26. Copland, *Secrete* (note 1), 378, 379, 381; *Phisionomi* (note 2), f. 4r, 5r.

27. Thorndike, "De complexionibus" (note 16), 403, 405.

Both passages are constructed as ideal types of certain complexions not or only very rarely to be found in real life. In the last decades of the fifteenth and the first of the sixteenth centuries, a considerable number of older physiognomic texts—from the pseudo-Aristotelian *Physiognomonica* to the *Secretum* and the treatises of Rhazes, Michael Scotus, and Pierre d'Ailly—had been printed.[28] More popular and widespread were the editions of so-called *Physiognomiae et chiromantiae compendium* or *Complexionenbüchlin,* based on the *Secretum Secretorum* and the writings of Michael Scotus, but usually (and falsely) ascribed to the Italian physician and astrologer Bartolomäus Cocles (d. 1504). Such compendia were repeatedly reprinted in Latin, German, English, and Italian versions in the first half of the sixteenth century. They offered the reader clues to detect the secret features of their enemies with the help of physiognomy, "a striking natural art" to identify "reckless disgraceful people through their bodily signs."[29]

Part of the attraction of these texts for medieval readers seems to have been their careful use and display of fragments from learned authorities (there is clearly a pleasure in lists and references at play here) combined with a new emphasis on portraits, physiognomics, and self-observation. Petrarch played a very similar game of literary quotations and allusions in his famous 1374 *Letter to Posterity,* describing himself as "though not blessed with a physique of the first order, I enjoyed the advantages of youth, sparkling eyes, the skin colour between white and dark [*inter candidum et subnigrum*]"[30] No description of a face can do without models and the protocols that developed along with them. The wording and vocabulary of the physiognomic literature reappeared in the popular descriptions of the face of Jesus as given in the alleged Letter of Publius Lentulus, which was extremely popular in the late fifteenth and early sixteenth centuries. The body and face of the Saviour, of course, showed no birthmarks or black spots; Christ was described as *makellos* or unblemished in the most literal sense. An Augsburg broadsheet published around 1512 combined a woodcut of the face of Christ by Hans Burgkmeier with the detailed physiognomic interpretation of Christ's ideal *complexio,* his nose, eyes, eye-

28. Ulrich Reisser, *Physiognomik und Ausdruckstheorie in der Renaissance* (Munich: Beiträge zur Kunstwissenschaft 69, 1997), 52f.

29. *In disem büchlin wirt erfunden von complexion der menschen* (Augsburg: Schönsperger, 1514), f. 4v; *Phisionomi* (note 2), f. 3r.

30. Francesco Petrarca, *Posteritate / Lettere ai posteri,* trans. Gianni Villani (Rome: Salerno 1990), 35f.

brows, and so on.[31] The ideal face points to that of the beholder: for the popular Franciscan Strassburg preacher and ardent moral reformer Johann Geiler von Kaisersberg, active in the last decades of the fifteenth century (d. 1511), acute and unrelenting self-observation became the key to perfection and salvation. "Only the one who scrutinizes himself to some degree and perceives his own body becomes aware of how far he is from perfection"—so goes an old topos from monastic spiritual literature. In the texts of Kaisersberg's sermons posthumously published by Johannes Pauli in 1517, however, the author urges his audience and readers to concentrate these efforts "on your own complexion [*auff die eygen complexion*]"— to take utmost care to decipher the signs of one's own nature on one's own body.[32]

Changing Colors

At the turn of the sixteenth century, the term "complexion" had detached from the framework of Galenic medicine and natural philosophy and moved into a broader sphere of the description of human bodies. Scholastic medicine had shaped *complexio* as a mode of establishing and interpreting complicated relations in a wide range of bodily phenomena—including tactile qualities (the firmness or softness of flesh, for example), temperature, and the effects of emotions, exercise, or medication—used to describe both mutable and unchangeable characteristics. The notion of "well-tempered" blood in the descriptions of seventeenth-century transfusion experiments (see Londa Schiebinger's essay in this volume) still derived its meaning from this classical set of categories. Yet the physiognomization and essentialization of the term in the vernacular literature and in the developing genre of treatises on signs narrowed its use to something more particular, determinative, individually innate, and visible—or at least traceable for the trained eye. Highlighting nature's diversity as well as nature's constraint, *complexio,* we may argue, shifted from the interior to the exterior with increasing emphasis on the skin, its colors and marks.

The use of such notions of diversity and constraint may remind us that

31. Philine Helas, "Lo 'smeraldo' smarrito, ossia il 'vero profilo di Cristo,'" in *Il volto di Cristo, Catalogo della mostra, Roma, Palazzo delle Esposizioni 2000–2001* (Milan: Electa, 2000), 215–26; Joseph Leo Koerner, *The Moment of Self-Portraiture in German Renaissance Art* (Chicago: Chicago University Press, 1993), 116.

32. Johannes Pauli, ed., *Die Broesaemlin doct. Keiserspergs uffgelesen von Frater Johannes Paulin* (Strassburg: Johann Grueninger, 1517), f. 12v, 17v, 42r.

the term "nature" itself always hovers ambiguously between the descriptive and the normative. Even on the descriptive level alone, there is an always-present tension between the Scholastic notion of the nature of the species that displays qualities inherent to the specific category, and the nature of the individual. In medieval and Renaissance texts, a number of qualities of certain groups were portrayed as problematic for the identification of the nature of their individual members. Medical theory conceptualized womens' alleged abilities of disguise and dissimulation as a function of their "cold" and "wet" nature. Similarly, gypsies, with particular reference to their dark skin, curly hair, and their Assyrian or Egyptian origins, were accused of treacherously changing their outward appearances as well as their names. Gypsies were the first group whose passports and safe conduct documents, regardless of who issued them, were declared invalid by Imperial legislation in the whole Holy Roman Empire in 1551 and again in the following decades. People of such a treacherous nature, it was implied, could only have forged identity papers.[33]

How, then, did the meaning of a person's *complexio* move from a balance of invisible interior liquids and their combination to exterior signs on skin and face? The ways skin colors were described between the fourteenth and sixteenth centuries seem to have less to do with the actual degree of reflection of light by the skin than with a complex set of notions of colors and signs informed by the theories of physiognomy and complexion outlined above. Descriptions of slaves bought and sold in Florence between 1366 and 1397 in the city's "Registro degli schiavi" (the overwhelming majority of them female, 329 out of a total 357), focus on their outward appearance in ways familiar to the readers of medical and physiognomic texts. They speak, for example, of an eighteen-year-old woman sold in July 1366 as "above medium height, olive-colored skin [*ulivigna*], with a big nose and a black mole above the nose and two scars on her left hand," another of "quasi black skin, with some marks of the left side of her nose" or "of

---

33. Reimer Gronemeyer, *Zigeuner im Spiegel früher Chroniken und Abhandlungen: Quellen vom 15. bis zum 18. Jahrhundert* (Giessen: Focus, 1987). For the Imperial legislation on gypsies, see Johann Schmauss and Heinrich Senckenberg, eds., *Neue und vollständigere Sammlung der Reichs-Abschiede,* vol. 2 (Frankfurt am Main, 1747), 609–32. On the gypsies' assumed Assyrian origin, cf. Polydorus Vergilius, *Beginnings and Discoveries (De rerum inventoribus,* 1521), trans. and ed. Beno Weiss and Louis Pérez (Niewkoop: De Graaf, 1997), 481. His account is not without tongue in cheek: Vergilius, a close friend and collaborator of Erasmus, uses his description of the treacherous tattooed gypsies for a polemic attack on mendicant orders and especially on the Antonines, who, he explains, are cheating and deceiting the faithful with begging and stealing, wearing on their chest the letter "T."

white skin, with pierced ears and a black mole on the left side of her forehead." Along with the detailed description of "natural" signs is the emphasis on marks figuratively written on the body: "With a great scar on her head by her left eyebrow, and a scar on her left cheek, by her nose"; a tattoo "like a cross, on the right finger"; "a great scar or brand, on the top of the right hand." Privately compiled notary records of sales and emancipations of such female slaves give similar descriptions of their appearances. They list their stature (small, rather small, medium, large), the shape of their faces (round, long, square), skin color (*flava* or yellow, bruna, *nigra, ulivigna, rossa,* or even *verdastro,* greenish), and provide extensive descriptions of birthmarks on hands and faces, scars and tattoos.[34] Their categories of skin colors are somewhat puzzling for modern readers. As olives take very different colors in the stages of their ripening and processing, from light brown to violet and pitch black, what could *ulivigna*—as opposed to, let's say, *bruna*—possibly mean? In other texts, descriptions of color and origin of slaves are turned into explicit puns. Alessandra Strozzi advised in one of her letters to her son Filippo in 1465 to look for a tatar slave ("qualche tartera di nazione") to buy, because they could better stand hard labor, but she admitted that "the red ones—that is, the Russians [*le rosse, cioè quelle di Rossia*] were more attractive and handsomer."[35]

The meaning of *rossa* is (deliberately?) vague here. Does it really refer to skin pigmentation? References to birthmarks, moles and signs, and to "red" or "black" as a person's characteristic color appear as well in a detailed 1464 description of the soldiers garrisoned in the Roman Castel S. Angelo. Practically all of the sixty-two soldiers, with the exception of three men—Jacobus Hungarus, Iohannes Albanesi, and Antonius Sclavus—are of Spanish, Italian, and German origin. Their descriptions are rather short, giving only name, age, and two or three distinctive marks. A certain "Johannes Scomel," *gallicus,* is described as red (*pinguis et rubeus*); another is simply presented as "Apricus Rubeus," without further specification if the indicated color referred to skin or hair. Yet the list provides its reader with very detailed descriptions of the signs these men are wearing on their

34. The "Registro degli schiavi" is edited in Ridolfo Livi, *La schiavitù domestica nei tempi di mezzo e nei moderni* (Padova: Milani, 1928), 141–217, here 146 and 149 (1366). Livi's edition is used by Iris Origo, "The Domestic Enemy: The Eastern Slaves in Renaissance Tuscany in the 14th and 15th Century," *Speculum* 30 (1955): 321–66, see esp. 337 and 333. Descriptions of runaway slaves, however, seem to concentrate less on scars and birthmarks than on the clothes the fugitives are wearing (ibid., 349).

35. Alessandra Macinghi Strozzi, *Lettere ai figliuoli,* ed. Angela Bianchini (Rome: Bulzoni, 1987), 247. The passage has a peculiar undertone: is she teasing the son for his sexual appetites?

hands, faces, and arms, of their scars, moles, and birthmarks: "with a black sign of his right jaw"; "with a sign of his small finger of the right hand"— not only refering to traces of their violent profession but echoing the detailed lists of marks from the literature of complexion and physiognomy, "with a black mole on his left hand." [36] When the list presents the German Michael de Maguntia as a man of medium height with a scar on his forehead and having a black appearance or a black face (*facie nigra*), or a certain Thomas de Trever (from Trier) having a *facie nigra,* or an Alfonsus de Salamancha as *homo nigris colore,* what notions of blackness are employed here?

Such descriptions of Reds and Blacks guarding the papal fortress in the centre of medieval Christendom in the West cast an interesting light on Renaissance notions of skin colors and bodily appearance in general. Christopher Columbus's first account in October 1492 of the inhabitants of the island of Guahani describes them as "well formed, with handsome bodies and good faces. Their hair is coarse—almost like the tail of a horse—and short. . . . Some of them paint themselves with black, and they are the color of the Canarians, neither black nor white, and some of them with red, and some of them with whatever they find." [37] The "neither black nor white" formula is clearly not the actual description of a person's appearance. If not simply a general figure of speech (Petrarcha's "*inter candidum et subnigrum*"), it might well be derived from the description of the Canary islands translated by Boccaccio from Italian to Latin— the skin color of the inhabitants of the "Happy Islands" as a borrowing of a Petrarchian metaphor. [38] Later sixteenth-century chroniclers of the European expansion to the Americas, however, seem eager to present a rather different story. Bartolomé de las Casas describes the very same encounter on a Caribbean beach in October 1492 in his *Historia de las Indias* thirty years later:

> The Indians, who witnessed these actions in great numbers, were astonished when they saw the Christians, frightened by their beards, their white-

36. Giuseppe Zippel, "Documenti per la storia del Castel Sant'Angelo IV: La guarnigione di Castel s. Angelo nel 1464," *Archivio della Società Romana di Storia Patria* 35 (1912): 196–200.

37. *The Diario of Christopher Columbus' First Voyage to America,* Oliver Dunn and James Kelley, eds. (Norman: Oklahoma University Press, 1989), 65, 67.

38. Rinaldo Caddeo, ed., *Le Navigazioni Atlantiche di Alvise di Ca da Mosto, Antoniotto Usodimare, e Niccoloso da Recco* (Milan: Edizioni Alpes, 1928); Giorgio Padoan, "Petrarcha, Boccaccio e la scoperta delle Canarie," *Italia Medioevale ed Umanistica* 7 (1964): 263–77.

ness, and their clothes; they went up to the bearded men, especially the Admiral since, by the eminence and authority of his person, and also because he was dressed in scarlet, they assumed him to be the leader, and ran their hands over the beards, marvelling at them, because they had none, and carefully inspecting the whiteness of the hands and faces.[39]

Commenting on this passage, Peter Hulme has dryly noted that, after several weeks at sea, the Spanish and Italian sailors might not have differed much at all in color from the Amerindian natives.[40] The whiteness Las Casas mentions is located precisely at those parts of the Europeans' bodies most exposed to tropical sun. I want to leave aside the importance of clothes and beards so prominent in Las Casas' account in contemporary European protocols of identification in the first half of the sixteenth century. Among other things, it was the right to wear a beard that distinguished free male laborers from slaves in sixteenth- and seventeenth-century Italy and Spain; and physiognomic compendia had much to say about the importance of "a well kept beard with thick and strong hairs" as a sign of a person of reason and "of good nature."[41]

Whereas blackness or a black appearance could be derived from moles and bodily marks, very few of the inhabitants of Renaissance Europe seem to have considered themselves to be white in any form—let alone in the all-too-familiar meaning of the term as a category signifying a central dominant "nonrace" position from which others can be seen to differ (see Londa Schiebinger's essay in this volume). On the contrary, European theories of climate and complexion handed down from the classics reserved a number of rather unflattering terms for the fair-haired and fair-skinned people to the north. Aristotle's *Politics* described them as "full of spirit" but lacking political organization, "incapable of ruling others . . . (they are)

---

39. Bartolomé de las Casa, *Historia de las Indias,* vol. 1, ed. Agustin Millares Carlo (Mexico Citya: Fondo de Cultura Económico, 1965), 202. The Hulme reference is to Peter Hulme, "Tales of Distinction: European Ethnography and the Caribbean," in Stuart B. Schwartz, ed., *Implicit Understandings: Observing, Reporting, and Reflecting on the Encounters between Europeans and Other People in the Early Modern Era,* 157–97 (Cambridge: Cambridge University Press, 1996), here 161.

40. Hulme, "Tales" (note 39), 162.

41. *Physionomi* (note 2), f. 8v. On the significance of beards for free laborers, see Raffaela Sarti, "Viaggiatrice per forza. Schiave 'turche' in Italia in età moderna," in Dinora Corsi, ed., *Altrove: Viaggi di donne dall'antichità al Novecento* (Rome: Viella, 1999), 241–96, here 274. On their role in early modern racialist discourse, see Londa Schiebinger, *Nature's Body: Gender in the Making of Modern Science* (Boston: Beacon Press, 1993), 120–25.

deficient in intelligence."[42] Widely read classical texts like Pliny's *Natural History*, Vitruvius's *Ten Books on Architecture*, and Flavius Vegetius's *Military Institutions* all agreed over the interactive relationship between the body's constitution and its environment. As the sun in the southern regions caused the moisture to evaporate, northern climates prevented this drying process. The cold northern climates therefore forced their inhabitants to be hardy, but this excess moisture also produced "sluggish minds." Albertus Magnus repeated in the thirteenth century that those who live in cold regions abounded in "blood and bodily spirit . . . their humour is thick and bodily spirit does not respond to the motion and receptivity of mental activity. Therefore, they are dull-witted and stupid."[43]

It has been repeatedly noted that in the making of the category "race" (in the sense of a group of persons or animals of common descent or origin and likeness, the term itself coined in the sixteenth century), early modern authors drew on material from both antique and medieval learned authorities to configure innate, permanent, "natural" categories suitable for the classification of humans.[44] Yet the texts they used not only portrayed Africa and Asia as both monstrous and noble, but did the same for the barbarous northern regions of Europe. Medieval humoral discourse had many negative things to say on the bodily consequences of northern climate. Cold external air clogged the body's pores, drawing heat and moisture inward and producing a particular complexion—white. Drawing on an wide range of Scholastic authorities, Jean Bodin in 1565 in his widely read *Method for the Easy Comprehension of History* urged historians to consider the temperaments of various peoples in the world so that they may write histories "drawn from nature." Southerners were cold, dry, and melancholic, he stated; northerners were hot, moist, and sanguine. In an

---

42. Quoted in Mary Floyd-Wilson, "Clime, Complexion and Degree Racialism in Early Modern England," Ph.D. diss., University of North Carolina, 1996, 31. Even less favorable is Hippocrates's account of the Scyths dwelling in northern regions: see John Friedman, *The Monstrous Races in Medieval Art and Thought* (Cambridge: Harvard University Press, 1981), 106ff.

43. Floyd-Wilson, "Clime" (note 42), 34; Clarence Glencken, *Traces on the Rhodian Shore, Nature and Culture in Western Thought from Ancient Times to the End of the Eighteenth Century* (Berkeley: University of California Press, 1967), 258, 429–60.

44. For a broader bibliographic survey, see Les Back and John Solomos, eds., *Theories of Race and Racism* (London: Routledge, 2000); and Michael Banton, *Racial Theories*, 2nd ed. (Cambridge: Cambridge University Press, 1998). I have found very useful Kim Hall, *Things of Darkness: Economies of Race and Gender in Early Modern England* (Ithaca: Cornell University Press, 1995); Peter Martin, *Schwarze Teufel, edle Mohren* (Hamburg: Junius, 1993); and Benjamin Braude, "The Sons of Noah and the Construction of Ethnic and Geographical Identities in the Medieval and Early Modern Periods," *William and Mary Quarterly* 54 (1997): 103–42.

attempt to reconcile ancient authorities with astrological and humoral models, Bodin suggested that cold, moist, phlegmatic peoples inhabited the extreme north, whereas, in general, shades of skin color varied with the latitude. "Under the tropics people are unusually black; under the pole, for the opposite reason, they are tawny in color . . . down to the sixtieth parallel, they become ruddy; thence to the forty-fifth they are white . . . to the thirtieth they became yellow, and when the yellow bile is mingled with the black, they grow greenish, until they become swarthy and deeply black under the tropics"—a specter of skin pigmentation allegedly due to nature that echoes the Italian categories we have already encountered. Such outward signs had their inward counterparts. The southerners "abound in black bile" and proved wise, weak, swarthy, and small, Bodin argues. Conversely, the hearthy, fair-skinned, white northerners had such an excess of "blood and humour" that their "mind was so weighed down (with moisture) that it hardly ever emerged." [45]

The Frenchman Bodin was not the only Renaissance writer claiming the ideal, temperate middle position "favoured by Nature" for his own countrymen. Italian, German, Spanish, German, and English authors did the same. Neither was this kind of argument reserved for Europe; seventeenth-century Chinese intellectuals and their nineteenth-century commentators established very similar rhetorical links between birthplace, skin color, and bodily features, as Fa-Ti Fan's essay in this volume reminds us. [46] Yet Bodin's spectrum of climate-induced skin colors went from white through yellow and green to black without invoking explicit moral statements. On the contrary, he concluded that certain natural modes of conduct were not altogether subject to human volition: the chastity of northerners was not to be particularly admired because it was caused by the weakness of their sexual appetites. Nor was the licentiousness of the southerners to be blamed, he added, for it was especially the mark of their melancholic complexion. The apparent virtuousness of the Germans came, he wrote, from their lack of imagination; statesmen, his lesson went, must frame their policy in accordance to these laws of nature. [47]

In this sixteenth-century polyphony of categories, the term "complexion" could serve quite different functions, as we have seen. Physiognomi-

45. Jean Bodin, *Method for the Easy Comprehension of History,* trans. Beatrice Reynolds (1566; New York: Columbia University Press, 1945), 111; Floyd-Wilson, "Clime" (note 42), 89. See also Marian Tooley, "Bodin and the Medieval Theory of Climate," *Speculum* 28 (1953): 64–83; and Schiebinger, *Nature's Body* (note 41), 126–34, 18 ff.

46. Floyd-Wilson, "Clime" (note 42), 85, 171, 175.

47. Tooley, "Bodin" (note 45), 78.

cal, astrological, and medical treatises continued to use the term to highlight individual "natural" inclinations. The afore-mentioned 1536 physiognomic *Complexionenbüchlein* ascribed to Cocles devoted long chapters to the inhabitants of Africa and Asia—yet ignored the new-found lands in the West. But apart from a short reference to the blackness of certain African tribes, its accounts of exotic India and Asia (providing, among other things, detailed descriptions of the soldiers of the legendary Christian Prester John in India wearing cross-shaped tattoes) do not mention skin pigmentation at all.

White as a skin color, however, appears only once in the *Complexionenbüchlein*. It is reserved for those conceived or born under the astrological sign of Andromeda. These people, the anonymous author wrote, have a pretty face, are lucky, and unchaste. They are usually not too interested in women but engage in sodomy; they are flatterers, servile, and "of white skin color"—and usually end their lives either bankrupt or burnt on the stake.[48] And it is only in the last chapter that the author introduces his readers to a particular people that, however, "lives under Nature's guidance only [*nach der natur leitung*]." These "Ichtiophagi Affrice" walked naked, knew of no personal property, had children, women, and all things in common, would not differentiate between right and wrong, and knew nothing of pleasure or mourning. Yet, the author adds, they lived happily and peacefully from fish and a kind of bread baked between hot stones, knowing of no scarcity or famine, eating together and merrily singing songs in dissonant melodies.[49] What immediately follows this exotic idyll is, as a conclusion to the whole compendium, an account of the barbaric origins and of the national character of the Germans, "so ingenuous in both good and bad things." Drawing on Tacitus's descriptions of their wildness and on the Germans' military prowess, the author praises their victories in innumerable battles against the infidels, their perseverance, reliability, and ambition. They supercede all other nations in artifice, mechanics, and inventions, he closes, in printing, artillery, and many other arts (*und vil ander künst*).[50] In contrast to the mild-mannered fish-eaters living under Nature's guidance and without any history, the Germans ap-

---

48. *Phisionomi* (note 2), f. 25v.

49. Ibid., f. 73r. The account seems to be based on the description of the Guinea coast by the Augsburg merchant Balthasar Springer (1505/1506). On its influence, see Beate Borowka-Clausberg, *Balthasar Sprenger und der frühneuzeitliche Reisebericht* (Munich: Iudicium, 1999).

50. *Phisionomi* (note 2), f. 74r and v.

pear as self-educating noble savages and successful engineers of their own ambition and historical progress—with no particular nature (nor skin color, we might add) attached to them.

The 1536 *Physionomi* guides its reader from deciphering the bodily signs of people's true nature to the wild nature of savages living under Nature's guidance only—and from them, bringing the message home, as it were, to their originally not less savage but ambitious, enterprising, and cultured counterparts.[51] The (German) reader is flatteringly invited to see his own achievements in overcoming nature highlighted by the sluggish innocence of the Ichtiophagi. The rhetorical scheme is identical to the one displayed in the anecdote of Hippocrates and Physionomyas/Philemon with which the same book on "Physionomi" literally opens: a paradoxical lesson on both the deciphering and the transformation of the unchangeable innateness of humans, or in Hippocrates's words, on "the true tendencies of my nature." In a kind of utopia of shifting gazes and switching definitions, both the strict normativity of Physionomyas and the flexible descriptive reading of nature by Hippocrates is offered to the reader. It could be argued that the combination of such ambiguous references on infallible "true signs" with well-placed hints on their uses for dissimulation organized most Renaissance writing on the practices of identification and classification, allowing for a particular *plaisir du texte*.[52]

Yet the closing chapter of the 1536 *Physionomi* as a sixteenth-century heterology may also be read as an example of a specific literary technique of contrast to be found in most early modern accounts on complexion. Richard Hakluyt's *Principall Navigations, Traffiques, and Discoveries of the English Nation* includes George Best's often-quoted ponderings from 1587 of on the wonders of diversity and persistency:

> I my selfe have seene an Ethiopian as blacke as cole brought unto England, who taking a faire English wife, begat a sonne in all respects as blacke as the father was . . . ; whereby it seemeth this blackness proceeded rather of some natural infection of that man, which was so strong, that neither the nature

51. To which the Frenchman Bodin in his *Method* thirty years later maliciously adds the "lack of imagination" as a natural category of Germanness, as we have seen.

52. Montaigne offers in the essay "On physiognomy," preceding the above-quoted passages, another and quite different reading of both Socrates's *complexion, au dernier poinct de vigueur,* and of interpretations of his own outward appearance by contemporaries in the strifes of civil war. Michael Taussig, "The Abduction of Montaigne," in *Defacement: Public Secrecy and the Labour of the Negative* (Stanford: Stanford University Press, 1999), 226–29.

of the clime, neither the good complexion of the mother concurring, could any thing alter.[53]

In this passage, neither a woman's "good complexion" nor the blessings of English weather can make up for the father's black "nature." Like medieval *complexiones,* early modern complexions as (facial) skin pigmentation are relational. They do not exist for themselves but come in packages of neatly preorganized juxtapositions. But the actual meanings of "white" and "black" (as well as of a couple of other colors) turn out to be much more flexible than their respective rhetorical function of establishing a system of presumably fixed co-ordinates.

I have outlined here some of the uses of the term "complexion" in its journey from medieval natural philosophy to colonial rhetoric and administration. Long before debates on human difference between polygenists and monogenists (and their echoes in twentieth-century paleontology, as Robert Proctor shows in his essay in this volume), long before the discussions of moral versus climatic causes of difference between humans, and long before the efforts of the eighteenth-century professor Blumenbach to detect blackness around the nipples and on the testicles of European underclass males as a sign of their closeness to "Ethiopians" (decribed more fully by Londa Schiebinger in the following essay), medieval writers of the thirteenth and fourteenth centuries established *complexio* as a key term in the emerging new paradigm of isolating and reading signs (humoral, medical, astrological ones) in and on human bodies. In the framework of interpretating nature as both diversity and constraint, *complexio* shifted from an abstract and highly complicated notion that included tactile qualities (firmness or softness of flesh) or differences in the temperatures of organs or limbs into a new realm of *evidentia* in the literal sense of the word. In the older reading of the term, nobody had the same complexion. When complexion stood for an invisible and internal disposition, you had to be a trained and experienced physician to describe and define it. In the worldly sphere of physiognomics and in the games of dissimulation, however, statements about a person's "true nature" had to be made at a glance. And even more so in a broader administrative context, when judging if a person's appearance fits his or her description in search warrants or identity documents.

Moreover, the interpretation of the signs of *complexio*/complexion in-

53. Quoted in Hall, *Things of Darkness* (note 44), 11.

vokes definitions of place as something constructed in relational coordinates. Where were you born? Under which constellation? In which climate? The politics of place are a politics of creating an "us." Without such placing, no successful moral, aesthetic, or administrative authority of nature can rhetorically unfold. We are back to the making of boundaries here. The changing use of *complexio,* no longer confined to the protocols of the individual but increasingly prominent in the description of classification marks of groups, illustrates the shift from an individual to a collective astrology to classify and explain other people's outward appearances. *Complexio* thus became complexion, signifying something that was purely visual, located on the outside, on the skin, on the skin of the face and its color. With nature's allegedly unchangeable signs becoming ever more visible, skin became the privileged screen on which somebody's "true" nature would be located. It is not by chance that the term "complexion" turns out to be closely linked to the descriptions of individuals in their identity documents and to the history of the passport. The efforts of placing individuals within the closed boundary of complexion may be intrinsically (and paradoxically) linked to their movement, to their abilities of crossing boundaries in space.

15

Human Experimentation in the Eighteenth Century: Natural Boundaries and Valid Testing

Londa Schiebinger

In the natural course of events, humans fall sick and die. Physicians, neighborhood herb women, colonial doctors, worried relatives, and a host of apothecaries and pharmacists concoct potions and regimen in efforts to cure and to protect life. The history of medicine bristles with attempts to find new and miraculous remedies, to work with and against nature to restore humans to health and well-being.

Finding new and effective medicines requires empirical observation of the effects of new drugs in living organisms. A perennial question for doctors, patients, and ethicists is: who will go first? On whose body will unknown and potentially dangerous drugs be tested? By whom, and for whose benefit? By what standards of the good, the just, and the valuable are these decisions to be made? Today, such questions are mediated through carefully crafted codes of patients' rights, tightly enforced procedures for informed consent, and legally approved medical protocols.[1] This essay explores how drugs were tested in the eighteenth century. It looks specifically at how human subjects were chosen for experiments and at notions

My thanks to the Alexander von Humboldt-Stiftung for supporting my research for this essay. My thanks also to Lorraine Daston for her gracious hospitality during my stay at the Max-Planck-Institut für Wissenschaftsgeschichte and to members of the Moral Authority of Nature group for their wit, wisdom, and helpful comments. This essay is drawn from my larger project on gender in the voyages of scientific discovery

1. Susan Lederer, *Subjected to Science: Human Experimentation in America before the Second World War* (Baltimore: Johns Hopkins University Press, 1995); Jay Katz, *Experimentation with Human Beings* (New York: Russell Sage Foundation, 1972); Irving Ladimer and Roger W. Newman, eds., *Clinical Investigation in Medicine* (Boston: Boston University, Law-Medicine Research Institute, 1963).

of uniformity and variability across living organisms. Did physicians imagine a natural human body that once tested held universally? Were tests done on male bodies thought to hold for female bodies (and vice versa)? Were white and black bodies considered interchangeable in this regard? Which of these distinctions were considered a product of cultural artifice and which were thought to be jealously guarded by Dame Nature herself? What role did the "moral authority of nature" play in the choice of subjects?

As we shall see, the choice of experimental subjects in the eighteenth century responded to both natural and social imperatives. In the early modern period, Europeans made exacting distinctions in social rank, allowing only persons of the highest rank to wear fine ermines or scratch with a fingernail grown for the purpose at the king's chamber door, yet in this same period physicians assumed far-reaching unity across basic human nature in the matter of drug testing. Differences of estate—and the differences in lifestyle and temperament that entailed—did not necessarily raise the notion of unbreachable physiological barriers between social classes. Drugs successfully tested on prisoners or charity patients were prescribed as "safe" for valued upper-class patients who typically paid a goodly sum for treatment. Even differences in *complexio*—formed by a peculiar configuration of internal liquids and external climatic and cosmic forces, as discussed in Valentin Groebner's essay in this volume—did not disrupt physicians' notion that generalizations concerning drug testing could be applied across large groups of humans. Apart from specific bodily differences, such as sex and age, taken by experimenters to be so basic as to have been dictated by nature, physicians assumed a certain unity of humanity, an interchangeability of bodies.

In this setting, the populations of slaves, concentrated on New World plantations, seemed a boon to European physicians. Here was a captive and controllable group available for experimentation—an ample supply of experimental subjects. Groebner characterized a shift in perceptions of race from the thirteenth to the sixteenth century as one from a highly individual notion of *complexio* to more modern notions of complexion as denoting skin color and serving as a marker of distinct human taxons. By the late eighteenth century, new questions had arisen about how deeply racial difference penetrated the body. Skin color, and the growing racism associated with it, began raising troublesome queries in naturalists' minds about the presumed unity of human nature and about which bodies could properly represent humankind. New concerns arose about whether the bodies of African slaves in the West Indies could adequately stand in for upper-class

Europeans and whether standard therapies for whites would prove effective in black bodies. By the late eighteenth century, race seemed to raise "natural" barriers—enforced by nature's own authority—to medical testing.

This essay explores, then, which bodies were considered sufficiently universal for medical testing. To some extent, the choice of subjects was simply arbitrary. As with dissection, physicians and surgeons used any bodies they could lay their hands on (perhaps legally and morally, perhaps not). But when they had a choice, they were swayed, as we shall see, by traditions inherited from the ancients (the use of prisoners, for example), by Galenic notions (for instance, that females must be part of the experimental design), by rising currents of scientific racism, legal restrictions, and much else besides. Let us investigate, then, which bodies were used for experimentation, how far the presumed unity of living organisms stretched, and where perceived bodily differences arose and came to be seen as imposing natural boundaries.

Embracing Unity: Animal Experimentation

The voyages of scientific discovery engendered controversy among early modern naturalists concerning how to order the bounty of nature being transported by the boat load into Europe. The number of identified plant species alone rose from six thousand known to Kasper Bauhin in 1623 to fifty thousand known to Georges Cuvier in 1800. Expeditions were mounted to look for new species of humans, including Carl Linnaeus's golden-eyed, nocturnal "troglodytes." Vigorous debate erupted over natural boundaries between and within the three great kingdoms of nature: were "stone mosses" (probably lichens) properly denizens of the vegetable or mineral worlds, were "erba viva" (sponges) plants or animals, were "orangutans" (the generic term in this period for the great apes) humans who had simply been deprived of education; were "Hottentots" truly the missing link between humans and brutes?[2]

The approach to questions of the unity of humankind in the world of medical testing was more practical. Medical testing is dangerous; naturalists working in Europe and its colonies prudently tested unknown foods and potential drugs on animals, not humans. Experimentalists in this regard assumed a gross unity of animal organisms, seeing humans as part of

2. See Allen Grieco, "The Social Politics of Pre-Linnean Botanical Classification," *I Tatti Studies* 4 (1991): 131–49, here 137.

the animal kingdom. Anton von Störck, physician at the city hospital in Vienna, was typical in first using an animal to test his newly developed nonsurgical cure for breast cancer. One can imagine Störck's excitement in attempting to develop a therapeutic drug for breast cancer in a period when the only other option was mastectomy. Störck's experiments in the 1760s involved feeding extracts of his wonder drug—a hemlock extract—to a little dog. It would be "criminal," he remarked in his published report, to make the first trial of this extract on a human. He gave the dog a "scruple" (1.3 grams) of the drug in a piece of meat, three times a day over the course of three days, and took it as a good sign that the dog remained healthy and eager for the food.[3]

Since ancient time, testing had begun with animals, especially testing for toxicity. Rhazes charted the effects of quicksilver on apes. Paracelsus experimented with ether-like substances on chickens. In the eighteenth century, the Italian abbot Felice Fontana employed three thousand vipers, four thousand sparrows, numerous pigeons, guinea pigs, rabbits, cats, and dogs in his experiments with snake venom.[4] Experimentalists in this period were especially keen to use "higher" animals to increase the likelihood that results could be generalized to humans. To this end even horses met their fate at the hands of naturalists. Since the Renaissance, however, the dog had emerged as the experimental animal of choice. Apes, bears, and lions, according to Renaldus Columbus, Vesalius's successor at Padua, are internally more similar to humans, but they become angry on being cut, which made vivisection difficult. Pigs were too fat and their squeals annoying.[5]

Animals were also on the front lines of blood transfusion experiments, some of the first experiments to be carried out in the 1660s by the new Royal Society in London and the Académie Royale des Sciences in Paris.[6] Locked in a struggle over priority, experimentalists in England and France

3. Anton Störck, *An Essay on the Medicinal Nature of Hemlock* (Dublin, 1760), 12–13.

4. Melvin Earles, "The Experimental Investigation of Viper Venom," *Annals of Science* 16 (1960): 255–69. Felice Fontana, *Traité sur le vénin de la vipère, sur les poisons americains, sur le laurier-cerise et sur quelques autres poisons végétaux* (Florence, 1781).

5. Roger French, *Dissection and Vivisection in the European Renaissance* (Aldershot: Ashgate, 1999), 207. See also Andreas-Holger Maehle, "The Ethical Discourse on Animal Experimentation, 1650–1900," in Andrew Wear, Johanna Geyer-Kordesch, and Roger French, eds., *Doctors and Ethics: The Earlier Historical Setting of Professional Ethics*, 203–51 (Amsterdam: Rodopi, 1993), 218, 225.

6. See Harcourt Brown, "Jean Denis and Transfusion of Blood, Paris, 1667–1668," *Isis* 39 (1948): 15–29.

had been bullish on transfusion since William Harvey's demonstration of the circulation of the blood. Physicians such as the Parisian Jean Denis had high hopes that, if techniques could be perfected, transfusion could rejuvenate the old and cure the sick by sending "fresh, well-tempered blood" directly to the affected part in the same way that "the maternal blood [flowing] into the umbilical vein of the infant" nourishes and vivifies all its parts.[7] To this end, French and English physicians transfused blood from calves to sheep, from sound dogs to mangy dogs, and from a young dog to an old dog ("which two hours after did leap and frisk").[8] In some cases, a greater unity of human and nonhuman animals was assumed than was wise. In the 1660s, lambs' and sheep's blood was transfused into humans in "large quantities" until the death of two men led the Parlement of Paris to halt these celebrated experiments.[9]

Much more could be said about animal experiments. As Johann Friedrich Gmelin emphasized in 1776, however, while experiments on animals were one of the few ways to test drugs that might prove useful to humans, "in the end, no other option remains except to experiment on the human body itself."[10]

Epistemologically Weighty Bodies

Historians have emphasized the use of the poor and displaced in human experimentation.[11] There are, however, several important exceptions: self-experimentation using the physician's own body, and popularizing risky health measures using the bodies of royals or the wellborn. Early modern Europe enjoyed a vibrant and growing culture of scientific experimentation. As part of this culture, natural philosophers began experimenting on their own bodies. The literary critic Julia Douthwaite has described this developing tradition as a form of "autobiographical empiricism," whereby credible subjects scrutinized minute effects in their own bodies and con-

7. Cited in A. D. Farr, "The First Human Blood Transfusion," *Medical History* 24 (1980): 143–62, here 151.

8. John Lowthrop, *The Philosophical Transactions and Collections,* 3 vols. (London, 1722), 3:229.

9. *Lettres à M. Moreau contre l'utilité de la transfusion* (Paris, 1667). See also Brown, "Jean Denis and Transfusion of Blood" (note 6).

10. Johann Friedrich Gmelin, *Allgemeine Geschichte der Gifte,* 3 vols. (Leipzig, 1776), 1:34.

11. William Bynum, "Reflections on the History of Human Experimentation," in Stuart Spicker et al., eds. *The Use of Human Beings in Research,* 29–46 (Dordrecht: Kluwer, 1988), 32.

veyed these as reliable data to other scientists.[12] The historian Stuart Strickland has described how natural philosophers, such as Johann Ritter, used their own bodies as "calibrated instruments" epistemologically equivalent to the voltaic columns, Leyden jars, thermometers, and other devices cluttering their laboratories. The body of the experimenter provided unique information not available through the use of other instruments, yet ideally it also became an instrument, simulating as nearly as possible inanimate objects' indifference to the "prepossessions" of errant humans.[13]

Medical research participated in this general culture of self-experimentation. And following Galen's injunction, physicians often felt obliged to test new treatments first on their own bodies. But unlike the physicists, physicians only rarely used their bodies as unique instruments, as did Albrecht von Haller, who in the course of dying recorded the effects of opium in the human body taken daily for two and a half years.[14] More commonly, the physician was simply the first human subject to try a cure. But unlike many patients, the medical autoexperimenter considered himself a "proficient" subject, able to provide reliable information.[15] The information was deemed credible: a medical expert could presumably distinguish effects relevant to the experiment from other "subjective" states of his own body and could be used, for instance, in respect to sleep, pain, or states of mind. The information was also considered "pure"—that is, gathered from a healthy body. The autoexperimenter tested for toxicity not discovered in animals and for effects in the healthy human body (which, it was assumed, could handle a dangerous drug better than a body already weakened by illness).[16]

Self-experimentation—the willingness of an experimenter to "go first"—also tended to exonerate physicians when they used the medica-

12. Julia Douthwaite, *The Wild Girl, Natural Man, and the Monster: Dangerous Experiments in the Age of Enlightenment* (Chicago: University of Chicago Press, 2002), 72; Simon Schaffer, "Self Evidence," *Critical Inquiry* 18 (1992): 327–62.

13. Johann Ritter, cited in Stuart Strickland, "The Ideology of Self-Knowledge and the Practice of Self-Experimentation" (Max-Planck-Institut für Wissenschaftsgeschichte, preprint 65, 1997), 25.

14. Albrecht von Haller, "Abhandlung über die Wirkung des Opiums auf den menschlichen Körper," *Berner Beiträge zur Geschichte der Medizin und der Naturwissenschaften* 19 (1962): 3–31.

15. Schaffer, "Self Evidence" (note 12), 336. Autoexperimenters were primarily men; see this essay below.

16. Georg Friedreich Hildebrandt, *Versuch einer philosophischen Pharmakologie* (Braunschweig, 1786).

tion, perhaps with fatal results, in others. The willingness to take a drug oneself, the Boston physician Zabdiel Boylston noted, was a measure of a physician's "faith" in the medication.[17]

By the mid eighteenth century medical self-experimenters had built in certain safety procedures.[18] Substances were first to be smelled, then touched to the skin, and finally tasted—first only with the tip of the tongue, and then, if appropriate, taken internally. These procedures can be seen in Dr. Störck's hemlock experiments. Encouraged by the benign effect of his extract on his dog, Störck took the next step and experimented on himself, and recorded that he took each morning and evening for eight days one "grain" of his hemlock extract with a cup of tea. Perceiving no ill effects, he increased the dosage to two grains. Again perceiving no "ill or unusual" effects, he felt justified "for the best reasons to try this on others."[19]

While Störck seems to have been experimenting alone in his chamber, physicians taught that experimenters should not, in fact, be alone during such experiments, for two different reasons. The first was that it is important for others to "witness"—to observe, learn from, and verify—the findings; the second was more immediately practical, namely, to aid in the event that the experimenter should fall unconscious (and also to continue to log the results of the trial).

Toward the end of the eighteenth century, self-experimentation became more systematic and organized. In efforts to overcome the idiosyncrasies of an individual experimenter's body (to calibrate his body against others), physicians and medical students tested potential drugs in groups. James Thomson, a medical doctor most likely born in Jamaica, reported that "some years ago, while at the University of Edinburgh . . . a few of us associated for the purpose of making experiments on various medicines, the active properties of which we had reason to question." According

17. Lawrence Altman, *Who Goes First: The Story of Self-Experimentation in Medicine* (New York: Random House, 1986), 12; Zabdiel Boylston, *An Historical Account of the Small-Pox Inoculated in New England upon all sorts of persons, Whites, Blacks, and of all Ages and Constitutions* (London, 1726), vi.

18. Rolf Winau, "Experimentelle Pharmakologie und Toxikologie im 18. Jahrhundert" (Mainz, Habil. Schrift, 1971); Winau, "Vom kasuistischen Behandlungsversuch zum kontrollierten klinischen Versuch," *Versuche mit Menschen in Medizin, Humanwissenschaft und Politik,* ed. Hanfried Helmchen and Rolf Winau (Berlin: Walter de Gruyter, 1986), 83–107; and Andreas-Holger Maehle, *Drugs on Trial: Experimental Pharmacology and Therapeutic Innovation in the Eighteenth Century* (Amsterdam: Rodopi, 1999).

19. Störck, *An Essay on the Medicinal Nature of Hemlock* (note 3), 12–14.

to his report, the physicians and medical students were each assigned specific drugs to test in his own healthy body. They recorded in minute detail "the state of the pulse, vomiting, dizziness, and every other circumstance." When a particular drug seemed of sufficient importance, it was taken by different individuals at the same time; the results were carefully compared. Thomson continued, "[W]e then inferred that, generally speaking, the same results will follow in a morbid state, and combat successfully certain symptoms which we wish[ed] to obviate."[20]

Physicians, then, first tested the efficacy of drugs in their own bodies. What did physicians do, though, when drugs, like emmenagogues or abortifacients, were uniquely destined for women? The Edinburgh group, because it was university based, was all male; how was this difficulty surmounted? To date, I have not found a single report of self-experimentation by a woman, though thousands of such experiments might have taken place. Nor have I come across accounts of the bodies of wives or female servants standing in for the physicians' own body. Women, many of whom prepared medicines from their kitchen gardens, no doubt experimented on their own bodies. Root or herb women must have doctored themselves when ill and observed the effects. Midwives may also have experimented on themselves or their patients, but these have all gone unrecorded. The closest such account I have uncovered is a report on a broom-seed remedy for dropsy published by an anonymous woman in 1783. She persisted in "trials of this remedy" in numerous cases, as she recorded, also with a gentleman and, significantly, on a woman with child. No doubt this anonymous researcher first took the remedy herself.[21]

A very different type of epistemologically significant bodies were those of royalty or the well born. Royal bodies in this period carried a significance beyond their own materiality. Kings in France and England were thought to have two bodies: a visible, corporeal, mortal body and an immaterial, symbolic, and sacred body on which rested royal prerogative.[22] Willingness to risk those bodies inspired trust in their subjects. To be sure, nostrums and innovative surgical procedures were well tested on people of "inferior" status before offered to royalty, but the willingness of nobility to

20. James Thomson, *A Treatise on the Diseases of Negroes, as they occur in the Island of Jamaica; with Observations on the Country Remedies* (Jamaica, 1820), 145–46.

21. *A Sovereign Remedy for the Dropsy* (London, 1783).

22. Ernst Kantorowicz, *The King's Two Bodies: A Study in Medieval Political Theology* (Princeton: Princeton University Press, 1957); Sara Melzer and Kathryn Norberg, eds., *From the Royal to the Republican Body: Incorporating the Political in Seventeenth- and Eighteenth-Century France* (Berkeley: University of California Press, 1998).

submit to a procedure did much to popularize the use of a new remedy. What was persuasive in Europe also held sway in Europe's colonies. Hans Sloane, while serving as physician to the governor of Jamaica early in his career, noted that "at first the Inhabitants would scarce trust me in the management of the least Distemper, till they had observed the good effects the European method had in the Duke's numerous family."[23]

Royals often offered their own bodies in order to popularize public health measures favored by mercantilist governments attempting to augment national wealth by producing growing and healthy populations, which would increase the production of crops and goods, fill the ranks of standing armies, and pay substantial taxes and rents.[24] A well-known example involved the introduction of smallpox inoculation into Europe. Inoculation was introduced into England from Turkey by Lady Mary Wortley Montagu—who might in this role be styled an international broker of knowledge.[25] Lady Mary learned of the procedure from an "old Greek woman" (unnamed, as was typical) while living in Adrianople, where Montagu's husband served as ambassador. Charles Maitland, Lady Mary's surgeon, put off inoculating Montagu's daughter while in Adrianople because he felt she was too young but also because he hoped to use the occasion of her inoculation "to set the first and great example to England, of the perfect safety of this practice, and especially to persons of the first rank and quality."[26] (The fanfare surrounding the inoculation of Lady Mary's six-year-old son and only male heir in Turkey had reached England even before the return of this noble family.) It was these children's successful inoculation that sparked interest within England. Two royal princesses were inoculated in 1722 without mishap (and after successful trials with the procedure on prisoners, discussed below). In his virulent attack on Maitland and his procedures, the ardent anti-inoculist William Wagstaffe was hard pressed to set aside the fact that inoculating the smallpox had been "admitted into the greatest families."[27]

23. Hans Sloane, *A Voyage to the Islands Madera, Barbadoes, Nieves, St Christophers, and Jamaica; with the Natural History . . .* , 2 vols. (London, 1707), 1:xc.

24. Guenter Risse, *Hospital Life in Enlightenment Scotland* (Cambridge: Cambridge University Press, 1986), 12.

25. Genevieve Miller, *The Adoption of Inoculation for Smallpox in England and France* (Philadelphia: University of Pennsylvania Press, 1957); Genevieve Miller, "Putting Lady Mary in Her Place," *Bulletin of the History of Medicine* 55 (1981): 3–16; Isobel Grundy, *Lady Mary Wortley Montagu* (Oxford: Oxford University Press, 1999).

26. Charles Maitland, *Mr. Maitland's Account of Inoculating the Small Pox* (London, 1722), 8.

27. William Wagstaffe, *A Letter to Dr. Freind; Shewing the Danger and Uncertainty of Inoculating the Small Pox* (London, 1722), 3.

Social Categories: Prisoners, Hospital Patients, Orphans, and Soldiers

Physicians' and royal bodies, while epistemologically weighty, were few in number. One must remember that drugs are not absolutes. Early modern physicians were acutely aware that what heals at one dosage can be a deadly poison at another. Discovering correct dosages for a wide range of patients required testing more than the bodies of a few privileged individuals. Who were to be the subjects of these larger trials? As today, where drug trials tend to overselect subjects from vulnerable populations, such as prisoners, the mentally ill, students, ethnic minorities, and welfare recipients, early modern experimentalists tended to draw subjects from the socially disadvantaged. Most came from the same groups used for dissection—prisoners, hospital patients, orphans, and soldiers—persons who had no next of kin to insist on Christian burials or, in the case of medical care, to seek out and pay for expensive cures.

From a medical point of view, there was nothing special about these bodies, except their availability. When physicians failed in treating prominent persons, they often met with severe consequences, losing patients, livelihoods, and perhaps even lives. Sir Richard Croft, for instance, shot himself after Princess Charlotte died in childbirth. Few complained, however, when things went wrong in treating the poor. William Withering, known for developing digitalis, discussed openly the practice of using the poor to test dosages of his new cure before prescribing it for his paying patients. Practicing in Birmingham in the 1770s and 1780s, Withering provided free advice to the poor one hour per day, as was customary. This practice, he remarked, "gave me an opportunity of putting my ideas into execution in a variety of cases; for the number of poor who thus applied for advice was between two and three thousand annually." Overwhelmed by so many patients, he despaired that "in this mode of prescribing . . . it will be expected that I could not be very particular, much less could I take notes of all the cases which occurred." Concerning these patients he continued, "I soon found the Foxglove to be a very powerful diuretic; but then, and for a considerable time afterward, I gave it in doses very much too large, and urged its continuance too long."[28] Withering copiously recorded his paying patients' cases, and one could argue that these patients, too, were subjects of his experiments with foxglove. Withering, however,

---

28. William Withering, *An Account of the Foxglove and its Medical Uses* (Birmingham, 1785), 2–3.

did not prescribe for his paying patients until he had refined his dosages using charity patients.

Charity patients were used freely to test all manner of cures: intravenous injection of laxatives to cure venereal disease (in men) and epileptic fits (in women), new styptics for amputation and mastectomy. Soldiers, orphans, and other wards of the state also fell prey to these types of experiments. It was condemned criminals, however, who were deemed experimentalists' most exquisite subjects. Doctor of the Medical Faculty of Paris, Denis Dodart, advocated in 1676 the use of bodies already condemned to death for the most extreme drug trials, namely, for testing remedies against poison.[29] Christian Sigismund Wolff, in a 1709 disputation at the University of Leipzig, went so far as to advocate the ancient practice of vivisecting criminals on the grounds that society would profit from the knowledge gained.[30] In the 1750s, Maupertuis, president of the Berlin Academy of Sciences, advocated the use of prisoners for such operations as the removal of kidney stones or uterine cancers, for which "neither Nature nor the Art" had provided a cure. Even with the condemned, Maupertuis, echoing Dodart, advocated that for "the sake of humanity," the physician should diminish as much as possible the pain and the peril of the procedure. Further, physicians should first perform operations on cadavers and then on animals before trying them on criminals.[31]

Maupertuis, however, went beyond most of his colleagues in this period, and thrilled at the thought of using condemned persons for purely speculative research. "Perhaps we will discover," he enthused, "the marvelous union of the soul and the body if we dare to search in the brain of a living human." Intoning again the theme of public utility, he continued, "[O]ne person is nothing compared to the human species."[32]

Not all physicians, however, considered criminal bodies reliable instruments for experimentation. Jean Denis, who performed the first blood transfusion on a human in 1667, rejected the solicitations of "divers persons of much gravity and prudence" who urged him "to beg some condemned Criminal, on whom to make the first essay." Denis feared that a condemned man might deem "transfusion a new kind of death" and

29. Denis Dodart, *Mémoires pour servir à l'histoire des plantes* (Paris, 1676), 10.

30. Christian Sigismund Wolff, *Disputatio philosophica de moralitate anatomes circa animala viva occupatae* (Leipzig, 1709), 28–40.

31. P.-L. Moreau de Maupertuis, *Lettre sur le progrès des sciences* (1752), in *Oeuvres,* vol. 2 (Hildesheim: Georg Olms Verlag, 1965), section 11, "Utilités du supplice des criminels."

32. Ibid.

that the fear of it "might cast him into faintings and other accidents" that would unfairly defame his own "great experiment."[33]

Why in societies so obsessed with class distinctions that marriage to a commoner dethroned an heir apparent were the bodies of persons of the lowliest estates considered representative of humankind in general? Medical theory generally spoke against this practice. A rich diet, coupled with abundance and sedentary leisures, was thought to form peculiarly delicate physiques in the upper classes. Johann Friedrich Blumenbach spoke to these issues in the 1770s, emphasizing, for example, that social class might influence skin color. "The face of the working man or the artisan," he wrote, "exposed to the force of the sun and the weather, differs as much from the cheeks of a delicate [European] female, as the man himself does from the dark American, and he again from the Ethiopian." Blumenbach pointed especially to the problem of anatomists generalizing about humankind from dissection because corpses were typically procured from what he described as the "lowest sort of men." These European men, he remarked, have skin around the nipples and on the testicles that comes nearer to the "blackness of the Ethiopians than to the brilliancy of the higher class of Europeans."[34]

For similar reasons William Wagstaffe objected to transferring cures from poor nations to rich nations. He argued that smallpox inoculation could not be safely transferred from Turkey into England because of what he described as the English "National Blood." Believing that smallpox attacked the body with a greater or lesser degree of ferocity according to the state of the blood at the time of infection, Wagstaffe urged that it would be impossible to "transplant to us with success, or naturalize to our advantage" a medical procedure deriving from a people [the Turks] who lived on a spare diet and in the lowest manner. The English "national" blood—even among the "meanest of our people"—is a "rich" blood, produced by the lushest diet in the world and abounding "with particles more susceptible of inflammation." Wagstaffe concluded by suggesting that the legislature might wish to take action "in order to prevent such an artificial way of depopulating a country."[35]

Although physicians of the Galenic school set great store (and often

33. J.-B. Denis, *Lettre écrite à Monsieur de Montmort* . . . (Paris, 1667); Farr, "The First" (note 7), 160.

34. Johann Friedrich Blumenbach, *On the Natural Varieties of Mankind* (1775 and 1795) trans. Thomas Bendyshe (1865; New York: Bergman, 1969), 108.

35. Wagstaffe, *Letter* (note 27), 5–8.

garnered great income) by careful attention to individual distinctions in temperament, diet, and regime as described by Groebner in his essay, experimentalists generally assumed that persons of the lower classes provided valid information about drugs and their effectiveness. In the early modern period, as today, experimental subjects were rare and much sought after. Physicians in this period were desperate to get bodies, preferably compliant bodies, for experimentation. For their part, the poor received free medical care or occasionally small monetary compensation. A poor fruit seller who dutifully took Störck's anticancer hemlock pills was rewarded with a few coins; orphans who successfully withstood smallpox inoculation were sometimes educated at state expense.

Natural Categories: Sex and Twins?

Interestingly enough, in this setting where bodies were assumed to be interchangeable across social classes, male and female bodies were kept distinct. Unlike in the twentieth century, the results of tests performed on male bodies were not automatically generalized to females. Surprisingly to modern eyes, females were a required part of experimental design in eighteenth-century medicine. As Joan Cadden has shown, Galenic medicine—still influential in the eighteenth century—taught that disease and medications often progress differently according to the age, sex, and temperament.[36] Groebner's essay also emphasizes that *complexio* was conceived as age- and sex-linked throughout the early modern period. Experiments in the eighteenth century were deliberately designed to take these factors into account.

Controlling for sex and age can best be seen in the experiments on prisoners at London's Newgate prison in 1721 to test the smallpox inoculation. Upon Montagu's return to England, experimentation began in earnest. Doctors wanted to know if the climate (England being cold and Turkey hot) affected the outcome, where in the body the inoculation had the greatest effect (the right or left arm, the thigh, or another part), whether the benefits of the operation lasted a lifetime, how the patient was to be prepared (whether bleeding or purging beforehand improved the outcome), and so forth.

Although it remains unclear who initiated the Newgate experiment,

---

36. Joan Cadden, *Meanings of Sex Difference in the Middle Ages: Medicine, Science, and Culture* (Cambridge: Cambridge University Press, 1993).

according to Hans Sloane's account, Queen Caroline "begged the lives of six condemned criminals" to undergo the experiment of inoculation "to secure her other children, and for the common good."[37] Lady Mary's surgeon Charles Maitland, now royal surgeon, was to carry out the experiment in consultation with two royal physicians, Johann Steigertahl and Sloane, the latter soon also to become president of the Royal Society. Six prisoners were selected—three women and three men matched as closely as possible for age—since it was desired to know how the operation acted in persons of "all Ages, Sexes, and different Temperaments":[38]

Anne Tompion, age 25	John Cawthery, age 25
Elizabeth Harrison, age 19	Richard Evans, age 19
Mary North, age 36	John Alcock, age 20

The experiments began at 9:00 on the morning of August 9, 1721. A German visitor reported, perhaps only to heighten the drama, that the criminals trembled when Maitland took out his lancet fearing that they were to be bled to death.

Maitland recorded the events with the immediacy of detail characteristic of experimental case histories in this period and in first person: "I made an incision in both arms and the right leg of all six."[39] While this was a "controlled" experiment, Maitland could not control John Alcock, who took it upon himself to open all his hot and itchy pustules with a pin, no doubt in search of relief. About two weeks after the operation, Maitland purged two of the men, Alcock and Cawthery, and pronounced them fit for release. He intended also to purge the women, but was prevented by their monthly periods which, he was surprised to hear, "seiz'ed them all about the same time."[40] He may have attributed this to the experiment, not knowing that females in close proximity often menstruate synchronously. On the sixth of September, all participants were pardoned by the king and council and released from Newgate prison.

---

37. Hans Sloane, "An Account of Inoculation," *Philosophical Transactions* 49 (1756): 516–20, here 517. Richard Mead also reported that the experiments were done "by order of his Sacred Majesty, both for the sake of his own family, and of his subjects" (*The Medical Works*, 3 vols. [Edinburgh, 1763], 2:145).

38. Emanuel Timonius, "An Account, or History, of the Procuring the Small Pox by Incision, or Inoculation; as it has for some time been practised at Constantinople," *Philosophical Transactions of the Royal Society of London* 29 (1714): 72–82, here 72.

39. Maitland, *Account* (note 26), 21–22.

40. Ibid., 24.

Thomas Fowler's experiment with tobacco in the 1780s on 150 persons offers another example of tests designed to include an equal mix of males and females. Fowler, physician to the infirmary in Stafford, England, found the diuretic effects of the alkaline fixed salt of tobacco, administered internally, effective against dropsie. The cure, however, depended on the correct dosage; too high a dosage might cause vertigo, nausea, or excessive purging. In the course of his testing, Fowler found that when age was held constant, sex, indeed, became the most important variable. He adjusted the dosages of his nicotine-based diuretic (to be taken daily two hours before dinner and at bed time) accordingly:

1st class, 21 cases (3 men, and 18 women) 35–60 drops.
2nd class, 57 cases (29 men and 28 women) 60–100 drops.
3rd class, 13 cases (9 men and 4 women) 100–150 drops.
4th class, 3 cases (3 men) 150–300 drops.

The extremes in this study were at one end a weak and nervous woman, Sarah Dudley, who tolerated but twenty drops of the infusion, and at the other an old man, Charles Nicols, who, accustomed to the use of tobacco, required four hundred drops. Fowler also tested his concoction on children, although not children under age five because, he wrote, "they could not so well describe the effects of so active a medicine."[41]

It is unclear when women were removed from mainstream medical testing. The growing propensity in the nineteenth century to use medical students for experimentation privileged all-male groups (characteristic of the Edinburgh University self-experiments discussed above). The military, used already in the eighteenth century, was another site of medical experimentation that was exclusively male. Prison populations, too, became more decidedly male in course of the nineteenth and twentieth centuries. But the use of males also seems to have resulted from a deliberate preference for them. Concentrated populations of women were available among prostitutes (those with three or more convictions were often used in dissection),[42] in the many lying-in hospitals founded in the eighteenth century, schools of nursing founded in the nineteenth century, and, in Catholic countries, in nunneries.

A second group of interest in medical experimentation of the eigh-

---

41. Thomas Fowler, *Medical Reports of the Effects of Tobacco* (London, 1785), 72–79.
42. *Erneurte Medicinalordnung vom 21. December 1767*, 477–78.

teenth century was twins. Though physicians did not address the point, the reasons for using twins seem similar to those that made studies of identical twins popular in the twentieth century: they ensured that the circumstances of an experiment were held as constant as possible.

The royal physician John Wreden, working in Hannover in the 1720s, conducted extensive research on smallpox inoculation; one test concerned the twin daughters of a common soldier. We do not know what incentive was given to the parents for this experiment, whether money was exchanged, for example, as was sometimes the practice. In the presence of Dr. Hugo, another of the king's physicians, Wreden inoculated the younger of the three-year-old twins. The girl was then laid into the same cradle as her sister, with the unspoken intention of infecting the second girl. The purpose of Wreden's experiment was to show that smallpox contracted through inoculation—called the "artificial smallpox"—was not as dangerous as the "natural" form contracted directly from a sick person. Children at this time were commonly brought to persons sick with smallpox in an effort to have them catch a light case through this contact. But Wreden's experiment with the older twin contradicted the accepted medical practices of the time in that she was not offered the safest known treatment. She became so dangerously ill that the physicians gave up all hope for her recovery (she did eventually recover).[43]

Wreden observed the twins for some time. Several months later they both were taken ill with measles at the same time but, Wreden recorded with enthusiasm for his inoculating techniques, "with a great difference": the inoculated child overcame the illness with greater facility than her sister. Wreden further concluded with the zeal of a partisan that inoculation contributed to the general health of the younger twin. The inoculated child, he wrote, grew so well that, having been before the smaller and weaker of the two, she now seemed "lustier, stronger, and older" than her sister.[44]

Women and twins, then, captured physicians' interest as natural groups required and useful in medical testing. While they apparently deemed it necessary that females be included in experiments in this period, I have not found any evidence of physicians who conducted tests within Europe attempting to control for class, ethnic, or national differences. As we will

43. John Wreden, *An Essay on the Inoculation of the Small Pox* (London, 1729), 30–33. Wreden does not say whether these girls were identical twins.

44. Ibid., 32–33.

see, however, while physicians seemed content to consider females in Europe as all belonging to one natural category, considerations of race soon intruded on this seeming natural unity.

Slaves in the West Indies

The voyages of scientific discovery were fueled by the assumption that plants, medicines, food stuffs, animals, and peoples could be transplanted and acclimatized—that nature was interchangeable globally. The great Carl Linnaeus, for example, hoped that he could "fool," "tempt," and "train" tropical plants to grow in Arctic lands and thereby create "Lapland cinnamon groves, Baltic tea plantations, and Finnish rice paddies."[45] Humans were seen as no less adaptable; physicians assumed that exotic cures could squelch recalcitrant fevers inside Europe as well as keep slaves, merchants, and planters alive in the colonies.

Patterns for experimentation in the colonies were similar to those in Europe—or at least this is what has survived that we can know about. If there were independent experimental traditions in the West Indies among the indigenous or slave populations, these have not been preserved. What has survived are reports on new substances sent by European-educated physicians working in the colonies to their colleagues in Europe. While medicines were tested according to European protocols, it should be kept in mind that new cures often had Amerindian or African origins—and, importantly, that in the early eighteenth century Europeans held these remedies in high esteem. Missionaries, planters, merchants, and soldiers in the West Indies were often afflicted by illnesses completely unknown to them. In these desperate situations, colonists discarded the costly, and also often old and ineffective, drugs shipped from Europe, employing instead tropical remedies offered by "the naturals of the country whom one calls savages" (the Tainos or Caribs) or the "marvelous cures" abounding in the islands that "only the negroes know how to use."[46]

Populations used for drug testing carried out by European-educated physicians in the West Indies were similar to those used in Europe: the well-born, hospital patients, soldiers, and physicians' own bodies. One popula-

---

45. Lisbet Koerner, *Linnaeus: Nature and Nation* (Cambridge: Harvard University Press, 1999), 121.

46. Jean-Baptiste-René Pouppé-Desportes, *Histoire des maladies de S. Domingue*, 3 vols. (Paris, 1770), 3:59; Nicolas Bourgeois, *Voyages intéressans dans différentes colonies Françaises, Espagnoles, Anglaises* (London, 1788), 460, 487, 503.

tion not available in Europe but used in colonial experiments was slaves. In many instances European physicians in the colonies did *not*—as might be expected—use slaves as guinea pigs. Slaves were considered valuable property of powerful plantation owners whom doctors were employed to serve. Physicians treated the slaves under their care, when they were sick, much as they did the free Africans, persons of mixed race and their white clientele (whether these were planters, sailors, soldiers, masons, prostitutes, or innkeepers) and recorded the effects of the drugs used.[47]

James Thomson, who learned the art of experimentation while a medical student in Edinburgh, returned to Jamaica where he studied the many new remedies the island had to offer. Thomson experimented, for example, with unroasted coffee, first on himself, then on free patients and slaves under his care. From suggestions found in a foreign medical journal, he wrote, "I was induced to try the effects of this substance in intermittent [malarial] fevers." He was hoping to find a substitute for the Peruvian bark, which made him and his patients nauseous. "Satisfied with the results" in his own healthy body, Thomson waited for an opportunity to use this new cure on some diseased subject, "who I little imagined would prove to be myself." It "worked well," so he also gave it to a young gentleman and also to a "negro" woman who had long been troubled with fever for whom neither the bark nor snake-root offered respite.[48] Thomson made "trials," as he called them, also with prickly yellow wood, quassia, the bark of the lilac or hoop tree, and the bullet tree.

Many slaves, then, were treated individually and with care. The Caribbean also saw, however, remarkable and unsavory experiments using unique and large populations: plantation slaves. These plantation trials were carried out by the English physician John Quier, a European-educated practitioner on the island of Jamaica, who, according to the son of his medical partner, Thomas Dancer, carried "the practice of inoculation to a much greater length than has been done by any in Europe."[49] Quier's study group of at least 850 and probably closer to a thousand persons was extraordinarily large for the eighteenth century.[50] Many physicians considered tests on five or six patients significant; large tests might include up to three hundred patients.

47. See, e.g., *Mémoires du Cercle des Philadelphes* 1 (1788): 35–143, here 104–15.

48. Thomson, *Treatise* (note 20), 153–54.

49. Thomas Dancer, *The Medical Assistant; or Jamaica Practice of Physic* (Kingston, 1801), 156.

50. Quier inoculated 700 persons in 1768 and 146 in 1774. He also inoculated another group in 1773 but did not record the number. John Quier, *Letters and Essays* (London, 1778), 8, 64.

It should be noted that Quier's experiments with inoculation took place after the College of Physicians in London had endorsed the procedure in 1755, but while heated debates on the subject still raged in France. In face of epidemics throughout the island of Jamaica, Quier and others in the colonies were persuaded to undertake inoculation because of the "extraordinary accounts of success received from England."[51] Inoculators in Jamaica followed the rules for inoculation laid down by Dr. Dimsdale in his treatise, "sent over by the agent for the island," even though they were calculated for a colder climate and thus possibly invalid in England's Caribbean colonies.[52]

Quier was employed by plantation managers and would have inoculated plantation slaves whether he was engaged in medical research or not. We can see from his reports, however, that he took what he considered a rare opportunity to answer questions about inoculation still pressing within European medical circles. Because his subjects were slaves, Quier was able to investigate questions that doctors back in Europe could not, including whether preparation (in terms of purging and diet) was necessary before inoculation, whether it was safe to inoculate menstruating women, whether one could safely inoculate a person already suffering from another disease, and so forth. In order to answer these questions, Quier reported that he (sometimes at his own expense) "made numerous repetitions of the inoculation in the same patient"—a practice rarely carried out on European subjects either in Europe or in the colonies.[53] He also inoculated populations (new infants and pregnant women) that physicians in Europe commonly did not. The information he gathered was sent in three detailed letters to Dr. Donald Monro of Saint George's hospital in London and read to the College of Physicians.

Pregnant women were Quier's special interest. Two special dangers made pregnancy a matter of some concern for both Quier in Jamaica and his colleagues in London: women who contracted smallpox generally miscarried; yet at the same time pregnant women who were not engrafted during a mass inoculation were in danger of contracting a more severe form of the disease naturally.

In his first letter of 1770, Quier wrote that "pregnancy is no obstacle to this process [inoculation] during the first six or seven months of gestation; afterward I think there is danger of abortion; not so much from the vio-

51. Ibid., 6.
52. Ibid., 7.
53. Ibid., 43, 56.

lence of the disease, as from the necessary method of preparation."[54] His colleagues in London questioned whether this conclusion, derived from women of African origins, was valid for European women given that they were more fragile than slaves, insisting in particular that women "of fashion" and "of delicate constitution" should not be inoculated unless absolutely necessary.[55]

Quier's work raised a new and potentially explosive question: whether medical experiments done on blacks were valid for whites. This was rarely an issue for naturalists in the seventeenth and early eighteenth centuries. If they did interrogate cures learned from slaves or Amerindians in the Caribbean, they were concerned that differences in climate—the West Indies being hot, England being cold—might render a drug ineffective. The question about physiological differences between the races took on steam only after the rise of scientific racism in the late eighteenth and nineteenth centuries. We see James Thomson, Quier's colleague in Jamaica, moving in this direction with his numerous "experiments regarding the differences of anatomical structure, observable in the European and Negro, but particularly those of the skin."[56] But for Quier the differences between blacks and whites were still primarily matters of habit and class. For him, results from slave women might not hold for European ladies because the bodies of slave women had been hardened and rendered coarse by their backbreaking labors.

By Quier's second letter of 1773, more queries had come from London concerning the validity of his experiments on slave populations for the women of Europe. In this letter, Quier backtracked on his claim concerning the hardiness of slave women and noted that the females of that class were excused from all kinds of labor for three or four weeks after lying-in and generally treated with great care. He judged his experiments on slave women to be valid, if not for upper-class European women, then at least for women of what he called "the rustic part."[57] Again, we see that Quier considered his experiments on blacks valid for Europeans of a similar class. Slave women and "the rustics" of Europe (peasants and anyone whose livelihood depended on physical labor), he judged, would have similar experiences in pregnancy.

In this second letter, Quier repeated his claim that his inoculations in

54. Ibid., 11.
55. Ibid., 12.
56. Thomson, *Treatise* (note 20), 3.
57. Quier, *Letters* (note 50), 54–56.

pregnant women had caused "not a single instance of abortion."[58] In his third letter of 1774, having been pressed again on the subject by his English colleagues, Quier "took pains" to make a strict inquiry into the matter. In so doing, he discovered that two slave women had miscarried shortly after their inoculations, but given that child bearing, miscarriage, and similar female complaints were generally taken care of by a slave midwife, he had not been called to attend them or—even more extraordinary—had not even been told about their miscarriages at the time.[59] By this final letter, Quier had to admit that inoculation caused miscarriage, but he had shown that, despite his London colleagues' presumptions concerning disunity among the races, neither racial nor class differences influenced women's response to inoculation, that inoculation, indeed, endangered pregnancy across the races. What is interesting is that social divides—his position as a European doctor to slave women—had distanced him from birthing practices in the colonies to such an extent that he had earlier missed data crucial for his experiments because he was not privy to information about miscarriages among slaves.

Conclusion

The eighteenth century fostered an extensive experimental culture. As we have seen, many new cures were tested on a variety of human subjects. Europeans in the eighteenth century inherited many notions from the past, such as the Hippocratic precept that medicine must do no harm. Physicians inherited notions that certain boundaries—thought to be introduced and patrolled by nature—existed in bodies such that proper experimentation required repeated "trials" on persons of "different Ages, Sexes, and Constitutions, in different seasons of the year, and in different climates."[60] They also inherited notions about which social groups were to be subjected to testing. It was generally accepted that wards of the state, such as condemned criminals and orphans; employees of the state, including soldiers and sailors; and persons devoid of personhood (such as slaves) could "benefit society" more generally by being used in medical testing. Lorraine Daston has shown, in her essay included in this volume, how

58. Ibid., 56.

59. Ibid., 67–69. On the politics of abortion among slave women, see Londa Schiebinger, *Exotic Abortifacients: Colonial Botony between Europe and the West Indies* (Cambridge: Harvard University Press, forthcoming).

60. Wagstaffe, *Letter* (note 27), 4.

seemingly trivial, even unsavory, nonhuman organisms such as insects or worms could become valorized as objects of "wonder" and "affection" through naturalists' persistent observations. In the case of human subjects used for medical experiments, devalued members of society were revalued by lending their bodies to medicine, an act seen as contributing to the greater "public utility."

Historians of medicine commonly urge that until the 1860s physicians prescribed for the individual patient, taking into account his or her age, sex, temperament, social, and physical environment.[61] This Galenic outlook, it has been suggested, would preclude generalizing the results of experimentation from one group of patients to another. This, however, was not true of empirical medicine in the seventeenth and eighteenth centuries. Harold Cook has argued that already in seventeenth-century England a conflict had emerged between royal and military physicians in this respect. While the College of Physicians insisted on treating patients according to their costly and individual needs, army physicians were pressed to develop a small number of effective drugs that could be employed in the field without observing the niceties of lengthy preparations and individual constitutions.[62] In the course of the eighteenth century, medical practitioners increasingly tested "specifics" (single drugs aimed to cure particular diseases) on individual patients with the notion that a successful treatment could be generalized to others, while still heeding the Galenic trio of age, sex, and temperament.

Experimentalists, then, generally assumed an interchangeability of bodies among Europeans, even of very different social classes. As Valentin Groebner explores in his essay, during the Renaissance *complexio* was thought to reveal the moral character of individuals; outward signs (moles, birth marks, skin color) allowed those schooled in their interpretation to read the inner disposition of both body and soul. Groebner intimates that toward the end of the sixteenth century, *complexion* took on modern connotations and more narrowly denoted skin color that was thought to mark out natural categories of human beings. But European naturalists in the eighteenth century, initially at least, continued to assume a certain unity among peoples of different "complexions." Naturalists working in the colonies recognized many bodily differences. They saw that Europeans and

61. John Warner, *The Therapeutic Perspective: Medical Practice, Knowledge, and Identity in America, 1820–1885* (Cambridge: Harvard University Press, 1986).

62. Harold Cook, "Practical Medicine and the British Armed Forces after the 'Glorious Revolution,'" 34 *Medical History* (1990): 1–26.

Africans suffered from different diseases; they noted that *Negres nouveaux* (slaves recently brought from Africa) were less hearty than creoles born in the islands. They noted that different people responded differently to different cures. But these observations were dependent mostly on place, climate, diet, and labor, not race.[63] West Indian creoles, for example, whether white or black, were seen as having the best survival rates. A number of European physicians and naturalists in this period subscribed to what might be called an environmentalist theory of race. Environmentalists were monogenists who believed that all humanity shared a common ancestry with Adam and Eve. They opposed the polygenists, who taught that the human races were immutable physical entities created separately at the beginning of time. Environmentalists saw all human bodies as potentially made of the same raw materials: physical differences resulted from the effects of diverse climates acting on an otherwise uniform human nature. For them, racial traits were not fixed but impressed into heredity by any number of external factors, including climate, diet, and customs; the vagaries of epidemics or disease; the crossing of different races; and the manipulative hands of midwives and mothers.[64]

As we have seen, notions of racial unity were challenged toward the end of the eighteenth century, as political struggles (to end slavery and the slave trade, to expand political and economic liberties) brought about the rise of what Michel Foucault has called "political anatomy."[65] The physical body—stripped clean of history and culture—came to ground political rights and social privileges. This peculiarly Western notion of race (see the essay by Fa-ti Fan in this volume) increasingly identified anatomists as possessing the most exact methods for uncovering internal "signs" of racial difference. As notions of race changed, experimentalists threw into question earlier practices based on the assumption that the bodies of Europeans and of enslaved Africans were interchangeable. This does not mean that

63. As Jorge Cañizares Esguerra has pointed out, Spanish creoles often did not share the views of their European colleagues. See his "New World, New Stars: Patriotic Astrology and the Invention of Indian and Creole Bodies in Colonial Spanish America, 1600–1650," *American Historical Review* 104 (1999): 33–67.

64. Londa Schiebinger, *Nature's Body: Gender in the Making of Modern Science* (Boston: Beacon Press, 1993), chaps. 4–6. See also George Stocking Jr., *Race, Culture, and Evolution: Essays in the History of Anthropology* (New York: Free Press, 1968); and Winthrop D. Jordan, *White over Black: American Attitudes toward the Negro, 1550–1812* (Chapel Hill: University of North Carolina Press, 1968).

65. Michel Foucault, *Discipline and Punish: The Birth of the Prison*, trans. Alan Sheridan (New York: Pantheon, 1977), 193.

European-educated physicians did not continue to use enslaved Africans as test subjects. By the nineteenth century, the use of slaves as medical specimens (for teaching anatomy, practicing surgical techniques, and testing new drugs) in teaching hospitals in Virginia and elsewhere had grown to such proportions that even popular songs warned unwary slaves away.[66]

The story of sexual differences in medical testing is somewhat different. As we saw above, females had long been a required element in experimental design. Testing drugs in the uniquely female body was thought crucial for women's continued health and well-being. Physicians in the eighteenth century also called for new and exacting studies of sexual differences, much in the same way that they had called for the scientific study of racial differences. The rise of what Cynthia Russett has called "sexual science" after the 1750s did not always work to women's advantage.[67] The academic study of sexual differences was rarely designed to promote women's health: its overriding purpose was to answer the troublesome "woman question," concerning political rights for women. These studies set on new scientific foundations ancient traditions that had portrayed women as the weaker sex and granted nature a newly sovereign authority to determine what constituted female nature. (Legacies of these developments are discussed by Michell Murphy in his essay in this volume.) In the course of the late eighteenth century, the results of studies of women's narrow crania but capacious pelvises, for instance, came to underwrite elaborate ideologies of motherhood and sexual complementarity that served to exclude women from university educations, the sciences, and the professions.[68] Jakob Ackermann was one of the few physicians who remained concerned about the implications of sexual difference for health care, but these concerns were confined to an appendix to his lengthy study of sexual difference.[69]

66. Todd Savitt, "The Use of Blacks for Medical Experimentation and Demonstration in the Old South," *Journal of Southern History* 48 (1982): 331–48; Todd Savitt, *Medicine and Slavery: The Diseases and Health care of Blacks in Antebellum Virginia* (Urbana: University of Illinois Press, 1978).

67. Cynthia Russett, *Sexual Science: The Victorian Construction of Womanhood* (Cambridge: Harvard University Press, 1989).

68. Londa Schiebinger, *The Mind Has No Sex? Women in the Origins of Modern Science* (Cambridge: Harvard University Press, 1989), chaps. 7–8. See also Joan Landes, *Women and the Public Sphere in the Age of the French Revolution* (Ithaca: Cornell University Press, 1988); and Geneviève Fraisse, *Reason's Muse: Sexual Difference and the Birth of Democracy*, trans. Jane Marie Todd (Chicago: University of Chicago Press, 1994).

69. Jakob Ackermann, *Über die körperliche Verschiedenheit des Mannes vom Weibe außer Geschlechtstheilen*, trans. Joseph Wenzel (Koblenz, 1788), appendix.

Questions about which bodies could be used as valid objects of medical experimentation took place in a broader context where natural rights, natural (in)equalities, natural kinds, and natural differences were much discussed. The Enlightenment conferred new authority on nature. Nature simultaneously stood for objectivity, the morally good, and was seen as providing the foundation for judging society and its institutions. In this regime, science (as the knower of nature) and the body (as the physical form of the newly sovereign individual) both took on new epistemological force. As we have seen, despite the notion that science is value free, sharpening political struggles surrounding race and sex molded thinking about which bodies were interchangeable, which bodies could truly represent humankind, and which bodies were reliable nature objects.

16

Nature and Nation in Chinese Political Thought: The National Essence Circle in Early-Twentieth-Century China

Fa-ti Fan

Barbarians and Chinese are born in different places, which are endowed with different atmospheres. Since the atmospheres are different, their habits must be different. Since their habits are different, their thinking and behavior must be different.

WANG FUZHI (1619–1692)

One of the most powerful discourses of nationalism is the argument that the components of a nation are "naturally" bound together. For the nation idealized in this way, its people, land, and sovereignty constitute an organic whole whose history is an analogy, or even an extension, of natural history. There is an ultimate, perfect, natural form of the nation, governed by teleological laws: just as water runs downward, the river into the valley, so the nation must follow a natural course. If a stream doesn't flow smoothly downward, it's not because nature perverts itself, but because there are obstacles—artificial, human-caused impediments that ought to be removed. Nation has its character, its soul, its spirit and will. A classical statement of this nationalism is racial nationalism of the late nineteenth and early twentieth centuries, which demanded that race be the essence of nation. Racial boundaries determined national boundaries. This view readily admitted culture into the natural history of the nation, but insisted that it was an organic component of nation—it was the manifestation of the

I am grateful for the comments of Jorge Cañizares-Esguerra, John Chaffee, Yuehtsen Juliette Chung, Fang-yen Yang, and the participants at the Moral Authority Workshop at the Max Planck Institute for the History of Science and at the Histories of Natural History in East Asia Workshop at Princeton University.

nature or character of the nation-organism. Often couched in scientific language, the discourse frequently drew upon biology, social Darwinism, anthropology, and other sciences. Similar assumptions can be found in other forms of nationalism. In geography, for example, vast terrain, woodlands, mountains, islands, or some other kind of geographic formation of a nation has been understood to determine national essence and national character.

This essay examines an influential strand of discourse of Chinese nationalism in the first decade of the twentieth century; its major voices included some of the foremost political thinkers of the time. We shall see how the historical actors defined and redefined history, tradition, and nationhood in relation to the transmutations of the concept of *nature*. The intense political controversies during that tumultuous decade, in which the nationhood of China was heatedly contested, contributed to the founding of republican China in 1912.[1] But the historical significance of this nationalist discourse was not only political. Activist Chinese intellectuals in the late Qing faced many challenges, not the least that of Western learning. It was a moment in Chinese intellectual history that has been seen as a transition, a response, and a crisis. During the last decades of the nineteenth century, Chinese intellectuals frequently had to revise their conceptual framework and invent new language to accommodate the foreign ideas—an intellectual enterprise full of political implications. The nationalist discourse discussed in this essay grew in this intellectual milieu; its proponents tried to write a national narrative that directly confronted these intellectual issues. The drive of the discourse was to define and recover the national essence of China, of which nature was a significant component. This essay thus interacts with Julia Thomas's essay in this volume, in which she examines how Japanese intellectuals in the twentieth century searched for meaning in the intersections among nature, nation, and modernity.

Like its Japanese counterpart, *shizen,* the now standard Chinese term for nature—*ziran*—did not receive its distinctly modern meaning until the late nineteenth century (possibly by way of Japan). In its traditional connotations, derived from ancient Daoist texts such as the *Daodejing,* the term suggested a state of harmonious relationship between human and cosmos rather than designated an objective ontological entity. This emphasis underscored, for example, the *Daoxue* notions of a universe in-

---

1. See, e.g., Prasenjit Duara, *Rescuing History from the Nation: Questioning Narratives of Modern China* (Chicago: University of Chicago Press, 1995).

formed by *li* (principle), as well as the *qi* model of cosmos, to which we shall soon return. Similarly, none of the other traditional Chinese concepts that have sometimes been equated to nature—such as *tiandi* (heaven and earth), *wanwu* (myriad things), and *zaohua* (creative transformation in the universe)—corresponded to either machine-like nature or nature as God's creation. *Ziran* became "nature" in the modern scientific sense only with the introduction of Western learning in the late Qing. Nevertheless, Chinese intellectuals saw analogous bodies of knowledge in the two intellectual traditions, and they believed that knowledge translation was possible—though they disagreed widely on the value and use of Western learning and on the methods of accommodating it.

Complex conceptual operations were required to interpret and translate the modern Western idea of nature and the knowledge built thereon. Not surprisingly, the Chinese typically approached them through traditional conceptual schemata. A striking example of how Chinese intellectuals wrestled with the new ideas was the reformist Tan Sitong's *Renxue*, "A Study of Benevolence" (1896). Tan proposed a scientific, moral, and political philosophy grounded in a universal theory of force and principle that cooked Christianity, Confucianism, Buddhism, pre-Qin schools of philosophy, and Western science in one pot. The common denominator for the different traditions was *ren*, the Confucian idea of benevolence, which was equated by Tan to love, *qi*, ether, force, gravity, heat, and electricity. Admittedly, Tan's project was unusual in its ambition, but it indicates the struggle the Chinese intellectuals experienced, and the creativity they possessed, in dealing with foreign ideas. Impressed with the power of Western learning, they tried to absorb that which they deemed valuable in the body of knowledge, and were often confronted with challenges to their previous *Weltanschauung*. Even such a basic notion as *tian* (heaven, skies) in cosmology required revision, and the intellectual processes were documented in the reading notes on Western science of Kang Youwei, Zhang Binglin, and other remarkable thinkers of the period.

By the first years of the new century, most of the better-informed young Chinese intellectuals came to accept the nebular theory in cosmology, and many described in their own writings the scenes of the formation of the planets from hot substances spun off the sun, the cooling off of Earth, and the formation of life on the planet, often in conjunction with some sort of evolutionary theory. This doesn't mean that these Chinese intellectuals quickly abandoned their intellectual tradition. Even Yan Fu (1853–1921), who had received a Western-style education in China and had studied at a naval academy in England, did not advocate total Westernization. In fact, the Western works he translated and made popular in China documented his efforts and difficulties in reconciling what he took to be traditional and what he took be modern—and, in multi-angle connections, Chinese and Western.[2] That same struggle is evident in the *guocui* (national essence) authors' endeavor to define the temporal and spatial dimensions of the Chinese nation.

By developing two schemes of national narrative, the *guocui* writers (more below) imagined a Chinese nationhood and found a *raison-d'être* for the nation's distinctive existence. They did so, first, by drawing boundaries between alien and kindred, immoral and righteous, unnatural and natural—that is, between "them" and "us." In their nationalist discourse, the "them" were, to echo Gregg Mitman's essay in this volume, like unwanted weeds growing in the wrong place that must be pulled out. The Chinese intellectuals claimed that primordial ties validated the existence and solidarity of the Chinese nation, and in this linear national history, nature promised its permanence. Second, their national narrative was rooted in the politics of place, of defining the natural or native place of the nation. The Chinese nationhood, grounded in moralized spaces, was legitimized by civil and natural history that wove together memory, cultural heritage, textual authority, geography, and biology. Not surprisingly, this complex configuration of ideas doesn't fit in the nature/culture dichotomy that has underlined much of Western notions of nature since the nineteenth century.[3]

The Chinese authors actually did not oppose or resist the global discourse of modernity; they welcomed science, nationalism, and political ideals like liberty and equality. Their thrust was, rather, to develop a native way of accommodating modernity. In the bewildering woods of new and old ideas wound tangled paths to the past and the future. The national essence writers cut from one path to another—from the *qi* cosmology of an early Qing philosopher to evolutionary theory, from natural history to the classics—all in the intellectual, moral, and political realm of nature and nation.

2. Benjamin Schwartz, *In Search for Wealth and Power: Yen Fu and the West* (Cambridge: Harvard University Press, 1964).

3. Interestingly, this rigid dichotomy might have steered historians of China toward emphasizing "culturalism" in interpreting the Chinese view of world order, hence oversimplifying different strands of traditional Chinese political thought. I cannot here deal extensively with the writings of historians of China. But see Fang-yen Yang, "Nation, People, Anarchy: Liu Shi-pei and the Crisis of Order in Modern China," Ph.D. diss., University of Wisconsin-Madison, 1999, esp. chap. 4.

The *Guocui* (National Essence) Circle

Reform-minded Chinese intellectuals had much to worry about at the dawn of the twentieth century. The country had been ravaged by rebellions, wars, and famines for decades, and it suffered a complete humiliation in the Sino–Japanese War in 1895. The Reform Movement in 1898 lasted little more than three months before the conservative *coup d'état* reversed the course; leaders of the movement either fled abroad or were executed. The Boxer Uprising, foolish and destructive in the eyes of the intellectuals, led only to ruthless retaliations by the Western powers and Imperial Japan. Many of the young intellectuals gave up any hope in the Qing dynasty, which had ruled China since 1644, and became revolutionaries. Despair over the Qing dynasty fueled a powerful discourse of nationalism. The dynasty was founded by Manchus, a Tungusic-speaking people from outside China proper, and in spite of the predominant presence of Han Chinese in the government, the imperial court remained Manchu. Drawing upon the anti-Manchu writings by Han Chinese in the early Qing and the racial theories popular in the West, many radical Chinese intellectuals campaigned for a nationalism that mandated the expulsion of Manchus as well as resistance to the Western imperial powers. This campaign quickly gained momentum in the first years of the twentieth century. Others, defining "nation" in different terms and weary of excessive bloodshed, defended the advantages of constitutional monarchism. Political journals and newspapers mushroomed in Shanghai and in Tokyo, where political refugees and overseas students from China concentrated. In their pages, young Chinese intellectuals engaged in heated debates over the fate of China and the path to a new nation.

Among them were a group of intellectuals associated with the Society for the Preservation of National Learning in Shanghai, founded in 1905. Most of the major figures were from the higher strata of the scholarly gentry in South China and pursued careers as artists, scholars, or journalists. Politically, the circle was predominantly prorevolutionary and had connections with the Revolutionary Alliance, a radical revolutionary association led by Sun Yat-sen. The National Learning circle shared a sense of crisis. The influx of Western learning triggered among them the fear of a permanent loss of the traditional intellectual heritage. They believed that actions had to be taken to preserve "national learning" *(guoxue)*. They compared their enterprise to two foreign examples: the Japanese *kokusui* movement that had begun several years earlier and, more loftily, the Renaissance in Europe. The *kokusui* movement urged a populist reaction to

Meiji Westernization and helped to shape many so-called traditional Japanese arts and ritual practices known to us today. Like their Japanese counterparts, the Chinese preservationists held an essentialist view of national culture and endeavored to protect the authenticity of the culture as they defined it. Unlike the *kokusui* circle, however, the Chinese group adopted a strictly scholarly approach. Their official organ, the *National Essence Journal—Guocui xuebao*—published monthly from 1905 to 1911, regularly printed recondite studies on esoteric subjects that even the contemporary literati found difficult.[4]

Nevertheless, some of the intellectuals associated with the circle wielded enormous influence on the Chinese revolutionaries through their sharp, fierce, and learned polemics against their equally brilliant rivals in the form of constitutional monarchists. Virulent anti-Manchuism quickly gathered support among the young intelligentsia after 1903. Exiled in Japan, Zhang Binglin (1869–1935), who was generally considered to be the intellectual leader of the revolutionaries, fought battles with Kang Youwei, Liang Qichao, Yan Fu, Yang Du, and a host of other political thinkers who did not approve the anti-Manchu revolution he championed. Similarly, Liu Shipei (1884–1919), another intellectual powerhouse of the *guocui* circle, poured out inflammatory propaganda for anti-Manchuism and radical social changes. Both Liu and Zhang were leading classical scholars, and their scholarship, as we shall see, formed an integral part of the national narrative they constructed. They were supported by a cast of talented authors who shared with them a holistic and preservationist view of nationhood.[5]

Belief in the *guocui* doctrine also required its adherents to take a stance

4. Although the term *guocui* is commonly translated as "national essence," it should be noted that the Chinese authors sometimes equated it to national learning or a particular body of intellectual heritage instead of some intrinsic nature shared by all Chinese, which the English term strongly suggests. Nevertheless, in their intellectual enterprise, the authors shared core assumptions about the intrinsic and essential characters of the Chinese nation, and it is these assumptions or configurations of nationhood that I try to examine in this essay. For a general study of the *guocui* circle, see Zhen Shiqu, *Wan Qing guocui pai wenhua sixiang yanjiu* (A study of the culture and thought of the *guocui* school in the late Qing) (Beijing: Beijing Shifan Daxue Chubanshe, 1993). See also the essays on "national essence" in Charlotte Furth, ed., *The Limits of Change: Essays on Conservative Alternatives in Republican China* (Cambridge: Harvard University Press, 1976), 57–170.

5. See, e.g., Kauko Laitinen, *Chinese Nationalism in the Late Qing Dynasty: Zhang Binglin as an Anti-Manchu Propagandist* (London: Curzon Press, 1990); Young-tsu Wong, *Search for Modern Nationalism: Zhang Binglin and Revolutionary China, 1869–1936* (Hong Kong: Oxford University Press, 1989); Hao Chang, *Chinese Intellectuals in Crisis: Search for Order and Meaning (1890–1911)* (Berkeley: University of California Press, 1987).

regarding the issue of Westernization, which was an extremely controversial issue among Chinese intellectuals in the late nineteenth and early twentieth centuries. Some of them were pragmatic. Their attitude was that anything that would increase the wealth and strength of China should be welcome, whether it was Western or Chinese in origin. This view found a powerful spokesman in Yan Fu, the most important translator of Western works in natural and social science at the time. The *guocui* group rejected this approach. Not that they saw Western learning as worthless. Huang Jie, one of the founders of the *National Essence Journal,* admitted that China had much to learn from Japan and the West.[6] Another major contributor agreed and found it necessary to explain that preserving national learning would not hinder the introduction of Western learning.[7] In fact, a majority of the articles published in the journal employed Western learning or at least nodded at it. Spencer, Darwin, Huxley, J. S. Mill, among others were frequently cited as authorities. These Chinese intellectuals were hardly cultural purists. Most of them, furthermore, did not brandish the talisman "East is spiritual" as a way to keep China from being engulfed by "Western materialism."

The National Essence group claimed that just as the Renaissance revived Western classics, what China needed was a revival of pre-Qin learning *(zhuzi xue),* which flourished in the late Zhou (ca. 480–220 B.C.), but largely perished with the bibliocaust in the Qin (221–200 B.C.)—whatever had survived the destruction was suppressed by the state-sponsored monopoly of Confucianism during the succeeding dynasty. In their view, the pre-Qin philosophers were comparable to those of ancient Greece. Indeed, the *guocui* authors were not unsympathetic to the notion then popular among some conservative sections of the literati that pre-Qin learning covered much of what was valuable in Western learning. These conservative scholars discovered, for example, optics and mechanics in the Mohist writings, logic in the School of Names, and chemistry in the Daoist writings. The National Essence preservationists believed that pre-Qin learning was solid, useful, practical, and empirical. It corresponded to Western learning in fundamental ways. Instead of matching superficial similarities in descriptions of natural phenomena, the National Essence authors drew parallels between the methodological and epistemological approaches of the two independent intellectual traditions. They argued

6. *Guocui xuebao* (hereafter *GCXB*), 1:11–15. All references are to the reprint edition (Taipei, n.d.) in thirteen volumes.

7. Ibid., 2:771–78.

that, like Western science, pre-Qin learning was grounded in observation *(muyan)* and experiential knowledge *(shiyan)*. Liu Shipei did not hesitate to add a Baconian twist to the teaching in the classic *Great Learning* on the way to true knowledge, *"zhizhi zai gewu."* The sentence was elusive and had received many different, even contradictory, readings, yet Liu interpreted it as a slogan saying that knowledge is based on sensory perceptions.[8]

This and similar interpretations of pre-Qin learning helped the National Essence authors work through the problems of Westernization and the conceptual transformations of nature. Their essentialist and holistic view of culture did not allow them to submit to Western learning, and they were convinced that the introduction of Western learning would work only on the foundation of national learning. Culture could not be directly transplanted; it withered in an unfavorable land. Put differently, in order to digest Western learning, China must have a strong Chinese stomach. And once Western learning was absorbed, it would no longer be Western learning, but Chinese.[9]

Strange as it might sound, the controversy between the reformists and the revolutionaries had a good deal to do with interpretations of the classics. Kang Youwei (1858–1927), a leading voice in the New Text school, elevated Confucius as a prophetic sage king who authored all the canonic classics—half a dozen ancient texts that had long become the core of the education of a Confucian scholar, including the *Book of Poetry* and the *Spring and Autumn Annals* (see below). Kang argued that the classics should be read as sacred scripts in which were hidden clues for the solutions to problems in the present world, and he actually had used this doctrine to promote the abortive Reform Movement. Zhang and Liu, on the other hand, had come from the Old Text school, and they tended to see ancient texts as historical records from which the history of ancient China could be reconstructed. The fact that controversy concerning major political problems of the day should have had so much to do with textual criticism tells us that the intellectual world of late-Qing scholars was very different from ours.[10]

---

8. Ibid., 6:4423–30. See also ibid., 2:1277–81.

9. See, e.g., Liu Shipei's treatment of *zhuzi* learning, "Zhou mao xueshu shi zongshu," which appeared in parts in ibid., vol. 1.

10. See, e.g., Joseph Levenson, *Confucian in China and Its Modern Fate,* vol. 1 (Berkeley: University of California Press, 1968), 79–94; and from a different perspective, Lionel M. Jensen, *Manufacturing Confucianism* (Durham: Duke University Press, 1997), 151–214.

The Nature of Chineseness

Ancient Chinese political geography mapped the world into inside or outside China *(nei/wai)* and, closely related to this division, near or distant from the center of civilization *(yuan* or *jin)*. This geography also suggested a gradation of similarity and implied a scale of humanity by which other peoples could be measured. The ideal ruler, perfectly accomplished in morality, would oversee *tianxia* (all under heaven) according to the Way of the Sage Kings, and peoples from near and afar would willingly submit themselves to, and benefit from, his benevolent governance. Ultimately, there would be no place left outside of this umbrella of benevolence and civilization, and there would be only a great harmonious unity. Much of the controversy over anti-Manchuism used this political geography as a point of reference and concerned a few sentences in the classic *Spring and Autumn Annals*. The message of the remarks seemed to some to be saying that barbarians may be admitted into China and that Chinese, if they violate moral principles, should be treated like barbarians. For these readers, notably Kang Youwei, this statement spelled out that civilization *(lijiao)* lay at the heart of the matter. Any barbarians who have accepted Chinese culture should be treated like Chinese. But a few sentences in an ancient text were easily open to different interpretations; besides, interpretations of the text were intertwined with the controversy between the New Text and the Old Text schools mentioned above. Not surprisingly, Zhang, Liu, and other anti-Manchu authors read the text very differently. They either took the remarks to be a warning to Chinese not to become assimilated by barbarians or argued that the peoples named in the text were actually Chinese located on the fringes of China.[11]

The recent rediscovery of Wang Fuzhi's works, which responded eloquently to the Manchu conquest, had much influence on the *guocui* writers. Saddened by the political upheaval, Wang Fuzhi (1619–92) chose to live a self-exiled life and reflected on the loss of China to the barbarians. Although the central message of his major political writings was a teaching on benevolent governance based on Confucian moral cosmology, his call for drawing boundaries between Chinese and barbarians attracted intense interest, more than two hundred years later, from the revolutionar-

---

11. See, e.g., *GCXB* (note 6), 4:2457; Liu Shipei, *Liu Shipei quanji* (Complete works of Liu Shipei), vol. 2 (Beijing: Zhonggong Zhongyang Dangxiao Chubanshe, 1997), 2–3 (*Liu Shipei quanji* is hereafter cited as *LSP*); Zhang Binglin, *Zhang Taiyan quanji* (Complete works of Zhang Binglin), vol. 4 (Shanghai: Shanghai Renmin Chubanshe, 1982–86), 174.

ies. In his writings, Wang used the theory of *qi* to explain the natural territory of China and its relationships to other lands and their inhabitants. In Qing Chinese cosmology, *qi* (usually translated as air, atmosphere, pneuma, or ether) was a kind of material substance that permeated the universe and mediated between the beings. It was also central to medical theories and geomancy of the time. The concept was used in medical literature to explain, for example, the supposedly different physiques of the southerners and the northerners of China.[12]

In Wang's geography, China proper was the Central Zone *(zhongqu);* outside this region were lands of wilderness and inferior *qi*. The surrounding barbarians lived in harsh, wild lands—barren deserts; dense jungles shrouded in hazy miasma; steep rugged mountains; arid plains swept across by howling winds. China proper, on the other hand, was a cultivated land. Its climate was mild, its land better endowed, its people settled, not nomadic. They farmed; they developed a writing system; they possessed refined rites, manners, literature, and categorically superior moral principles that were in resonance with heaven and earth. The northern barbarians wore pelts and drank the blood of animals. They were hunting tribes. They knew no morals and behaved like beasts. Other barbarians were hardly any better.[13] Superficially, Wang's view appears similar to the romantic idea of nature and wilderness, reinforced in the American frontier myth, but the emphases and value systems underlying them can't be more different. In Wang's cosmo-geography, nature (or *qi*) and civilization were not opposite to each other; they were in concentric unity. The Central Zone enjoyed *both* superior nature and superior civilization.

In a way, Wang used the concept of *qi* to challenge the legitimacy of the Manchu dynasty in China. Since *qi* varies from place to place, the inhabitants must vary accordingly. Water birds have webbed feet; mountain birds don't. Each thing has its right place in the universe. The ancient sages observed the order of things and categorized them. Manchus in China are out of their cosmic-ecological niche, subverting the cosmic order at their own peril. Simultaneously, Wang also sketched out moral reasons that the Chinese ought to draw boundaries between themselves and the barbarians: to transgress or neglect the boundaries is to violate the order of heaven and earth.

12. Marta Hanson, "Robust Northerners and Delicate Southerners: the Nineteenth-Century Invention of a Southern Medical Tradition," *Positions* 6 (1998): 515–50.

13. Wang Fuzhi, *Chuanshan quanshu* (Complete works of Wang Fuzhi), vol. 12 (Changsha: Yuelu Chubanshe, 1992), 532–37.

Wang speculated that before the Yellow Emperor of the archaic age, China was a land of barbarians, and that before Tai Hau, another mythic hero, it was simply a land of beasts. Beasts differ from humans in that they are not complete with innate qualities *(zhi);* barbarians from the Chinese in that they are not complete with civilization *(wen).* Barbarians do not have knowledge, they cannot tell right from wrong. They cry when they are hungry, they eat as much as they can at a time and abandon what is left. They are just beasts that walk erect. The progress from bestiality to civilization is not irreversible, however. When the barbarians rule China, its civilization suffers. When China loses its civilization, its people will return to the time before the Yellow Emperor. They will eat what's not for them to eat, wear what's not for them to wear. Their blood and *qi* will alter; their appearance will change; their innate qualities will deteriorate. They will be worse than beasts. They will have no language, no writing, no history. There will be only chaos.[14]

Thus, Wang classified Chinese, barbarians, and beasts into three hierarchic categories according to innate qualities (which included physical attributes) and civilization. The boundary between innate qualities and civilization was porous and could be transgressed. It was precisely because of the possibility of this transgression that Wang eagerly admonished the Chinese to guard the boundaries against barbarians, especially now as the Manchus ruled China. There was only a thin line between humans and beasts. It is not difficult to see that Wang's theory could be modified to accommodate some versions of evolutionary theory, or vice versa, as many of the *guocui* writers were prone to do.

The traditional theory of *qi* appeared only occasionally in the writings of the *guocui* group. Liu Shipei, for instance, employed the concept of pernicious miasma *(nieqi)* to assert the distinction between Chinese and the beast-like barbarians—this application came straight out of Wang's theory.[15] Sheng Weizhong, another frequent contributor to the *National Essence Journal,* accepted the explanation based on the notion of *diqi* (local or earth *qi*) of why certain animals were confined to certain regions.[16] By this time, however, traditional cosmology and the theory of *qi* had lost much of their currency, and they existed mostly in hybridization with Western science, whose prestige was growing rapidly. Meanwhile, evolutionary theory became widely known among young Chinese intellectuals

14. Ibid., 12:467.
15. *LSP* (note 11), 2:7.
16. *GCXB* (note 6), 10:2551.

through Yan Fu's free translation (1898) of Huxley's *Evolution and Ethics,* which he managed to render into a tract of Spencerian social Darwinism. An ardent admirer of Spencer, Yan also translated the British philosopher's *A Study of Sociology.* These and other similar works in Chinese and Japanese translation had an immediate impact, and evolutionary theory enjoyed esteem and popularity among Chinese intellectuals.[17] Witnessing a China trampled by imperial powers and humiliated in the Sino–Japanese War, they found in the theory of evolution a familiar picture: the strong triumphed; the weak suffered and faced an imminent danger of subjugation, enslavement, and ultimately extinction. In the eyes of the Chinese intellectuals, the world was in a continuous fierce competition between nations and races: stronger, fitter, and higher races won out.

The anti-Manchu writers quickly appropriated Western racial discourse and evolutionary theory for their political cause.[18] In the Qing, the most common way to refer to Manchus and Han Chinese had been *Manren* (Man people) and *Hanren* (Han people).[19] This usage was discarded by the *guocui* authors. They argued that evolutionary theory had proved the biological distances between Chinese and neighboring barbarians, including Manchus, and produced evidence from ancient texts. Ancient Chinese named the surrounding peoples with derogatory names implying their bestiality. The Chinese characters in the names typically consisted of radicals reserved for animals. This practice earned the approval of Liu Shipei and Huang Jie. They praised the ancients' vigilance on racial boundaries and justified it on the ground of evolutionary theory. Huang simply declared that these "races" were not humans: the Qiang was a sheep race (because the character had the radical of the sheep); the Man, a snake race; the Di, a dog race. The breakdown of the boundary between the Yi (barbarians) and the Xia (Han Chinese) would mean the breakdown of the boundary between humans and animals. Interracial marriage was bestial. Evolutionary theory had demonstrated that the ancestors of humans were animals. The barbarians were far behind Chinese on the evolutionary ladder, and they still retained the nature of animals.[20] This view largely con-

---

17. James Pusey, *China and Charles Darwin* (Cambridge: Harvard University Press, 1983).

18. Frank Dikötter, *The Discourse of Race in Modern China* (Stanford: Stanford University Press, 1992). The book is controversial and should be read with critical care.

19. It is noteworthy, though, that both the Manchu court and the leaders of the Taiping Rebellion (1851–64) had tried to encourage "racial consciousness" among their own peoples. See Pamela Kyle Crossley, "Thinking about Ethnicity in Early Modern China," *Late Imperial China* 11 (June 1990): 1–34.

20. GCXB (note 6), 1:171–78.

curred with Liu's, though he did not go so far as to banish the barbarians from the realm of humanity altogether.[21]

Increasing contact with Western natural history did not necessarily damage the authority of ancient texts; in fact, it prompted some to defend the credibility of the intriguing *Shanhaijing* (Classic of mountains and seas), which was regarded by many as the earliest geographic work in China. (Modern scholars consider it a compilation of mythological writings over a period of hundreds of years from ca. 400 B.C., but there had been an influential, though contested, tradition of attributing its authorship to the heroes in the archaic age.) The book gave many colorful descriptions of places, peoples, plants, and animals in foreign lands as well as in China, including stories of giants, dwarfs, and a tribe of one-eyed people. The more cautious among Chinese scholars had treated these accounts as fantasies. The introduction of Western natural history opened up a new world of nature to the Chinese. Those who visited natural history museums in Shanghai or abroad all marveled at the collections, and the less well-traveled encountered descriptions of extinct or exotic animals in translated books. The authority of Western science inspired them to take a new look at *Shanhaijing*. If giant beasts and other bizarre animals described in Western books actually existed at some point of time, then the strange creatures in ancient Chinese texts might well have been real: couldn't the dragon have been a kind of dinosaur? It was simply that they had been wiped out in the process of evolution. Impressed with translated Western zoological books, Sun Zhongrun tried to draw the correspondences between them and Chinese works.[22] Perhaps it wasn't strange after all that the giants described in ancient Chinese texts seemed to have disappeared from this part of the world. Drawing on Western works, Sun speculated that they had been driven out of Asia by the Chinese and had migrated to Patagonia. Similarly, Liu Shipei suggested that the centaur-like creatures in *Shanhaijing* were not imaginary.[23] According to the theory of evolution, humans had evolved from animals, so the mysterious creatures, half-human and half-beast, might have been the ones in transitional form. *Shanhaijing* was probably written at the time when the struggle between humans and beasts was still intense. This interpretation of *Shanhaijing* underscored Liu's racial theory: some of the creatures, recorded in high antiquity by the ancients, eventually evolved into the barbarian tribes living around China.

21. *LSP* (note 11), 2:3–4.
22. *GCXB* (note 6), 10:2927–31.
23. Ibid., 2:1257–58. See also Ma Xulun's comments on *Shanhaijing* in ibid., 3:787.

However, neither a general theory of *qi* nor that of evolution provided the radical anti-Manchu authors with the techniques of drawing boundaries between Han Chinese and Manchus. They had found in Western science, and nature, a theoretical ground for excluding Manchus: Manchus were subhumans or inferior humans. This general judgment needed to be supplemented by a set of practical techniques to mark the boundaries. Western racial discourse focused on physical attributes defined by certain technology of representation (as demonstrated by Valentin Groebner and Londa Schiebinger in their contributions to this volume). In contrast, this kind of discussion was largely absent in the writings of the anti-Manchu authors. In spite of their strident ideology and language, they said little about physical differences between Chinese and the barbarian tribes they so relentlessly dehumanized. One possible explanation is that, physically, the Chinese were not noticeably different from their neighbors, including the Manchus, and this plain fact forestalled any serious attempt to classify them in physical terms. But this convenient explanation bypasses the real issue. Surely, physical similarities among peoples in Europe had not prevented nineteenth-century European scientists from producing strict theories, detailed research, and voluminous treatises precisely for the purpose of separating them *physically*. Physical similarities and differences were more the outcome than the empirical foundation of racial thinking.

A more pertinent explanation, then, is that the focus of Western racial discourse was still new to the Chinese intellectuals. Even in recent traditional works on other peoples—such as *Huang qing zhi gong tu,* an eighteenth-century ethnological study of tributary peoples, including Westerners from the trading nations, and similar works of the nineteenth century—descriptions of the different peoples were extremely sketchy, referring to their dresses, social customs, and sometimes visible physical attributes (for example, hair color and complexion), usually all in a few sentences. Physical identifications were included primarily as part of the overall appearances of the peoples described or as exoticism.[24] Detailed specifications of human features existed in physiognomic discourse, but they had not thrived in ethnological literature. The anti-Manchu writers were hard pressed to come up with physical criteria to mark out the bar-

---

24. Laura Hostetler mentions the characteristics used to distinguish ethnic groups in Qing ethnographic writings in her *Qing Colonial Enterprise: Ethnography and Cartography in Early Modern China* (Chicago: University of Chicago Press, 2001), chaps. 5 and 6. But see also my review of the book in *Metascience* 10 (November 2001): 458–61.

barian tribes around China, and they typically recycled the age-old derogatory comments on supposedly primitive behaviors—not much an improvement from Wang Fuzhi's work two hundred years before. On a few
occasions, Liu Shipei claimed that the Li in ancient texts were a black
people on the ground that the character *li* meant black. He did not spec-
ify any further the physical features of the people, however, and at one
point he suggested that the Li, who were enslaved by the ancient Chinese,
gained a dark skin color from laboring in the sun.[25] Alarmed by the lack
of scientific data, Huang Jie lamented the underdevelopment of racial
science *(zhongxue)* in China.[26] It was, in his eyes, a cause of the current
misery of the Chinese. Actually, he himself knew little about Western
racial theories and had to console himself with the thought that there were
still written records and indigenous practices to fall back on. Huang, Liu,
and Zhang derived from traditional literature methods that would enable
them to write a grand national narrative and to police racial boundaries
(zhongjie).

It is important to note here that the Chinese used several words, often
indiscriminately, to translate the concept of race. This promiscuity or flu-
idity of terms reflects the struggle of the Chinese intellectuals in trying
to comprehend the meaning of race through the conceptual frameworks
available to them. Different as they were in Chinese, *zhong, zu,* and *lei*
all had been adopted to translate "race." Traditionally, *zhong* and *lei* were
often used to classify animals and plants, as in the great herbal *Bencao
gangmu* (1596), whereas *zu,* often used in linkage with clan *(shi),* stressed
the kinship among the individuals of a group. Although all of them had
appeared in the context of describing peoples, they had not had the same
connotations. To complicate the matter, Chinese intellectuals of the late
Qing frequently used them in various combinations—*zhongzu, zulei,* for
example—and slid quietly from one to another. In his forceful revolutionary tract, *The Revolutionary Army* (1903), Zhou Ron drummed hard to
raise his people's racial consciousness and tried to educate them about
racial differences. In a diagram, he listed "Han *zu*" under "yellow *zhong*,"
but in the accompanying text, he frequently used the term "Han *zhong*"
instead of "Han *zu.*" The very incoherence of Western racial discourse
also contributed to the confusion of the Chinese intellectuals on the subject. Are there only five races in the world or are there dozens of them?

25. *LSP* (note 11), 2:4–5, 3:241, 4:289.
26. *GCXB* (note 6), 1:52–53, 173.

Kinship or blood lineage was the most familiar way for the Chinese to understand the biological connections between individuals in a community. The Qing literati had ample knowledge of mapping lineage lines. They had developed a lineage discourse that encouraged the determination of one's ancestry—ideally back to the great and the good in history. The practice flourished in the Qing, and the literati of the late Qing were all familiar with the ideas and practice of tracing one's genealogy. The *guocui* writers adopted this lineage discourse in interpreting the concept of race, and they promoted the practice of tracing surnames in historical records as the way to discriminate the barbarians. Correspondingly, they developed a national narrative based on the idea of blood lineage.[27]

Drawing on ancient texts and Western Orientalist scholarship, they narrated that the Chinese people had originated on the Pamirs in central Asia and subsequently migrated to the Central Plains, by the Yellow River. As it happened, another people, the Miao or Li, had settled in the place first. What followed was an intense struggle between the two races, and the stronger race, Han Chinese, finally won. After the conquest, the Han Chinese enslaved the Li people and chased the resisting remnants away. The process was an imperial expansion of ancient China. China thus had its origin as a colony of the Hua Xia or Han Chinese from central Asia, who were closely related to the Babylonians, and it broke away from the Western nations only later at the time of the Yellow Emperor (fl. 2700 B.C.)— the starting point of the Chinese nation.[28]

Evidently, Western racial theory and imperialism had influenced this version of the history of ancient China. The surging interest in the mythic Yellow Emperor had several origins. One of them was that his name conveniently corresponded to the color of the yellow race. There was also the influence of the recent example of Japan's invention of the imperial ancestry. Although none of the *guocui* authors under discussion were sympathetic to monarchism and theocracy, they believed that this national nar-

---

27. Kai-wing Chow, *The Rise of Confucian Ritualism in Late Imperial China: Ethics, Classics, and Lineage Discourse* (Stanford: Stanford University Press, 1994). My analysis of lineage discourse in anti-Manchuism is indebted to Chow's essay, "Imagining Boundaries of Blood: Zhang Binglin and the Invention of the Han 'Race' in Modern China," in Frank Dikötter, ed., *The Construction of Racial Identities in China and Japan: Historical and Contemporary Perspectives* (London: C. Hurst, 1997), 34–52.

28. See, e.g., Liu Shipei's *Rangshu* (The book of expulsion), in *LSP* (note 11), 2:1–17; Huang Jie, "Huangshi" (Yellow history), which appeared in parts in *GCXB* (note 6), esp. 1:41–60.

rative would excite and consolidate national consciousness. It conformed to the Chinese sentiment and tradition of ancestral worship: the Chinese were descendants of the Yellow Emperor, and China was a large family nation.[29]

On the authority of Spencerian evolutionary sociology, Liu argued that the formation of the Chinese race-lineage *(minzu)* grew from family. The Chinese term for nation-state, *guojia,* consisted of two characters, one of which was family (the other one being kingdom or principality), so the origin of nation was family and clan *(jiazu).* A race-nationality *(minzu)* naturally shared common character and had its foundation in a common blood lineage. Since nationality grew from the amalgamation of families and clans, the foundation of nation was family.[30] He also insisted that the ancients had deep racial consciousness, and cited a statement from an ancient text to the effect that the Yellow Emperor distinguished among different clans.[31] In one of his more petulant moments, Liu Shipei went as far as to suggest that people in China should be divided into three classes according to their surnames.[32] Those who had surnames recorded in ancient texts were to be honored as the descendants from the original Hua or Han race-lineage; those who had later authentic Chinese names belonged to the second class; at the bottom were the ones with surnames of barbarian origins. The anti-Manchu authors repeatedly stressed the importance of genealogical research, thereby distinguishing between the Chinese and the barbarians.[33]

The focus of the lineage discourse of nationalism was on kinship rather than biological types, though the two obviously overlapped The nation

29. Huang Zhonghuang [pseud.], ed., *Huangdi hun* (The soul of the Yellow Emperor) (Taipei: Zhonghua Minguo shiliao congbian, 1968 [1904]); Wang Zhongfu, "Zhongguo minzu xilai shuo' zhi xingcheng yu xiaoji di fengxi" (On the rise and decline of the theory of the western origins of the Chinese nation), *Zhongguo lishi xuehui shixue jikan* 8 (1976): 309–46; Shen Songquiao, "Wo yi wo xue jian Xuanyuan: Huangdi shenhua yu wan Qing di guozu jiangou" (The myth of the Yellow Emperor and the construction of Chinese nationhood in the late Qing), *Taiwan shehui yanjiu jikan* 28 (December 1997): 1–77.

30. *LSP* (note 11), 4:145. Many other authors also employed this argument, see, e.g., *GCXB* (note 6), 2:1031–32.

31. *LSP* (note 11), 2:6. Other *guocui* writers also frequently cited it as an example of the racial consciousness of the ancient Chinese, see, e.g., *GCXB* (note 6), 1:47–48.

32. *LSP* (note 11), 2:7.

33. On Zhang Binglin's role in propagating this notion, see Chow, "Imagining Boundaries of Blood" (note 27). Zhang's view influenced many young Chinese intellectuals. See, e.g., *LSP* (note 11), 2:6–9; *GCXB* (note 6), 1:58–60, 1:283–86.

they envisioned derived its authority from both nature and history. It should be emphasized, however, that the Chinese authors' understanding of Western science, including Darwin's theory of evolution, was fragmentary, and that their notion of race-lineage was based on traditional lineage discourse. They also tended to return to the argument of cultural and emotional bond. In his debate with the reformist Kang Youwei, Zhang Binglin argued that the notion of nationality was rooted in human nature, in human consciousness.[34] Drawing on historical records, Kang had made the point that Han Chinese and Manchus were of the same race *(zongzu)*. Zhang, on the other hand, insisted that they were not and found evidence in Western racial science. But the thrust of his argument lay elsewhere. Even if they were of the same origin, Manchus had long lived far way from China, in a wild land, and their language, political system, and customs all differed greatly from those of Chinese. How could they be called the same race *(zhong)?* He emphatically distinguished historical race-nationality *(lishi minzu)* from natural race-nationality *(tianran minzu)*. If we use natural race as the classifying category, he contended, since humans of all colors and on all continents had come from the same origin, what would be the point of arguing? What Zhang eagerly called for was a kinship-based ethnic community.[35] He assumed that common culture began with kinship, and this belief led him to refute the argument put forth by Yang Du that China was a geographic and cultural concept, not an ethnic nationality *(minzu guojia)*.[36] However, Zhang did not oppose intermarriage and immigration; he considered these private matters to be left to individuals to decide. The foundation of his view was some form of ethno-nationalism: the Manchus had invaded China and had refused to be assimilated, so they should be given only the choice of leaving China altogether or surrendering their sovereignty and accepting sinicization (that is, assimilation by acculturation and intermarriage). The bonds that tied China together would be culture and kinship.

34. Zhang, *Zhang Taiyan quanji* (note 11), 4:173–84, here 173.

35. The National Essence authors' version of the Chinese nation comes close to the ethnic nation defined by Anthony D. Smith, which assumes, among other things, "a myth of common ancestry, shared historical memories, an association with a 'homeland,' and one or more differentiating elements of common culture." Anthony D. Smith, *National Identity* (Reno: University of Nevada Press, 1991), 21. See also his *The Ethnic Origins of Nations* (Oxford: Blackwell, 1986). Of course, this doesn't necessarily mean that nation is of ethnic origins or that ethnicity, however one defines it, fundamentally shapes the nation.

36. Zhang, *Zhang Taiyan quanji* (note 11), 4:253.

Nation's Nature

The fever of anti-Manchuism decreased noticeably by 1910, as an increasing concern about national construction in the future blended into the immediate cause of overthrowing the Qing dynasty. Not surprisingly, in a period of political and intellectual turmoil, many sensitive minds frequently revised themselves. Zhang Binglin stepped back from the vehement anti-Manchu stance he had held since about 1908, and influenced by Buddhism began to question evolution, science, nation, and even nature itself. He did not forsake nationalism, however, and remained a staunch champion for national essence. He believed that the nation was what the Chinese urgently needed in order to resist Western and Japanese imperialism. Liu Shipei's intellectual and political moves were even more extreme. He at one point defected from the revolutionary camp and went to teach at an academy for classical learning sponsored by the Qing government. Beginning in 1907, he also toyed with anarchism. Like Zhang, however, he never relinquished the notion of national essence. The anarchist vision of tearing down national and racial boundaries did not prompt him to discard the idea of a nation bound together by history, culture, and primordial ties like race or ethnicity (zhongzu, minzhu). Of course, the nation included not only the people and the sovereignty, both of which, as we have seen, were major concerns of the *guocui* authors, but also the territory or, rather, the land. Liu Shipei, for example, wrote several geography textbooks and local histories. Geography was an essential component of the *guocui* writers' national narrative, which sought to define the space of the nation. Where is the homeland of the nation? Where are the boundaries of the nation? And what are the relationships between the local and the national? These were questions of paramount importance at a time that witnessed the disintegration of the Qing empire and growing regionalism.[37]

According to Zhang Binglin, the territory of China was defined in the Han dynasty, about two thousand years ago, a glorious period in the eyes of many Chinese. It did not include Mongolia, Xinjiang (eastern Turkestan), and Tibet. (These places had become part of the Qing empire.) On the other hand, the Han territory included Korea and Vietnam, both of which were now under the rule of foreign imperial powers. In fact, Zhang very much repeated Wang Fuzhi's map of the territory of China. As has

37. Bryna Goodman, in *Native Place, City, and Nation: Regional Networks and Identities in Shanghai, 1853–1937* (Berkeley: University of California Press, 1995) discusses similar issues in a different context.

been mentioned, Wang used the notion of *qi* to describe the natural territory of China. The *qi* of the Central Zone was better than that of the surrounding lands, lands of wilderness. He argued that Vietnam and Burma
received the *qi* from the Central Zone and were natural extensions of
China—it was "*ziran zhi he*" (natural integration).[38] As late as the early
1890s, Kang Youwei used the notion of *qi* to explain why Confucianism
was confined to China and the other East Asian countries. The high
mountains of the Tibetan plateau blocked the *qi* coming from China. On
the other side, the rivers and mountains of China all ran toward the east,
so the earth *qi* also flowed in that direction and carried with it Chinese
influences, including Confucianism.[39]

The *qi* model persisted into the twentieth century, but at the same time,
Western geographic determinism was also influential among the Chinese
intellectuals. In a journal published by the revolutionary students based in
Tokyo, for example, an article on the Yellow River opened with the assertion that "the history of this river was the most honorable history of the
nation."[40] The author continued, "Since human geography was deeply
influenced by physical geography, research into the history of our nation
must first study the history of the Yellow River." Like many of his contemporaries, the author viewed the Yellow River as the cradle of Chinese
civilization. Referring to scientific ideas common at the time, he asserted
that people in the tropics were apt to grow lazy. . The location of the Yellow River, though arable, was dry, cold, and windy; moreover, the river
often flooded and caused disasters. Challenging environments like this
shaped the excellent character of the Chinese, and they became diligent
and frugal. Equipped with these good hereditary qualities, the Chinese
were on the same footing as the white race in the race war.

This description of China's natural environment differed greatly from
Wang Fuzhi's. The climate of the Central Plains, where Chinese civilization originated, was no longer mild, but rigorous, and it was no longer *qi*
that had produced the superior qualities of Chinese, but the environment
as defined in Western scientific terms. Both, however, emphasized the importance of agriculture, and both assumed strong ties between people and
land: Chinese were superior because their land was favored by nature. One

38. Wang, *Chuanshan quanshu* (note 13), 12:533.

39. Kang Youwei, *Kang Youwei quanji* (Complete works of Kang Youwei), vol. 1 (Shanghai: Shanghai Guji Chubanshe, 1987), 193–95.

40. *Hubei xueshengjie*, vol. 1 (Taipei: Zhonghua Minguo shiliao congbian, 1968 [1904]),
248–50.

depicted a scenario of struggle; the other, order. The qualities that constituted the collective identity of Chinese *vis-à-vis* other peoples were no longer a higher level of civilization; in fact, the author of the article on the Yellow River painfully admitted the ignorance of his people in comparison with Westerners. Instead, frugality and diligence—which the Chinese had hitherto hardly used to distinguish themselves from other peoples in spite of their worth as personal and social virtues, and which had been selected from Western characterizations of the Chinese—were incorporated into the nationalist discourse and became part of the essence of the Chinese.

Freely drawing on new and old theories, the Chinese intellectuals maintained a sense of place in constructing the nation's history and geography. The Yangzi River had long been recognized as the north/south divide in Chinese history and geography, in part because it had been the border between a Chinese and a "barbarian" state a number of times, in part because the noticeable change of landscapes and crop across the line. Many Qing scholars believed that the center of Chinese civilization had been shifting southward. Frequently trampled by barbarians, including the Manchus, northern China had turned into a wasteland inhabited by semi-civilized, ignorant people. The ancient center of Chinese civilizations in the north had sunk into dark oblivion. The Yangzi provinces, on the other hand, had grown to be a prosperous region blessed with intellectual energy and refined culture. Wang Fuzhi had attributed this phenomenon to the gradual shifting of "the *qi* of the heavens and the earth," or the center of cosmic energy, from north to south.[41] Writing on the same phenomenon, Liu Shipei refrained from making such a general statement. Although he evidently still kept the *qi* notion, he employed historically specific reasons, such as migration, communication, foreign invasions, and other similar historical phenomena. However, the main reason he cited was *fengtu*.[42]

The traditional notion of *fengtu* (literally wind and earth), and a cluster of other similar concepts, proved extremely valuable to the National Essence writers' discussions of place and people. The concept was inclusive and elastic. Although it usually rested on the theory of *qi*, it was compatible with different geographic theories. Without the assumption of *qi*, it could simply mean the relationships between the inhabitants and the climate, physical geography, and other environmental factors of a particular

41. Wang, *Chuanshan quanshu* (note 13), 12:467–68.
42. His comparative study of the northern and southern schools of learning appeared in parts in *GCXB*, starting with a general introduction in *GCBX* (note 6), 1:205–6.

place. In a noted essay, Liu marked the differences between the northern and the southern schools of pre-Qin philosophy. The land in the north was mountainous and barren, and its communication was cut off by natural obstacles. Living in this environment, the inhabitants were firm, solid, and practical. The southerners, on the other hand, resided in a rich land full of lakes and waterways, so they were idealistic and lively. These differences in regional culture shaped the respective philosophical traditions.[43] Along a similar line of argument, he remarked elsewhere that the natural environment of a place—whether it was located in the mountains, on a plain, or by a river—would determine the occupations of its inhabitants.[44] The environment and the occupation would in turn determine the characters of the people—hardy, solid, frugal, and diligent or smart, calculating, and frivolous. Peasants in a village in a northern province, for example, were bound to be different in character and physique from shopkeepers in a big southern city.

With this interest in people and their environment, the *guocui* authors promoted the study of local history. Local history itself was hardly new. The Chinese had a long tradition of producing works in local history. The *fangzhi* or gazetteer, for example, was a publication issued by the county or prefectural government that described the climate, geography, society, main events, plants and animals, dialects, customs, and other aspects of the place. The National Essence preservationists imbued this form of literature with a new meaning. The ultimate objective was to write a grand history of the Chinese people *(minshi)*. Commenting on some social customs in his area, Sheng Weizhong lamented the dearth of this kind of history in traditional historiography; existing gazetteers or local histories, formulaic as they were, omitted from the record important social customs, such as the cricketfight.[45]

In a series of essays, Liu Shipei urged a massive collective endeavor to compile new, substantial county gazetteers.[46] He believed that this enterprise would be instrumental to the preservation of national essence. For national essence could be retrieved not only from written records, but also from language, folkways, collective memory, and closely related to all these, the natural environment. Liu traced the practice of recording local

---

43. Ibid., 1:207–8.

44. Ibid., 4:2915–16.

45. A game popular in some parts of China. It basically was like the cockfight, but using insects instead of birds. Ibid., 6:4673.

46. Ibid., 4:2539–43.

ethnographic and environmental data back to the Zhou dynasty, the golden age of Chinese civilization that the preservationists tried to recover. The ancients, while they paid great attention to each local society and recorded its history, customs, dialects, the occupations of the natives, and so on, did not neglect the natural environment, and particular offices were created to collect information about the rivers, mountains, birds, animals, and other natural resources.

A good local history would inspire people's love for their land and could be used to educate them. More than once, Liu compared the counties and prefectures to the principalities or kingdoms of feudal China of the Zhou. The principalities of the Zhou compiled local studies, and by analogy, the counties of modern China should do the same. The amalgam of local histories was the history of the nation. In this vision, which was shared by many *guocui* authors, the nation recognized regional differences; it grew from local families and societies, and was not created by the authority of the state. (This is not the place to dwell on the political underpinning of this proposal. Let it suffice to say that this model of the nation-state helped the authors, as republicans, to dispute the constitutional monarchists' top-down statist model.) They maintained that the endeavor of compiling local histories would prove the unity beneath the superficial diversity. China would not disintegrate under republicanism, for despite its regional differences, the Chinese nation had a history and nature that reflected its permanence and unity.

The *guocui* writers believed that much of the power of local history lay in its ability to summon the collective memory of the people about the land. A common phrase for the land, or territory, of China was "rivers and mountains" *(he shan)*. In the view of the *guocui* authors, landscapes were littered with visible memories of the people; history was inscribed in the landscapes, the temples, the monuments and memorials, the fields on which battles against barbarian invaders had been fought, the paths the ancestors had trodden. The strong Chinese literary tradition in evoking history and landscapes provided a rich lore of representations of this kind. All of them, according to Liu, could be used to play on the chords of memory, to induce *"si gu you qing,"* a deep feeling for the past.[47] Nature was clothed in symbolism; it became the geo-body of the nation.[48]

Another important dimension of the project of compiling gazetteers

47. Ibid., 4:2542, 2546.
48. My use of the notion of "geo-body" is adapted from Tongchai Winichakul, *Siam Mapped: A History of the Geo-body of a Nation* (Honolulu: University of Hawaii Press, 1994).

involved investigations into objects of nature. Liu explained that the ancients of the Three Dynasties gathered information about the natural environment, identified the uses and properties of the plants and animals, and determined the ways to maximize natural resources.[49] Each kingdom or principality had its own work on its natural resources, because the ancients understood that the wealth of a nation depended on agriculture, manufacture, and commerce. Using the same historical framework, Huang Jie studied ancient Chinese works on soils and concluded their knowledge to be superior to the modern understanding.[50] The ancients employed experts to examine the lands; the branch of knowledge reached its peak during the Spring and Autumn Period of the Zhou dynasty. The ancient text *Guangzi* classified earth into eighteen different types, which were further divided into nearly one hundred subtypes. Their properties were carefully specified. The information was a great help to crop choice, water control, and other considerations that directly influenced farmers' lives. Huang sighed that the knowledge had been lost.

Thus the *guocui* authors argued that ancient Chinese had substantial knowledge about the natural environment. In a simple society, the ancients lived among animals and plants, and they observed them daily. The lore on natural history was embodied in folksong and documented in the classic *Book of Poetry*. The knowledge was augmented, refined, and systematized by pre-Qin philosophers, and the study of nature's objects was part of the education of a gentleman scholar in the Zhou. Confucius himself encouraged his disciples to "learn widely the names of birds, beasts, grasses, and trees," so that they could study the *Book of Poetry*. Like everything else, the knowledge diminished and was eventually lost as the scholarly tradition swung further and further away from experiential knowledge *(shiyan)*. Nowadays, the learned could not tell wheat from barley *(bubian shuji)*.[51]

The history reconstructed by Liu and Huang amplified the common opinion that, without the necessary knowledge, China was wasting its natural riches. Liu admitted that Western-style schools offered courses on natural history. But it was hardly enough. "Isn't it a shame," he asked, "that

---

49. *GCXB* (note 6), 4:2911–16.

50. Ibid., 6:3977–82.

51. See, e.g., ibid., 2:783–87, 2:1277–81, 7:63–64. I took some liberty in translating the phrase "*shuji,*" which originally referred to two different kinds of millets familiar to the ancient Chinese. To preserve the force of the comment, I substituted them with "wheat and barley."

we still do not have a work on the natural history of China?"[52] To compile such a work, one must start from the local, as part of the project of collecting local history. Each county was to be divided into several districts, each of which would have a researcher whose primary responsibilities included listing nature's objects in the district and collecting specimens. Compiling these results would produce a collective work on the natural history of the nation.[53]

Clearly, Liu Shipei's proposal of collecting specimens and other data had been influenced by his understanding of Western natural history, which involved extensive fieldwork. However, he and others in the *guocui* circle did not entertain the idea of simply replicating Western science. Even Xue Zhilong, who taught natural history at a Western-style school, thought it important to recover the lost ancient knowledge about the living world.[54] What the Chinese intellectuals had in mind was the tradition of *bowu* learning. The modern Chinese term for the science of natural history is *bowu xue* (the study of *bowu* or a wide range of things). The term was adopted in the second half of the nineteenth century (possibly by way of Japan) to denote the Western learning of natural history. Traditional *bowu* studies did look like the earlier phase of Western natural history in its extensive coverage of plants, animals, and other objects or phenomena of nature, in its attention to exotic things and objects, and in its inclusion of anecdotes, stories, quotations, miscellaneous information, and occasionally personal observations. Obviously, this apparent similarity convinced the translator to adopt it for the science of natural history, although the traditional meaning of *bowu* did not have the connotation that the branch of knowledge was primarily about nature. By the end of the nineteenth century, however, the term *bowu* had been very much contaminated by the Western meaning, and the National Essence writers viewed it as a Chinese counterpart to natural history.

Yet, *bowu* meant much more to the Chinese preservationists than gathering specimens. In his study of ancient polity, Ma Xulun, a frequent contributor to the *National Essence Journal,* placed *bowu* studies among genealogy, political geography, and other branches of learning crucial to government.[55] Ma had gone to a Western-style school and surely had basic notions of Western natural history. He pointed out that natural history was

52. Ibid., 4:2912.
53. Ibid., 4:2911–16.
54. Ibid., 7:63–64.
55. Ibid., 2:783–87.

closely related to all the other branches of science, such as physiology and sociology. He did not stop here, however, and went on to expound the political and moral significance of *bowu* studies. A true scholar must take *bowu* studies seriously, for he is supposed to examine things *(wu)* in order to thoroughly comprehend the universal principles *(gewu qunli)* as it is stated in the *Great Learning*. Hence, the true scholar observes the changes and transformations of all things, and he will not let any detail, however small it is, escape his notice. Only those who are accomplished in morality can reach the roots of things. To be a true *bowu* scholar, therefore, is more than possessing book learning. A scholar who is good at tracing and annotating phrases in old books may be called *bogu* (learned in old books), but a true *bowu* scholar has the ability to discover the principles of things, elucidate the most obscure, and explain the most difficult. With such broad knowledge, a true scholar will be able to maintain or restore the order of a civilization. Along a similar line and citing the classic *Book of Changes,* Liu Shipei argued that humankind and all creatures each has its own innate nature *(xing)*.[56] They have particular places in the natural environment and follow certain patterns of change. Only the morally accomplished can discover these fundamental principles.

Although this categorization of *bowu* was somewhat grandiose, the deep concern about the political and moral aspects of studying nature was shared by the other contributors to the *National Essence Journal*. While the ancients studied nature to cultivate virtue and learn the Way of the Heavens and the Earth, the modern Chinese pursued *bowu* studies as part of the enterprise of preserving their intellectual heritage, saving the nation, and maintaining a cultural identity. Sheng Weizhong took great trouble to identify the insect "*cuzhi*" in old texts because of the association between the name and the image of a woman diligently weaving. (The image was a traditional symbol of the domestic virtues of the ideal woman.) Sheng called the insect "a special product of our land" and "a symbol of the agricultural nation." He added, "It cannot be neglected by the learned gentleman."[57] A tiny insect thus embodied a gendered nationhood.

The National Essence writers had common notions about how to revive the classical or national learning of *bowu*. One of their central ideas was *huitong gujin*, to connect or bridge the past and the present. A Chinese student of *bowu* not only should study nature based on extensive observa-

---

56. *LSP* (note 11), 4:205.
57. *GCXB* (note 6), 6:4665–74, 6:4668.

tion, but he should use the knowledge thus obtained to explicate traditional works.[58] In his proposal of collecting natural historical data for gazetteers, Liu Shipei insisted that the investigator consult old dictionaries, the classics, and traditional *bowu* works.[59] If any items collected were also found in the old texts, the references must be cited and the vernacular names changed to the ancient. Thus, he assumed that nature or nature's objects provided a path to reaching back to the past when the ancients observed and wrote about them. An accomplished *bowu* scholar would find that path.

In practice, the authors typically employed three kinds of evidence: folklore, personal observation, and texts. The task was to accurately identify the animals and plants in the classics. It was closely connected to the Confucian tradition of rectifying names—that is, determining the correct correspondences between things and names. The objective was to clarify confusion, expose usurpers, and maintain the order of things among the moral, political, and natural worlds.[60] The lineage discourse of nationalism also drew on this intellectual tradition. It was a moral act to identify usurpers (that is, the Manchus), denounce them, and send them back to their proper position. Only then would the order of the ethicopolitical universe be restored. In spite of their repeated appeal to observation, the *guocui* authors' starting point in *bowu* research was annotating the classics rather than studying "nature" as such. It was basically an archaeology of texts.

As serious classicists, the authors were familiar with the rigorous techniques of *kaozheng* philology, which had been developed by scholars in the early Qing (about the late seventeenth and early eighteenth centuries) in response to the abstract neo-Confucianism of the Song and Ming.[61] Philological scholarship, Liu Shipei argued, had much to contribute to natural history and was connected to empirical research.[62] As one of the most accomplished philologists of his time, Liu discovered certain phonological rules to help identify animals, plants, and even artifacts in ancient texts. The ancients did not name things arbitrarily. Objects received their par-

---

58. Ibid., 7:63–64.

59. Ibid., 4:2913.

60. The *guocui* authors wrote extensively on the importance of the rectification of names. See, e.g., *LSP* (note 11), 2:15–17; *GCXB* (note 6), 3:1879–80, 4:2295, 4:2701–9.

61. Benjamin Elman, *From Philosophy to Philology: Intellectual and Social Aspects of Change in Late Imperial China* (Cambridge: Harvard University Press, 1984).

62. *LSP* (note 11), 3:247–49, 3:250–54, 3:255–56; *GCXB* (note 6), 5:3635.

ticular names mostly because of their forms; the form of an object often determined the sound or pronunciation of the name given to the object. Although the vernacular names of objects are widely differing in China, it is possible to trace the original pronunciations of the names and identify the forms of the objects in question. The apparent diversity in the nation dissolved when one reached far enough into the past to come near to the nation's origin.

Therefore, the goal of *bowu* research was not simply to make new discoveries, collect every item in the land, classify them, identify their properties and usage, demonstrate the natural riches of the nation, and investigate the relationships between the people and the land—though all of these considerations were important. In their version of *bowu* studies, the core of the natural world of China was the one described in antiquity. In fact, in their own research, the *guocui* authors focused exclusively on identifying the plants and animals *already* recorded in the classics. The first task of a *bowu* scholar was to unfold the records of nation's nature.

Conclusion

The *guocui* authors constructed a national narrative that designated the cultural and natural essence of China. This version of nationalism proved to be a powerful one to early-twentieth-century Chinese intellectuals (though it should be noted that there were other influential rival discourses of Chinese nationalism at the time and that not everyone who embraced this nationalism shared the *guocui* writers' enthusiasm for national learning). The force of their discourse derived in part from its mobilization of the traditional symbolism of community, its invocation of the primordial ties of nation. This discourse projected a national identity that, at an agonizing moment in their nation's history, many Chinese found familiar and persuasive. In defining the nature of Chineseness, the National Essence writers inherited the traditional sinocentric anthropological map by way of Wang Fuzhi's exposition, yet they reinterpreted it by appropriating evolutionary theory and Western racial science, which in turn were inserted into traditional lineage discourse. Nature was constantly redefined throughout this process. The shaping force in nature was no longer *qi,* but competition among nations and races. While the Manchus and other "barbarians" were to be put in their right places in nature all the same, the reason was no longer to maintain the cosmic order, but to preserve the nation.

Nature was no longer a moral cosmos as such, and it had gained a different form. However, there were still order and boundaries that demanded respect. In fact, its voice and authority were becoming sterner. To follow the rhythm of nature was now less a matter of cultivating virtue, as required in the traditional moral cosmology, than struggling for survival. But to the *guocui* authors, nature was not simply an objective ontological entity devoid of values. Not all nature's objects were the same in the nation's body. Under the authority of the classics, of the text, some of them were more valuable than others because they were embodiments of national essence. Valuable nature was not wilderness, not the unknown. Only when a nature's object carried history and memory did it become a cherished subject for investigation and appreciation. And when one studied nature, the primary goal was not to discover its secrets, nor simply to experience it as an individual, but to articulate one's bonds with the nation.

The Chinese intellectuals thus imagined a nationhood that echoed distant voices from a utopian antiquity of moral and ethnic purity. The nation-state they envisioned professedly recast the two thousand years of degeneration in Chinese history and returned to the authentic, the original, to the fountain head. They asserted that the Chinese nation was firmly anchored in place, in evolution, in ecology, in plants and animals—that is to say, in nature. Yet, just like the traditional cosmology they no longer adhered to, their new conceptions of nature continued to elude the nature/ culture opposition prevailing in Western thought since the late nineteenth century. Their intellectual conviction and interpretive strategy allowed them to invent a distinctive modernity, one that aspired to be at the same time universal and local or national. Under the moral authority of an intricately configured nature, they charted a meandering path into the misty terrain of nation and modernity.

17

When Pollen Became Poison: A Cultural Geography of Ragweed in America

Gregg Mitman

At the Atlantic City meetings of the American Public Health Association in October 1947, the Committee on Air Pollution of the Engineering Section presented a report on atmospheric pollution that defied the boundaries between pure nature and impure civilization that historians have taken to be central to the story of modern environmentalism in America. Of the many plants that discharged smoke, dust, radioactive isotopes, obnoxious odors, and other industrial hazards into the atmosphere, one stood out among the list of smelters, cement plants, nuclear plants, and oil refineries: ragweed.[1] These "pollen factories," each "operating without Federal permission," annually produced an estimated million tons of "toxic dust," 275,000 tons of which made its way into the air. Ninety-three percent of the population of the United States was exposed to this "atmospheric contamination" produced in the ragweed belt east of the Rocky Mountains.[2] For the 2 to 4 percent of the U.S. population suffering from hay fever,

I owe special thanks to Maureen McCormick for her invaluable research assistance. The generous support of the Max Planck Institute for the History of Science in Berlin and the Alexander von Humboldt Foundation, as well as the intellectual camaraderie of the Moral Authority of Nature group, were indispensable in creating a stimulating atmosphere for research and writing. For archival assistance, many thanks to the staff at the Bancroft Library and the National Library of Medicine. This project was supported by National Science Foundation grant SES-0196204.

1. Committee on Air Pollution, Engineering Section, American Public Health Association, "Report on Air Pollution," *American Journal of Public Health* 38 (1948): 761–69.

2. Oren C. Durham, *Your Hay Fever* (New York: Bobbs-Merrill Co., 1936), 154; R. P. Wodehouse, "Weeds, Waste, and Hay Fever," *Natural History* 43 (1939): 150; "Hay Fever and How to Escape It," *Popular Mechanics* 70 (July 1938): 56.

pollen pollution resulted in lost time from work, "reduced efficiency," and aggravating symptoms that led to sleep deprivation and irritability. The fourth leading chronic illness in the United States in the 1930s, hay fever became a national public health concern and ragweed public health enemy number one.[3]

How could a native plant species of North America prove itself to be such a troubling problem? The question goes far beyond the watery eyes and sneezing fits that plagued hay fever sufferers. While auto-camping had become a popular middle-class recreation in 1920s America, ragweed followed a less traveled road than the one taken by sagebrushers who headed to America's national parks seeking a return to nature. The history of ragweed is a story of travel, not from the city to the country, but from the wilds of nature to the wastelands of civilization. Its unexpected movement across defined boundaries—country and city, natural and engineered—marked it as a dissident. Just as increased travel and mobility accentuated administrative concern with the problem of identification in Renaissance Europe (see Valentin Groebner's essay in this volume), so too did ragweed's unexpected movement across seemingly fixed categories make it the object of increasing scrutiny by citizens, botanists, physicians, public health officials, and engineers in modern America.

A "slum dweller, preferring to live in . . . city dumps," a "river rat," a "squatter on vacant property," or "nature's reply to man's destructive and wasteful exploitation of natural resources," ragweed's identities became marked by the cultural geographies of the places where it thrived.[4] This act of placing, as Groebner notes in his essay, is one means by which nature accretes moral, aesthetic, or administrative authority. To understand the how, why, and what of nature's authority, then, we must attend to the material, social, and symbolic relations in which such authority is situated. We must, in short, also consider the production of place.

This essay investigates the shifting geographies of place, disease, and moral authority of nature in ragweed's nomadic wandering across the boundaries of rural and urban, native and immigrant, pure and polluted, nature and civilization. In transgressing its natural boundaries, ragweed had become, in the urban spaces it inhabited, a symptom of the moral

3. Committee on Air Pollution, Engineering Section, American Public Health Association, "Public Health Significance, Distribution, and Control of Air-borne Pollens," *American Journal of Public Health* 39 (May 1949): 86–102.

4. Durham, *Your Hay Fever* (note 2), 145; Wodehouse, "Weeds, Waste, and Hayfever" (note 2), 162.

depravity and waste brought about by modern civilization. Neither wild nor domesticated, ragweed merited no place in the cultivated spaces of the city. Through great effort and expense, public health officials, sanitary engineers, and biomedical researchers sought to control or eliminate it from the urban environment, to push it back to its benign, even valued, place in nature.

A Native in the Land

Roman wormwood, hog-weed, or bittersweet, as *Ambrosia artemesiifolia* was commonly known, prospered in places of freshly disturbed soil— along river banks, and on flood plains, deltas, and erosion gills and gullies—in the native landscape of North America. Botanists believed South America to be its place of origin. It then traveled along the foothills of the Andes, made its way through Central America, and settled in North America, where it established itself in patchy areas east of the Rockies, long before the alien invasion of Europeans. To the Harvard botanist Asa Gray, *Ambrosia artemesiifolia* was an "extremely variable weed" that flourished in "waste places" and flowered from July to September. In the nineteenth-century American south, in particular, it was commonly used as a domestic remedy for nosebleeds and flourished in cultivated pastures and fields.[5]

A relatively innocuous plant species in nineteenth-century America, ragweed attracted increasing prominence with the rise of hay fever resorts. By the 1880s hay fever had become the pride of America's leisure class and the base of a substantial tourist economy that catered to a culture of escape. In mid-August, thousands of hay fever sufferers each year fled to the White Mountains of New Hampshire, to the Adirondacks in upper New York State, to the shores of Lake Superior, and to the Colorado plateau. Seeking refuge from the watery eyes, flowing nose, sneezing fits, and attacks of asthma that developed with the "regularity of a previously calculated eclipse," these "accomplished tourists" also sought refuge from the "desk, the pulpit, and the counting room" of the city.[6] It was in such ur-

---

5. Asa Gray, *Manual of the Botany of the Northern United States,* 3rd ed. (New York, 1862), 212. Roger P. Wodehouse, *Hay Fever Plants: Their Appearance, Distribution, Time of Flowering, and Their Role in Hayfever* (New York: Hafner Publishing, 1971), 137–44.

6. "Hay Fever Day," *The White Mountain Echo and Tourist's Register* 7 September 1878, 3; Harrison Rhodes, "American Holidays: Springs and Mountains," *Harper's Monthly Magazine* 129 (1914): 545; George M. Beard, *Hay-Fever; or, Summer Catarrh: Its Nature and Treatment* (New York: Harper & Bros., 1876), 82.

ban spaces that a nervous predisposition, which physicians deemed necessary for the development of their ailment, prevailed.

Hay fever, in the popular opinion of physician George Beard, was a functional nervous disease that bore a close relation to the much-celebrated American malady of the late nineteenth century: neurasthenia. In *American Nervousness,* published in 1881, Beard pointed to modern civilization, and particularly American civilization, as the source of nervous exhaustion, which included among its many symptoms sensitivity to climatic change and "special idiosyncracies in regard to food, medicines, and external irritants." Rapid technological progress, including the development of steam power, the telegraph, and the periodical press; expanded civil, religious, and social freedoms; the extremely variable climate of the United States; and the tendency of Americans to work hard and hurriedly, particularly among "brain-workers"—these and other causes were implicated in the development of a nervous temperament, found prominently among the educated and well-to-do classes in the northern and eastern regions of the United States. An extremely sensitive nervous system, coupled with the depressing influences of heat, made a particular class of individuals susceptible during the dog days of summer to a host of external irritants that ranged from dust, to sunlight, to plant pollens. In the absence of effective drugs, removing oneself from the cause to exempt places became the preferred remedy among the country's afflicted bourgeoisie.[7]

The various causes that triggered hay fever were well-known to hay feverites, although there was much individual variation, requiring each patient to become "his own physician." Dust and the smoke of a railway train were the bane of many a sufferer, a cruel irony given the necessity to travel. Strong sunlight, fruits of various kinds, particularly peaches, and the fragrances of flowers could also trigger an attack during the sneezing season. But the precise onset of the disease each year also led Boston physician Morrill Wyman to suspect a vegetative origin for the general cause of paroxysms. More than one sufferer reported on the irritations provoked

7. George M. Beard, *American Nervousness: Its Causes and Consequences* (New York: G. P. Putnam's Sons, 1881), 7. On Beard and neurasthenia, see Charles Rosenberg, "Pathologies of Progress: The Idea of Civilization as Risk," *Bulletin of the History of Medicine* 72 (1998), 714–30; George Frederick Drinka, *The Birth of Neurosis: Myth, Malady and the Victorians* (New York: Simon & Schuster, 1984); F. G. Gosling, *Before Freud: Neurasthenia and the American Medical Community, 1870–1910* (Chicago: University of Illinois Press, 1987); Barbara Sicherman, "The Uses of a Diagnosis: Doctors, Patients, and Neurasthenia," *Journal of the History of Medicine and Allied Sciences* 32 (1977), 33–54.

when walking along a road where Roman wormwood was present. *Ambrosia artemisiifolia* flowered in the middle of August, grew abundantly along the seashore and in catarrhal regions, and was rarely present in the mountains. In the fall of 1870, Wyman gathered specimens of the flowering plant in Cambridge, Massachusetts, sealed them in a parcel, and carried them along on his railway journey to Glen House, a favorite place of residence in the White Mountains of New Hampshire for hay feverites seeking relief. On September 23, he opened the package and the contents were inhaled by he and his son, upon which they promptly were seized with sneezing and itching of the eyes, nose, and throat. When a portion of the plant Wyman had sent arrived at the Waumbec House at Jefferson Hill, fifteen miles northwest of Glen House, eight persons who sniffed the plant developed symptoms that ranged from sneezing and watering eyes to "asthma and stricture in the chest." Eight other hay fever sufferers staying at Waumbec House who did not inhale the pollen remained free of their usual hay fever symptoms. Wyman was unwilling to attribute the "cause of the whole disease" to ragweed. It occurred, for example, in exempt places, although Wyman believed this could be explained on the basis of the plant having different properties in different regions. It also failed to produce the same symptoms when grown in a pot indoors and prompted to flower one month earlier in July. Although Wyman refused to attribute the disease's origin to ragweed, locating it instead in an individual predisposition to exciting causes that acted on the nervous system during particular seasons of the year, ragweed, as a primary exciting cause, became a focal point of concern among hay feverites residing in America's most popular hay fever resorts.[8]

In Bethlehem, New Hampshire, for example, home to approximately five hundred seasonal hay fever residents and the U.S. Hay Fever Association, established in 1874, the appearance of ragweed along the railroad in 1878, just north of Bethlehem near the village of Littleton, was cause for alarm. Railroads brought hay fever sufferers to this natural sanctuary, but the building of the branch line of the White Mountains Railroad into Bethlehem Station, and an additional narrow-gauge line added in 1881 to the town center, had also brought a most unwelcome guest. Captain Farr of the Oak Hill House first dispatched someone to root it out on its

---

8. Morrill Wyman, *Autumnal Catarrh (Hay Fever)* (New York: Hurd & Houghton, 1872), 127, 101, 103. For a history of hay fever resorts, see Gregg Mitman, "Hay Fever Holiday: Health, Leisure, and Place in Gilded-Age America," *Bulletin of the History of Medicine* 77 (2003) (forthcoming).

appearance. Littleton residents launched a community effort to extermi-
nate the "baneful weed" that season. Despite their attempts, Dr. Morrill
Mackenzie observed eight years later that "the spread of this pest is simply
marvelous."[9]

By the turn of the century, ragweed embodied all that hay fever tourists
sought to escape. Like hay fever, it came to be regarded as a product, not
so much of nature, but of civilization. Ragweed's appearance in Bethlehem
with the coming of the railroad only pointed to the failure of hay feverites
to keep the cause of their ailment—civilization—from defiling the purity
of place. Other pollen-bearing plants that followed the plow also came
under attack. "Everything around the patient" was "saturated with . . .
poisonous emanations," of "corn, peas, fodder, and other farm products,"
wrote James Bell in the U.S. Hay Fever Association's prize essay of
1887.[10] The proliferation of vegetable and flower gardens in the early
1890s prompted the U.S. Hay Fever Association to pass a resolution urging Bethlehem citizens to restrict "the planting of corn and other pollen-
bearing vegetables" on the north side of town at some distance from the
street.[11] "Improvements and other civilizing changes" in Bethlehem, were,
in the opinion of some, "diminishing its immunity from Hay Fever."
"Even in the very best of resorts," observed Professor Samuel Lockwood,
"unless Nature has been left to her virgin forms and moods," complete relief could no longer be found.[12]

Lockwood, a New Jersey naturalist, was a devoted Bethlehem pilgrim,
who understood well that if this natural sanctuary were to remain a favored
refuge, more than individual testimonies to the power of place were
needed to attract hay fever exiles, particularly as dissenting voices grew
alongside Bethlehem's weeds. Other places to the west, Petoskey and
Mackinac Island, and Denver and Colorado Springs vied for the hay fever
tourist trade. In 1896, Denver's Chamber of Commerce and Board of
Trade invited the U.S. Hay Fever Association to move their annual meeting to the "Queen City of the Plains," a place known for its curative power
over asthma, backed not only by individual testimonies but also by scien-

9. "Ragweed on the Railroad," *White Mountain Echo,* 14 September 1878, 3; "U.S. Hay
Fever Association. Annual Report, 1886," *White Mountain Echo,* 18 September 1886, 6.

10. James Eugene Bell, "Prize Essay. II.," *Manual of the United States Hay Fever Association
for 1887* (Lowell, Mass.: Vox Populi Press, 1887), 30.

11. "Hay Fever Talk," *White Mountain Echo,* 10 September 1892, 8.

12. "The United States Hay Fever Association," *White Mountain Echo,* 8 September 1900,
19. Samuel Lockwood, "The Comparative Hygiene of the Atmosphere in Relation to Hay
Fever," *Journal of the New York Microscopical Society* 5 (1889): 50.

tific studies on Colorado's climate and health conducted by Charles Denison, president of the American Climatological Association.[13] Lockwood himself had turned the efforts of the U. S. Hay Fever Association in 1888 to what he hoped would become a comprehensive scientific study of the hygiene of the atmosphere in exempt and nonexempt regions. Enlisting the support of laymen, Lockwood endeavored to gather meteorological records of temperature, wind velocity and direction, humidity and barometric changes, along with microscopic analysis of atmospheric particles and experiences of patients under their local influence to arrive at "trustworthy" results "on the line of comparative pathology."[14] After three consecutive seasons of scientific studies comparing the air around three White Mountain resorts to his nonexempt home in Freehold, New Jersey, Lockwood concluded that the tonic mountain air, coupled with the smaller quantity of vegetable matter and comparative absence of pollen in Bethlehem, went far in explaining the hygienic qualities of the White Mountains as a hay fever resort.[15]

Health and leisure combined to shape an urban vision of the landscape in America's most luxurious hay fever resorts in which the purity of nature sustained body, soul, and a thriving tourist trade. These resorts and their seasonal residents attributed great importance to the distinction between country and city in fashioning hay fever as a disease of the urban bourgeoise and the lifestyle of leisure that their class membership afforded. But just as hay fever transgressed the boundaries of social class, ragweed propagated across the boundaries between pure nature and impure civilization. As a native plant species, ragweed was a product of nature's nation. On "virgin" land, botanists believed ragweed to be confined to areas where natural disturbances such as fire or flooding had occurred. But such events were regarded as so rare that the presence of ragweed was considered relatively scarce and ephemeral until axe and plow enabled it to establish a firm foothold in the North American landscape. In nature's econ-

13. "Hay Fever Experiences," *White Mountain Echo*, 12 September 1896; "Not To Be Sneezed At," *Denver Rocky Mountain News*, 25 August 1896. On Denison and medical climatology, see Billy M. Jones, *Health-seekers in the Southwest, 1817–1900* (Norman: University of Oklahoma Press, 1976), and Sheila Rothman, *Living in the Shadow of Death: Tuberculosis and the Social Experience of Illness in American History* (New York: Basic Books, 1984).

14. "U.S. Hay Fever Association," *White Mountain Echo*, 17 September 1887; "A Warning to Bethlehem," *White Mountain Echo*, 14 September 1889, 7. See also "Hay-Fever: A Five Year's Resume," *White Mountain Echo*, 5 September 1891, 8.

15. Lockwood, "The Comparative Hygiene of the Atmosphere in Relation to Hay Fever" (note 12).

"AND WITH CLEARANCE OF THE LAND UP POPPED THE DEVIL—RAGWEED. IN A SHORT SEASON OR TWO SUSPICIOUS SNEEZES BEGAN TO VEX THE GUESTS. COMMERCE HAD AGAIN OVERSHOT THE MARK."

FIGURE 17.1. With the development of America's hay fever resorts, hay feverites were unsuccessful at keeping the cause of their ailment—civilization—at bay, as illustrated in this 1940s article on ragweed that appeared in *Nature Magazine* 35 (1942): 17.

omy, ragweed's place was limited. But the economy driving the progress of civilization enabled ragweed to proliferate beyond its limited geographic range in the natural landscape (fig. 17.1). Humans had tipped nature's balance, and thereby became subjects of it's revenge. Hay fever, like onanism in the Enlightenment, came to be seen as a consequence of civilization's progress and a symptom of its discontents (see Fernando Vidal's essay in this volume).

An Immigrant in the City

Prospering in vacant lots and waste places in American cities, ragweed migrated from nature to the metropolis with the tidal wave of immigration sweeping the nation. Its ability to flourish in soils disturbed by humans became an important passage point through which it became a weed of civilization.

The urban landscapes in which ragweed settled and thrived were spaces of increasing importance to Progressive reform efforts of public health officials, social workers, and charity and civic organizations. In adapting to areas in the city where few other plants could survive, ragweed's identity became shaped by the physical, moral, and social geography of the city. Immigrant neighborhoods had been a specific target of public health officials, who looked suspiciously on the health and hygiene practices of the many newcomers, particularly those from southern and Eastern Europe, who came to American cities in the late nineteenth century seeking their fortunes or at least a better quality of life. To the sanitation efforts aimed at improving the unhealthy conditions of lower-working-class neighborhoods, Detroit mayor Hazen S. Pingree added a novel experiment in moral and social reform in 1894. Responding to the industrial slowdown and large-scale unemployment following the economic depression of 1893–94, which had strained city relief efforts and fueled labor unrest, Pingree proposed turning municipally owned and privately donated vacant land over to the poor and unemployed for the purpose of raising their own food. Pingree's Potato Patch scheme exceeded all expectations. In the first year, 945 families turned 430 acres of vacant land in the city into productive gardens that yielded fourteen thousand bushels of potatoes, in addition to beans, cucumbers, corn, tomatoes, and turnips. An initial investment of three thousand dollars reaped an estimated twelve thousand dollars in produce the first year. Within three years, twenty-five cities had adopted vacant-lot cultivation as a novel form of work relief. Civic reformers and charitable organizations believed vacant-lot gardens restored a sense of independence and self-respect to the urban poor, and helped assimilate immigrants into the American environment. In turning the soil, urban immigrants became homesteaders, those pioneers who had forged a nation, regardless of the fact that vacant-lot gardeners were never entitled to ownership of the quarter-acre lots or cooperative farms that they worked. But nature in the city did more than impart a sense of the "Jeffersonian agrarian ideal." To reformers, "fresh air and moderate exercise" greatly benefited the "physical and moral health" of those living where concrete, refuse, and squalor seemed to proliferate.[16]

While the return of prosperity by the early 1900s diminished vacant-lot gardening programs, despite continued interest among the urban poor, the playground and school garden movements extended reform efforts to revamp the wasteland of city slums. Through their active involvement in settlement houses, civic organizations, and domestic reform institutions, women played a critical role in the establishment of children's playgrounds and gardens on vacant lots and schoolyards in cities across the United States beginning in the 1890s. Believing that play on public streets contributed to vagrancy and juvenile delinquency, women representing organizations

16. Thomas J. Bassett, "Reaping on the Margins: A Century of Community Gardening in America," *Landscape* 25 (1981): 2; Allan Sutherland, "Farming Vacant City Lots," *American Monthly Review of Reviews* 31 (1905): 569. See also R. F. Powell, "Vacant Lot Gardens vs. Vagrancy," *Charities* 13 (1904): 25–28; Freder W. Speirs, Samuel McCune Lindsay, and Franklin B. Kirkbride, "Vacant-Lot Cultivation," *Charities Review* 8 (1898): 74–107.

FIGURE 17.2. Vacant city lots, such as this one in Chicago, were a focal point of civic reform efforts in late-nineteenth- and early-twentieth-century America and a prime habitat for ragweed. Adapted from *Charities and the Commons* 21 (1908): 50.

like the Massachusetts Emergency and Hygiene Association and the Mothers Club of Cambridge persuaded public officials to clear vacant lots of trash, garbage, and rubble so they could be converted into sand gardens. In Chicago, for example, clean-up efforts were funded by the Relief and Aid Society, which employed homeless men, representing fourteen different nationalities, to clear vacant lots of "dead dogs, tin cans, wire springs, and all sorts of rubbish" and to fill in bad holes with cinder, ashes, and sand (fig. 17.2). Once the filth had been removed, lots were equipped with swings, sandboxes, and see-saws, trees and shrubs were planted by city park commissions, and playground supervisors employed. Suzanne Spencer-Wood suggests that "in creating playgrounds, parks, and other green spaces, women's organizations used the domestic reform argument that women's high morality, closeness to nature, and domestic values of community and cooperation were needed to physically re-form men's immoral capitalist urban landscapes of unnatural stone." [17]

Nature, however, could work both ways in the city. While it could readily offset the physical, mental, and moral depravity of the concrete jungle, it could also be used to teach children lessons in efficiency, productivity, and civic virtue of importance to industrial progress and successful government. School gardens, part of America's nature study move-

---

17. B. Rosing, "Chicago's Unemployed Help Clean the City," *Charities and the Commons* 21 (1908): 51; Suzanne M. Spencer-Wood, "Turn of the Century Women's Organizations, Urban Design, and the Origin of the American Playground Movement," *Landscape Journal* 13 (1994): 134.

ment, offered the child much more than instruction in the principles of biology. "I did not start a garden simply to grow a few vegetables and flowers," wrote Mrs. Fannie Parsons, who established a school garden in New York City's DeWitt Clinton Park. "The garden was used as a means to show how willing and anxious children are to work, and to teach them in their work some necessary civic virtues; private care of public property, economy, honesty, application, concentration, self-government, civic pride, justice, the dignity, and the love of nature." All this from a vegetable garden on a vacant lot! But there was more. Henry Parsons suggested in his book *Children's Gardens for Pleasure, Health, and Education,* that children could learn the principles of scientific management through study and experiments that demonstrated how energy was wasted in carrying water cans inefficiently or through the improper selection and use of garden tools. By expending less energy, children would see, Parsons reasoned, how to "gain larger net returns in vegetables, flowers, health, strength, and knowledge for himself and for his fellow man." Through gardening, the future workers of America would become sound citizens schooled in the principles of Taylorism.[18]

To the city's middle and upper classes, vacant lots were places of waste, inefficiency, poverty and filth. Rarely, if at all, were vacant lots and nearby residents considered a by-product of the inefficiencies and injustices produced by industrial capitalism. To many, those who lived in the wastelands of the city were individuals whose ethnicity, race, or mental or moral attributes made them unable to successfully accommodate themselves to the American environment. Transients, eking out an existence on the economic margins, they were urban weeds. "We all know," wrote the physician Robert Hessler in 1911, "how large cities with a river front are infested by a class of people known as 'river rats,' a highly undesirable class, human weeds, so to speak." In Hessler's botanical and social survey of the city, the plant and human residents of "Shanty Town"—where waste lots prevailed—were nearly all foreign born. Hessler's meditations delivered on behalf of the Indiana Academy of Sciences brought together themes of disease, nativism, and back to nature that were a staple part of anti-immigrant and eugenic sentiments common during the Progressive era. But contrary to Hessler, and to the observations of an *Outing Magazine* au-

18. Quoted in Bassett, "Reaping on the Margins" (note 16), 3. On the nature-study movement, see Sally Gregory Kohlstedt, "Nature Study in North America and Australasia, 1890–1945: International Connections and Local Implementations," *Historical Records of Australian Science* 11 (June 1997): 439–54.

thor in 1906 that "scarcely any of our native weeds are especially trouble-some," at least one native, which flourished in vacant lots and waste places of the city, had become of increasing concern to urban hay fever sufferers and physicians in the early decades of the twentieth century: ragweed.[19]

Ragweed's visibility in the city, and its specific association with the ge-ography of waste, owed much to the rise of allergy clinics in major met-ropolitan centers in the 1910s. The clinical specialty of allergy, which fell under various headings including "protein sensitization," "human hyper-sensitiveness," and "clinical anaphylaxis," first developed around a set of practices that included preparation of pollen extracts, skin testing, and hyposensitive injection treatments. These experimental treatments were themselves a result of a major shift in the interior geography of hay fever as a disease from the nervous system to the body's immune system. The work of French physiologist Charles Richet on anaphylaxis in the early 1900s, demonstrating the hypersensitivity of animals to certain proteins, which could result in fatal shock, led the German physician Wolff-Eisner to speculate in 1906 that hay fever was an anaphylactic reaction caused by an increased sensitivity to atmospheric pollens. Pollen extracts quickly be-came an important diagnostic tool for physicians, since only hay fever suf-ferers would exhibit skin reactions to small doses of these protein extracts applied through scarification of the skin or intradermal injections. By 1920, at least a half dozen clinics located in Boston, Chicago, New York City, New Orleans, and San Francisco were in operation where hay fever sufferers could be tested and desensitization treatments, which often in-cluded a combination of pollen and vaccine therapies, administered.[20]

The therapeutic strategy of desensitization involved gradually acclima-tizing the body to the environment. Warren Vaughan, one of the founders

19. Robert Hesler, "Weeds and Diseases," *Survey* 26 (1911): 54; Clifton Johnson, "Our Im-ported Pests," *Outing Magazine* 48 (1906): 39. For more on the context of nativism as applied to germs, plants, and people, see Alan M. Kraut, *Silent Travelers: Germs, Genes, and the "Immi-grant Menace,"* (New York: Basic Books, 1994); Philip J. Pauly, "The Beauty and Menace of the Japanese Cherry Trees: Conflicting Visions of American and Ecological Independence," *Isis* 87 (1996): 51–73.

20. On the history of anaphylaxis and asthma research, including hay fever, see M. B. Emanuel and P. H. Howarth, "Asthma and Anaphylaxis: A Relevant Model for Chronic Dis-ease? An Historical Analysis of Directions in Asthma Research," *Clinical and Experimental Allergy* 25 (1995): 15–26. On Richet, see Kenton Kroker, "Immunity and Its Other: The Anaphylactic Selves of Charles Richet," *Studies in the History and Philosophy of Biological and Biomedical Sciences* 30 (1999): 273–96. On the history of allergy clinics in America, see Sheldon B. Cohen, "The American Academy of Allergy: An Historical Review," *Journal of Al-lergy and Clinical Immunology* 64 (1979): 332–466.

of the American Association for the Study of Allergy, explained hyposensitization in the following terms:

> The tissues of the person allergic to the pollen . . . are exposed to gradually increasing quantities of this allergen by means of hypodermic injections, until they reach a stage where the inhalation of previously damaging quantities produces no symptoms. . . . [T]he enemy has become naturalized. The workers find him in the community most of the time, but in such small numbers that they pay little attention to him. As the number increases, long familiarity leads to mutual readjustments so that the original antagonism no longer exists. A so-called foreign protein becomes less foreign, to the extent that the native cells are acclimated to its presence in their neighborhood.[21]

Assimilation was an important strategy in the relief efforts of both allergists and civic reformers in American cities throughout the early decades of the twentieth century. Those deemed incapable of adjustment or a threat to the health of the body politic were marked as weeds. Be they people or plants, such urban weeds had no place in the cultivated landscapes of urban America. They deserved only one fate: removal.[22]

Although weed ordinances existed in many municipalities and states, dating back to the 1860s, rarely were such bills enforced. In the early 1910s, the U.S. Hay Fever Association began lobbying its members to write to their local boards of health encouraging more active policing of city weeds. Most antiweed laws, however, applied strictly to those weeds along highways and on private property that were deemed a nuisance to agriculture. Seeking a noxious weed law regulated not by committees of forestry, agriculture, or highways, but by committees on public health, the U.S. Hay Fever Association drafted a bill modeled after Michigan's noxious weed law in 1915 that singled out ragweed, Canadian thistle, milkweed, wild carrots, oxeye daisies, goldenrod, and other noxious weeds believed aggravating to hay fever sufferers. Highway commissioners were responsible for cutting such weeds along roads twice each season, while property owners within the city limits were responsible for destroying weeds on their premises. Those that disregarded the ordinance risked hav-

21. Warren T. Vaughan, *Primer of Allergy: A Guidebook for Those Who Must Find Their Way through the Mazes of This Strange and Tantalizing State* (St. Louis: C. V. Mosby, 1939), 55.

22. The most prominent example during this period was Mary Mallon. See Judith Waltzer Leavitt, *Typhoid Mary: Captive to the Public's Health* (Boston: Beacon Press, 1997).

ing the city do the weed removal for them and a lien placed on their property until the city's expenses were paid. In 1916 an amendment to New York State's noxious weed law was added that charged person's failing to cut hay fever weeds such as ragweed and Canadian thistle with a misdemeanor and a fine of not less than five and not more than twenty-five dollars for each violation.[23]

The earliest, most systematic, municipal weed eradication effort in the name of hay fever prevention took place in New Orleans in the summer of 1916. The principal organizer behind this campaign was the laryngologist and president of the American Hay Fever Prevention Association, William Scheppegrell. Scheppegrell had established one of the earliest allergy clinics in the country at New Orleans Charity Hospital and began working closely with botanists, the U.S. Department of Agriculture, and the U.S. Public Health Service in extensive local and regional studies and surveys of North American plants found to cause hay fever. In May of 1916, he boldly asserted to members of the Louisiana State Medical Association that the number of hay fever cases could be reduced by 50 percent if common and giant ragweed were eliminated from the city of New Orleans.[24]

Mapping the location of ragweed throughout the city was the responsibility of a topographic committee that relied on a variety of informants. In its treatment of hay fever, Scheppegrell's clinic furnished patients with a nine-block map of their neighborhood on which they were to locate lots "infected with weeds." Considerable assistance was also furnished by the Women's Civic League, which appointed a committee to report on the location of vacant lots where ragweed grew in abundance. Through their active efforts in urban relief, from vacant-lot gardening to school playgrounds, women knew the geography of the city in ways that proved invaluable to the weed eradication programs of New Orleans and cities that later followed.[25]

23. U.S. Hay Fever Association, "Forty-second Anniversary Report," (1915), 7–8, 26; U.S. Hay Fever Association, "Forty-third Anniversary Report," (1916), 14.

24. See W. Scheppegrell to Miss L. B. Gachus, 7 August 1916, in U.S. Hay Fever Association, "Forty-fourth Anniversary Report," (1917), 24. See also William Scheppegrell, "The Seasons, Causes, and Geographical Distribution of Hay Fever and Hay Fever Resorts in the United States," *Public Health Reports* 35 (September 24, 1920): 2241–64; William Scheppegrell, *Hayfever and Asthma: Care, Prevention, and Treatment* (Philadelphia: Lea & Febiger, 1922).

25. William Scheppegrell, "Anaphylaxis Due to Pollen Protein, With a Report of the Results of Treatment in the Hay-Fever Clinic of the New Orleans Charity Hospital," *Laryngoscope* 28 (1918): 859; William Scheppegrell, "Hay-Fever: Its Cause and Prevention," *Journal*

In vacant-lot gardening programs, weeding was viewed by advocates as a form of physical and moral reform. R. F. Powell, superintendent of the Philadelphia Vacant Lot Cultivation Association, described how a "helpless paralytic" man "pulling the weeds from among the little plants" was transformed into a healthy, productive citizen, who in a matter of five years, became manager of a nine-acre farm. We can only wonder whether city officials looked on ragweed eradication efforts in similar terms. To clear the streets and sidewalks of ragweed in the outlying districts of the city, for example, twenty convicts were placed at the service of the American Hay-Fever Prevention Association. Did they see prisoners weeding the city of ragweed as a form of rehabilitation, or did they view their participation simply as a form of cheap labor? We can't be certain. But the destitute of society were often used in local ragweed eradication efforts as a form of social and inevitably moral uplift (fig. 17.3). During the Great Depression, for example, New York City put fifteen hundred unemployed men to work through the Work Progress Administration to rid the urban landscape—132,600,000 square feet to be exact—of ragweed. It was a form of work relief that had a history dating back to vacant-lot cultivation. The moral space vacant lots and their inhabitants occupied was one prevalent in the geographic imagination of the city's most upstanding citizens.[26]

Ragweed did not get its reputation as a "vegetable criminal," in botanist Oren Durham's words, simply from hanging out in seedy parts of town. This "undesirable citizen," even though it could "boast of being one hundred percent American," had characteristics that distinguished it from more notable patriotic plants.[27] In a national poll conducted by the *Independent* in 1918, the public selected goldenrod and columbine as the two most favored candidates for the national flower. When a concerned reader wrote to *Science* that goldenrod accounted for 15 percent of hay fever's symptoms, William Scheppegrell came to the plant's defense. In the north-

of the American Medical Association 66 (March 4, 1916): 707–12. In El Paso in 1927, an extensive hay fever weed eradication program was mobilized by the Women's Auxiliary of the El Paso County Medical Association. See N. S. Ives, "Weed Eradication by Community Effort Reduces Hay Fever," *American City* 37 (1927): 214. In 1938, a massive statewide "war on ragweed" in Michigan was assisted by the Boy Scouts, Camp Fire Girls, women garden clubs, and Daughters of the American Revolution. See, e.g., R. Ray Baker, "Boy Scouts Launch County Ragweed War," *Daily News,* 21 May 1936; R. Ray Baker, "Butler Named General of Washtenaw Ragweed Army," *Daily News,* 12 June 1936.

26. Powell, "Vacant Lot Gardens vs. Vagrancy," 27–28. On New York City efforts, see "Allergy, Pollen, and Ah-choo Time," *Literary Digest* 123 (June 12, 1937): 17.

27. Durham, *Your Hay Fever* (note 2), 145–46.

FIGURE 17.3. Ragweed eradication programs, such as this one in New York City, were part of a history of social relief and moral reform efforts dating back to urban vacant-lot cultivation programs of the 1890s. Adapted from *Life*, 24 August 1942, 52.

ern, eastern, and southern states, the pollens of ragweeds *(Ambrosiaceae)* were the principal causes of fall hay fever. In the pacific and Rocky Mountain states, the wormwoods *(Artemisia)* were the main culprits. Spring hay fever was largely caused, Scheppegrell reported, by the pollens of grasses.[28] And there were specific reasons why goldenrod should not be included in the "Rogue's gallery of the plant world."[29] All hay fever plants, Scheppegrell argued, displayed four characteristics that goldenrod did not share: "(1) they are wind-pollinated; (2) very numerous; (3) the flowers are inconspicuous, without bright color or scent; (4) the pollen is formed in great quantities." All were characteristics of "plants which occur as weeds in empty lots, neglected gardens, sidewalk and waste land generally." The inefficiency of hay fever plants, which produced a far greater excess of pollen grains than would be successful in reproduction, made them appear wasteful, not simply because of the places where they thrived, but also because of their reproductive traits. Neither ragweed, nor the particular

28. Horace Gunthorp, "Hay-Fever and a National Flower," *Science* 49 (1919): 147–48; William Scheppegrell, "Hay Fever and the National Flower," *Science* 49 (1919): 284–85.

29. "The Rogue's Gallery of the Plant World," *Better Homes and Gardens*, August 1934, 32, 51.

groups targeted by American eugenicists, practiced reproductive moderation. It was its alleged lack of reproductive restraint that made ragweed,
along with eastern and southern Europeans, the target of the American
justice system. In its migration from country to city, ragweed, despite its
American heritage, became part of the "immigrant menace" that threatened the health of the nation.[30]

Ecological Pioneers

In the urban spaces of clinics, public health, and reform, ragweed "produce[d] nothing of utility or beauty" and thus "deserve[d] only one fate—
extermination."[31] But the study and treatment of hay fever relied on
knowledge in both the clinic and the field. In the latter, ragweed and other
hay fever plants were valued products of nature. Almost every allergy clinic
either employed or worked closely with local botanists steeped in the taxonomy, ecology, and seasonality of regional flora in their diagnosis and
treatment of hay fever. On the West Coast, the limited amount of appropriate pollen samples available to allergists for clinical testing proved difficult in the teens and twenties. The prevalence of hay fever east of the
Rockies, and its believed absence in the West, meant that the few commercial drug houses involved in the preparation of pollen extracts had limited samples available that were representative of Western vegetation. Purity of extracts was also a problem, since incorrect species identification or
contamination of samples with other plant pollens could greatly affect testing and treatment. Questions about the specific sensitivity of hay fever sufferers to plant pollens also bore on problems of taxonomy, since needless
splitting of species proved unproductive to pollen collection if patients
could only distinguish between larger species or species groups.[32]

To Harvey Monroe Hall, a botanist at the University of California-
Berkeley, hay fever plants of the West, particularly the sagebrushes *(Artemisia),* offered abundant material for phylogenetic study and a lucrative
side income to support his family. In 1916, and again in 1917, he was hired
by Dr. Grant Selfridge, a San Francisco otorhinolaryngologist, to conduct
the first botanical and pollen survey of western hay fever plants. Funded by
a grant from the Southern Pacific Railroad, Hall traveled along the rail-

30. Scheppegrell, *Hay Fever and Asthma* (note 24), 38.
31. Durham, *Your Hay Fever* (note 2), 146.
32. See, e.g., Harvey M. Hall, "Hay-Fever Plants of California," *Public Health Reports* 37 (1922): 803–22.

road's right-of-way through Utah, Nevada, Oregon, and California collecting pollens from the most prevalent wind-pollinated species such as
cockleburs, sagebrushes, mugworts, and lambs-quarters, which Selfridge,
a hay fever sufferer, tested on himself. The Southern Pacific regarded Hall's
survey to be of practical importance, since the productivity of a significant
number of its outdoor railway men was affected by lost work time to hay
fever. "It is very attractive work," Hall wrote to his family from the field,
"like panning for gold, and the pollen is more valuable than gold by
weight." In his field journals, Hall commonly wrote down "gold" when
he meant pollen. For mugwort pollen, he received more than twenty dollars an ounce—the going rate for gold—from the pharmaceutical firm
Park-Davis Co. Other pollen collectors also commonly made the comparison to gold prospecting. But hay fever plants yielded Hall more than
just pollen and additional income. His botanical survey resulted in a major
publication on the ecological and evolutionary taxonomy of *Artemisia* and
a research position at the Alpine Laboratory in Manitou Springs under the
directorship of the plant ecologist Frederic Clements.[33]

While Hall's study of hay fever plants was limited to largely unpopulated regions in the West, during the 1920s extensive urban field surveys
were conducted by botanists in association with allergy clinics. Vegetation
maps of the city provided physicians with detailed knowledge about the
relative abundance of local hay fever plants, but they provided no information about daily, monthly, and seasonal pollen counts in the atmosphere.
Scheppegrell was the first to begin regular aerial sampling of atmospheric
pollen using glass slides coated with glycerin exposed for twenty-four
hours outside the eighth floor of the Audubon office building in down-
town New Orleans. In 1925, Oren Durham, chief botanist at Abbott Laboratories and a major influence behind the standardization of pollen sampling, counting, and survey techniques, developed a pilot pollen survey

33. Hall to Family, 11 August 1916, Harvey Monroe Hall Papers, Bancroft Library, University of California–Berkeley. On Selfridge's involvement and that of the Southern Pacific
Railroad, see Grant Selfridge, "Spasmodic Vaso-Motor Disturbances of the Respiratory Tract,
with Special Reference to Hay Fever," *California State Journal of Medicine* 16 (1918): 164–70;
and Cohen, "The American Academy of Allergy" (note 20), 333–34. For a later but similar
account of a pollen collector, see Steven M. Spencer, "Pollen for the Sneezers," *Saturday Evening Post*, 26 August 1944, 26–27, 105. Hall's phylogenetic study appears in Harvey M. Hall
and Frederic E. Clements, *The Phylogenetic Method in Taxonomy: The North American Species
of Artemisia, Chrysothamnus, and Atriplex* (Washington: Carnegie Institution of Washington,
1923). For biographical information on Hall, see Ernest Brown Babcock, "Harvey Monroe
Hall," *University of California Publications in Botany* 17 (1934): 355–68.

of Chicago in collaboration with University of Chicago physician Karl Koessler. Utilizing the quadrat sampling method developed in plant ecology by Clements, Durham and Koessler divided the city into 171-mile-square blocks. For each square, the percentage of area occupied by vegetation, use of the area (for example, industrial, residential), and relative abundance of plants were noted. Atmospheric pollen plates were exposed on a daily basis at various sampling sites in the city from June through September and pollen grains of grasses, ragweeds, chenopods, amaranths, wormwoods, composites, docks, plaintains, and miscellaneous other plant groups were counted. Based on field and aerial surveys, Durham and Koessler calculated that the ragweeds accounted for 65 percent of the total pollen load in the city of Chicago and 80 percent during the fall hay fever season. Intensity of hay fever symptoms on a daily basis furnished by clinical data correlated closely with ragweed pollen counts in the atmosphere.[34]

In 1929, Durham arranged the first national atmospheric pollen survey through the cooperation of the U.S. Weather Bureau. Through its services, Durham distributed uniform sampling devices, materials, and instructions to local meteorologists in twenty-eight cities throughout the United States. At weekly intervals, Durham received a shipment of slides, which by 1936 totaled eighteen thousand. He personally identified and counted pollen from all species of the ragweed family on every slide. Pollen counts were correlated with meteorological factors including temperature, sunshine, and rainfall for each city. In 1936, the number of sampling locations exceeded one hundred and extended from Winnipeg, Manitoba to Tampico, Mexico. The following year the New York *World-Telegram* was one of the first newspapers to begin publishing a daily pollen count in its weather section, today a standard part of television weather forecasts.[35]

At the local level, ragweed's identity took root in the urban wasteland, a place markedly shaped by the reform efforts of city public health officials

34. Karl K. Koessler and O. C. Durham, "A System for an Intensive Pollen Survey, with a Report of Results in Chicago," *Journal of the American Medical Association* 86 (April 17, 1926): 1204–9. For a later analysis of urban vegetation mapping directed not toward public health but toward species conservation, see Jens Lachmund, "Mapping Urban Nature: Bio-Ecological Expertise and Urban Planning in Germany, ca. 1979–1999," in Gerd Gigerenzer and Elke Kurtz, ed., *The Expert in Modern Society* (forthcoming).

35. Oren C. Durham, "Incidence of Ragweed Pollen in United States during 1929," *Journal of the American Medical Association* 94 (June 14, 1930): 1907–11; Durham, *Your Hay Fever* (note 2), 110; "Allergy, Pollen, and Ah-choo Time" (note 26), 17.

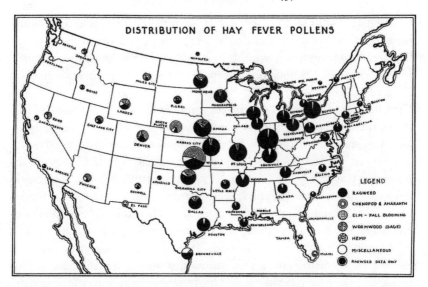

FIGURE 17.4. Pollen maps of the United States, such as this one from the 1930s, placed ragweed firmly within the ecological crisis of the Dust Bowl and the immorality of thoughtlessness and greed. *Journal of Allergy* 6 (1935): 129.

and social scientists. But Durham's national surveys, which relied on the knowledge of botanists and meteorologists, introduced another scale and another identity. The pollen maps of the United States produced in Durham's 1936 popular book *Your Hay Fever,* with the largest circles of ragweed pollen dominating the corn and wheat belt of the Midwest, placed ragweed firmly within an agricultural crisis gripping the nation (fig. 17.4).

For the American public, the mental image of 275,000 tons of ragweed pollen, "an invisible cyclone," as George Kent described it, blowing "like a storm" from the Great Plains to the eastern seaboard was not difficult to imagine or comprehend. In 1934, just two years previous to Kent's article in *American Magazine,* residents in eastern cities including Boston, New York, Washington, and Atlanta, watched as more than 12 million tons of midwestern dirt darkened the skies. By the time the dust storms subsided in the late thirties, 2.5 million people had abandoned their farms on the Great Plains. While many plains people blamed this national disaster on six years of drought, federal scientist began to attribute its cause to ecologically destructive land-use practices.[36]

36. George Kent, "Sneezes on the Breezes," *American Magazine* 122 (September 1936): 74. For a history of the Dust Bowl, see Donald Worster, *Dust Bowl: The Southern Plains in the 1930s* (New York: Oxford University Press, 1979).

To writer Paul Horgan, the Dust Bowl signified a struggle with nature lost to human arrogance and desire. "The men are taken away by life or death; and their houses stay until the weather and the weed render dust," he wrote in 1936. "A government report has called them, these abandoned places of human passage, visible evidences of failure. They are monuments. They lie ruined, as memories of audacities long-gone, and poor judgements; the almost inscrutable remains of aspiration wedded to tragedy."[37] But Horgan was not the only one to see a moral lesson in the "return of the weed." Roger Wodehouse, scientific director of the Hayfever Laboratory at the Arlington Chemical Company, attributed the increase in ragweed and hay fever to "man's recent land abuses." "Hay fever," Wodehouse told Natural History readers in 1939, "is nature's reply to man's destructive and wasteful exploitation of natural resources just as much as is soil erosion, wind erosion and floods. It is less spectacular than the great gullies carved out of hillsides by running water or the disasterous dust storms that bury farm buildings and move whole farms into the next state, or the floods that sweep away bridges. These are nature's answer in her boisterous mood. In her more subtle mood the answer is hay fever. And so softly it comes that few of us ever suspect that it is the answer to our thoughtlessness or greed."[38]

Amidst the topography of ecology and dust, ragweeds became "pioneers, not invaders." Ragweed was a symptom of human exploitation of the land. "Weeds," wrote ecologist Paul Sears, looking out on the blown out, baked out land in Oklahoma, "like wild-eyed anarchists, are the symptoms, not the real cause of a disturbed order."[39] Western hay fever plants like prairie sagewort (Artemisia frigida) were valuable signs to plant ecologists of overgazing on the Great Plains. Their presence signified an earlier stage in the development of the climax community, a stage in community succession, when the environment was more conducive to plant species that could gain a firm foothold in denuded soil. But native weed species, like ragweed, were not only good indicators of disturbed conditions. They were also important pioneers in regenerating the health of the land. "Ragweeds . . . perform a useful service," Wodehouse noted, "in holding the soil against wind and water erosion until it is taken over by other more

37. Paul Horgan, The Return of the Weed (New York: Harper & Bros., 1936), 3.
38. Wodehouse, "Weeds, Waste, and Hayfever" (note 2), 162.
39. Ibid.; Paul B. Sears, Deserts on the March, 4th ed. (Norman: University of Oklahoma Press, 1980), 113.

permanent plants." In the art of "land doctoring," weeds were both diagnostic tools of plant ecologists and nature's therapy.[40]

Ragweed's identity as a native pioneer species and its presence in the nation's upper atmosphere existed at odds with its place in the city, where its immigrant status and prolific reproductive habits made it the target of removal. Oren Durham looked skeptically on such municipal weed eradication programs, as did Wodehouse. Given the prodigious quantities of ragweed pollen in the upper atmosphere, Durham argued that local efforts, no matter how widespread and successful, would fail to affect what was essentially a national weather problem. Wodehouse offered a different reason, but it fit readily within the common space shared by botanists, meteorologists, and ragweed. "Ragweed was here thousands of years before the *Mayflower* touched the shores of the New World," Wodehouse told his fellow Rotarians, "and I predict that it will be here as many thousands of years after the human race has ceased to trouble this planet, faithfully doing its job of soil conservation." Citizens needed to follow nature's lead and practice sound soil conservation if ragweed's presence in the city was to diminish. Appropriate plantings in cleaned-up vacant lots would be far more effective in the long run, Wodehouse argued, than the usual "root-and-burn" method, which attacked the "symptom, not the malady itself." Pulling ragweed up by the roots only created more favorable conditions for the plant the following year. The newly disturbed soil offered an ideal environment for the germination of ragweed seeds, which could lie dormant in the ground and viable for forty years. In Wodehouse's urban ecology, ragweed may have still been a "vagrant riffraff of the plant world." Nevertheless, it held value as nature's instructor, saving "billions of dollars" in soil lost to erosion. Ragweed was nature's ecological pioneer and subtle revenge *against* the ravages of human civilization.[41] Only if humans were to become responsible stewards of the land would ragweed return to its proper place in nature. This was the moral lesson gleaned by ecologists, who had taken on the authoritative voice of Nature in the twentieth century (see Katharine Park's essay in this volume).

40. Wodehouse, "Weeds, Waste, and Hayfever" (note 2), 162. On the ecology of *Artemisia* in range management, see Hall and Clements, *The Phylogenetic Method in Taxonomy* (note 33), 110. The notion of ecologists as land doctors is from Aldo Leopold, *A Sand County Almanac with Essays on Conservation from Round River* (New York: Ballantine Books, 1970).

41. Durham, *Your Hay Fever* (note 2), 160–61; Roger P. Wodehouse, "Hold that Sneeze!" *Rotarian* 57 (July 1940): 43–44.

Controlling an Industrial Plant

While botanists and plant ecologists appealed to a conservation ethic in the control of ragweed, an engineering ethos dominated ragweed control measures after World War II. Cautionary tales of limits, spun from the depression, Dust Bowl, and a rationed economy, were dimly heard amid the renewed promises of science and technology in the war's aftermath. But warning signs of the environmental consequences of the nation's booming postwar industrial economy did not go completely unnoticed. In October of 1948, a thermal inversion blanketed the northeast and left many residents and virtually all the asthmatics in the small, industrialized town of Donora, Pennsylvania, gasping for breath. The Donoran smog, as it came to be known, was linked to emissions, particularly sulfur dioxide, from the town's zinc works and initiated a new era in air pollution research and regulation. Air pollution, previously regarded as the domain of industrial engineers, became after the Donoran incident an issue of environment and health.[42]

Environmental historians have looked to the Donoran smog episode as a crucial event in moving air pollution concerns out of the federal Bureau of Mines into the Public Health Service, where it became the focus of research in industrial hygiene, sanitary engineering, and medicine.[43] But regard for the health effects of air pollution did not emerge solely from the invisible toxins spewing out of industrial smokestacks like those of the American Steel and Wire Company. Exactly one year before the Donoran smog killed seventeen people, the Committee on Air Pollution of the American Public Health Association concluded that of all contaminants in the atmosphere regarded as public nuisances, only two were proven health hazards: radioactive substances and pollen. "The control of atmospheric pollution," committee members recommended, "is an administrative and technical problem that can be solved by engineering means."[44] Ragweed, a source of air pollution threatening the public's health, suddenly

---

42. See, Lynne Page Snyder, "'The Death-Dealing Smog over Donora, Pennsylvania': Industrial Air Pollution, Public Health, and Federal Policy, 1915–1963," Ph.D. diss., University of Pennsylvania, 1994.

43. Both Snyder, "The Death-Dealing Smog" (note 42) and David Stradling, *Smokestacks and Progressives: Environmentalists, Engineers, and Air Quality in America, 1881–1951* (Baltimore: Johns Hopkins University Press, 1999), cite the Donoran smog episode as a crucial event in this shift.

44. H. A. Whittaker et al., "Report of the Committee on Air Pollution," *American Journal of Public Health* 38 (1948): 768.

become part of industrial regulation and control. Some industrial scientists like Louis McCabe and Robert Kehoe strongly objected to blurring the boundaries between natural and "localized man-made metropolitan air pollution," which the Committee on Air Pollution had done, but the municipal control of ragweed after World War II only reinforced ragweed's boundary transgressions between the natural and engineered.[45]

An industrial plant like ragweed required industrial strength control measures. Luckily, for sanitation engineers determined to eliminate one of the city's most notorious polluters, the nation's military-industrial complex had produced an ideal weapon of destruction: the herbicide, 2, 4-D. The development of 2, 4-D derived from interwar research in plant physiology on growth hormones that became part of a top-secret, National Academy of Sciences' wartime committee on biological and chemical warfare. Although a major priority dispute and a prolonged legal battle between the American Chemical Paint Co. and Dow Chemical erupted after the war over the discovery of 2, 4-D as an effective herbicide, the chemical quickly made its way into the marketplace. Sold under varying brand names including Weedone, Weedex, Weed-be-Gone, and Weed-No-More, 2, 4-D was initially met with great enthusiasm by suburbanites and golf course owners who sought to keep their lawns and fairways free of broad-leafed weeds. Within a few years, agricultural scientists successfully sold its benefits to farmers. In the field of public health, 2, 4-D spraying campaigns against ragweed exceeded the use of DDT for mosquito control. By 1949, production estimates of the weed killer had reached 20 million pounds.[46]

The earliest, largest, and most sustained effort at ragweed eradication using 2, 4-D began in New York City in the summer of 1946. A plan submitted by the Health Department's Bureau of Sanitary Engineering met with the approval of the mayor and city council, which released funds for the project. During the first year, approximately 850,000 gallons of the herbicide were sprayed on three thousand acres of public properties, roads, and sidewalks, and on privately owned vacant land where ragweed was found. Maps that furnished information on the location and abundance of ragweed were provided to spraying crews by local police department

45. Louis C. McCabe, "National Trends in Air Pollution," in *Proceedings of the First National Air Pollution Symposium* (Los Angeles, 1949), 52. See also Robert A. Kehoe, "Air Pollution and Community Health," *Proceedings of the First National Air Pollution Symposium* (Los Angeles, 1949), 115–20.

46. Nicolas Rasmussen, "Plant Hormones in War and Peace: Science, Industry, and Government in the Development of Herbicides in 1940s America," *Isis* 2 (2001): 291–316.

precincts. The involvement of police in vegetation mapping only reinforced the moral space vacant lots occupied within the historical geography of urban reform. Mosquito control units, road oiling and chloride distributing equipment, street flushers, and tree-sprayers furnished by the U.S. Public Health Service, Borough President's Offices, the Department of Sanitation, and the Department of Parks were converted into thirty-three spraying units. By the early 1950s, municipal demand prompted manufacturers to design spray units that could be driven and operated by one person to cover "infected" areas of ragweed with this new "hayfever vaccine." A widespread public educational campaign was also launched that included exhibits, posters, pamphlets, and leaflets distributed to and enlisting the aid of civic groups, chambers of commerce, hayfever and ragweed societies, and educational institutions.[47]

Although nine years of spraying had reduced the ragweed population in the New York City metropolitan area by half, and also reduced the atmospheric pollen count, hay fever sufferers continued to be plagued by ragweed pollen blowing from the west side of the Hudson River. The control of pollen pollution was an interstate problem that required a large-scale coordinated effort. By 1955 more than 150 communities in the state of New Jersey alone had adopted municipal ragweed spraying campaigns in an effort to reduce atmospheric contamination. Such programs helped push the sale of 2, 4-D to over $50 million that year.[48]

Despite the objections of industrial scientists, who wished to maintain a distinction between the products of nature and civilization, it was difficult to differentiate between the cultural geography of ragweed and air pollution in the immediate postwar years. Alfred Fletcher, director of the Division of Environmental Sanitation for the New Jersey State Department of Health, believed that a new air pollution law and the appointment of an air pollution control commission legislated by the 1953 statute, which included the control of airborne pollen, added considerable legal muscle to the rather weak municipal antiweed ordinances that were difficult, time-consuming, and expensive to enforce. Pollen studies were

47. See, "H-F Day," New Yorker, 24 August 1946, 17–18; Philip Gorlin, "Planning and Organizing a Ragweed Control Program," American City 62 (June 1947): 88–90; Israel Weinstein and Alfred Fletcher, "Essentials for the Control of Ragweed," American Journal of Public Health 38 (May 1948): 664–69; "Noncompetitive," New Yorker, 26 August 1950, 17–18.

48. Philip Gorlin, "Ragweed Control in New York City," Hay Fever Bulletin 6, no. 1 (1955): 4–6; Alfred H. Fletcher, "Ragweed Control in New Jersey," Hay Fever Bulletin 6, no. 2 (summer 1955): 7–8, 12.

also an important component of the air pollution research activities of the U.S. Public Health Services' newly established Robert A. Taft Sanitary Engineering Center in Cincinnati. And Philip Gorlin's move from supervisor of New York City's Weed & Insect Control Unit, to technical director and coordinator of its Ragweed Control Program, and finally to supervisor of the city's Department of Air Pollution Control is further evidence of how easily ragweed moved across the divide separating nature from civilization.

Getting Back to Nature

Weed eradication measures signified postwar American optimism in the complete technological control of nature (fig. 17.5), an optimism that was embraced equally by public health officials, environmental engineers, and segments of postwar feminism (see Michelle Murphy's essay in this volume). Those species deemed menaces to humans were readily dispensed with. When ragweed moved from the wild to the metropolis, it was no longer subject to the laws of nature, but to those of society. Or so sanitary engineers and public health officials readily believed. But the arrogance and optimism seen in weed spraying campaigns were tempered by another moral lesson ragweed imparted, one that became increasingly prominent in the United States after World War II. It was a lesson, not of nature transgressing boundaries, but of humans stepping outside their place in nature. Rachel Carson's *Silent Spring,* one of the classic texts of America's environmental movement, would bring into question whether humans ultimately would pay the price for such intellectual hubris seen in engineering nature. Unlike the nature of archaic Greek poetry, nature in 1960s American environmentalism took on a foundational moral authority, the ultimate arbiter of justice (see Laura Slatkin's essay in this volume). Cancer, genetic deformations, and dying songbirds were just a few of the specters Carson evoked in demonstrating nature's revenge against the ravages of human civilization. Just as Wodehouse saw in ragweed's proliferation nature's reply to human arrogance and greed, so too did Carson play on an ecology of fear in her writings. Nature did impose constraints that humans ignored at their peril. And allergy came to be regarded as one such price of civilization in the twentieth century.

At the time of 2, 4-D's release, another weapon was added to the arsenal of hay fever sufferers that would once again alter the topographic space that ragweed occupied in American culture. Antihistamines, in particular,

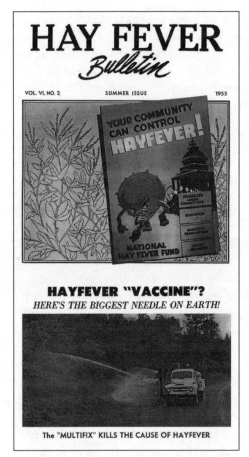

FIGURE. 17.5. During the 1950s an industrial plant like ragweed required industrial strength control measures. Herbicides, like 2,4-D, used in municipal ragweed control programs signified postwar American optimism in the complete technological control of nature.

benadryl, first became available as a prescription drug in the mid-1940s. Concerns about side effects, which included drowsiness and convulsions, and their long-term impact, particularly on children, made the public cautious about their use in the treatment of hay fever compared to the initial enthusiasm expressed for 2, 4-D. Such concerns were short-lived. Today, antihistamines represent the first line of attack Americans adopt in their struggles with allergy.

Pharmaceuticals targeted at chronic allergy sufferers point to a different nature than that from which ragweed emigrated—not to the nature outside us, but to that within us. By biochemically altering the allergic body, biomedicine has freed individuals from the constraints of the external environment. And nature is also made safe from civilization. In contemporary drug ads, like those for the new and improved antihistamine Claritin, allergy sufferers can now "get back to nature," represented by a meadow of beautiful wildflowers in hyperreal colors. Through antihistamines, pollen is once again made pure; ragweed is returned to the wild. After a long and arduous journey, ragweed has, through the technological wonders of biomedicine, once again founds its place in nature.

18

Three Roots of Human Recency: Molecular Anthropology, the Refigured Acheulean, and the UNESCO Response to Auschwitz

Robert N. Proctor

That Neanderthals are thought of in terms of a "problem" or a "question" is remarkably similar to the way in which Germans thought about Jews prior to World War II. In both instances, the objects of such treatment were cast in the role of a collective "other" whose differences have been assumed to indicate the extent of their failure to qualify for fully human status.

    C. LORING BRACE

When did humans become human? Did this happen 5 million years ago or fifty thousand years ago? How sudden was the transition, and is this even a meaningful question? Strange as it may seem, there is radical disagreement over the timing of human evolution, understood as the emergence of the language-using cultural creature of today. No one knows whether speech, consciousness, or the human aesthetic sense is a fairly recent phenomenon—ca. 50,000 years ago—or ten or even a hundred times that old, though it seems that recency currently enjoys the upper hand.[1]

This is an early version of a longer paper that appeared, with comments, in *Current Anthropology* 44 (2003): 213–39.

    1. The sentiment can be found in many different sciences of prehistory: witness the recent skepticism with regard to the formerly rock-hard claims for human-generated fire at the five-hundred-thousand-year-old site at Zhoukoudian, south of Beijing (S. Weiner et al., "Evidence for the Use of Fire at Zhoukoudian," *Science* 281 [1998]: 251–53.) Hominid control of fire had been pushed back to 1.5 or even 1.8 million years ago in Britain and elsewhere, though there is the perennial problem of how to distinguish human-made from accidental fire. Scholars have looked for, but not yet found, the tell-tale carbon signatures of anthropogenic fire in the seabeds downwind from the hominid sites of Africa; see M. I. Bird and J. A. Cali, "A Million-year Record of Fire in Sub-Saharan Africa," *Nature* 394 (1998): 767–69.

For many years, it was fashionable to project "humanness" (whatever that might mean) into any and every hominid scratched out by a paleontologist; Lucy was "our oldest ancestor," an Australopithecine "woman" (versus "female"); and even older hominids were sometimes granted humanity. Today, however, it is more common to see the *Australopithecines* as far more chimp-like; "humanness" is often not even granted to *Homo erectus,* the earliest in our genus (itself an arbitrary designation), and there are those who do not want to ordain the Neanderthals or even early *Homo sapiens* as "fully human."

What is going on here? What makes us want to grant or withdraw humanity from a given or presumptive ancestor? What is the evidence one way or another, and what larger prejudices are at stake?

In this essay I explore some of the separate lines of evidence leading to the idea that humanness in the *cultural* sense is a relatively recent phenomenon—no more than 100,000 years old, and perhaps even more recent than this, since that is when you get the first clear signs of representational art, kindled fire, deliberate ritual, compound tools, and other things we like to think of as signs of human intelligence. To avoid becoming bogged down in definitions, I want to equate "humanness" for the moment with language and culture—recognizing that these categories are no more secure, no less in flux than *Menschlichkeit.* Witness the recent work on chimpanzee "material culture,"[2] which casts the traditional Boasian concept in an altogether different light from how American anthropologists have regarded this category. I simply want to black box some of these definitional issues to make sure I get across the novelty implicit in recent thinking with regard to human recency.

For example, it was widely thought several decades ago that the 2-, 3-, and then 4-million-year-old hominid fossils being found in Africa had "culture" in the Boasian sense—including folkways and mores, fables and religion, and so forth. Humanness in the wake of the 1950 UNESCO *Statement on Race* was pushed back even into the middle Miocene—as when Louis Leakey suggested that *Ramapithecus* ca. 14 million years ago was a "hominid" and "tool-user"—both of which were taken to mean that the creature was human in some deep and inclusive sense.[3] The equa-

2. Frans B. M. de Waal, "Cultural Primatology Comes of Age," *Nature* 399 (1999): 635–36; William C. McGrew, *Chimpanzee Material Culture: Implications for Human Evolution* (New York: Cambridge, 1992).

3. John Gribbin and Jeremy Cherfas, *The Monkey Puzzle: Reshaping the Evolutionary Tree* (New York: Pantheon Books, 1982), 12. Leakey was not unusual in this regard; Pat Shipman,

tion of hominid and humanity fit with the older tradition of humans as an evolutionary *Sonderweg:* only humans use tools, tool-use implies language, language implies culture, language and culture are unique to humanity, and so forth. It also had certain advantages for career-conscious fossil-finders, since it was surely preferable to have found some kind of human rather than some kind of chimp (or worse). It was not until the 1960s that Vince Sarich and Allan Wilson showed that humans shared a common ancestor with chimps only 5 – 6 million years ago — and not until the 1980s that this idea was widely accepted. (A few maverick evolutionists as recently as the 1960s could maintain that humans and apes had not shared an ancestor since the Eocene — roughly 50 million years ago by modern counts.) It is also noteworthy that it took a racial inegalitarian (Vincent Sarich) to discover the more recent split — more on that below.

Much of that consensus — equating hominid and humanity — has been broken in the past couple of decades, and here I explore how and why that came to pass. It has partly to do, of course, with Jane Goodall's celebration of nonhuman tool use and, to a lesser extent, the rise of "pop ethology," evolutionary psychology, and sociobiology; but there are several other key transitions that warrant an accounting. I focus on three of these transformations, or "crises," all of which have given force to the idea that humanness may be relatively recent:

(1) *Archaeology.* First is the crisis in interpretation of the *oldest tools*— specifically the Oldowan and Acheulean assemblages of the Lower Paleolithic, the oldest tools to have epochal names attached, and the oldest to count as evidence of hominid or human "culture." (Chimpanzee cultural traditions have thus far been treated only ahistorically, since there is almost no "archaeological" evidence of chimpanzee tool use — though Frédéric Joulian has recently found evidence of stone anvils being used by chimps for at least two hundred years and we should, in theory, be able to find these going back millions of years).[4] The key question here is whether Oldowan and Acheulean artifacts can be considered evidence of a cul-

---

who did her Ph.D. on the Kenya *Ramapithecus* site, notes that in the mid 1970s "everybody thought *Ramapithecus* was a hominid" (personal communication).

4. Frédéric Joulian in his "Techniques du corps et traditions chimpanzières" (*Terrain,* March 2000, 37–54) describes sites where nut-cracking has gone on for at least two hundred years, judging from the number of nutshells and wear on the stone anvils at such sites. William H. Calvin speculates that early hominids may have begun deliberately chipping stone after having accidentally caused stones to flake in the course of chimp-like hammering. See his *The Throwing Madonna: Essays on the Brain* (New York: McGraw-Hill, 1983), 27; see also his *A Brain for All seasons* (Chicago: University of Chicago Press, 2002), 133–46.

tural "tradition" in any interesting sense—part of human "history." An argument can be made that they cannot, or at least cannot in the conventional Boasian sense, given their apparent stability and uniformity over vast stretches of time and space. Oldowan tools persist for roughly a million years in Africa (from 2.5 to 1.5 million years ago), and Acheulean tools last even longer, from about 1.5 to .2 million years ago. The argument has been made that one reason these tools are so stable is that their users were not transmitting knowledge of their use by abstract symbols, and that some other mechanism must account for their endurance. One possible implication is that their inventors were not yet human in any interesting sense (for example, not linguate creatures); some kind of nonlinguistic transmission may have been involved (imitation, for instance), the way Japanese macaques copied Imo the inventive one, who sorted grain from sand by tossing them both into the water (grain floats).[5] The apparent stability of the Acheulean remains a puzzle.

(2) *Paleontology.* This crisis derives from the recognition of *fossil hominid phyletic diversity*—another innovation of the 1960s and 1970s, following spectacular south and east African hominid fossil finds (Leakey's *Homo habilis,* Johanson's *Australopithecus* "Lucy," and so on) showing that more than one species of hominid must have existed at many points in the course of hominid evolution. Many paleoanthropologists today place the total number of hominid species at about twenty—in four or five distinct genera: *Australopithecus* and *Homo,* of course, but perhaps also *Paranthropus* and *Ardipithecus,* and now *Kenyanthropus, Orrorin,* and *Sahelanthropus.* Hominid diversity seems to have peaked about 2 million years ago, when three, four, or even more separate hominid species coexisted on the planet (and all in central or east Africa). The present situation, in fact, where there is only one surviving species—*Homo sapiens*—seems to be an unusual state of affairs in the 5-million-year span of "human" evolution. There may have been other periods with only one hominid, but the last thirty thousand years or so—since the extinction of the Neanderthals—is certainly unusual in having only one living representative of the hominid family. Fossil hominid diversity was not accepted without a struggle, however. There was a certain degree of ideological resistance stemming from the liberal antiracialist climate of the post-Auschwitz era, when it was

---

5. Calvin, *Throwing Madonna* (note 4), 22–23. One could also argue that the stability of Acheulean tools implies a tremendous efficiency of that tool type, given the needs of protohumans over that vast stretch of time. Acheulean stability in this sense would be evidence not of a lack of inventiveness, but rather of absence of more prolific needs.

dogmatically assumed that only one hominid species could exist at any given time (the "single species hypothesis"). This is interestingly tied to the reevaluation of *race* in the early post–World War II era, when a broad anthropological consensus emerged that the humans living today are more or less equal in terms of cultural worth and standing in the Family of Man— culminating in the 1950 UNESCO *Statement on Race,* which branded race an "unscientific" category and "man's most dangerous myth" (Ashley Montagu's epithet).

(3) *Molecular anthropology.* A third crisis or turning point stems from the recognition that all living humans have descended from a small group of Africans who lived roughly 150,000 years ago. "Modern humans" are therefore relatively recent in a *biological* sense, though nothing is necessarily implied about *cultural* recency (since earlier hominids could still have been linguate/complex cultural creatures). This "Out-of-Africa II" scenario has received immense coverage in the popular press—through its vivid emblem of an "African Eve," of course, but also through the clarity and simplicity of its opposition to the "multiregional" or "Regional Continuity" hypothesis, according to which the diverse local *Homo erectus* populations in different parts of the world didn't go extinct (as proposed by the out-of-Africanists), but gave rise to the *Homo sapiens* that eventually evolved in those regions. The opposing molecularist, sequence-based recency thesis has become the dominant view, partly through the strength of its molecular methods,[6] but also by successfully tarring the multiregional model (originally proposed by Franz Weidenreich in the 1930s) with older polygenist traditions, which presumed deep and usually invidious racial divisions.

All three of these transformations—archaeological, paleontologic, and genetic—have been important in the rising stock of human recency. Of course, the factors I have mentioned and will elaborate in a moment are not the only elements at work; there are others,[7] including the triumph of Gould and Eldredge's punctuated equilibrium, or efforts by paleoanthro-

---

6. There have also been important physical anthropologists championing "out-of-Africa"—for example, Christopher Stringer of London's Natural History Museum, who had already made a good case for Neanderthal "replacement" when the molecular evidence became available. See Erik Trinkaus and Pat Shipman, *The Neandertals* (New York: Vintage, 1992), 360–419.

7. The argument—spurious in my view—has been made that human linguistic diversity is too shallow to be very old, and that human languages diverged from a common origin only about thirty thousand years ago. See Merritt Ruhlen, *The Origin of Language: Tracing the Evolution of the Mother Tongue* (New York: Wiley, 1994).

pologists like Richard Klein, who argues that the explosive growth of human innovativity ca. fifty thousand years ago—Pfeiffer's "creative explosion" or Diamond's "great leap forward"—may be traceable to some sort of "neural mutation."[8] Recency is not the same as suddenness, however, and the idea of recency has become (interestingly) at least as popular among antigradualists as among Gouldians. Indeed, it was two *anti*-Gouldian aspects of the thesis that first piqued my own interest in human recency: (1) the idea that language capacities may have developed relatively late in human evolution, long after the *sapiens* speciation event, contra what one might expect from punctuated equilibrium; and (2) the awkward fact that the human cultural "Big Bang" seems perilously close to the point of human racial differentiation and dispersal (in the extreme recency model), raising the specter that some "races" may actually have become "human" earlier than others—a common idea among segregationalists and polygenists as late as the 1950s and 1960s.[9] Both of these are rather non-Gouldian concerns and avoidable; both, I would say, can be rectified with a proper chronicling of events consistent with both racial egalitarianism and punctuated equilibrium.

Before I turn to these crises, let me make two methodological points about opportunities for historical inquiry in this area. The first is simply a call for historians of science and technology to entertain paleoanthropology and the Paleolithic. Paleoanthropology is a fascinating and understudied area of modern technoscience, full of adventure and ideology; but so, too, at least in this latter aspect, is the Paleolithic itself. Prehistoric tools have generally not become the objects of analysis by historians of technology—and the explanation is fairly obvious (if moronic), given the founding mytho-myopia of our discipline (history) that "historical" events are those that postdate the invention of writing ca. 3000 B.C. The parochialism of such an approach has long been obvious to oral historians, archaeologists, and historians of material culture; but the history of tools prior to text

8. Jared Diamond, *Guns, Germs, and Steel: The Fates of Human Societies* (New York: Norton, 1997); John E. Pfeiffer, *The Creative Explosion: An Inquiry into the Origins of Art and Religion* (New York: Harper & Row, 1982). Richard G. Klein points out that it is not until about fifty thousand years ago that you find the first evidence of religion, representational art, ornamentation, new and changing tool styles, perhaps even the first boats; see his *The Human Career: Human Biological and Cultural Origins*, 2nd ed. (Chicago: University of Chicago Press, 1999), 512–17, 590–91; see also Ian Tattersall, *Becoming Human: Evolution and Human Uniqueness* (New York: Harcourt Brace, 1998), and Margaret W. Conkey et al., eds., *Beyond Art: Pleistocene Image and Symbol* (Berkeley: University of California Press, 1997).

9. See, for example, Carleton S. Coon, *The Origin of Races* (New York: Knopf, 1962).

remains remarkably undertheorized—by historians, at least.[10] I would therefore like to make a pitch for "deep history of technology," closer collaborations with archaeologists and prehistorians, a serious reckoning with that 99.9 percent of hominid experience that predates what historians define as "history proper" (since the invention of script), perhaps even an increased attention to human evolution as part of our understanding of history (and vice versa). The textual turn in anthropology[11] in this sense needs to be complemented by a nontextual (or pretextual) turn; we need to problematize the disciplinary divide that has tended to isolate prehistorians from historians of technology.

A second point is that we need to look for the political good in the technically bad and vice versa, the politically bad in the technically good. The point is not that tools may be used for good or for ill (think of Tom Lehrer's version of Wernher von Braun), but rather that political evil may be creative, and political good-will stifling. Nazi tobacco research is an obvious case of the former—the fertile face of fascism[12]—whereas the UNESCO *Statement on Race* is, I'll argue, a heretofore unnoticed example of the latter, since one of my claims in this essay is that the racial liberalism of the 1950s and 1960s was partly responsible for delaying the recognition of fossil hominid diversity by ten or twenty years. Let me turn now, though, to archaeology, moving then to paleontology, and finally to race and genetics.

Refiguring the Acheulean

In 1797 John Frere, an English country squire and former high sheriff of Suffolk, discovered a number of curious artifacts in a brick-clay pit in the parish of Hoxne. In a letter published three years later in *Archaeologia,* the journal of the Society of Antiquaries, Frere described the implements as "evidently weapons of war, fabricated and used by a people who had not the use of metals." The situation under which they had been found led him to conclude that they had to be very old, having been buried under

10. Among archaeologists there is a great deal of theory in this sense. For a sampling, see especially the work of Clive Gamble, Tim Ingold, Margaret Conkey, Pierre Lemonnier, and André Leroi-Gourhan.

11. I mean more in the sense explored by Brinkley Messick in his *Calligraphic State: Textual Domination and History in a Muslim Society* (Berkeley: University of California Press, 1993), and not so much in the sense intended by Clifford Geertz, Hayden White, and others that "all the world is text."

12. See my *The Nazi War on Cancer* (Princeton: Princeton University Press, 1999).

ten feet of well-stratified vegetable earth and "Argill" clay. Frere concluded that this particular manner of burial, plus their association with the bones of animals no longer found in England, meant that these artifacts must date from "a very remote period indeed; even beyond that of the present world." [13]

Historians have often commented on the failure of Frere's contemporaries to recognize the antiquity of human artifacts; his paper went more or less unnoticed for more than fifty years, until the Prehistoric Revolution of 1859, when from diverse angles—and fairly suddenly—it was recognized that humans have a profound antiquity.[14] Paleontologists even in Frere's time had by and large abandoned Archbishop Ussher's estimate of six thousand years since creation (Buffon had calculated seventy-five thousand years based on experiments with cooling bodies), but the absence of human remains in geologic deposits had made it unfashionable to argue for the existence of humanity beyond the more miserly biblical chronology. Paleontologic time-markers were introduced in the early decades of the nineteenth century, but even George Cuvier, the primary architect of such markers, died in 1832 believing that there was no such thing as fossil man.[15] It was not until the late 1850s that human antiquity was widely recognized, the key event being the acceptance by English and French geologists of the authenticity of the Acheulean "hand-axes" found by Boucher de Perthes in the gravels south of St. Acheul, now a suburb of Amiens, northwest of Paris.[16] The discovery was interestingly coincident with the publication of Darwin's *Origin of Species,* though the latter book seems to have had little or no immediate impact on the question of human antiquity. The leading architect of the revolution, Boucher de Perthes, was in fact a biblical catastrophist and antitransformationist who argued that humans were probably created and destroyed several times before Adam was called into being.

13. John Frere, "Account of Flint Weapons Discovered at Hoxne in Suffolk," *Archaeologia* 13 (1800): 204–5.

14. Donald K. Grayson, *The Establishment of Human Antiquity* (New York: Academic Press, 1983); Claudine Cohen and Jean-Jacques Hublin, *Boucher de Perthes, 1788–1868: Les origines romantiques de la préhistoire* (Paris: Belin, 1989).

15. Martin Rudwick, *Georges Cuvier, Fossil Bones, and Geological Catastrophes* (Chicago: University of Chicago Press, 1997), 232–34. Cuvier's denial of human antiquity can be found in his 1812 *Discours préliminaire,* translated into English in 1813 as *Theory of the Earth* with an elaborate preface by Robert Jameson, who gave the book a natural theologic flavor, equating Cuvier's last revolution with the biblical flood.

16. Grayson, *Establishment* (note 14), 168–223.

Acheulean tools are remarkable in several different respects—quite apart from their stunning beauty and symmetry, earning for them recognition as the very first traces of a human aesthetic sense (Oldowan tools, dating back to ca. 2.5 million years ago, look more like crudely broken rocks). The oldest "hand-axes" are about 1.5 million years old, and the youngest about one hundred thousand.[17] That brings us to one of the most remarkable features of such "tools" (if that is what they are): their relative uniformity over vast reaches of time and space. Acheulean "hand-axes" are found for a span of 1.4 million years—more than fifty thousand generations—over most of the range occupied by *Homo erectus,* from the Pleistocene gravels of England (though not in Ireland, scraped clean by later glaciers) to the open-air sites of northern Spain, Algeria, and Morocco, to the famous *erectus* sites of Africa and as far east as the Urals and occasionally beyond, as indicated by the recent finds at Bose Basin in Nihewan southern China. (Much debate has swirled over whether early hominids ever crossed any significant body of water; most paleoanthropologists have said that this does not occur until fairly late in human evolution, ca. forty or fifty thousand years ago, when humans voyaged into Australia, though the seven-hundred-thousand-year-old tools recently found on Flores Island in Indonesia may make this harder to argue. The question is: did people "boat" to Flores, or just drift and land as castaways on accidental rafts? The question in a sense divides two different theories of wondrous capacity: of early hominids to build, and of wind and waves to mobilize log jams and peat mats.)

A second remarkable fact about hand-axes is how difficult it has been to come up with an adequate sense of how to interpret them. This is partly due to the lack of contemporary ethnographic evidence (Acheulean tools have not been used for at least a hundred thousand years), but also due to the difficulties of understanding what life was like for creatures that may have been very different from us. *Homo erectus* is generally assumed to have made these tools, but there are several other (albeit closely allied) candidates, including *Homo habilis, Homo rudolfensis, Homo antecessor, Homo heidelbergensis,* and *Homo ergaster.* The idea of the same type of tool being made and used by entirely different species is not one that many cultural anthropologists are generally comfortable with—which is one reason to doubt whether the Acheulean is a "culture" or "tradition" in any interest-

17. An interesting visual display can be found at http://www.personal.psu.edu/users/w/x/wxk116/axe.

ing sense. The term itself blends paleontologic and ethnic categories— nature and culture—since there are some who talk about the "Acheulean" as a "people" (strange, since it may have embraced three or four different species), and others who treat it more as a chronologic or periodizing category (like "Pleistocene"), and still others who regard it as simply a formal tool-type designation ("Acheulean hand-axe"). The ambiguity is reflected in how and where such artifacts are found, since they were produced for so many hundreds of millennia as to have become distributed as quasi-geologic objects. They can almost be used as index fossils, for example, to date a sediment. They are also fairly "plastic" when it comes to imputable theories of function, which means they can be used as "inadvertent Rorschach tests," blank slates onto which different conceptions of antiquity have been inscribed.

Frere is credited as the first to recognize the profound antiquity of stone tools, but he was by no means the first to have collected and marveled at them. Stone tools of various sorts have been picked up since time immemorial, a common understanding being that these were the work of fairies or some other natural or magical cause. (The oldest image of a paleolithic artifact may be the medieval French painting depicting St. Etienne holding a typical Acheulean flint "hand-axe.") Agricola and Gesner suggested that chipped-flint implements were the traces left by thunderbolts; there was also the idea that such artifacts had originally been made of iron and had converted into stone by their long continuance in the earth.[18] Stone oddities of various sorts must have been picked up wherever people are curious; some people may even have recognized some as "prehistoric"— but in the absence of a well-founded belief in profound human antiquity, prehistoric artifacts seem rarely to have been written about as such. Aldrovandi in his posthumous 1648 *Musaeum metallicum* classed stone points along with *glossopetrae* ("tongue stones"), or what we today would recognize as sharks' teeth; Mercatis's *Metallotheca* (first published in 1719) included stone points under the general category of *ceraunia*—Pliny's grouping that included belemnites. Stone points were often confused with fossils—it being not at all obvious whence either of these had come. Even some of the early utilitarian explanations strike us today as quaintly comical—for example, Buckland's 1823 characterization of "a small flint,

18. Charles W. King, *The Natural History of Gems, or Semi-Precious Stones* (London: Bell & Daldy, 1870), 80, reporting on the views of Anselm Boetius de Boodt in his 1609 *De Gemmis et Lapidibus*.

the edges of which had been chipped off, as if by striking a light."[19] If axes were projections of woodsmen, flints here were presumed to have something to do with the flintlock or strike-a-light.

For many years after Frere's discovery—and subsequent work by Boucher de Perthes—it was argued that (what we now call) Acheulean tools were "axes"—a not-implausible suggestion, given their symmetry and size and cutting edge, which generally extends around the entirety of the tool. In the nineteenth century, it was often suggested that these were weapons of some sort (recall Frere's account) used by primitives to defend themselves against ferocious beasts, perhaps also to wage war against one another. Typical is Louis Figuier's 1870 *L'homme primitif,* which shows club and axe-wielding savages from "the period of extinct animals" fending off an attacking cave bear.[20] Axes in the nineteenth-century European ethnographic imagination were often accoutrements of medieval armor, supplemented by non-Western images of dress or habit. This was consonant with older images of early man as Adam or Hercules (with club and skin) or the more or less noble savage—all of which were recycled for use in "man of the Stone Age" representations.[21] The idea of a "Stone Age man with hafted axe" was also consistent with the nineteenth-century urbanist equation, primitive = woodsman, an equation visible in countless early illustrations: Boitard's "Fossil Man" (1861), *Harper Weekly*'s "Neanderthal" (1873), Henri du Cleuzieu's "Pithecanthropus" (1887), Faivre's "Deux mères" (1888), Forestier's "Modern Man, the Mammoth Slayer" (1911), and many others.

The problem with this view, as subsequent studies showed, was that none of the "axes" used in the Lower Paleolithic show any evidence of having ever been hafted (no polish from cords or handles, for example). This was already recognized in the nineteenth century, when Gabriel de Mortillet (1821–98), an early French Darwinian, identified Acheulean hand-axes as *coups de poing*—"blows of the fist"—the idea being that such instruments would be held in the hand to chop or dig or to butcher large animals.[22] The absence of hafting or any other kind of compound tool use prior to the Paleolithic (hook with string, hoe with handle, knife with wooden grip, spear with stone point, and so on) has been used to argue

19. William Buckland, *Reliquiae deluvianae* (London: Murray, 1823), 83.

20. Louis Figuier, *L'Homme Primitif,* 3rd ed. (Paris: Hachette, 1873), fig. 16; Stephanie Moser, *Ancestral Images: The Iconography of Human Origins* (Ithaca: Cornell University Press, 1998), 127.

21. Moser, *Ancestral Images* (note 20), 135.

22. Gabriel de Mortillet, *Formation de la nation francaise* (Paris: Alcan, 1897).

that something changed in the cognitive regimen of hominids in the last hundred thousand years or so—a cognitive falling-into-place that allowed some new type of inventive, recombinatorial capacity. Does the invention of compound tools mark the beginning of new linguistic capacities?

Kathy Schick and Nicholas Toth in the 1990s provided experimental archaeological support for the idea that Acheulean "hand-axes" were used to process large animal carcasses (hunted or scavenged),[23] though that is only one of many theories put forward. J. Desmond Clark of Berkeley has suggested their use as bark-stripping tools, to allow feeding on the cambium layer of trees; others have proposed a digging scenario to extract plant roots or water or burrowing animals. William Calvin has argued that they may have been thrown for hunting ("killer frisbees"); the idea has also been floated that hand-axes were designed for myriad diverse uses—cutting, digging, scraping, hammering, chopping, and so on. "Hand-axes" in this view were the Swiss army knives of the Paleolithic.[24]

Evolutionary psychologists have also thrown their hats into the ring. In the spring of 1998, University of Reading archaeologist Steven Mithen proposed that "hand-axes" may actually be sexual lures, bragging points made by men to attract the opposite sex, the Ferraris or Armani suits of an earlier age. The rather macho (yet thin) theory here is that females were attracted to handsome stone-axe makers, which enabled those who made the more perfect forms to have more offspring.[25] This could presumably help explain why many "hand-axes" never seem to have been used (there is often no edge wear); it might also explain why some sites contain more such tools than would seem to have been needed—in some cases thousands scattered over a very small area. (More likely a symptom of the fact that there aren't very many feminists in paleoanthropology!) The theory is part of Mithen's larger view that the rise of modern consciousness involved a (relatively recent) onset of communication between different parts of the brain—"multi-tasking"—from which we get art, language, religion, and the rest of the show.[26]

A rather different approach has been to argue that Acheulean tools are not so uniform as they might first appear. There are different shapes

23. Kathy D. Schick and Nicholas Toth, *Making Silent Stones Speak: Human Evolution and the Dawn of Technology* (New York: Simon & Schuster, 1994), 258–60.

24. Ibid., 258–59; Eileen O'Brian in 1981 proposed the "killer frisbee" idea later taken up by Calvin.

25. Bob Holmes, "The Ascent of Medallion Man," *New Scientist*, 9 May 1998.

26. Steven Mithen, *The Prehistory of the Mind* (London: Thames & Hudson, 1996).

and sizes (some as large as fifty centimeters), and of course different kinds of materials, the earliest African assemblages being more often basalt or quartzite, with subsequent European tools being more often flint, chert, or jasper. There are Acheulean sites without hand-axes (Clacton-on-Sea, in England), and Acheulean-like hand-axes that persist into the Late Mousterian that were probably used by Neanderthals. The selection of appropriate materials may have involved a measure of skill and connoiseurship lost to us today; the fact that to many of us such tools "all look alike" may be simply an artifact of distance and lack of familiarity. There is also the growing recognition that a given tool can have had multiple uses, or may even have once been one kind of tool and later cannibalized for a novel use. Large flakes (axes?) thus become cores, cores are refigured as choppers, choppers are used as cores for smaller flakes, and so forth. Ancient hominids in this sense may have been rather more like us—opportunistic and flexible—than is sometimes thought.

This last-mentioned prospect has made it much harder to say for sure what is a core (waste or resource) and what is a flake (tool)—giving rise also to the suggestion that many so-called hand-axes may actually be discarded cores from which flakes were taken. Nicholas Toth has argued that most Oldowan "tools" are actually remnant cores, the idea being that suitable pebbles would be carried around and then struck whenever needed to produce a thin, sharp flake. Such flakes are effective cutting tools, and would serve very well for rapid butchery and excision of flesh.[27] The same could well be true for many of the "hand-axes" found in Europe, the Middle East, and Africa: their marvelous symmetry may simply indicate that the core has been exhausted, flakes having been taken from all around the edge. Archaeologists for more than two centuries may have been celebrating the earliest preserved form of human waste: not tools, in short, but trash.

To return to the question of human recency, one interesting explanation for the consistency of Acheulean artifacts over such vast stretches of time and distance might be that humans were not making them. Richard Klein at Stanford and Alan Walker and Pat Shipman at Penn State have put forward this hypothesis, the basic claim of which is that whoever or whatever made them was culturally and intellectually more like a creative chimp than a modern human. Nonlinguistic creatures may have been the makers—hominids without the use of symbolic language, in other

27. Nicholas Toth, "The Oldowan Reassessed: A Close Look at Early Stone Artifacts," *Journal of Archaeological Science* 12 (1985): 101–20.

words.[28] It is hard to say what to think about this view, that the first hominid creations that are genuinely beautiful, displaying symmetry and undeniable skill (albeit perhaps only trash, or perhaps first trash and then exapted to be some kind of tool), may have been produced by people that were not yet fully people. It could even be that it was in perfecting such things that humans became more fully human—though this latter idea ("more fully human") may make no more sense than the idea of a creature being "fully cockroach" or "fully chimpanzee." Narratives of arrival are pervasive in paleoanthropology, reflecting not just our (understandable?) sapiento-centrism, but also the (questionable?) sense that we alone have managed to leave some important part of nature's authority behind. The difficulty is compounded, as we shall see, by the fact that more than one species of human may have walked the earth, at several different and simultaneous points in hominid history.

Racial Liberalism, the UNESCO Statement, and the Single Species Hypothesis

Understandings of hominid diversity have undergone a profound shift in recent decades, from a conception that there could be only one kind of hominid at any given time to a view that the past thirty thousand years or so is unusual in having only one. Many paleoanthropologists now believe there may have been as many as twenty different species of hominids since our last common ancestor with chimps,[29] with a peak sometime around 2 million years ago when three, four, or possibly more different hominid species coexisted in Africa, just prior to the *Homo erectus* exodus. That is a dramatic change from the dominant view of the 1960s, defended by C. Loring Brace and others, that the human culturo–ecologic "niche" was so narrow that only one kind of hominid could exist at any given time.[30]

28. Richard Klein, *The Dawn of Human Culture* (New York: John Wiley & Sons, 2000); Alan Walker and Pat Shipman, *The Wisdom of the Bones* (New York: Knopf, 1996). The key anatomic evidence here, developed by Walker and Ann MacLarnon, is the narrowness of the *erectus* spinal column, which suggests that *erectus* did not have the chest-cavity nervous links and musculature required for language. The implication is that *Homo erectus* was unable to speak. See Ann MacLarnon, "The Vertebral Canal," in Alan Walker and Richard Leakey, eds., *The Nariokotome Homo Erectus Skeleton* (Cambridge: Harvard, 1993).

29. Ian Tattersall, "Once We Were Not Alone," *Scientific American,* January 2000, 38–44; also Ian Tattersall and Jeffrey H. Schwartz, *Extinct Humans* (New York: Westview, 2000).

30. C. Loring Brace, "The Fate of the 'Classic' Neanderthals: A Consideration of Hominid Catastrophism," *Current Anthropology* 5 (1964): 3–43. Milford Wolpoff, Brace's student and then colleague at Michigan, was another exponent of the single species hypothesis;

This older idea was partly a political outcome of the fear of excluding extinct hominid species from the ancestral Family of Man,[31] but it was also interestingly consistent with older, gradualist, ladder-like phylogenies deriving from the Great Chain of Being. The "single species hypothesis" popular in the 1950s and 1960s championed a linear nonbranching evolutionary sequence according to which *Australopithecus* begat *erectus*, *erectus* begat *Neanderthal*, *Neanderthal* begat *sapiens*, and so forth. The newer family trees, by contrast, are more often "bushy," with false starts and dead ends (extinctions), and for long stretches of palaeontologic time showing more than one species living concurrently.

What accounts for the rise of the single species hypothesis? Ecological concerns à la Brace were only one of several impulses; another was a growing worry over the out-of-control proliferation of hominid taxa. "Lumpers" such as Ernst Mayr, for example, contributed to the hypothesis with his effort to reduce the clutter of hominid generic names. Mayr in 1950 maintained that the proliferation of hominid generic names made little taxonomic sense, and proposed that the zoo of names circulating at that time—*Australopithecus, Plesianthropus, Paranthropus, Pithecanthropus, Sinanthropus, Paleoanthropus,* etc.—be reduced to a single genus, *Homo,* defined by upright posture. Mayr also maintained, though, following Dobzhansky, and with race clearly on his mind, that "never more than one species of man existed on the earth at any given time."[32] Taxonomists were by no means univocal in classing all fossil finds as either *Homo* or *Australopithecus* in the period 1950–70, or even in eschewing "bushes,"[33] but it is worth noting that the resurgence of interest in hominid diversity since the 1970s has been accompanied by the sprouting of several new (or newly revived) genus-level branches on the hominid taxonomic tree—notably *Ardipithecus* and *Paranthropus,* and most recently *Orrorin* and *Sahelanthropus,* though one can hardly expect these to be the final words on the topic.

The single species hypothesis began to come undone in the 1960s, as

see his "Telanthropus and the Single-Species Hypothesis," *American Anthropologist* 70 (1968): 477–93.

31. This idea is echoed in Jonathan Marks's argument that the Neanderthals should be considered to have bred with us for political reasons; see his "Systematics in Anthropology: Where Science Confronts the Humanities (and Consistently Loses)," in G. A. Clark and C. M. Willermet, eds., *Conceptual Issues in Modern Human Origins Research* (New York: Aldine De Gruyter, 1997), 46–59.

32. Ernst Mayr, "Taxonomic Categories in Fossil Hominids," *Cold Spring Harbor Symposium on Quantitative Biology* 15 (1950): 109–18.

33. See, e.g., Gerhard Heberer, ed., *Menschliche Abstammungslehre* (Stuttgart: Gustav Fischer Verlag, 1965).

new fossil finds made it obvious that human evolution was not a single-stalk, nonbranching developmental process. Prominent among these was *Homo habilis* ("handy man"), a fossil found in the early 1960s by the Leakeys working at Olduvai Gorge in Tanzania. This creature was clearly more "human-like" than *Australopithecus,* yet quite a bit older than had previously been imagined for our genus (about 1.75 million years, based on volcanic ash-fall dates), and the first real evidence that *Homo* must have lived contemporaneous with the *Australopithecines.* The reigning assumption had been that ape-like hominids had been fully replaced by more human-like hominids—but here was a new and disturbing idea that shook the world of the single species hypothesizers.[34] Could this be possible? Might two different kinds of ape-men have lived at the same time? If so, how did they interact? Could they have conversed with one another? Traded with one another? Fought with one another? The idea of multiple coexisting human lineages seemed a rather unsettling prospect—fertile ground for sci-fi, though, as more than one author recognized.[35]

The story was made still more complex when it became clear that there was more than one kind of *Australopithecus.* The key discovery here came in 1974 when Donald Johanson, a graduate student working at a dig near Hadar, in Ethiopia, discovered a 3.2 million-year-old hominid soon regarded as the first-found member of a new species, dubbed *Australopithecus afarensis* (southern ape of Afar), but better known as "Lucy" from the fact that the paleoanthropologists were rocking to the Beatles song "Lucy in the Sky with Diamonds" (LSD) as they put the pieces together. The skeleton was only 40 percent complete, but clearly showed that "humans" walked upright more than 3 million years ago.

34. The nail in the coffin came in 1975, when Leakey and his colleagues announced the discovery of a *Homo erectus* skull old enough to have clearly coexisted with *Australopithecus boisei;* see Walker and Shipman, *Wisdom* (note 28), 140–47.

35. Three of the best multiple hominid species novels are Vercors, *You Shall Know Them* (1953, from the French); William Golding, *The Inheritors* (1955); and Bjorn Kurten, *Dance of the Tiger* (1995). Vercors' book imagines the discovery of a band of *Australopithecines* ("Tropis") living in a remote Javanese jungle—a sort of paleoanthropologic version of Conan Doyle's *Lost World* with added subplots of hominid enslavement and interspecies breeding. The popular feature film *Quest for Fire* was based on J.-H. Rosny's *La guerre du feu: Un roman des ages farouches* (Paris: E. Fasquelle, 1911), a World War I–era novel with Neanderthal/Cromagnon love and war themes. Desmond Morris orchestrated the gestures for the film, which includes a lot of sniffing and head-smashing. Clive Gamble suggests that the finding of Neanderthal sites near the Somme in the early part of the twentieth century prompted associations of these creatures with a bestial and violent past that humans had supposedly transcended (personal communication).

Subsequent years would see many other fossil finds, representing as many as a dozen different hominid species. It eventually became clear that there were several different species of these southern ape-like men of Africa, the oldest known as of this writing being *Australopithecus anamensis,* a fossil hominid found in 1995 by Meave Leakey of the National Museum of Kenya and Alan Walker, then at Johns Hopkins University. The creature is 4.2 million years old, which cannot be very long after the point when (what are now) chimps and humans must have branched off from one another. Since the year 2000, even older hominids have been found: *Orrorin tugenensis* from the Tugen Hills of Kenya lived about 6 million years ago, and *Sahelanthropus tchadensis* from Chad is even older. No one knows for sure whether any of these are our direct ancestors: they almost certainly are not, given the bushiness of the hominid lineage and the fact that most lines eventually perish. Vince Sarich points out that you always know that a fossil had parents, but you never know whether it left any offspring.[36]

How have such finds affected theories of human recency? I earlier mentioned the strong professional pressures in favor of "splitters": it is surely better for your career to have found a new hominid species than yet another example of some other scholar's already-discovered species. Taxonomic modesty favors lumping, hubris sanctions splitting. Similar pressures influence the "humanity" of one's finds: it is clearly better to have found an early "man" than a rather later or precocious ape. The pressure to speak in such terms is enormous: witness Ian Tattersall's most recent book, *Extinct Humans,* whose very title brandishes a concept he himself has cautioned against. I assume his agent cautioned him that a book titled *Extinct Hominids* would not sell as well.

The trend since the 1970s, however, has been to argue that hominids prior to *Homo sapiens* were not as "human" as once was thought. I've already noted several causes for this shift in thinking, to which we can add the following factors: (1) a retreat from some of the more optimistic assessments of chimpanzee cognitive capacities of the 1960s and 1970s; and

---

36. Vince Sarich, personal communication, via Alan Walker. The odds that any given organism will contribute genes to subsequent generations become vanishingly small as time marches on; the use of the term "ancestor" with reference to any of the hominid fossils under discussion here is therefore misleading, since there is only a remote possibility that any would have contributed anything to the genetics of modern *Homo sapiens*. Sarich's point in this comment was primarily to contrast the kinds of evidence provided by paleontology versus molecular genetics: you never know if a particular fossil reproduced or not, but molecular information is always about lineages known to have survived (since you sample from people alive today).

(2) acceptance of the view that it was not such a bad thing to be "not fully human."

There has also been the growing sense, though, that it is not necessarily "racist" to believe that nonsapient hominids were radically different from us. Here, it is important to appreciate the ideological obstacles faced by those who wanted to emphasize fossil hominid diversity. The most prominent among these obstacles was the liberal antiracist sentiment of many postwar anthropologists, especially in the Anglo-American world, where shock and horror over the events of Nazi Germany combined with concerns that in other parts of the world as well, racial prejudice was still a potent force. Such concerns mounted in the first decade after the war and led liberal activists in the newly founded United Nations Educational, Scientific and Cultural Organization (UNESCO) to draft a "Statement on Race" denouncing racial theory and racial prejudice. The resulting document, published in 1950 and in various revised versions ever since, became the canonical liberal resolution of the race issue: race as usually conceived did not exist; people were equal throughout the world in terms of intellectual and cultural worth; the most important differences that you find among peoples are due to nurture rather than nature, and so forth. The Boasian position was vindicated and strengthened: Boas had said that race, language, and culture were separate and independent variables; the new view was essentially that race does not exist at all.[37]

Historians of science are familiar with the obstructive impact of ill-willed ideologies on science, less familiar are examples of political goodwill stifling science. In the question of fossil hominid phyletic diversity, however, the impact of the UNESCO statement on race and the larger population-genetics critique of racial typology must be regarded as somewhat stifling. The most common fear seems to have been that by allowing multiple lineages of humans, one would open the door to racism by excluding one or another lineage from the mainline ancestral sequence leading to modern humans. This was clearly the case in C. Loring Brace's rejection of multiple lineages, one of his fears being that Neanderthals would be dehumanized (and excluded from the human ancestral line) by what he called "hominid catastrophism."[38] Antilinearity, in his view, was also tantamount to "anti-evolution." Ian Tattersall has suggested that the

37. UNESCO, *The Race Concept: Results of an Inquiry* (Paris: UNESCO, 1952).

38. Brace, "Fate of the 'Classic' Neanderthals" (note 30). Brace's was an ecologic argument: he maintained that since Neanderthals possessed culture, they could not have been overwhelmed by another species practicing culture.

emphasis on population thinking in these peak prestige years of the New Synthesis also helped foster the idea that "no amount of variation" was too great to be contained within a single species.[39] The emphasis on genetic diversity in this sense may have retarded the acceptance of new hominid lineages; it may also have made it difficult to believe that some lineages had perished without issue. The seeds for this myopia were already sown in 1944, when Theodosius Dobzhansky argued that "no more than a single hominid species existed at any one time level," a view that was taken to an extreme in 1959, when Emil Breitinger argued that hominid evolution was punctuated by "only one single *a priori* certain case of a complete speciation and splitting"—the divergence of hominids from tertiary primate species.[40] Implicit in such assertions of hominid unity was also the idea that "our" branching point from the other apes was very remote—11 or 12 million years even in the most conservative (radical?) estimates.[41]

Morphologists also had their blinders, albeit coming from quite different technical and conceptual traditions. Paleoanthropologists at Cambridge in the 1960s, for example, could be heard muttering about how "there simply wasn't enough 'morphological space' between *Australopithecus africanus* and *Homo erectus* to shoehorn in a new species."[42] Stephen Jay Gould would later argue for a more "bush-like" hominid lineage, in harmony with his model of punctuated equilibrium.[43] Gould was an important transitional figure in that he was one of the first to clearly accept the UNESCO redefinition (or abandonment) of race, while also maintaining that an overly ladder-like phylogeny had straitjacketed human evolution and underestimated the morphological diversity of human (and other) lineages. Multilinearity after Gould finally became acceptable again, when purged of its earlier racialist overtones.

The liveliness of this issue has to be understood in light of the fact that, even as late as the 1960s, human *racial* diversity was still being routinely

39. Ian Tattersall, *The Fossil Trail: How We Know What We Think We Know about Human Evolution* (New York: Oxford, 1995), 116.

40. Richard Delisle, "Human Paleontology and the Evolutionary Synthesis," in Raymond Corbey and Bert Theunissen, eds., *Ape, Man, Apeman: Changing Views since 1600* (Leiden: Leiden University, Department of Prehistory, 1995), 217–28.

41. See, for example, Günther Bergner, "Geschichte der menschlichen Phylogenetik seit dem Jahr 1900," in Heberer, *Menschliche Abstammungslehre* (note 33), 49.

42. Tattersall, *Fossil Trail* (note 39), 116.

43. Stephen Jay Gould, "Bushes and Ladders in Human Evolution," in his *Ever Since Darwin* (New York: Norton, 1977).

characterized by many physical anthropologists as taxonomically significant. In 1962, for example, Carleton Coon—then president of the American Association of Physical Anthropologists—claimed that African *Homo
erectus* populations ("Congoids") had actually crossed the threshold to fully
human *Homo sapiens* two hundred thousand years later than other hominid populations (Europeans, of course, led the way). Africa, as he put it,
was the "kindergarten" of humanity.[44] Coon also used this prejudice to
work secretly behind the scenes to undermine the *Brown v. Board of Education* civil rights ruling of the U.S. Supreme Court, which declared that
separate was *not* equal and mandated desegregation.[45] Franz Weidenreich,
a Jewish emigré anthropologist from Germany, had carried an implicit
polygeny over into American physical and paleoanthropology, one in
which humans were understood to have diverged into separate racial
groups prior even to the transition from *erectus* to *sapiens*. *De facto* polygeny
continued also in Germany: a 1965 book edited by the former SS officer
Gerhard Heberer, for example, features a chart showing racial differentiation beginning at the end of the Pleistocene, about 1 million years ago.[46]

The changing graphic conventions used to portray human phyletic development are interesting in this context, especially when compared with
countervailing trends in the portrayal of human racial diversity. Hominid
species diversity has increased at the same time that *racial* diversity has been
progressively downplayed. The concurrence is not a coincidence: the same
stress on genetic diversity that helped put an end to the idea that human
"races" constitute separate species also helped eclipse the idea that the humans must also have been diverse in the distant past (whence the single
species hypothesis). Human racial unity seemed to preclude hominid
phyletic diversity.

Molecular Anthropology

The idea of modern humans developing slowly and separately in different
parts of the world is today known as "multiregionalism"; this is the infa-
mous alternative to what is often called the "replacement" or "out-of-
Africa" model, the idea that fully modern humans emerge rather suddenly

44. Coon, *Origin of Races* (note 9), 656.

45. William H. Tucker, *The Science and Politics of Racial Research* (Urbana: University of Illinois Press, 1994), 162–68; Coon, *Origin of Races* (note 9); John Jackson, "In Ways Unacade-
mical: The Reception of Carleton S. Coon's *The Origin of Races,*" unpublished manuscript
(2000).

46. Bergner, "Geschichte" (note 41), 49.

in Africa about 135,000 years ago and spread from there throughout the world, replacing (without interbreeding with) the *Homo erectus* populations they encountered. The two sides are represented by different instrumental traditions: multiregionalists, led by Milford Wolpoff of Michigan, tend to be physical anthropologists; the most prominent out-of-Africanists, by contrast, have been molecular geneticists—notably the diaspora from Allan Wilson's Berkeley lab in the 1980s, including Mark Stoneking, Rebecca Cann, and Svante Pääbo, to name some of the more distinguished. Multiregionalists tend to stress continuities in physical type as evidence of regional continuity; out-of-Africanists emphasize rates of nucleotide divergence as evidence of bottlenecks and human biorecency.

Apart from these disciplinary differences, however, there are also interesting ideological divides—though not always those that make it into the popular press. The tendency has been to gloss the debate as one between "we're all Africans" and "racial divisions are really deep," when that is not necessarily the most interesting or accurate fracture plane in the debate (Wolpoff is not Carleton Coon—and even Coon is more complex than commonly imagined). One thing going on is a very deep difference over how to grant the Neanderthals dignity. Wolpoff and the multiregionalists basically maintain the UNESCO line that to deny them a biological link to the present is to exclude them from the Family of Man, a move that smacks of racism. (The question also concerns the "specialness" of modern humans—with the anthropologists of the so-called Michigan school cautioning against an exaggerated conception of "human uniqueness.")

Critics of this view—like Tattersall—say that the Neanderthals are no less respectable for having gone extinct, or for not having been able to breed with *Homo sapiens;* their dignity should not hinge on their biocompatibility with successor populations. What is also interesting, though, are the different rhetorical strategies used by the two groups—multiregionalists and out-of-Africanists. These are interesting since each has tried, at various points, to accuse the opposing group as being more "racist." Out-of-Africa theorists have accused multiregionalists of exaggerating racial divisions (conceived as going back as long as a million years in some of the still-used "candelabra models"); multiregionalists in turn have accused out-of-Africa advocates of implying a total and perhaps violent (genocidelike?) replacement of *Homo erectus* or the Neanderthals by *Homo sapiens*[47]—

47. Milford H. Wolpoff and Rachel Caspari, *Race and Human Evolution: A Fatal Attraction* (New York: Westview Press, 1997).

a misconception fueled by silly and sensationalist articles in the popular press (like *Der Spiegel*).[48]

Each side has also managed to brand the opposing camp as "old-fashioned"—the Wolpoffians seeing in the out-of-Africa idea echoes of a rather imperial racist/progressivist "replacement model" going back to W. J. Sollas ca. 1911 or even Nicolaas Witsen in the 1600s;[49] while the "African Eve" or "Garden of Eden" supporters find in Wolpoff's multiregional model vestiges of the hoary specter of polygenesis—the idea occasionally expressed in the nineteenth century, for example, that white people descended from chimps, Africans from gorillas, and Asians from the orangs. Multiregionalists see the debate in terms of "cooperation versus violence"; the mitochondrialists see a world of recent unity versus "deep divisions."

So far, it seems that the geneticists are winning the field. Multiregionalists have no comparable technical wonder, and the original mitochondrial evidence for recency has been joined by nonmitochondrial evidence—for example, from the Y chromosome. Critics of progressivist teleology have also successfully argued that it makes no biological sense that populations in widely differing parts of the world would have converged in becoming "fully modern"—since "modernity" in this sense is not just a form of culture that diffuses. Molecular anthropology has also gained authority from its many other spectacular successes: the extraction and sequencing of DNA from Neanderthal bones, suggesting a last common ancestor with humans ca. five hundred thousand years ago; the sequencing of the Ice Man of the Alps and the Tsar family's remains; the use of sequences to establish dates for the origin of birds, the colonization of land by plants, and so forth. Molecular techniques have been used to determine what archaic Native Americans ate, distinguished by sex and by season, to help determine whether the Wistar Institute caused the AIDS epidemic, and to determine the "gender" of dried blood on Virgin Mary statues. Molecular anthropology is riding high: you really can no longer do phylogeny, for

48. See, for example, the March 20, 2000, cover story in *Der Spiegel,* "Der Krieg der Ersten Menschen: Wie der *Homo sapiens* den Neandertaler verdrängte," where the *Homo sapiens* "Ausbruch" out of Africa and confrontation with the Neanderthals reads very much like the description of a World War II battle scene. See Matthias Schulz, "Todeskampf der Flachköpfe," *Der Spiegel,* 20 March 2000, 240–55.

49. Wil Roeboeks, " 'Policing the Boundary'? Continuity of Discussions in 19th and 20th Century Palaeoanthropology," in Corbey and Theunissen, eds., *Ape, Man, Apeman* (note 40), 173–80.

example, without paying at least some attention to molecular sequences, and few molecularists have sided with the multiregionalists.

The politics of human genetic recency, though, have been complex. On the inegalitarian Right, Vince Sarich showed in the late 1960s that humans and apes shared a common ancestry with chimps only 5−6 million years ago. Meanwhile, on the egalitarian Left, Richard Lewontin showed that racial differentiation was relatively recent.[50] Sarich's innovation can be seen as part of an effort to reemphasize the animal in our humanity, and Lewontin's the opposite—to deemphasize whatever biological differences may divide us. Sarich moved up the break with apes, Lewontin moved up (and trivialized) the separation of races from one another.

Confusion and Projection

The idea of human recency comes from many different directions, only a few of which have been mentioned here.[51] The ideological aspects are interesting, because people seem to be getting different things out of recency. Some seem to like the fact that "we are all Africans"; there is a kind of "Black Athena" resonance in the molecularist account of the paleolithic, especially in its popularization by the media. But there is also support for recency from those who reject the single species hypothesis. Hominid bushiness seems to reopen one of the questions at the root of the UNESCO statement: how deep can human biodiversity go? Hominid bushiness raises the difficult question of what it must have been like to have multiple species of humans living at the same time; it also forces us to realize that our "humanness" in the sense of biological specificity must be distinguished from the "human condition." Everything we do we do as animals, but the part in this that is distinctly human remains at least partly up for grabs.[52]

Two things we can be sure of: the history of science is often a history of confusion, and ideologies often come in cumbersome packages. Argu-

50. Richard C. Lewontin, "The Apportionment of Human Diversity," *Evolutionary Biology* 6 (1972): 381–98; see also his *The Genetic Basis of Evolutionary Change* (New York: Columbia, 1974), 152–57.

51. Margaret W. Conkey and S. H. Williams, "Original Narratives: The Political Economy of Gender in Archaeology," in M. di Leonardo, ed., *Gender at the Crossroads of Knowledge: Feminist Anthropology in the Postmodern Era* (Berkeley: University of California, 1991), 102–39.

52. Tim Ingold, "Humanity and Animality," in Ingold, *Companion Encyclopedia of Anthropology* (London: Routledge, 1994), 14–32.

ments developed for dealing with racial differences and racial prejudices have been projected onto dealings with fossil hominid diversity; that was true before the UNESCO statement on race, but it is also true afterward. There are those who feel that it is morally wrong to claim that the Neanderthals, for example, were anything less than fully human.[53] No one can deny that the bestial impregnation of this species in the early part of this century was wrong in many respects, but their refitting with flowers (in 1969 at the Shanidar site in Iraq, where pollen was found in a grave, whence the "flower child of Shanidar") may eventually seem just as quaint, if rather more pleasant. The Neanderthals may or may not have bred with "us" (the molecular evidence suggests they didn't), their replacement by "us" may have been peaceful or bloody (there is no evidence one way or another). What we can safely assume, though, is that no matter how much evidence we get, the prehistory of tools, bodies, and beliefs will forever remain a fertile field for projection and wishful thinking.

The question with which I began this essay—when did humans become human?—is ultimately a moral question, for the simple reason that one really cannot say when "they" became "us." Species are never well defined over time in any event, since you cannot do the breeding experiment. And for humans, there is the complicating fact that reproductive compatibility is not really what people are concerned about when they talk about human origins. Several different species may have been "human" in some of the commonsense ways we use that term. The notion of fully modern humans emerging at some well-defined point in time helps buttress theories of human uniqueness, but it doesn't face up to the fact that (apart from our species specifically) humanity remains a category in flux. Sartre is famous for his quip that the only thing we can say about human nature is that we have no nature; self-definition is one of our defining fea-

53. My personal view as of this writing is that humanness is a linguistic rather than a biological (or phyletic-typological) concept, and that if intelligent creatures are discovered in some other part of the universe, they should probably be accorded some kind of "human rights." Humanity in this sense is a moral category that transcends biological specifics. There are obvious ethical conundrums in such a view (for instance, with regard to the "humanity" of nonlinguistic *Homo sapiens*); there is also the intriguing question of what kind of answer we should give if and when machines of human construct begin to ask for "rights" of one sort or another. Lorraine Daston suggests that "personhood" is perhaps a more appropriate category than "humanness" in this respect (personal communication), the point being to distinguish our species-being (humanness) from our moral status (personhood). The question then would be whether "human" rights are sufficient to cover whatever rights are demanded for all intelligent creatures—on this planet or some other.

tures. How far back we are willing to go and see ourselves in this sense is a moral choice; nature provides a guide but the human part must decide. When Gandhi was asked what he thought about Western civilization, he answered, "It would be a good idea." Holding up the mirror, humanity often appears just as unrealized.

❋ CONTRIBUTORS ❋

DANIELLE ALLEN is associate professor in the classics and political science departments and the Committee on Social Thought at the University of Chicago. She is author of *The World of Prometheus: The Politics of Punishing in Democratic Athens* (2000) and *Talking to Strangers: On Rhetoric, Trust, and Citizenship* (forthcoming).

JOAN CADDEN is professor of history and of science and technology studies at the University of California, Davis. Her current research concerns medieval natural philosophers' explanations of male homosexual desire, and centers on the work of the late medieval physician Pietro d'Abano. She is author of *Meanings of Sex Difference in the Middle Ages: Medicine, Science, and Culture* (1993) and has written articles on the medical and scientific ideas of medieval women such as Hildegard of Bingen and Christine de Pizan.

LORRAINE DASTON is director of the Max Planck Institute for the History of Science and honorary professor at the Humboldt-Universität. She has taught at many other universities as well, including Harvard, Princeton, Brandeis, Göttingen, and Chicago. Her publications include *Classical Probability in the Enlightenment* (1988) and *Wonders and the Order of Nature, 1150–1750* (1998, with Katharine Park), both of which were awarded the Pfizer Prize of the History of Science Society, as well as articles on the history of quantification, objectivity, and scientific attention.

FA-TI FAN is assistant professor of history at the State University of New York at Binghamton. He is author of *British Naturalists in Qing China: Science, Empire, and Cultural Encounter* (in press) and is currently working on a study of science and nationalism in twentieth-century China.

ECKHARDT FUCHS is assistant professor in the Department of Education at the University of Mannheim. His main field of research is modern American and European intellectual history, and he is currently working on the history of scientific internationalism. He is author of *Henry Thomas Buckle: Geschichtsschreibung und Positivismus in England und Deutschland* (1994); coauthor of *"J'accuse": Zur Affäre Dreyfus in Frankreich* (with Günther Fuchs, 1994); editor of *Weltausstellungen im 19. Jahrhundert* (1999); and coeditor of *Across Cultural Borders: Historiography in Global Perspective* (with Benedikt Stuchtey, 2002), *Macht und Geist: Intellektuelle in der Zwischenkriegszeit* (with Wolfgang Bialas, 1995), and *Geschichtswissenschaft neben dem Historismus* (with Steffen Sammler, 1995).

VALENTIN GROEBNER teaches medieval and Renaissance history at the University of Basel, Switzerland. He is author of *Liquid Assets, Dangerous Gifts* (2002) and *Defaced. The Visual Culture of Violence at the End of the Middle Ages* (forthcoming). He is currently working on a new book on identification practices and identity papers in Europe between the thirteenth and seventeenth centuries.

A. J. LUSTIG is a postdoctoral fellow at the Dibner Institute for the History of Science and Technology at the Massachusetts Institute of Technology. She is currently working on a book titled *Calculated Virtues: Altruism and the Evolution of Society in Modern Biology.*

GREGG MITMAN is professor of the history of science, medical history, and science and technology studies at the University of Wisconsin, Madison. He has written widely on the intellectual, cultural, and political history of ecology and nature in twentieth-century America. His most recent works include an *Osiris* volume coedited with Michelle Murphy and Christopher Sellers on the history of environment and health (forthcoming) and *Reel Nature: America's Romance with Wildlife on Film* (1999). He is currently completing a book on the ecological history of allergy in American that brings together his interests in environmental history, history of science, and medical history.

MICHELLE MURPHY is assistant professor of history and women's studies at the University of Toronto. Her essay in this volume is connected to a larger project on the roles of technoscience in women's health movements and its intersections with the transnational history of popular control. She is working on a book titled *Toxic Privilege: Chemical Exposure, Gender, and the Politics of Imperception in the United States.*

KATHARINE PARK is Zemurray Stone Radcliffe Professor of the History of Science and Women's Studies at Harvard University, where she teaches courses on the history of early science and medicine and on the history of women, gender, and the body. She is author of *Doctors and Medicine in Early Renaissance Florence* (1985) and coauthor of *Wonders and the Order of Nature, 1150–1750* (1998, with Lorraine Daston). Her current book project, *Visible Women: Gender, Generation, and the Origins of Human Dissection,* focuses on late medieval and early modern Italy.

MATT PRICE received his Ph.D. in history and philosophy of science from Stanford University in 1998 and currently teaches in the history department at the University of Toronto. He is writing a history of ecological economics and has also published on the histories of prosthetic devices and physical rehabilitation in the twentieth century.

ROBERT N. PROCTOR was born in 1954 in Corpus Christi, Texas, the great-grandson of a Baptist missionary in China and grandson of a Klansman. He grew up on the King Ranch and later in Kansas City, where he was raised on a diet of Dewey's liberalism and Veblenian economics. He studied biology/chemistry at Indiana University, following which he obtained a doctorate in history of science at Harvard University, writing on the origins of the idea of value-neutrality. His most recent book is *The Nazi War on Cancer* (1999), and he's currently writing on human origins, agates, expert witnessing, agnatology ("ignorance lore"), and molecular coproscopy.

HELMUT PUFF is associate professor in the Department of Germanic Languages and in the Department of History at the University of Michigan. His research on early modern Europe focuses on German-speaking regions; the culture of the Reformation; the history of sexuality; gender history; the history of literacy, reading, and printing; and early modern visual culture. He is author of *Deutsch im lateinischen Grammtikunterricht* (1995) and *Sodomy in Reformation Germany and Switzerland* (2003) and editor of *Lust, Angst und Provokation: Homosexualität in der Gesellschaft* (1993) and *Zwischen den Disziplinen? Perspektiven der Frühneuzeitforschung* (2003).

ROBERT J. RICHARDS is professor of history, philosophy, and psychology at the University of Chicago, where he is director of the Fishbein Center for the History of Science. He is author of *Darwin and the Emergence of Evolutionary Theories of Mind and Behavior* (1987), winner of the Pfizer Prize of the History of Science Society; *The Meaning of Evolution* (1992); and *The Romantic Conception of Life: Science and Philosophy in the Age of Goethe* (2002). Currently he is working on a book tentatively titled *The Tragic Sense of Life: Ernst Haeckel and the Battle over Evolution in Germany.*

LONDA SCHIEBINGER is Edwin E. Sparks Professor of the History of Science and codirector of the Science, Medicine, and Technology in Culture program at Pennsylvania State University. She is author of *The Mind Has No Sex? Women in the Origins of Modern Science* (1989), the prize-winning *Nature's Body: Gender in the Making of Modern Science* (1993), and *Has Feminism Changed Science?* (1999); editor of *Feminism and the Body* (2000); coeditor of *Feminism in Twentieth-Century Science, Technology, and Medicine* (2001, with Angela Creager and Elizabeth Lunbeck); and section editor of the *Oxford Companion to the Body* (2001).

LAURA M. SLATKIN teaches classics at New York University and the University of Chicago. Her research centers on poetry and poetics in archaic and classical Greece. Her current book project, *Figure, Measure, and Social Order in Early Greek Poetry,* explores the problematics of reciprocity and its representations in Homer, Hesiod, and Solon.

JULIA ADENEY THOMAS is associate professor of history at the University of Notre Dame. She is author of *Reconfiguring Modernity: Concepts of Nature in Japanese Political Ideology* (2002), winner of the 2002 John K. Fairbank prize for the outstanding book in Asian history from the American Historical Association. She has published several essays from her current research on the aesthetics of democracy in postwar Japan, including the award-winning "Photography, National Identity and the 'Cataract of Times'" in *American Historical Review* 103, no. 5 (1998). Professor Thomas was educated at Princeton, Oxford, and the University of Chicago.

FERNANDO VIDAL is a research scholar at the Max Planck Institute for the History of Science. He has published widely on the history of psychology and the human sciences since the eighteenth century. He is author of *Piaget before Piaget* (1994) and *Analyse et sauvegarde de l'âme au siècle des Lumières* (forthcoming), the latter focusing on the emergence of psychology as a discipline in the eighteenth century. He is currently investigating the history of personal identity in connection with the debates about the resurrection of the body.